VOLUME SEVENTY

CURRENT TOPICS IN MEMBRANES

Co-Transport Systems

CURRENT TOPICS IN MEMBRANES, VOLUME 70

Series Editors

ROBERT BALABAN
National Heart, Lung and Blood Institute,
National Institutes of Health,
Bethesda, Maryland, USA

SIDNEY A. SIMON
Department of Neurobiology,
Duke University Medical Centre,
Durham, North Carolina, USA

VOLUME SEVENTY

CURRENT TOPICS IN MEMBRANES

Co-Transport Systems

Edited by

MARK O. BEVENSEE
Department of Cell, Developmental and Integrative Biology
University of Alabama at Birmingham
Birmingham, AL, USA

ELSEVIER

AMSTERDAM • BOSTON • HEIDELBERG • LONDON
NEW YORK • OXFORD • PARIS • SAN DIEGO
SAN FRANCISCO • SYDNEY • TOKYO
Academic Press is an imprint of Elsevier

Academic Press is an imprint of Elsevier
525 B Street, Suite 1900, San Diego, CA 92101-4495, USA
225 Wyman Street, Waltham, MA 02451, USA
32, Jamestown Road, London NW1 7BY, UK
The Boulevard, Langford Lane, Kidlington, Oxford, OX51GB, UK
Radarweg 29, PO Box 211, 1000 AE Amsterdam, The Netherlands

First edition 2012

ISBN: 978-0-12-394316-3
ISSN: 1063-5823

For information on all Academic Press publications
visit our website at store.elsevier.com

Printed and bound by CPI Group (UK) Ltd, Croydon, CR0 4YY

Transferred to digital print 2013

DEDICATION

Both this volume on cotransporters and the upcoming volume on exchangers are dedicated to Dr Dale J. Benos, Endowed Professor and Chair of the Department of Physiology and Biophysics at the University of Alabama at Birmingham (UAB) at the time of his unexpected death in October of 2010. Dale's dedication to the areas of research, teaching, and service to UAB and beyond was inspirational. He instilled a passion for pursuing excellence, and led by example with an emphasis on support, encouragement, and compassion for others.

CONTENTS

CONTRIBUTORS

Jeffrey Adijanto
Department of Pathology, Anatomy, and Cell Biology, Thomas Jefferson University, Philadelphia, PA, USA

Jürg Biber
Institute of Physiology and Zurich Center for Integrative Human Physiology, University of Zurich, Winterthurerstrasse, Zurich, Switzerland

Marc Bürzle
Institute of Biochemistry and Molecular Medicine, and Swiss National Centre of Competence in Research, NCCR TransCure, University of Bern, Bern, Switzerland

Inyeong Choi
Department of Physiology, Emory University, Atlanta, Georgia, USA

Barbara Döring
SLC10 family research group, Institute of Pharmacology and Toxicology, Justus Liebig University Giessen, Biomedical Research Center (BFS), Giessen, Germany

Ian C. Forster
Institute of Physiology and Zurich Center for Integrative Human Physiology, University of Zurich, Winterthurerstrasse, Zurich, Switzerland

Joachim Geyer
SLC10 family research group, Institute of Pharmacology and Toxicology, Justus Liebig University Giessen, Biomedical Research Center (BFS), Giessen, Germany

Matthias A. Hediger
Institute of Biochemistry and Molecular Medicine, and Swiss National Centre of Competence in Research, NCCR TransCure, University of Bern, Bern, Switzerland

Nati Hernando
Institute of Physiology and Zurich Center for Integrative Human Physiology, University of Zurich, Winterthurerstrasse, Zurich, Switzerland

Peter Hinterdorfer
Institute for Biophysics, Christian Doppler Laboratory of Nanoscopic Methods in Biophysics, Johannes Kepler University of Linz and Center for Advanced Bioanalysis GmbH (CBL), Linz, Austria

Ken-ichi Inui
Kyoto Pharmaceutical University, Kyoto, Japan

Rolf Kinne
Max Planck Institute of Molecular Physiology, Dortmund, Germany

Patrick B. Knight
Department of Molecular & Cellular Physiology, University of Cincinnati College of Medicine, Cincinnati, Ohio, USA

Thomas Lütteke
Institute of Veterinary Physiology and Biochemistry, Justus Liebig University Giessen, Giessen, Germany

Bryan Mackenzie
Department of Molecular & Cellular Physiology, University of Cincinnati College of Medicine, Cincinnati, Ohio, USA

Bryan D. Maliken
Department of Molecular & Cellular Physiology, University of Cincinnati College of Medicine, Cincinnati, Ohio, USA

Daniel Markovich
Molecular Physiology Group, School of Biomedical Sciences, University of Queensland, Brisbane St Lucia, QLD, Australia

Heini Murer
Institute of Physiology and Zurich Center for Integrative Human Physiology, University of Zurich, Winterthurerstrasse, Zurich, Switzerland

Eric J. Niespodzany
Department of Molecular & Cellular Physiology, University of Cincinnati College of Medicine, Cincinnati, Ohio, USA

John A. Payne
Department of Physiology and Membrane Biology, School of Medicine, University of California, Davis, California, USA

Ernst Petzinger
SLC10 family research group, Institute of Pharmacology and Toxicology, Justus Liebig University Giessen, Biomedical Research Center (BFS), Giessen, Germany

Nancy J. Philp
Department of Pathology, Anatomy, and Cell Biology, Thomas Jefferson University, Philadelphia, PA, USA

Theeraporn Puntheeranurak
Department of Biology and Center of Nanoscience, Faculty of Science, Mahidol University, Ratchathevi, Bangkok, Thailand

Mobeen Raja
Max Planck Institute of Molecular Physiology, Dortmund, Germany

Eva S. Schweikhard
Department of Structural Biology, Max Planck Institute of Biophysics, Frankfurt, Germany

Ali Shawki
Department of Molecular & Cellular Physiology, University of Cincinnati College of Medicine, Cincinnati, Ohio, USA

Tomohiro Terada
Department of Pharmacy, Shiga University of Medical Science Hospital, Shiga, Japan

Christine M. Ziegler
Department of Structural Biology, Max Planck Institute of Biophysics, Frankfurt, Germany

PREFACE

Carrier proteins or transporters in biological membranes play vital roles in the physiology of cells, thereby contributing to the proper function of tissues and organs. Transporters establish and maintain ion/solute gradients across biological membranes, regulate chemical signals such as Ca^{2+} and neurotransmitters, control cellular properties including pH and cell volume, and contribute to ion/solute reabsorption or secretion by epithelial cells. Many of these vital roles involve a close functional association between transporters and channels. A classical example is the role of the Na^+ pump (or Na^+-K^+ ATPase) in actively pumping Na^+ from inside to outside the cell, and K^+ in the opposite direction, to balance the passive leaks of Na^+ into and K^+ out of the cell through ion channels. This "pump-leak" model helps explain a cell's ability to establish a resting membrane potential and to maintain steady-state intracellular concentrations of these important cations (see An Introduction to Membrane Transport and Bioelectricity, 2nd Ed. by JH Byrne and SG Schultz, New York: Raven Press, Ltd., 1994).

Transporters are a rich and diverse group of proteins, and their categorization deserves some attention. A transporter that engages in facilitated diffusion uses thermal energy to promote conformational changes that drive a solute down its chemical gradient; no additional energy is required. In contrast, an active transporter uses energy released from either ATP hydrolysis (primary active transporter), or another solute moving down its electrochemical gradient (secondary active transporter) to drive transport of a solute against its electrochemical gradient. One could also define a tertiary active transporter as one that harnesses the energy from a solute gradient established by a secondary active transporter. Many secondary active transporters use the large electrochemical Na^+ gradient across the plasma membrane (established by the Na^+ pump) to fuel transport. Secondary active transporters can be further classified as either exchangers (also known as antiporters) or cotransporters (also known as symporters). Exchangers move one or more solutes in one direction across the membrane, and one or more different solutes in the opposite direction. On the other hand, cotransporters move all solutes in the same direction. The current volume focuses on cotransporters.

I was both honored and excited when the late Dr Dale Benos and Dr Sidney Simon as the Editors of this series approached me to cover the topic of *Co-Transport Systems*. My excitement stemmed from my longstanding interest in transporters in general, both from a research and teaching perspective. For the last ~20 years, my research has focused on both the cellular and molecular physiology of acid–base transporters, with a particular emphasis on cation-coupled bicarbonate transporters. In addition, every year for the past 9 years, I have been pleasantly reminded of the vast array of other important transporters when teaching first-year medical students the mechanisms of reabsorption and secretion along the nephron of the kidney. This volume was a golden opportunity to expand my knowledge of some of these transporters. I was further excited by the opportune timing of this volume. Now that the cloning era has come to an end and many genomes have been fully sequenced, we are left with the important task of understanding what we have identified. We must now unravel the structure, function, and regulation of the cloned transporters in detail.

In organizing this volume, I decided to focus on the SoLute Carrier (SLC) gene series as defined by the Human Genome Organization (HUGO). However, guilt quickly set in as I systematically marched through a list of the SLC families and selectively began picking out cotransporters. I detested the thought of discriminating against the exchangers, even though I knew a single volume covering all the SLCs would be too large and overwhelming (and difficult to pick up!). As an aside, there is evidence that a cotransporter can function as an exchanger depending on experimental conditions, so the distinction may not always be straightforward. I fired off an email to Dr Benos with my concern. His solution was simple; edit two volumes—one on cotransporters and the other on exchangers. That plan is in place, and readers should expect to see a subsequent volume in this series on exchangers shortly.

The current volume is divided into chapters written by leading experts on specific SLC cotransporter families. Because the chapters are organized by transported species, some chapters necessarily cover more than one SLC family. For instance, Schweikhard and Ziegler in Chapter 1 on amino acid transporters cover the Na^+-coupled amino acid and neurotransmitter transporters (SLC6), the H^+-coupled amino acid transporters (SLC36), the Na^+-coupled glutamine transporters (SLC38), the cationic amino acid transporters (SLC7), and the heterodimeric amino acid transporters (SLC7/SLC3). Raja et al. in Chapter 2 on sugar transporters cover the Na^+-independent glucose transporters (SLC2) and the Na^+/glucose cotransporters (SLC5). Forster et al. in Chapter 10 on phosphate transporters primarily cover both the type II Na^+/P_i cotransporters (SLC34) and the type III Na^+/P_i cotransporters (SLC20). Although there is some variability, each chapter covers some history and physiology of the specific SLC families, and then details where possible transport biophysics, important protein regions/residues, and structural considerations, e.g., from mutagenesis, cysteine-scanning accessibility, and crystallography studies. Our current knowledge about these transporters is impressive. Results from structure-function and cysteine-scanning mutagenesis studies and crystallography/modeling have identified specific residues and domains that contribute to functional properties such as solute binding, selectivity, and translocation across the membrane. Techniques such as epitope tagging and cysteine-scanning mutagenesis have also elucidated transporter topology. In addition, crystallography and homology modeling have provided insight into the detailed structure of many transporters. Based on modeling with a limited number of known, high-resolution structures of related proteins such as the bacterial Na^+/leucine transporter (LeuT), many different transporters appear to have the same core design with an inverted structural repeat of a set of transmembrane domains (also see Abramson and Wright, Current Opinion in Structural Biology, 19:425-432, 2009). Overall, this volume will be particularly appealing to those interested in both molecular function and structure of cotransporters. Additional information about these and other SLC transporters can be found in Dr Matthias Hediger's special issue in *Pflügers Arch.* published back in 2004 that contains mini-reviews of the SLC families (see the Introduction by Hediger et al. in Vol. 447:465–468). This series is being updated with a focus toward the biology and physiology, with specific attention toward clinical and therapeutic aspects of the SLC families [special issue in Molecular Aspects of Medicine entitled "The ABCs of membrane transporters in health and disease (SLC series)", Guest Editor: Matthias A. Hediger, publication in 2012/2013].

I am grateful to many individuals for making this volume possible. First and foremost, I thank all the authors for their outstanding and comprehensive contributions, and in taking the time to consider my suggestions in revising their chapters. I owe special thanks to Drs Eva Schweikhard and Christine Ziegler for presenting the particularly larger number of amino acid cotransporters in a well-organized and informative Chapter 1. Drs Matthias Hediger and Rolf Kinne were particularly supportive and encouraging. I also thank the Editors of this series including Drs Sidney Simon, the late Dale Benos, and Robert Balaban for providing me the opportunity to compile and edit this volume. Finally, I am grateful to my contacts at Elsevier (particularly Radhakrishnan Lakshmanan, Sarah Latham, Sarah E. Lay, and Mary Ann Zimmerman) for working with both me and the authors, and for always being organized and helpful.

Mark O. Bevensee

PREVIOUS VOLUMES IN SERIES

Current Topics in Membranes and Transport

Volume 23 Genes and Membranes: Transport Proteins and Receptors* (1985)
Edited by Edward A. Adelberg and Carolyn W. Slayman

Volume 24 Membrane Protein Biosynthesis and Turnover (1985)
Edited by Philip A. Knauf and John S. Cook

Volume 25 Regulation of Calcium Transport across Muscle Membranes (1985)
Edited by Adil E. Shamoo

Volume 26 Na^+–H^+ Exchange, Intracellular pH, and Cell Function* (1986)
Edited by Peter S. Aronson and Walter F. Boron

Volume 27 The Role of Membranes in Cell Growth and Differentiation (1986)
Edited by Lazaro J. Mandel and Dale J. Benos

Volume 28 Potassium Transport: Physiology and Pathophysiology* (1987)
Edited by Gerhard Giebisch

Volume 29 Membrane Structure and Function (1987)
Edited by Richard D. Klausner, Christoph Kempf, and Jos van Renswoude

Volume 30 Cell Volume Control: Fundamental and Comparative Aspects in Animal Cells (1987)
Edited by R. Gilles, Arnost Kleinzeller, and L. Bolis

Volume 31 Molecular Neurobiology: Endocrine Approaches (1987)
Edited by Jerome F. Strauss, III, and Donald W. Pfaff

Volume 32 Membrane Fusion in Fertilization, Cellular Transport, and Viral Infection (1988)
Edited by Nejat Düzgünes and Felix Bronner

Volume 33 Molecular Biology of Ionic Channels* (1988)
Edited by William S. Agnew, Toni Claudio, and Frederick J. Sigworth

Volume 34 Cellular and Molecular Biology of Sodium Transport* (1989)
Edited by Stanley G. Schultz

---

*Part of the series from the Yale Department of Cellular and Molecular Physiology

Volume 35 Mechanisms of Leukocyte Activation (1990)
Edited by Sergio Grinstein and Ori D. Rotstein

Volume 36 Protein–Membrane Interactions* (1990)
Edited by Toni Claudio

Volume 37 Channels and Noise in Epithelial Tissues (1990)
Edited by Sandy I. Helman and Willy Van Driessche

Current Topics in Membranes

Volume 38 Ordering the Membrane Cytoskeleton Trilayer* (1991)
Edited by Mark S. Mooseker and Jon S. Morrow

Volume 39 Developmental Biology of Membrane Transport Systems (1991)
Edited by Dale J. Benos

Volume 40 Cell Lipids (1994)
Edited by Dick Hoekstra

Volume 41 Cell Biology and Membrane Transport Processes* (1994)
Edited by Michael Caplan

Volume 42 Chloride Channels (1994)
Edited by William B. Guggino

Volume 43 Membrane Protein–Cytoskeleton Interactions (1996)
Edited by W. James Nelson

Volume 44 Lipid Polymorphism and Membrane Properties (1997)
Edited by Richard Epand

Volume 45 The Eye's Aqueous Humor: From Secretion to Glaucoma (1998)
Edited by Mortimer M. Civan

Volume 46 Potassium Ion Channels: Molecular Structure, Function, and Diseases (1999)
Edited by Yoshihisa Kurachi, Lily Yeh Jan, and Michel Lazdunski

Volume 47 Amiloride-Sensitive Sodium Channels: Physiology and Functional Diversity (1999)
Edited by Dale J. Benos

Volume 48 Membrane Permeability: 100 Years since Ernest Overton (1999)
Edited by David W. Deamer, Arnost Kleinzeller, and Douglas M. Fambrough

Volume 65 Claudins
Edited by Alan S. L. Yu

Volume 66 Structure and Function of Calcium Release Channels
Edited by Irina I. Serysheva

Volume 67 Advances in Adrenergic Receptor Biology
Edited by Qin Wang

Volume 68 Membrane Fusion
Edited by Leonid V. Chernomordik and Michael M. Kozlov

Volume 69 Metal Transporters
Edited by Svetlana Lutsenko and José M. Argüello

CHAPTER ONE

Amino Acid Secondary Transporters: Toward a Common Transport Mechanism

Eva S. Schweikhard, Christine M. Ziegler[1]
Department of Structural Biology, Max Planck Institute of Biophysics, Frankfurt, Germany
[1]Corresponding author: E-mail: Christine.ziegler@biophys.mpg.de

Contents

Abstract

Solute carriers (SLC) that transport amino acids are key players in health and diseases in humans. Their prokaryotic relatives are often involved in essential physiological processes in microorganisms, e.g. in homeostasis and acidic/osmotic stress response. High-resolution X-ray structures of the sequence-unrelated amino acid transporters unraveled a striking structural similarity between carriers, which were formerly assigned to different families. The highly conserved fold is characterized by two inverted structural repeats of five transmembrane helices each and indicates common mechanistic transport concepts if not an evolutionary link among a large number of amino acid transporters. Therefore, these transporters are classified now into the structural amino acid–polyamine–organocation superfamily (APCS). The APCS includes among others the mammalian SLC6 transporters and the heterodimeric SLC7/SLC3 transporters. However,

Current Topics in Membranes, Volume 70
ISSN 1063-5823
http://dx.doi.org/10.1016/B978-0-12-394316-3.00001-6

it has to be noted that the APCS is not limited entirely to amino acid transporters but contains also transporters for, e.g. amino acid derivatives and sugars. For instance, the betaine–choline–carnitine transporter family of bacterial activity-regulated Na^+- and H^+-coupled symporters for glycine betaine and choline is also part of this second largest structural superfamily. The APCS fold provides different possibilities to transport the same amino acid. Arginine can be transported by an H^+-coupled symport or by antiport mechanism in exchange against agmatine for example. The convergence of the mechanistic concept of transport under comparable physiological conditions allows speculating if structurally unexplored amino acid transporters, e.g. the members of the SLC36 and SLC38 family, belong to the APCS, too. In the kidney, which is an organ that depends critically on the regulated amino acid transport, these different SLC transporters have to work together to account for proper function. Here, we will summarize the basic concepts of Na^+- and H^+-coupled amino acid symport and amino acid–product antiport in the light of the respective physiological requirements.

1. CLASSIFICATION OF AMINO ACID TRANSPORTERS

Amino acid transporters play very important roles in nutrient uptake, neurotransmitter recycling, cell redox balance, cell signaling and regulation of cell volume (Broer and Palacin, 2011). Providing the building blocks in protein synthesis, the 20 proteogenic amino acids are particularly vital for cell survival and growth in mammalian cells. Not all the proteogenic amino acids essential for mammals can be synthesized in the organism (Verrey et al., 2009), and their uptake is required from the environment. Some amino acids also play an important role as substrates for energy production and as precursors for the synthesis of other molecules, such as compatible solutes (osmolytes) and/or neurotransmitters (Verrey et al., 2009). For instance, the neutral amino acids leucine, isoleucine, methionine and valine participate in the production of brain neurotransmitters such as glutamate or acetylcholine (Broer, 2006) and proline can be converted directly into glutamate (Zhang et al., 2004). Aromatic amino acids serve as precursors for serotonin, noradrenaline and dopamine, and branched-chain amino acids provide tricarboxylic acid intermediates for the production of glutamate and γ-aminobutyric acid (GABA) (Murai, Saito, Masuda, & Itoh, 1988).

Until recently, amino acid transporters were assigned to a large number of different, sequence-unrelated families (http://www.tcdb.org.superfamily.php/). They were initially classified according to their sequences (Wong et al., 2012). Members of the sequence-related families can share the affinity for a substrate class and/or for the driving force. For instance, the neurotransmitter–sodium symporter (NSS) family comprises only sodium-coupled symporters that transport a broad range of neurotransmitters and amino acids. The betaine–choline–carnitine transporter (BCCT) family displays variety in coupling mode as we can find sodium- and proton-coupled symporters as well as antiporters, which import preferably substrates with a trimethylammonium group (Ziegler, Bremer, & Krämer, 2010). The amino acid–polyamine–organocation (APC) families (Saier, 2000; Wong et al., 2012) function either as H^+-coupled symporters or as substrate–product antiporters covering a large spectrum of cationic but also neutral, aromatic anionic and branched amino acids (Jack, Paulsen, & Saier, 2000).

Amino acids are very diverse in chemical nature and structure and apparently the combination of a certain amino acid and a coupling substrate is not exclusive, e.g. the same amino acid can be transported by different mechanisms. Each amino acid transport system adapts to the environmental conditions by choosing exactly the coupling mode, which allows to achieve the affinity required for certain physiological conditions (Broer, 2008; Verrey et al., 2009; Wong et al., 2012). Although very little sequence identity exists between amino acid transporters of different families (even between the ones with common substrate specificity), structural biology has lately contributed significantly unifying different amino acid transporter families. Recently, the increasing number of archaeal and bacterial amino acid transporter structures (Fang et al., 2009; Gao et al., 2009, 2010; Kowalczyk et al., 2011; Ma et al., 2012; Ressl, Terwisscha van Scheltinga, Vonrhein, Ott, & Ziegler, 2009; Shaffer, Goehring, Shankaranarayanan, & Gouaux, 2009; Yamashita, Singh, Kawate, Jin, and Gouaux, 2005) have revealed a similar overall fold (Chang, Lin, Keith Studley, Tran, & Saier, 2004; Rudnick, 2011) (Fig. 1). This fold is referred to as the APC superfamily (APCS) fold (Wong et al., 2012) and is characterized by two structural repeats of five transmembrane (TM) helices each (Fig. 2). One repeat is inverted with respect to the other, thereby related by a twofold axis of symmetry that runs parallel to the membrane plane (Fig. 2, solid arrow). The five helices of each repeat, denominated often as A1–A5 in repeat 1 and B1–B5 in repeat 2, are closely intertwined and form three structural domains, which have been shown to exhibit very distinct roles in the alternating-access mechanism of secondary transport (Forrest, Kramer, & Ziegler, 2011). A1–A2 and B1–B2 form a four-TM helix bundle, which lines the substrate translocation pathway from one side. Adjacent to the bundle, a helix scaffold in the form of a hash-sign is located that comprises A3–A4 and B3–B4. The last helices in each repeat (A5 and B5) are named arms (Fig. 2). All three elements (bundle, hash and arms) play distinct roles during transport as described below. Additional

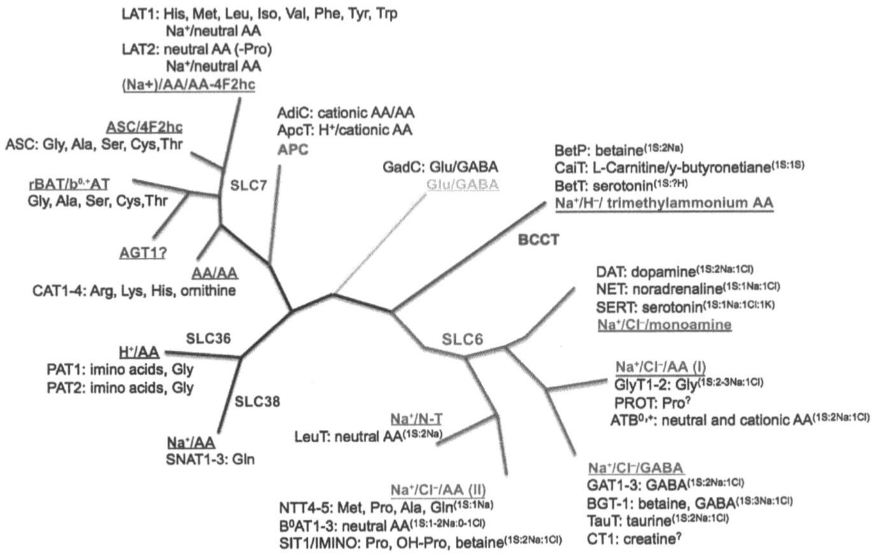

Figure 1 Phylogenetic tree of secondary transporters specific for amino acids (AA), polyamines and amino acid derivatives that share a common structural fold (adapted from Rudnick, 2011; Broer and Gether, 2012). The different branch colors indicate the sequence-related family assignments. The amino acids are shown in the three-letter code. For each transporter, the literature name (e.g. SERT for serotonin transporter), the substrates and the coupling ion or the counter-substrate (Na^+/H^+-coupled symport or substrate–product antiport) are listed. Additional ions required for transport are mentioned. The stoichiometry, if known, is given in brackets next to the substrate. See the color plate.

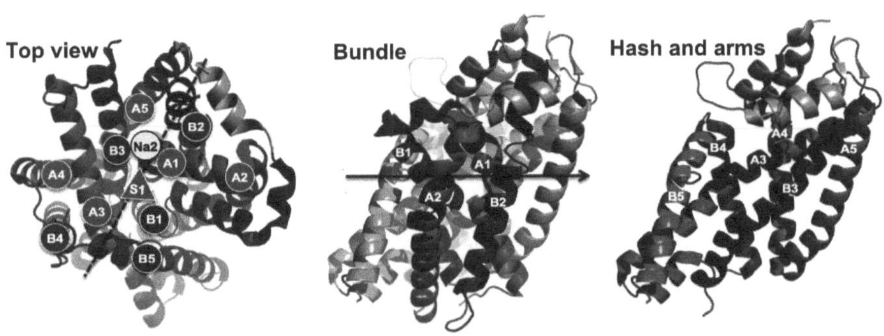

Figure 2 Structural repeats in LeuT (pdb entry 3F3A). Top view and side views on the LeuT structure. Repeat 1 is colored in red, and repeat 2 is in blue. The position of the central substrate-binding site is shown as green triangle, the conserved Na2 ion-binding site is colored in yellow. The bundle helices (A1–A2 and B1–B2) and the scaffold helices (A3–A5 and B3–B5) are depicted separately. The bundle helices are removed in the scaffold picture. Arms (A5 and B5) are colored in green. The twofold pseudosymmetry axis is shown as solid arrow in the picture of the bundle. See the color plate.

helices at the N- or C-terminal end often flank the three domains of the 10 TM transporter core. Interestingly, the appearance of the APCS fold seems to be independent of the sequence, coupling mode, and the nature of the substrates and thereby provides the basis for a more general mechanistic description of amino acid transport. It is important to note here that several of amino acid transporter families do not belong to the APCS, e.g. the family of sodium-dependent glutamate/aspartate transporters (excitatory amino acid carriers; Jiang and Amara, 2011) exhibit a very different fold (Yernool Boudker, Jin, & Gouaux, 2004).

2. HUMAN SLCS INVOLVED IN AMINO ACID AND NEUROTRANSMITTER TRANSPORT

Crystal structures of their prokaryotic homologs unambiguously showed that transporters of different mammalian SLC families share the APCS fold (Gao et al., 2009; Ma et al., 2012; Shaffer et al., 2009; Yamashita et al., 2005). Here we focus on the following two amino acid SLC families SLC6 and SLC7. Moreover, homology modeling, multi-sequence alignment procedures based on hydropathy blots and secondary structure predictions (Khafizov, Staritzbichler, Stamm, & Forrest, 2010), suggests a possible structural relation of SLC6 and SLC7 transporters to members of the SLC36 and SLC38 families (Schlessinger et al., 2010), for which no structural information is available to date.

2.1. Sodium-Coupled Amino Acid and Neurotransmitter Transport (SLC6)

The NSS family comprises sodium-coupled transporters specific for a broad range of amino acids, neurotransmitters, osmolytes and related nitrogenous substances (Broer and Gether, 2012). Eukaryotic genes that encode for NSS transporters are denominated as SLC6 (Kristensen et al., 2011), which is one of the largest transporter families in the human genome and currently has 20 members (Fig. 1, see recent review by Broer and Gether, 2012). Based on sequence similarity, the SLC6 family was grouped into the following four subfamilies: the GABA transporter branch, the monoamine transporter branch and the amino acid transporter branches (I) and (II) (Fig. 1). Although SLC6 transporters exhibit partly overlapping substrate specificity, they differ in cellular locations. GABA is the major inhibitory neurotransmitter in the brain. The transporters BGT-1 and TauT of the GABA subfamily facilitate uptake of the osmolytes glycine betaine and taurine, respectively (Lang, 2007). The creatine transporter (CT1) is specific for creatine, which is a high-energy phosphate storage compound found in muscle and brain (Wallimann, Tokarska-Schlattner, & Schlattner 2011). The monoamine transporter branch comprises the neurotransmitter transporters for dopamine, serotonin (5-HT) and noradrenaline, all of which play modulatory roles in many pathways affecting the central nervous system. The amino acid transporter branch (I) is grouped into highly specific transporters for glycine, GlyT1 and GlyT2 (Aragon and Lopez-Corcuera, 2005; Javitt, 2009), and proline (PROT) (Velaz-Faircloth et al., 1995). Glycine is the major inhibitory neurotransmitter in the spinal cord and imported exclusively by GlyT2 at this location. PROT is almost solely expressed in vesicular compartments in the brain, although its physiological role remains unclear. The amino acid transporter branch (I) comprises also the broadly specific transporter $ATB^{0,+}$, which transports neutral (0) and cationic (+) amino acids (Sloan, Grubb, & Mager, 2003). The amino acid transporter branch (II) covers the neutral amino acid transporters B^0AT1 and B^0AT3 (Parra et al., 2008). These transporters are found in the apical membrane of the epithelial cells in the kidney and intestine. Together with the proline- (hydroxyproline) and betaine-specific transporter IMINO, they remove amino acids from the lumen (Romeo et al., 2006; Vanslambrouck et al., 2010). The branch contains two other very similar transporters to IMINO, namely, SIT1 (IMINOB) and XT3 (IMINOK) localized in the kidney proximal tubule brush border membrane (Romeo et al., 2006).

SLC6 transporters are all Na^+ dependent, but with varying stoichiometry (1–3 Na^+). Stoichiometry can differ even among very homologous transporters, e.g. GlyT1 requires 2 Na^+, while GlyT2 uses 3 Na^+ and therefore accumulates glycine to higher amounts (Supplisson and Roux, 2002). GlyT1 and GlyT2 are expressed in locations that differ with respect to the substrate concentration and, evidently, the sodium coupling is adapted to the specific conditions (Raiteri and Raiteri, 2010).

Members of the GABA, monoamine and amino acid (I) branch seem to be all Cl^- dependent, while some transporters of the amino acid (II) branch (B^0AT1, B^0AT2, and NTT4) and the bacterial NSS transporters are Cl^- independent (Fig. 1). The monoamine transporter SERT is unique in its dependence in potassium antiport (Rudnick, 1998). Some prokaryotic homologs of the SLC6 family are stimulated by an inversely oriented proton gradient, indicating an additional proton-coupled antiport (Zhao et al., 2010). Interestingly, this antiport characteristic was only evident in the absence of chloride dependence consistent with an alternative role of both ions in the transport cycle (Broer and Gether, 2012).

2.2. Proton-Coupled Amino Acid Transporters (SLC36)

The SLC36 family comprises proton-coupled transporters mainly for proline and glycine. Proton motif force is considered as the evolutionarily older energizing mechanism (Anderson and Thwaites, 2010), although sodium motif force is the major transport driving force at the plasma membrane of mammalian cells (Boll, Daniel, & Gasnier, 2004). Therefore, proton-coupled transport is often facilitated in cells such as the small intestines that exhibit an acidic microclimate. The proton-coupled amino acid transporters (PAT) of the SLC36 family were initially classified as a lysosomal amino acid transporter (LYAAT1) exporting lysosomal proteolysis products (Boll et al., 2004). Their classification has been rendered more difficult by the fact that the proton gradient across the luminal membrane of proximal tubule cells depends on the activity of the sodium–proton exchanger NHE3, such that in many assays their transport function first appeared to be Na^+ dependent (Donowitz et al., 2009). PAT1 and PAT2 facilitate symport of protons, glycine or proline with a stoichiometry of 1:1, but different affinities (Zebisch and Brandsch, 2012). Proline binds with higher affinity than glycine to both PAT transporters that are also able to transport short-chain fatty acids (Foltz, Boll, Raschka, Kottra, & Daniel, 2004). PAT2 shows higher activity at neutral pH and the substrate affinity is strongly affected by the pH. Another notable characteristic is the additional affinity in PAT2 for GABA and glycine betaine (Foltz, Oechsler, Boll, Kottra, & Daniel, 2004), which may act as inhibitors (Edwards et al., 2011). It was suggested that the SLC36 transporters are distantly related to other PAT, such as the vesicular SLC32 neurotransmitter transporter VIAAT/VGAT and to the sodium-coupled SLC38 transporters (Boll et al., 2004).

In the absence of a structure, hydropathy profile alignments, e.g. by the AlignMe server (http://www.bioinfo.mpg.de/AlignMe; Khafizov et al., 2010), suggest some structural similarities with Na^+-coupled SLC6 transporters. Therefore, SLC36 transporters might share the APCS fold (Schlessinger et al., 2010). The similar substrate spectrum of PAT2, which is also able to facilitate GABA and glycine betaine transport, might be an additional argument. Structural data are required to clarify the fold of SLC36 transporters.

2.3. Sodium-Coupled Glutamine Symport (SLC38 Family)

The sodium-coupled neutral amino acid transporters (SNAT) of the SLC38 family mediate uptake of small, aliphatic amino acids by the System A subtypes (SNAT1, SNAT2, and SNAT4) in a pH-dependent manner. The

System N subtypes SNAT3 and SNAT5 also counter-transport protons and exhibit a narrower substrate spectrum compared to the System A subtypes. Glutamine represents the favored substrate throughout the SLC38 family. Glutamine is the most abundant circulating amino acid in the body (Broer, 2008). It is involved in the synthesis of the neurotransmitters GABA and glutamate. SLC38 transporters expressed in blood–brain barrier endothelial cells control the local glutamine gradients (Mackenzie and Erickson, 2004). SNAT1 expression is restricted to larger cortical microvessels, whereas SNAT3 is additionally expressed in blood–brain barrier capillary membranes (Ruderisch, Virgintino, Makrides, & Verrey, 2011) together with a phosphate-activated glutaminase (Solbu et al., 2005). SNAT transport systems are found also in the intestine (Sugawara et al., 2000), where glutamine transport is upregulated by amino acid depletion (Franchi-Gazzola, Visigalli, Bussolati, Dall'Asta, & Gazzola 1999). Thus, the intestine removes significant amounts of glutamine from the circulation for its metabolism (Sugawara, Nakanishi, Fei, Martindale et al., 2000). Only SNAT5 transports glycine, alanine, serine, and cysteine in addition to glutamine, asparagine, and histidine (Baird et al., 2004). The cellular and subcellular localization of the SNAT5 and its physiological role have not been determined to date.

SNAT3 mediates glutamine transport in cotransport with sodium in a 1:1 stoichiometry. Interestingly, the SNAT3 simultaneously counter-transports protons (Mackenzie and Erickson, 2004), a feature also observed in prokaryotic homologs of SLC6 transporters (Zhao et al., 2010). Similar to SLC36 transporters, no structural information is available for members of the SLC38. A profile-based sequence analysis approach predicted a distant relationship with SLC6 transporters (Zhang, Albers, Fiumera, Gameiro, & Grewer, 2009) and the homology modeling suggested conservation of one of the sodium-binding sites.

2.4. Cationic Amino Acid Transporter (SLC7) and Heterodimeric SLC7/SLC3

The SLC7 family comprises sodium-independent, proton-coupled symporters and amino acid antiporters and was therefore considered unrelated to the previously discussed SLC6 and SLC38 families. This distinction changed instantly, when structural data revealed the shared APCS fold of SLC7 transporters. The SLC7 family consists of the following two subfamily branches: the cationic amino acid transporters represented by the CAT system and the subfamily of heterodimeric amino acid transporters (HAT) (Palacin et al., 2005). The CAT branch comprises four members, which all exhibit a nearly identical substrate pattern (Verrey et al., 2004). One of them, CAT4, however, does not exhibit any transport activity and its physiological function is unclear. CAT1–3 all exchange cationic amino acids independent of Na^+ and the transport is stimulated by substrate concentration (Verri et al., 2005). CAT2A, CAT2B, and CAT3 display rather low affinity (Closs, Boissel, Habermeier, & Rotmann, 2006). CAT1 shows affinity for Arg, Lys and L-ornithine and is ubiquitously expressed in adult tissues except in liver (Broer, 2008; Closs et al., 2006).

The second branch of the SLC7 family comprises the so-called light chains of heterodimeric amino acid transporters (HAT) (Fotiadis Kanai, & Palacin, 2012). These transporters only reach the plasma membrane when they form a complex with a glycoprotein (Palacin and Kanai, 2004; Verrey et al., 2004) and are referred to as glycoprotein-associated transporters (Fort et al., 2007) (pdb entry 2DH2) (Fig. 3). The glycoproteins include

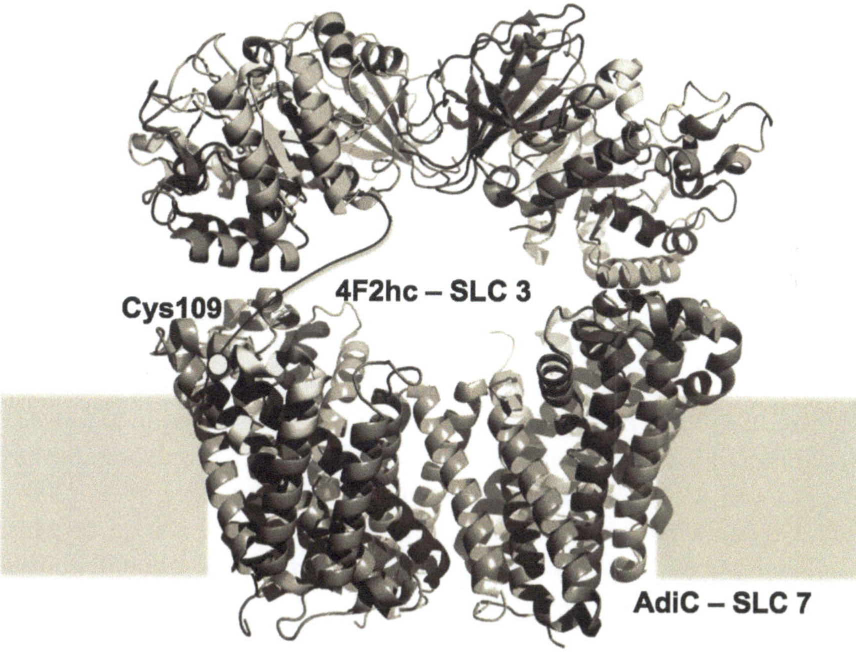

Figure 3 Side view on a putative complex between SLC3 and SLC7 showing a possible arrangement of both components to the HAT. Bundle helices of the AdiC structure (pdb entry 3BO6) are colored in red (A1 and A2) and the first hash helix A3 is colored in blue. It is suggested that 4F2hc SLC3 binds to the extracellular loop between A3 and A4 via disulfide bonds. Interestingly, cysteines can be found at the periplasmic side at the hash–membrane interface. See the color plate.

the two SLC3 family members, the rBAT protein (also called D2 or NBAT) and the 4F2 heavy-chain cell surface antigen (CD98hc or FRP) (Palacin and Kanai, 2004). SLC3 members comprise an intracellular N-terminus, a single TM domain and a large extracellular C-terminal domain that is related to bacterial α-glycosidases (Fort et al., 2007; Gabrisko and Janecek, 2009). rBAT and the 4F2hc protein are heavily glycosylated. The interaction between the trafficking heavy-chain subunit and the SLC7 transporter core is one of the intriguing characteristics of these heterodimeric transporter complexes.

The heterodimeric large amino acid transporter 4F2hc/LAT (Rossier et al., 1999) transports branched-chain (valine, leucine, and isoleucine) and aromatic (tryptophan and tyrosine) amino acids (del Amo, Urtti, & Yliperttula, 2008). LAT1 is expressed in the brain capillaries and in the basolateral membrane of the intestine and the kidney (del Amo et al., 2008). While the LAT1 system is rather specific for hydrophobic amino acids, the LAT2 has broad substrate specificity (Broer, Wagner, Lang, & Broer, 2000; Segawa et al., 1999). In addition, it was suggested that 4F2hc/LAT2 has a more specialized role exchanging intracellular cysteine for extracellular neutral amino acids (Bauch, Forster, Loffing-Cueni, Summa, & Verrey, 2003; Fernandez et al., 2003; Nemoto et al., 2003). 4F2hc/ASC-1 mediates the Na^+-independent transport of the small neutral amino acids (Gly, Ala, Ser, Cys, and Thr) with high affinity in the brain and lung. Transcripts are also observed in the kidney and intestine (Fukasawa et al., 2000; Nakauchi et al., 2000).

The cystinuria transporter rBAT/$b^{0,+}$AT is the major transporter for cationic amino acids and cysteine in the apical membrane of the kidney, small intestine, liver and placenta (Calonge et al., 1994; Chillaron et al., 1996; Fernandez et al., 2006). Cysteine and the cationic amino acids are transported with higher affinities compared to the neutral amino acids (Reig et al., 2002). Thus, physiologically, rBAT/$b^{0,+}$AT acts as an absorption mechanism for cationic amino acids and cysteine. rBAT is only stable when co-expressed with $b^{0,+}$AT transporter complex (Fernandez et al., 2002). $b^{0,+}$AT is stable but remains in the endoplasmic reticulum (Broer, 2008).

4F2hc/y^+LAT1 is the mediator of cationic amino acid efflux in epithelial cells (Broer, 2008). Importantly, 4F2hc guides y^+LAT1 to the basolateral membrane (Bauch and Verrey, 2002). The y^+LAT1 transporter accepts neutral and cationic amino acids with high affinity; however, the affinity of neutral amino acids increases by about two orders of magnitude in the presence of Na^+ (Kanai et al., 2000). In the absence of Na^+, it appears that the protons are cotransported (Kanai et al., 2000). The transport of cationic amino acids, in contrast, remains Na^+ independent (Kanai et al., 2000).

3. THE STRUCTURAL APCS: MECHANISTIC CONCEPTS

3.1. LeuT: A Paradigm for Na^+-Coupled Amino Acid Transporters

The conserved APCS fold was described for the first time for the bacterial Na^+-coupled symporter LeuT from *Aquifex aeolicus* (Yamashita et al., 2005) (pdb entry 2A65), a member of the NSS family. The structure of LeuT, which transports a broad spectrum of different amino acids, represented the first structural information on an SLC6 transporter. The superior resolution of the LeuT structure allowed for an unambiguous assignment of the position of the two coupling sodium ions, which are bound to an outward-facing occluded conformation in the center of the transporter. Very recently, the inward-facing structure of LeuT was reported (Krishnamurthy and Gouaux, 2012) (pdb entry 3TT3) and allowed for a molecular description of the alternating-access mechanism of symport (Jardetzky, 1966). During the alternating-access cycle, conformational changes modify the accessibility of a central binding site from both sides of the membrane upon binding and release of the substrates.

In LeuT, the following two types of conformational changes were observed: (1) a hinge-like gating movement of A1 and B1 (Fig. 2) and (2) a rigid-body rotation of the bundle against the hash domain (Krishnamurthy and Gouaux, 2012). A similar although slightly more pronounced rigid-body movement was predicted for LeuT by the homology modeling based on the conformational asymmetry between both repeats (Forrest et al., 2008). Key players in the transition from outward-facing state to inward-facing state are the formation and disruption of sodium- and substrate-binding sites that link movements of the bundle and scaffold helices and therefore drive the conformational changes. To allow for substrate binding, the conserved Na2 site (Fig. 2, yellow sphere in top view) has to be occupied with a sodium ion. As the carboxyl group of the substrate coordinates the second sodium ion in the Na1 site, the LeuT structure demonstrates the basic concepts of secondary transport (Forrest et al., 2011). Substrate and co-substrate have to be present simultaneously to enable the transporter to isomerize to the alternating conformation. Biophysical and bioinformatics studies have largely confirmed the structurally described mechanism of sodium-coupled transport via an NSS transporter (Claxton et al., 2010; Zhao et al., 2011; Zdravkovic, Zhao, Lev, Cuervo, & Noskov, 2012).

The structural data on the APCS fold shed light into the partly overlapping affinities of individual SLC6 transporters for some neutral amino acids, e.g. for alanine, which is a substrate in the GABA branch (TauT, GAT-2 and GAT-3) and in the amino acid (I) branch ($ATB^{0,+}$). Glycine is a substrate of GlyT and $ATB^{0,+}$, as well as for the amino acid (II) transporters B^0AT1 and B^0AT3. Glycine betaine is transported by BGT-1 and also by IMINO and SIT1. This shared affinity most likely results from substrate coordination being maintained by the carboxyl or amino group of the substrate (Broer and Gether, 2012). The size of an amino acid plays a role in substrate specificity. Several homology models on SLC6 transporters have allowed for the characterization of both substrate- and sodium ion-binding sites (Beuming et al., 2008; Jorgensen et al., 2007). Chloride-binding sites have been identified in the chloride-dependent members of the SLC6 family (Forrest, Tavoulari, Zhang, Rudnick, & Honig, 2007; Zomot et al., 2007). In chloride-independent SLC6 transporters, a negatively charged glutamate residue mimics binding of chloride close to the Na1 site. It was further suggested that in chloride-independent transporters, a proton might bind and unbind during the transport cycle and thus be counter-transported in an H^+ antiport fashion. H^+ involvement may facilitate the return step of the "empty" transporter (Zhao et al., 2010). This ensures a charge balance among the SLC6 transporters with similar mechanistic features but different molecular solutions (Zhao et al., 2010). However, the molecular mechanism of chloride coupling and potassium or proton antiport is not entirely provided by the structure of the bacterial homolog LeuT and requires a structural investigation of the respective SLC6 transporters.

3.2. Na^+-Coupled Glycine Betaine Transport: A New Flavor of the LeuT Mechanism?

The trimeric Na^+-coupled glycine betaine symporter BetP from *Corynebacterium glutamicum* is a member of the BCCT family (Ziegler et al., 2010). BCC transporters share the LeuT-like fold and are assigned to the APCS (Wong et al., 2012). However, members of the BCCT family have no human homologs and therefore are not related to an SLC family. BetP is exclusively specific for the zwitterionic glycine betaine, which is a trimethylammonium derivative of glycine. Residues in the bundle helices A1–A2 and B1 coordinate glycine betaine in a closed state and in an inward-facing state (Perez, Koshy, Yildiz, & Ziegler, 2012) (pdb entry 4DOJ). The trimethylammonium group of glycine betaine is caged by three tryptophans, which are highly conserved in the BCCT family. They are located in the middle of B1 forming a prism-like structure and contribute to the specificity of BetP (Ziegler et al., 2010). Although being sodium coupled, BetP shows some similarities to the proton-coupled SLC36 glycine transporter PAT2, both of which contain a glycine-rich motif in A1 and conserved aromatic residues in B1. Recent data showed that exchange of only one residue in A1 switches BetP from sodium coupled to proton coupled (Perez, Koshy et al., 2011) (pdb entry 3PO3). To achieve proton-coupled transport in the BetP, an aspartate was introduced, which might correspond to a lysine located at a similar position in PAT2, thereby supporting a possible relationship between BCC transporters and SLC36 transporters despite the different coupling modes.

Recently, it was shown in computational and functional studies that residues in B1–A3 and A1–B3 form the binding sites of sodium (Na1 and Na2), respectively, in a symmetry-related fashion (Khafizov et al., 2012). While the Na2 sodium site is highly conserved in APCS fold transporters, the Na1 sodium site in BetP has a very different location compared to LeuT (Khafizov et al., 2012). Six different conformational states in the transport cycle of BetP were observed in crystal structures (Perez et al., 2012). Compared to LeuT, the alternating-access mechanism in the BetP represents a hybrid of a rigid-body movement of the bundle and scaffold domains relative to each other and individual flexing of symmetry-related helices. The observation of a unique closed transition state reveals that the extra- and intracellular gates undergo uncoupled hinge movements. A new property that emerges for the BetP is the spring-like movement of the unfolded region of A1 (Perez et al., 2012). Partly unfolded TM helices are suggested to provide structural flexibility for ion coupling (Screpanti and Hunte, 2007). The structural data on BetP further develop an important concept raised already for LeuT (Krishnamurthy and Gouaux, 2012), suggesting that hinge motions trigger the formation and disruption of the substrate- and sodium-binding sites. However, the rigid-body movements are much less evident for the BetP than for LeuT and might be related to a tight intratrimeric ionic network that restricts the movement of the entire bundle. The unique coordination of the substrate by the Na1 sodium ion in LeuT and in BetP might be one of the reasons for the varying contributions of rigid-body rotation versus gating hinge movements (Perez et al., 2012). Comparison between LeuT and BetP indicated that the conservation of the Na2 site is a common element in the transport mechanism of symporters sharing the APCS fold. A similar hybrid "rocking bundle" and "gating" mechanism is conceivable for other APCS fold symporters, too, although the presence of glycines and prolines in the midsection of TM helices will probably dictate to what extent hinge-like gating can occur in a particular transporter (Perez et al., 2012).

Conformational changes reported for another APCS fold transporter, the benzyl-hydantoin sodium symporter Mhp1 from *Microbacterium liquefaciens*, are quite distinct from those reported for LeuT and BetP (Weyand et al., 2011) (pdb entry 2JLN). Mhp1, which requires only one sodium ion for transport, keeps the bundle helices static during the transport cycle and shows a flexible hash domain. Hinge-like gating movements are not evident in the bundle helices A1 and B1 but were observed in A5 and B5. These differences in conformational changes in APCS fold transporters raise the question on the role of the coupling substrate in rigid-body and flexing movements (Forrest and Rudnick, 2009; Forrest et al., 2011).

3.3. H^+ Symport and Antiport: Structures of APC Transporters

SLC7 transporters are homologs to members of the APC family that comprises proton-coupled symporters and amino acid antiporters. Surprisingly, the sodium-independent APC transporter structures revealed the same APCS fold as observed for sodium-coupled SLC6 symporters (Wong et al., 2012). The following three transporters of the APC family were crystallized in different conformational states, namely, ApcT from *Methanocaldococcus jannaschii* (Shaffer et al., 2009) (pdb entry 3GIA), AdiC from *Escherichia coli* (Fang et al., 2009; Gao et al., 2009, 2010; Kowalczyk et al., 2011) (pdb entries 3NCY, 3LRB, 3LIL, 3OB6) and GadC from *E. coli* (Ma et al., 2012) (pdb entry 4DJI) (Fig. 4A).

Direct comparison of the sodium-coupled LeuT and BetP contributed significantly to our understanding of the amino acid transport in general, and revealing the functional relationship between SLC6 and SLC7 transporters.

The antiporters AdiC and GadC are both crucially involved in the acidic stress response in enteric bacteria, which have developed sophisticated systems to maintain intracellular pH under acidic conditions (Gong, Richard, & Foster, 2003). One strategy is the import and decarboxylation of amino acids, thereby enabling enteric bacteria such as *E. coli* to grow even in the extremely acidic environment ($pH < 2$) of the stomach (Marteau and Chaput, 2011).

The glutamate–GABA antiporter GadC facilitates the exchange of extracellular glutamate with the intracellular reaction product GABA

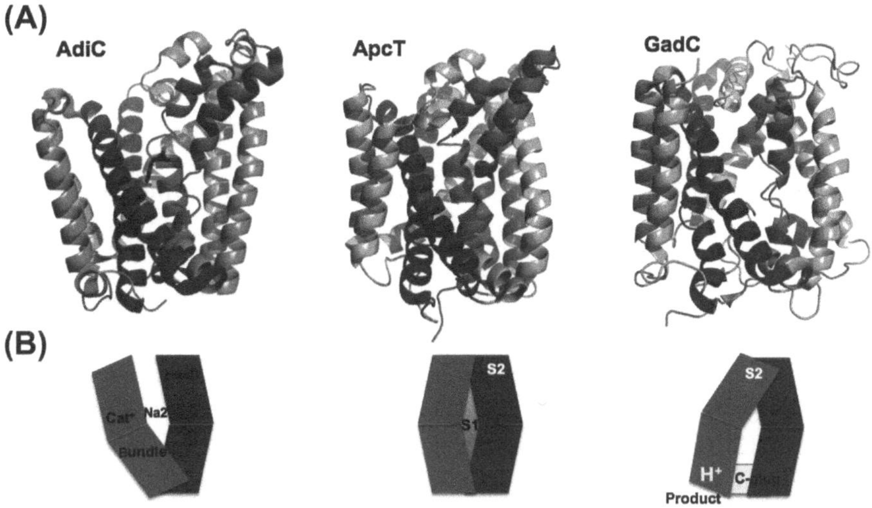

Figure 4 (A) Side views on the X-ray structures of AdiC (pdb entry 3OB6) in an outward-open conformation, ApcT (pdb entry 3GIA) in an occluded conformation and GadC (pdb entry 4DJI) in an inward-open conformation. The bundle helices A1 and B1 (in blue) show a pronounced unfolded stretches in their midsections. Hash helices A3 and B3 are colored in red. (B) Schematic drawing of conformational states in the APCS fold indicating the potential second binding site, the H^+ antiport and a possible role of C-terminal domains acting as a plug in an inward-facing state. See the color plate.

(De Biase, Tramonti, Bossa, & Visca 1999), while the AdiC exchanges extracellular arginine for intracellular agmatine (Gong et al., 2003). During decarboxylation of the transported glutamate or arginine for example, a proton in the cytoplasm replaces the COOH group. AdiC and GadC therefore promote the biochemical neutralization of the intracellular protons under acidic conditions acting as the virtual proton pump (Fang et al., 2009).

GadC is highly selective in substrate transport and shows only very reduced transport of other amino acids, e.g. Asp or Asn (Ma et al., 2012). Glutamate transport by GadC is strictly pH dependent. At pH values below 5.5, GadC shows elevated activity, which enables proton removal under acidic conditions. At neutral pH, transport activity is reduced and counteracts proton efflux (Ma et al., 2012). In contrast to GadC, the AdiC shows transport at all pH levels, although very moderate under basic conditions and elevated transport activities at pH 4 and 6 (Tomitori, Kashiwagi, & Igarashi, 2012). Interestingly, the SLC7 transporter CAT2 exhibits decreased transport with decreasing pH (Closs, Basha, Habermeier, & Forstermann,

1997; Vekony, Wolf, Boissel, Gnauert, & Closs, 2001), which is reciprocal to what was reported for AdiC and GadC.

GadC was crystallized in an inward-facing apo state (Ma et al., 2012). The available AdiC crystal structures capture the following three states of the transport cycle: the outward-open apo state (Fang et al., 2009), an Arg-bound outward-open (Kowalczyk et al., 2011) and an Arg-bound outward-occluded state (Gao et al., 2010).

In contrast to AdiC and GadC, ApcT is able to function under certain conditions as sodium-independent, PAT with broad substrate specificity (Shaffer et al., 2009). ApcT facilitates transport of Ser, Gln, Ala, and Gly, which is suppressed at neutral pH and is the archaeal ortholog of the SLC7 glutamate/cysteine antiporter xCT that supplies arginine for nitric oxide synthesis (Kanai and Endou, 2001). ApcT was crystallized in an occluded apo state and at low external pH (Shaffer et al., 2009).

Structural alignment between outward-facing states of AdiC, the occluded state of ApcT and the inward-facing state of GadC (Fig. 4) reveals conformational changes in the four-TM bundle, which undergoes a rigid-body clockwise rotation against the adjacent hash domain, but displays only reduced movements in A3, B3, and B4 (Ma et al., 2012). In addition, individual residues occlude the substrate-binding site during the transitions from outward- to inward-facing states.

These conformational changes are largely in agreement to what was reported for LeuT (Krishnamurthy and Gouaux, 2012) and BetP (Perez et al., 2012), but distinct from the changes observed for Mhp1 (Weyand et al., 2011), which although sharing the APCS fold is not an amino acid transporter.

One straightforward explanation for the pronounced hinge-like movements is that A1 and B1 in amino acid transporters possess a significant intrinsic flexibility around their midsection resulting in unfolded stretches, which are more pronounced in LeuT, AdiC and GadC than in BetP and ApcT. These stretches enable one or both bundle helices to undergo the described hinge-like bending movement during isomerization from one state to the other. The periplasmic and cytoplasmic end points of A1 and B1 or A5 and B5 open and close the periplasmic and cytoplasmic pathways, respectively. In the glycine betaine symporter BetP, this movement is the most pronounced (Gartner, Perez, Koshy, & Ziegler, 2011), as the flexible stretch of A1 bends away from the scaffold–bundle interface. Apparently, amino acid transporters conserve an intrinsic flexibility of A1 and B1 that promotes a common gating–rigid body movement. The degree of plasticity of the unfolded stretches is characteristic for each transporter. This property of intrinsic flexibility in A1 and B1 can be exploited to some extent in assigning members to the structural APCS (Schlessinger et al., 2010).

AdiC forms a homodimer in the crystal structure, while SLC7 transporters form heterodimeric transporter complexes with SLC3 (Palacin and Kanai, 2004). SLC3 is covalently bound through disulfide bonds to the membrane-embedded transporter core (Fig. 3) (Broer, 2008). The main role of these glycoproteins is the trafficking of the complex to the membrane. In fact, both subunits have to be co-expressed to reach the plasma membrane. As interaction site, the extracellular loop connecting A3 and A4 is proposed (Fort et al., 2007). The AdiC structure shows at least two cysteines in the scaffold helices A4 and B3 that would provide an interface for this interaction (Fig. 3). Based on the structural data for conformational changes in APCS fold transporters, a similar interaction might not directly affect the transport process.

4. IONS AND DIRECTIONALITY IN AMINO ACID TRANSPORT

4.1. Sodium and Proton Coupling: Role of Cationic Charges

As described independently for the leucine transporter LeuT and the glycine betaine transporter BetP, sodium binding has a predominant role in transducing and linking the scaffold and bundle helices, thereby causing the alternating movement of both repeats against each other (Krishnamurthy and Gouaux, 2012; Perez, Koshy et al., 2011). A sodium ion bound to the conserved Na2 position is essential to trigger the side chain rearrangement that renders the central binding site accessible. Subsequently, binding of an amino acid provides at least partly the energy necessary to overcome the potential barrier for the conformational change from outward to inward via a closed state.

In ApcT, the amine moiety of a conserved lysine residue in the gating helix A5 is located in a position equivalent to the Na2 site (Shaffer et al., 2009). The amine group when protonated evidently performs the same functional role as the Na2 sodium ion and triggers the opening of the extracellular pathway. Therefore, it can be concluded that during the alternating-access cycle in APCS fold transporters, a positive charge at the Na2 site created either through protonation of a charged residue or by sodium binding results in a close interaction of A1 with residues in B3. This interaction controls the conformation of the unfolded stretch in A1 and subsequently the opening of the central binding site toward the extracellular space pointing toward a unifying mechanism between proton and sodium coupling (Perez, Koshy et al., 2011; Shaffer et al., 2009). The conservation of the Na2 site is again a strong indicator that additional SLC transporters share the APCS fold. A homology model using LeuT as the template predicted the location of a potential Na2 site in the SLC38 transporter SNAT2, and this site was confirmed by the functional studies (Zhang et al., 2009).

The LeuT structure reveals that the second sodium ion located in the Na1 position coordinates the carboxyl group of leucine (Yamashita et al., 2005). The coordination by the carboxyl group of the substrates in LeuT (Yamashita et al., 2005) implies that the Na1 sodium ion dictates to some extent the specificity toward the substrate, e.g. a tendency to transport neutral amino acids. This specificity can be modulated by the additional ions such as chloride (Henry, Meiler, & Blakely, 2007) or potassium (Rudnick, 1998), which are required for some highly specific SLC6 transporters. However, this second Na1 ion is not required for all APCS fold symporters, such as the Mhp1 symporter (Shimamura et al., 2010) and the glucose symporter vSGLT (Faham et al., 2008). In comparison, H^+-coupled APCS fold transporters exhibit broad substrate specificity (Wong et al., 2012), most likely due to fixed side chains that are responsible for initiating the transition between states.

4.2. Symport versus Antiport

It is safe to assume that once the substrate fits into the central binding site, the conformational changes required to release the substrate on the other side of the membrane are independent on the nature of the motif force, e.g. proton- and sodium-coupled symport follow a similar mechanism. The question remains how an antiporter opens the extracellular pathways toward the central binding site without binding of the coupling ions.

In AdiC, A1 and B3 are interconnected directly by hydrogen bonds (Kowalczyk et al., 2011). Similarly, in the carnitine–γ-butyrobetaine BCCT antiporter CaiT (Schulze, Koster, Geldmacher, Terwisscha van Scheltinga, & Kuhlbrandt 2010) (pdb entry 2WSX), an arginine is located at a position equivalent to Na2 site and bridges an interaction between A1 and B3, making the presence of a sodium ion obsolete.

However, it has to be taken into account that in substrate–product exchange, the antiporter cycles back in complex with the counter-substrate.

It is very likely that the release of the counter-substrate leaves the antiporter already in an outward-open conformation that allows the substrate to bind without an additional step. Molecular dimension (MD) simulation on AdiC (Zomot and Bahar, 2011) showed that the protonation of a glutamate residue in B1 selectively causes the dissociation and the release of agmatine, but not of arginine to the extracellular side.

Evidently, opening of the central binding site toward the extracellular space requires a binding or release event, which might be in antiporters the release of the product counter-substrate. Therefore, the major difference to symport is that antiporter does not require a binding event to render the central substrate-binding site accessible, because there is no occluded apo state; the central binding site is occupied by the intracellular product counter-substrate.

For both symport and antiport, it is assumed that the subsequent substrate binding if necessary together with the second sodium or additional ions initiates conformational changes that occlude the central substrate-binding site. Recent site-directed spin labeling and electron paramagnetic resonance analysis on purified LeuT (Claxton et al., 2010) demonstrated the formation of an open outward-facing state with accessible substrate-binding site and the subsequent occlusion of this site after binding of leucine (Claxton et al., 2010). However, we have to ask the question which event will trigger now both in symport and antiport the conformational change from a closed state as observed for BetP to inward-facing state (Perez et al., 2012).

It was suggested that the free energy of closed conformation might be such that the transition occurs by thermal movement (Broer and Gether, 2012). However, an alternative possibility would be the binding of a second substrate to an extracellular binding site. This site was observed structurally in the BCCT family (Schulze et al., 2010) and for LeuT steered molecular dynamics have suggested the existence of a second extracellular substrate-binding site (Shi, Quick, Zhao, Weinstein, & Javitch, 2008). The existence of such a site has been further supported by single molecule Fluorescence resonance energy transfer (FRET) studies of conformational changes in LeuT (Zhao et al., 2011). The functional role of this binding site is still a matter of debate (Piscitelli, Krishnamurthy, & Gouaux 2010).

5. REGULATION IN APCS FOLD TRANSPORTERS

Regulation of amino acid transport is a crucial requirement to guarantee the fine-tuned balance of amino acid concentration in different cellular compartments and organs (Verrey et al., 2009). Some amino acids display similar concentrations within and outside of the cells, while others display an asymmetric distribution, and the amino acid transporters are in charge to maintain the amino acid concentration asymmetry.

To perform this task, amino acid transport is highly regulated via protein–protein interactions or by posttranslational modification such as phosphorylation (Chen, Reith, & Quick, 2004) or ubiquitination (Miranda and Sorkin, 2007) mostly at the terminal domains. Regulation occurs by regulated trafficking and/or insertion (Chen et al., 2004) rather than by changes of the catalytic activity. This process can be affected by activation of the kinases (Melikian, 2004) and binding of the substrate (Binda et al., 2008). Interaction of syntaxin1A with members of the SLC6 family, e.g. with GlyT1 and GlyT2 (Broer and Gether, 2012), can even promote reverse transport of substrate (Geerlings, Lopez-Corcuera, & Aragon, 2000). Evidently, oligomerization plays an important role in the SLC7 family and examples are also found for members of the SLC6 family, e.g. for the serotonin transporter (Kilic and Rudnick, 2000).

Although these regulatory interactions are very specific for the mammalian system, the bacterial structures of regulated transporters involved in the cellular stress response as the antiporter GadC in acidic stress or the symporter BetP in osmotic stress might shed some light into the basic principles of transport activity regulation.

The crystal structure of GadC (Ma et al., 2012) was determined under inactive conditions. In the inward-facing open transporter state, the C-terminal domain plugs the cytoplasmic pathway. The C-terminal domain is partly folded and occludes the central substrate-binding site from the cytoplasm. A blocked transport path under inactive conditions suggests that under active conditions, the C-terminal plug will move to give way to the central binding site. A C-terminal truncated GadC mutant ΔC34 showed a significant level of Glu accumulation at a pH of 6.5 where transport activity was strongly decreased for the wild-type (WT) protein. Deletion of the C-terminal domain in the GadC shifted its pH-dependent substrate transport toward a higher pH.

The regulation of the trimeric glycine betaine symporter BetP shows a similar pattern. Glycine betaine is an efficient osmolyte that is accumulated to high intracellular amounts by BetP in response to an upshift in the external osmolality (Ziegler et al., 2010). BetP is activated above ~200 mosmol/kg in *C. glutamicum* cells (Farwick, Siewe, & Kramer, 1995) triggered by an increase in internal K^+ concentration as a consequence of immediate cell shrinkage (Rubenhagen, Morbach, & Kramer, 2001).

The positively charged C-terminal domain was identified as a K^+ sensor (Schiller, Rubenhagen, Kramer, & Morbach, 2004). In the trimeric assembly, the C-terminal domain of each protomer is interacting with the cytoplasmic loops of the adjacent protomer (Ressl et al., 2009) (pdb entry 2WIT). Positively charged residues located in the C-terminal domain and the negatively charged residues located in the loop in front of A1 and B1 maintain this ionic interaction. Therefore, C-terminal interactions are transduced to bundle helices that are key players in the alternating-access mechanism of the APCS fold transporters (Perez et al., 2012). Nearly all residues in the α-helical C-terminal domain are critical in K^+ sensing and activation and any point mutations disturbing the helical motif result in deregulation of BetP (Perez et al., 2011). Truncation of the C-terminal domain resulted in deregulation and a constitutively active mutant. Although structural and functional data reveal the impact of the C-terminal domain, only little is known on how these elements dynamically change their conformation upon activation (Ge, Perez, Waclawska, Ziegler, & Muller, 2011; Nicklisch et al., 2012).

In light of a terminal regulation of these two APCS fold transporters, it seems very likely that similar concepts are found also in the regulatory mechanisms of, e.g., SLC6 and SLC7 transporters. The basic concept is that either in a monomeric state the intracellular pathway is blocked or an inward-facing state is locked by an interaction with the terminal domain. In an oligomeric state, terminal domains might interact with each other or with the transporter core of the adjacent protomer. Again, intensive structure–function studies on SLC transporters are required to understand the molecular basis of transport regulation.

ACKNOWLEDGMENTS

We are grateful to Manuel Palacin and Reinhard Krämer for critical reading of the manuscript, corrections and many important suggestions.

REFERENCES

Anderson, C. M., & Thwaites, D. T. (2010). Hijacking solute carriers for proton-coupled drug transport. *Physiology (Bethesda, Md.), 25*, 364–377.

Aragon, C., & Lopez-Corcuera, B. (2005). Glycine transporters: crucial roles of pharmacological interest revealed by gene deletion. *Trends in Pharmacological Sciences, 26*, 283–286.

Baird, F. E., Beattie, K. J., Hyde, A. R., Ganapathy, V., Rennie, M. J., & Taylor, P. M. (2004). Bidirectional substrate fluxes through the system N (SNAT5) glutamine transporter may determine net glutamine flux in rat liver. *The Journal of Physiology, 559*, 367–381.

Bauch, C., Forster, N., Loffing-Cueni, D., Summa, V., & Verrey, F. (2003). Functional cooperation of epithelial heteromeric amino acid transporters expressed in madin-darby canine kidney cells. *Journal of Biological Chemistry, 278*, 1316–1322.

Bauch, C., & Verrey, F. (2002). Apical heterodimeric cystine and cationic amino acid transporter expressed in MDCK cells. *American Journal of Physiology. Renal Physiology*, *283*, F181–F189.

Beuming, T., Kniazeff, J., Bergmann, M. L., Shi, L., Gracia, L., Raniszewska, K., et al. (2008). The binding sites for cocaine and dopamine in the dopamine transporter overlap. *Nature Neuroscience*, *11*, 780–789.

Binda, F., Dipace, C., Bowton, E., Robertson, S. D., Lute, B. J., Fog, J. U., et al. (2008). Syntaxin 1A interaction with the dopamine transporter promotes amphetamine-induced dopamine efflux. *Molecular Pharmacology*, *74*, 1101–1108.

Boll, M., Daniel, H., & Gasnier, B. (2004). The SLC36 family: proton-coupled transporters for the absorption of selected amino acids from extracellular and intracellular proteolysis. *Pflügers Archiv European Journal of Physiology*, *447*, 776–779.

Broer, A., Wagner, C. A., Lang, F., & Broer, S. (2000). The heterodimeric amino acid transporter 4F2hc/y+LAT2 mediates arginine efflux in exchange with glutamine. *Biochemical Journal*, *349*(Pt 3), 787–795.

Broer, S. (2006). The SLC6 orphans are forming a family of amino acid transporters. *Neurochemistry International*, *48*, 559–567.

Broer, S. (2008). Amino acid transport across mammalian intestinal and renal epithelia. *Physiological Reviews*, *88*, 249–286.

Broer, S., & Gether, U. (2012). The solute carrier family 6. *British Journal of Pharmacology*.

Broer, S., & Palacin, M. (2011). The role of amino acid transporters in inherited and acquired diseases. *Biochemical Journal*, *436*, 193–211.

Calonge, M. J., Gasparini, P., Chillaron, J., Chillon, M., Gallucci, M., Rousaud, F., et al. (1994). Cystinuria caused by mutations in rBAT, a gene involved in the transport of cystine. *Nature Genetics*, *6*, 420–425.

Chang, A. B., Lin, R., Keith Studley, W., Tran, C. V., & Saier, M. H., Jr. (2004). Phylogeny as a guide to structure and function of membrane transport proteins. *Molecular Membrane Biology*, *21*, 171–181.

Chen, N. H., Reith, M. E., & Quick, M. W. (2004). Synaptic uptake and beyond: the sodium- and chloride-dependent neurotransmitter transporter family SLC6. *Pflügers Archiv European Journal of Physiology*, *447*, 519–531.

Chillaron, J., Estevez, R., Mora, C., Wagner, C. A., Suessbrich, H., Lang, F., et al. (1996). Obligatory amino acid exchange via systems bo,+-like and y+L-like. A tertiary active transport mechanism for renal reabsorption of cystine and dibasic amino acids. *Journal of Biological Chemistry*, *271*, 17761–17770.

Claxton, D. P., Quick, M., Shi, L., de Carvalho, F. D., Weinstein, H., Javitch, J. A., et al. (2010). Ion/substrate-dependent conformational dynamics of a bacterial homolog of neurotransmitter: sodium symporters. *Nature Structural & Molecular Biology*, *17*, 822–829.

Closs, E. I., Basha, F. Z., Habermeier, A., & Forstermann, U. (1997). Interference of L-arginine analogues with L-arginine transport mediated by the y+ carrier hCAT-2B. *Nitric Oxide: Biology and Chemistry/Official Journal of the Nitric Oxide Society*, *1*, 65–73.

Closs, E. I., Boissel, J. P., Habermeier, A., & Rotmann, A. (2006). Structure and function of cationic amino acid transporters (CATs). *Journal of Membrane Biology*, *213*, 67–77.

De Biase, D., Tramonti, A., Bossa, F., & Visca, P. (1999). The response to stationary-phase stress conditions in Escherichia coli: role and regulation of the glutamic acid decarboxylase system. *Molecular Microbiology*, *32*, 1198–1211.

del Amo, E. M., Urtti, A., & Yliperttula, M. (2008). Pharmacokinetic role of L-type amino acid transporters LAT1 and LAT2. *European Journal of Pharmaceutical Sciences: Official Journal of the European Federation for Pharmaceutical Sciences*, *35*, 161–174.

Donowitz, M., Mohan, S., Zhu, C. X., Chen, T. E., Lin, R., Cha, B., et al. (2009). NHE3 regulatory complexes. *Journal of Experimental Biology*, *212*, 1638–1646.

Edwards, N., Anderson, C. M., Gatfield, K. M., Jevons, M. P., Ganapathy, V., & Thwaites, D. T. (2011). Amino acid derivatives are substrates or non-transported inhibitors of the amino acid transporter PAT2 (slc36a2). *Biochimica et Biophysica Acta, 1808*, 260–270.

Faham, S., Watanabe, A., Besserer, G. M., Cascio, D., Specht, A., Hirayama, B. A., et al. (2008). The crystal structure of a sodium galactose transporter reveals mechanistic insights into Na^+/sugar symport. *Science, 321*, 810–814.

Fang, Y., Jayaram, H., Shane, T., Kolmakova-Partensky, L., Wu, F., Williams, C., et al. (2009). Structure of a prokaryotic virtual proton pump at 3.2 Å resolution. *Nature, 460*, 1040–1043.

Farwick, M., Siewe, R. M., & Kramer, R. (1995). Glycine betaine uptake after hyperosmotic shift in Corynebacterium glutamicum. *Journal of Bacteriology, 177*, 4690–4695.

Fernandez, E., Carrascal, M., Rousaud, F., Abian, J., Zorzano, A., Palacin, M., et al. (2002). rBAT-b(0,+)AT heterodimer is the main apical reabsorption system for cystine in the kidney. *American Journal of Physiology. Renal Physiology, 283*, F540–F548.

Fernandez, E., Jimenez-Vidal, M., Calvo, M., Zorzano, A., Tebar, F., Palacin, M., et al. (2006). The structural and functional units of heteromeric amino acid transporters. The heavy subunit rBAT dictates oligomerization of the heteromeric amino acid transporters. *Journal of Biological Chemistry, 281*, 26552–26561.

Fernandez, E., Torrents, D., Chillaron, J., Martin Del Rio, R., Zorzano, A., & Palacin, M. (2003). Basolateral LAT-2 has a major role in the transepithelial flux of L-cystine in the renal proximal tubule cell line OK. *Journal of the American Society of Nephrology: JASN, 14*, 837–847.

Foltz, M., Boll, M., Raschka, L., Kottra, G., & Daniel, H. (2004). A novel bifunctionality: PAT1 and PAT2 mediate electrogenic proton/amino acid and electroneutral proton/fatty acid symport. *FASEB Journal: Official Publication of the Federation of American Societies for Experimental Biology, 18*, 1758–1760.

Foltz, M., Oechsler, C., Boll, M., Kottra, G., & Daniel, H. (2004). Substrate specificity and transport mode of the proton-dependent amino acid transporter mPAT2. *European Journal of Biochemistry/FEBS, 271*, 3340–3347.

Forrest, L. R., Kramer, R., & Ziegler, C. (2011). The structural basis of secondary active transport mechanisms. *Biochimica et Biophysica Acta, 1807*, 167–188.

Forrest, L. R., & Rudnick, G. (2009). The rocking bundle: a mechanism for ion-coupled solute flux by symmetrical transporters. *Physiology (Bethesda, Md.), 24*, 377–386.

Forrest, L. R., Tavoulari, S., Zhang, Y. W., Rudnick, G., & Honig, B. (2007). Identification of a chloride ion binding site in Na^+/Cl -dependent transporters. *Proceedings of the National Academy of Sciences of the United States of America, 104*, 12761–12766.

Forrest, L. R., Zhang, Y. W., Jacobs, M. T., Gesmonde, J., Xie, L., Honig, B. H., et al. (2008). Mechanism for alternating access in neurotransmitter transporters. *Proceedings of the National Academy of Sciences of the United States of America, 105*, 10338–10343.

Fort, J., de la Ballina, L. R., Burghardt, H. E., Ferrer-Costa, C., Turnay, J., Ferrer-Orta, C., et al. (2007). The structure of human 4F2hc ectodomain provides a model for homodimerization and electrostatic interaction with plasma membrane. *Journal of Biological Chemistry, 282*, 31444–31452.

Fotiadis, D., Kanai, Y., & Palacin, M. (2012). The SLC3 and SLC7 families of amino acid transporters.

Franchi-Gazzola, R., Visigalli, R., Bussolati, O., Dall'Asta, V., & Gazzola, G. C. (1999). Adaptive increase of amino acid transport system A requires ERK1/2 activation. *Journal of Biological Chemistry, 274*, 28922–28928.

Fukasawa, Y., Segawa, H., Kim, J. Y., Chairoungdua, A., Kim, D. K., Matsuo, H., et al. (2000). Identification and characterization of a Na(+)-independent neutral amino acid transporter that associates with the 4F2 heavy chain and exhibits substrate selectivity for small neutral D- and L-amino acids. *Journal of Biological Chemistry, 275*, 9690–9698.

Gabrisko, M., & Janecek, S. (2009). Looking for the ancestry of the heavy-chain subunits of heteromeric amino acid transporters rBAT and 4F2hc within the GH13 alpha-amylase family. *FEBS Journal, 276*, 7265–7278.
Gao, X., Lu, F., Zhou, L., Dang, S., Sun, L., Li, X., et al. (2009). Structure and mechanism of an amino acid antiporter. *Science, 324*, 1565–1568.
Gao, X., Zhou, L., Jiao, X., Lu, F., Yan, C., Zeng, X., et al. (2010). Mechanism of substrate recognition and transport by an amino acid antiporter. *Nature, 463*, 828–832.
Gartner, R. M., Perez, C., Koshy, C., & Ziegler, C. (2011). Role of bundle helices in a regulatory crosstalk in the trimeric betaine transporter BetP. *Journal of Molecular Biology, 414*, 327–336.
Ge, L., Perez, C., Waclawska, I., Ziegler, C., & Muller, D. J. (2011). Locating an extracellular K^+-dependent interaction site that modulates betaine-binding of the Na^+-coupled betaine symporter BetP. *Proceedings of the National Academy of Sciences of the United States of America, 108*, E890–E898.
Geerlings, A., Lopez-Corcuera, B., & Aragon, C. (2000). Characterization of the interactions between the glycine transporters GLYT1 and GLYT2 and the SNARE protein syntaxin 1A. *FEBS Letters, 470*, 51–54.
Gong, S., Richard, H., & Foster, J. W. (2003). YjdE (AdiC) is the arginine: agmatine antiporter essential for arginine-dependent acid resistance in I. *Journal of Bacteriology, 185*, 4402–4409.
Henry, L. K., Meiler, J., & Blakely, R. D. (2007). Bound to be different: neurotransmitter transporters meet their bacterial cousins. *Molecular Interventions*, 7, 306–309.
Jack, D. L., Paulsen, I. T., & Saier, M. H. (2000). The amino acid/polyamine/organocation (APC) superfamily of transporters specific for amino acids, polyamines and organocations. *Microbiology, 146*(Pt 8), 1797–1814.
Jardetzky, O. (1966). Simple allosteric model for membrane pumps. *Nature, 211*, 969–970.
Javitt, D. C. (2009). Glycine transport inhibitors for the treatment of schizophrenia: symptom and disease modification. *Current Opinion in Drug Discovery & Development, 12*, 468–478.
Jiang, J., & Amara, S. G. (2011). New views of glutamate transporter structure and function: advances and challenges. *Neuropharmacology, 60*, 172–181.
Jorgensen, A. M., Tagmose, L., Jorgensen, A. M., Topiol, S., Sabio, M., Gundertofte, K., et al. (2007). Homology modeling of the serotonin transporter: insights into the primary escitalopram-binding site. *ChemMedChem., 2*, 815–826.
Kanai, Y., & Endou, H. (2001). Heterodimeric amino acid transporters: molecular biology and pathological and pharmacological relevance. *Current Drug Metabolism, 2*, 339–354.
Kanai, Y., Fukasawa, Y., Cha, S. H., Segawa, H., Chairoungdua, A., Kim, D. K., et al. (2000). Transport properties of a system y+L neutral and basic amino acid transporter. Insights into the mechanisms of substrate recognition. *Journal of Biological Chemistry, 275*, 20787–20793.
Khafizov, K., Perez, C., Koshy, C., Quick, M., Fendler, K., Ziegler, C., et al. (2012). Identification of the sodium binding sites in the sodium-coupled betaine transporter BetP.
Khafizov, K., Staritzbichler, R., Stamm, M., & Forrest, L. R. (2010). A study of the evolution of inverted-topology repeats from LeuT-fold transporters using AlignMe. *Biochemistry, 49*, 10702–10713.
Kilic, F., & Rudnick, G. (2000). Oligomerization of serotonin transporter and its functional consequences. *Proceedings of the National Academy of Sciences of the United States of America, 97*, 3106–3111.
Kowalczyk, L., Ratera, M., Paladino, A., Bartoccioni, P., Errasti-Murugarren, E., Valencia, E., et al. (2011). Molecular basis of substrate-induced permeation by an amino acid antiporter. *Proceedings of the National Academy of Sciences of the United States of America, 108*, 3935–3940.
Krishnamurthy, H., & Gouaux, E. (2012). X-ray structures of LeuT in substrate-free outward-open and apo inward-open states. *Nature, 481*, 469–474.

Kristensen, A. S., Andersen, J., Jorgensen, T. N., Sorensen, L., Eriksen, J., Loland, C. J., et al. (2011). SLC6 neurotransmitter transporters: structure, function, and regulation. *Pharmacological Reviews, 63*, 585–640.

Lang, F. (2007). Mechanisms and significance of cell volume regulation. *Journal of the American College of Nutrition, 26*, 613S–623S.

Ma, D., Lu, P., Yan, C., Fan, C., Yin, P., Wang, J., et al. (2012). Structure and mechanism of a glutamate-GABA antiporter. *Nature, 483*, 632–636.

Mackenzie, B., & Erickson, J. D. (2004). Sodium-coupled neutral amino acid (System N/A) transporters of the SLC38 gene family. *Pflügers Archiv European Journal of Physiology, 447*, 784–795.

Marteau, P., & Chaput, U. (2011). Bacteria as trigger for chronic gastrointestinal disorders. *Digestive Diseases (Basel, Switzerland), 29*, 166–171.

Melikian, H. E. (2004). Neurotransmitter transporter trafficking: endocytosis, recycling, and regulation. *Pharmacology & Therapeutics, 104*, 17–27.

Miranda, M., & Sorkin, A. (2007). Regulation of receptors and transporters by ubiquitination: new insights into surprisingly similar mechanisms. *Molecular Interventions*, 7, 157–167.

Murai, S., Saito, H., Masuda, Y., & Itoh, T. (1988). Rapid determination of norepinephrine, dopamine, serotonin, their precursor amino acids, and related metabolites in discrete brain areas of mice within ten minutes by HPLC with electrochemical detection. *Journal of Neurochemistry, 50*, 473–479.

Nakauchi, J., Matsuo, H., Kim, D. K., Goto, A., Chairoungdua, A., Cha, S. H., et al. (2000). Cloning and characterization of a human brain Na(+)-independent transporter for small neutral amino acids that transports D-serine with high affinity. *Neuroscience Letters, 287*, 231–235.

Nemoto, T., Shimma, N., Horie, S., Saito, T., Okuma, Y., Nomura, Y., et al. (2003). Involvement of the system L amino acid transporter on uptake of S-nitroso-L-cysteine, an endogenous S-nitrosothiol, in PC12 cells. *European Journal of Pharmacology, 458*, 17–24.

Nicklisch, S. C., Wunnicke, D., Borovykh, I. V., Morbach, S., Klare, J. P., Steinhoff, H. J., et al. (2012). Conformational changes of the betaine transporter BetP from Corynebacterium glutamicum studied by pulse EPR spectroscopy. *Biochimica et Biophysica Acta, 1818*, 359–366.

Palacin, M., & Kanai, Y. (2004). The ancillary proteins of HATs: SLC3 family of amino acid transporters. *Pflügers Archiv European Journal of Physiology, 447*, 490–494.

Palacin, M., Nunes, V., Font-Llitjos, M., Jimenez-Vidal, M., Fort, J., & Gasol, E. (2005). Pflügers Archiv European Journal of Physiology. In The genetics of heteromeric amino acid transporters *Physiology (Bethesda, Md.) 20*, 112–124.

Parra, L. A., Baust, T., El Mestikawy, S., Quiroz, M., Hoffman, B., Haflett, J. M., et al. (2008). The orphan transporter Rxt1/NTT4 (SLC6A17) functions as a synaptic vesicle amino acid transporter selective for proline, glycine, leucine, and alanine. *Molecular Pharmacology, 74*, 1521–1532.

Perez, C., Khafizov, K., Forrest, L. R., Kramer, R., & Ziegler, C. (2011). The role of trimerization in the osmoregulated betaine transporter BetP. *EMBO Reports, 12*, 804–810.

Perez, C., Koshy, C., Ressl, S., Nicklisch, S., Kramer, R., & Ziegler, C. (2011). Substrate specificity and ion coupling in the Na+/betaine symporter BetP. *EMBO Journal, 30*, 1221–1229.

Perez, C., Koshy, C., Yildiz, Ö., & Ziegler, C. (2012). Alternating-access mechanism in conformationally asymmetric trimers of the betaine transporter BetP. *Nature Letter.*

Piscitelli, C. L., Krishnamurthy, H., & Gouaux, E. (2010). Neurotransmitter/sodium symporter orthologue LeuT has a single high-affinity substrate site. *Nature, 468*, 1129–1132.

Raiteri, L., & Raiteri, M. (2010). Functional 'glial' GLYT1 glycine transporters expressed in neurons. *Journal of Neurochemistry, 114*, 647–653.

Reig, N., Chillaron, J., Bartoccioni, P., Fernandez, E., Bendahan, A., Zorzano, A., et al. (2002). The light subunit of system b(o,+) is fully functional in the absence of the heavy subunit. *EMBO Journal, 21,* 4906–4914.

Ressl, S., Terwisscha van Scheltinga, A. C., Vonrhein, C., Ott, V., & Ziegler, C. (2009). Molecular basis of transport and regulation in the Na(+)/betaine symporter BetP. *Nature, 458,* 47–52.

Romeo, E., Dave, M. H., Bacic, D., Ristic, Z., Camargo, S. M., Loffing, J., et al. (2006). Luminal kidney and intestine SLC6 amino acid transporters of B0AT-cluster and their tissue distribution in Mus musculus. *American Journal of Physiology. Renal Physiology, 290,* F376–F383.

Rossier, G., Meier, C., Bauch, C., Summa, V., Sordat, B., Verrey, F., et al. (1999). LAT2, a new basolateral 4F2hc/CD98-associated amino acid transporter of kidney and intestine. *Journal of Biological Chemistry, 274,* 34948–34954.

Rubenhagen, R., Morbach, S., & Kramer, R. (2001). The osmoreactive betaine carrier BetP from Corynebacterium glutamicum is a sensor for cytoplasmic K^+. *EMBO Journal, 20,* 5412–5420.

Ruderisch, N., Virgintino, D., Makrides, V., & Verrey, F. (2011). Differential axial localization along the mouse brain vascular tree of luminal sodium-dependent glutamine transporters Snat1 and Snat3. *Journal of Cerebral Blood Flow and Metabolism: Official Journal of the International Society of Cerebral Blood Flow and Metabolism, 31,* 1637–1647.

Rudnick, G. (1998). Bioenergetics of neurotransmitter transport. *Journal of Bioenergetics and Biomembranes, 30,* 173–185.

Rudnick, G. (2011). Cytoplasmic permeation pathway of neurotransmitter transporters. *Biochemistry, 50,* 7462–7475.

Saier, M. H., Jr. (2000). Families of transmembrane transporters selective for amino acids and their derivatives. *Microbiology, 146*(Pt 8), 1775–1795.

Schiller, D., Rubenhagen, R., Kramer, R., & Morbach, S. (2004). The C-terminal domain of the betaine carrier BetP of Corynebacterium glutamicum is directly involved in sensing K^+ as an osmotic stimulus. *Biochemistry, 43,* 5583–5591.

Schlessinger, A., Matsson, P., Shima, J. E., Pieper, U., Yee, S. W., Kelly, L., et al. (2010). Comparison of human solute carriers. *Protein Science, 19,* 412–428.

Schulze, S., Koster, S., Geldmacher, U., Terwisscha van Scheltinga, A. C., & Kuhlbrandt, W. (2010). Structural basis of Na(+)-independent and cooperative substrate/product antiport in CaiT. *Nature, 467,* 233–236.

Screpanti, E., & Hunte, C. (2007). Discontinuous membrane helices in transport proteins and their correlation with function. *Journal of Structural Biology, 159,* 261–267.

Segawa, H., Fukasawa, Y., Miyamoto, K., Takeda, E., Endou, H., & Kanai, Y. (1999). Identification and functional characterization of a Na^+-independent neutral amino acid transporter with broad substrate selectivity. *Journal of Biological Chemistry, 274,* 19745–19751.

Shaffer, P. L., Goehring, A., Shankaranarayanan, A., & Gouaux, E. (2009). Structure and mechanism of a Na^+-independent amino acid transporter. *Science, 325,* 1010–1014.

Shi, L., Quick, M., Zhao, Y., Weinstein, H., & Javitch, J. A. (2008). The mechanism of a neurotransmitter: sodium symporter–inward release of Na^+ and substrate is triggered by substrate in a second binding site. *Molecular Cell, 30,* 667–677.

Shimamura, T., Weyand, S., Beckstein, O., Rutherford, N. G., Hadden, J. M., Sharples, D., et al. (2010). Molecular basis of alternating access membrane transport by the sodium-hydantoin transporter Mhp1. *Science, 328,* 470–473.

Sloan, J. L., Grubb, B. R., & Mager, S. (2003). Expression of the amino acid transporter ATB 0+ in lung: possible role in luminal protein removal. *American Journal of Physiology. Lung Cellular and Molecular Physiology, 284,* L39–L49.

Solbu, T. T., Boulland, J. L., Zahid, W., Lyamouri Bredahl, M. K., Amiry-Moghaddam, M., Storm-Mathisen, J., et al. (2005). Induction and targeting of the glutamine transporter

SN1 to the basolateral membranes of cortical kidney tubule cells during chronic metabolic acidosis suggest a role in pH regulation. *Journal of the American Society of Nephrology: JASN*, *16*, 869–877.
Sugawara, M., Nakanishi, T., Fei, Y. J., Huang, W., Ganapathy, M. E., Leibach, F. H., et al. (2000). Cloning of an amino acid transporter with functional characteristics and tissue expression pattern identical to that of system A. *Journal of Biological Chemistry*, *275*, 16473–16477.
Sugawara, M., Nakanishi, T., Fei, Y. J., Martindale, R. G., Ganapathy, M. E., Leibach, F. H., et al. (2000). Structure and function of ATA3, a new subtype of amino acid transport system A, primarily expressed in the liver and skeletal muscle. *Biochimica et Biophysica Acta*, *1509*, 7–13.
Supplisson, S., & Roux, M. J. (2002). Why glycine transporters have different stoichiometries. *FEBS Letters*, *529*, 93–101.
Tomitori, H., Kashiwagi, K., & Igarashi, K. (2012). Structure and function of polyamine-amino acid antiporters CadB and PotE in *Escherichia coli*. *Amino Acids*, *42*, 733–740.
Vanslambrouck, J. M., Broer, A., Thavyogarajah, T., Holst, J., Bailey, C. G., Broer, S., et al. (2010). Renal imino acid and glycine transport system ontogeny and involvement in developmental iminoglycinuria. *Biochemical Journal*, *428*, 397–407.
Vekony, N., Wolf, S., Boissel, J. P., Gnauert, K., & Closs, E. I. (2001). Human cationic amino acid transporter hCAT-3 is preferentially expressed in peripheral tissues. *Biochemistry*, *40*, 12387–12394.
Velaz-Faircloth, M., Guadano-Ferraz, A., Henzi, V. A., & Fremeau, R. T., Jr. (1995). Mammalian brain-specific L-proline transporter. Neuronal localization of mRNA and enrichment of transporter protein in synaptic plasma membranes. *Journal of Biological Chemistry*, *270(26)*, 15755–15761.
Verrey, F., Closs, E. I., Wagner, C. A., Palacin, M., Endou, H., & Kanai, Y. (2004). CATs and HATs: the SLC7 family of amino acid transporters. *Pflügers Archiv European Journal of Physiology*, *447*, 532–542.
Verrey, F., Singer, D., Ramadan, T., Vuille-dit-Bille, R. N., Mariotta, L., & Camargo, S. M. (2009). Kidney amino acid transport. *Pflügers Archiv European Journal of Physiology*, *458*, 53–60.
Verri, T., Dimitri, C., Treglia, S., Storelli, F., De Micheli, S., Ulianich, L., et al. (2005). Multiple pathways for cationic amino acid transport in rat thyroid epithelial cell line PC Cl3. *American Journal of Physiology. Cell Physiology*, *288*, C290–C303.
Wallimann, T., Tokarska-Schlattner, M., & Schlattner, U. (2011). The creatine kinase system and pleiotropic effects of creatine. *Amino Acids*, *40*, 1271–1296.
Weyand, S., Shimamura, T., Beckstein, O., Sansom, M. S., Iwata, S., Henderson, P. J., et al. (2011). The alternating access mechanism of transport as observed in the sodium-hydantoin transporter Mhp1. *Journal of Synchrotron Radiation*, *18*, 20–23.
Wong, F. H., Chen, J. S., Reddy, V., Day, J. L., Shlykov, M. A., Wakabayashi, S. T., et al. (2012). The amino Acid-polyamine-organocation superfamily. *Journal of Molecular Microbiology and Biotechnology*, *22*, 105–113.
Yamashita, A., Singh, S. K., Kawate, T., Jin, Y., & Gouaux, E. (2005). Crystal structure of a bacterial homologue of Na^+/Cl–dependent neurotransmitter transporters. *Nature*, *437*, 215–223.
Yernool, D., Boudker, O., Jin, Y., & Gouaux, E. (2004). Structure of a glutamate transporter homologue from Pyrococcus horikoshii. *Nature*, *431*, 811–818.
Zdravkovic, I., Zhao, C., Lev, B., Cuervo, J. E., & Noskov, S. Y. (2012). Atomistic models of ion and solute transport by the sodium-dependent secondary active transporters. *Biochimica et Biophysica Acta*, *1818*, 337–347.
Zebisch, K., & Brandsch, M. (2012). Transport of L-proline by the proton-coupled amino acid transporter PAT2 in differentiated 3T3-L1 cells. *Amino Acids*.

Zhang, M., White, T. A., Schuermann, J. P., Baban, B. A., Becker, D. F., & Tanner, J. J. (2004). Structures of the Escherichia coli PutA proline dehydrogenase domain in complex with competitive inhibitors. *Biochemistry, 43*, 12539–12548.

Zhang, Z., Albers, T., Fiumera, H. L., Gameiro, A., & Grewer, C. (2009). A conserved Na(+) binding site of the sodium-coupled neutral amino acid transporter 2 (SNAT2). *Journal of Biological Chemistry, 284*, 25314–25323.

Zhao, Y., Quick, M., Shi, L., Mehler, E. L., Weinstein, H., & Javitch, J. A. (2010). Substrate-dependent proton antiport in neurotransmitter:sodium symporters. *Nature Chemical Biology, 6*, 109–116.

Zhao, Y., Terry, D. S., Shi, L., Quick, M., Weinstein, H., Blanchard, S. C., et al. (2011). Substrate-modulated gating dynamics in a Na^{+}-coupled neurotransmitter transporter homologue. *Nature, 474*, 109–113.

Ziegler, C., Bremer, E., & Krämer, R. (2010). The BCCT-family of carriers: from physiology to crystal structure. *Molecular Microbiology*.

Zomot, E., & Bahar, I. (2011). Protonation of glutamate 208 induces the release of agmatine in an outward-facing conformation of an arginine/agmatine antiporter. *Journal of Biological Chemistry, 286*, 19693–19701.

Zomot, E., Bendahan, A., Quick, M., Zhao, Y., Javitch, J. A., & Kanner, B. I. (2007). Mechanism of chloride interaction with neurotransmitter: sodium symporters. *Nature, 449*, 726–730.

CHAPTER TWO

SLC5 and SLC2 Transporters in Epithelia—Cellular Role and Molecular Mechanisms

Mobeen Raja*,2, Theeraporn Puntheeranurak‡,2, Peter Hinterdorfer†, Rolf Kinne*,1

*Max Planck Institute of Molecular Physiology, Dortmund, Germany
‡Department of Biology and Center of Nanoscience, Faculty of Science, Mahidol University, Ratchathevi, Bangkok, Thailand
†Institute for Biophysics, Christian Doppler Laboratory of Nanoscopic Methods in Biophysics, Johannes Kepler University of Linz and Center for Advanced Bioanalysis GmbH (CBL), Linz, Austria
[1]Corresponding author: E-mail: rolf.kinne@mpi-dortmund.mpj.de

Contents

[2] The two authors contributed equally to the work.

Current Topics in Membranes, Volume 70
ISSN 1063-5823
http://dx.doi.org/10.1016/B978-0-12-394316-3.00002-8

Abstract

Members of the SLC5 and SLC2 family are prominently involved in epithelial sugar transport. SGLT1 (sodium-glucose transporter) and SGLT2, as representatives of the former, mediate sodium-dependent uptake of sugars into intestinal and renal cells. GLUT2 (glucose transporter), as representative of the latter, facilitates the sodium-independent exit of sugars from cells. SGLT has played a major role in the formulation and experimental proof for the existence of sodium cotransport systems. Based on the sequence data and biochemical and biophysical analyses, the role of extramembranous loops in sugar and inhibitor binding can be delineated. Crystal structures and homology modeling of SGLT reveal that the sugar translocation involves operation of two hydrophobic gates and intermediate exofacial and endofacial occluded states of the carrier in an alternating access model. The same basic model is proposed for GLUT1. Studies on GLUT1 have pioneered the isolation of eukaryotic transporters by biochemical methods and the development of transport kinetics and transporter models. For GLUT1, results from extensive mutagenesis, cysteine substitution and accessibility studies can be incorporated into a homology model with a barrel-like structure in which accessibility to the extracellular and intracellular medium is altered by pinching movements of some of the helices. For SGLT1 and GLUT1, the extensive hydrophilic and hydrophobic interactions between sugars and binding sites of the various intramembrane helices occur and lead to different substrate specificities and inhibitor affinities of the two transporters. A complex network of regulatory steps adapts the transport activity to the needs of the body.

1. INTRODUCTION

About 25% of the human genome code for membrane proteins; many are transport proteins that facilitate the stereospecific translocation of solutes across cellular membranes. This article deals mainly with the transporters involved in sugar uptake into cells and its transport across epithelial cells. It will focus on the solute carrier (SLC) 5 family and the SLC2 family. More emphasis is put on the SLC5 transporter with which the authors are most familiar. This group comprises the sodium cotransport systems, one of the most common and most important classes of transport systems that nature has developed during evolution. They encompass symport and antiport systems, which use the sodium gradient across the cell membrane as the electrochemical driving force to accumulate inorganic and organic solutes within the cells or to move them against concentration gradients out of the cells.

This review describes the discovery, the experimental confirmation, and the molecular identification and characterization of these systems. In 1960, the first hypothesis for cotransport was developed. In 1970, the first model of transepithelial active sugar transport was proposed by arranging a secondary active sodium–solute cotransporter (of the SLC5 family) in series with a sodium-independent transporter (of the SLC2 family). The two are located at different surface membranes of the epithelial cell. One of the authors of this review was a theoretical biologist applying thermodynamics to the system—as many of the investigators did at that time. In the next decade, the understanding of epithelial transport was refined, mainly in terms of carrier properties and transport kinetics. In the 1990s, finally, cloning allowed the identification of the transporter genes and their manipulation by mutagenesis.

The sodium-independent glucose transporter was isolated first as protein from the rich source of erythrocytes by biochemists and analyzed by theoretical and kinetically oriented scientists. The molecular identity allowed investigators to apply the techniques of molecular biology. But for some time, the intense interactions of membrane proteins with lipids and their decisive asymmetry prevented successful attempts to crystallization. Even today, only crystals of bacterial sodium–sugar transporters are available—the crystallization of mammalian glucose transporters is still a tantalizing challenge. The review traces these historical developments. In addition, it provides insights into the events at the cellular and molecular levels.

Recently, glucose-transporting molecules have become potential targets for the treatment of diabetes. Their regulation and specific inhibition by drugs have become a new focus of investigations. Thus, after 50 years, "translation" of findings "from bench to bedside" takes place, stimulating an exciting new phase in the understanding of sugar transport across cell membranes in health and disease.

2. PHYSIOLOGICAL SIGNIFICANCE AND MOLECULAR CHARACTERISTICS OF EPITHELIAL SUGAR TRANSPORT

The two epithelia commonly associated with transport of sugar in vertebrates are the small intestine and the kidney. In the small intestine, sugar is absorbed from ingested food and transferred into the blood stream to provide many cells with a necessary energy source. In the kidney, sugar is recovered from primary urine generated by filtration in the glomerulus to salvage this important metabolic compound and recycle it into the blood stream.

2.1. Sugar Transport in the Small Intestine

The general characteristics of these transport processes were first studied extensively in the intestine due to the availability of suitable in vitro and in vivo preparations, such as inverted sacs, mucosal rings or isolated villi. Epithelia could generate a concentration difference, the sugar concentration in the lumen being lower than the concentration in the surrounding medium or blood. Transport was also inhibited by inhibitors of cellular energy production. Energy dependence and transport against a concentration difference pointed to active transport (Crane, 1960a). Sugar transport was also shown to follow Michaelis–Menten kinetics, i.e. to display saturation kinetics and stereospecificity. In his review in 1960, Crane lists 14 sugars that are transported against a concentration difference, notably D-glucose (but not fructose or L-glucose) and D-galactose (but not 2-deoxy-D-glucose) (for chemical structures, see Fig. 1). From these results, the structural requirements for a sugar molecule to be actively transported were identified: a hexose ring and a free OH group at C2 are absolutely necessary, whereas constituents at C1, C3, C4 and C6 affect the affinity to the transporter and the rate of translocation (Crane, 1960b).

Apparent affinities for sugars in the millimolar concentration range were found. Because competition for uptake was observed between the actively transported sugars, a single transport system was postulated. This single transport system was later identified as the sodium–glucose cotransporter SGLT1, a member of the SLC5 family (Hediger, Coady, Ikeda, & Wright, 1987; Wright & Turk, 2004). In experimental animals and in humans, the rate of active transport in the lower ileum is less than in duodenal and jejunal regions. This correlates with the expression

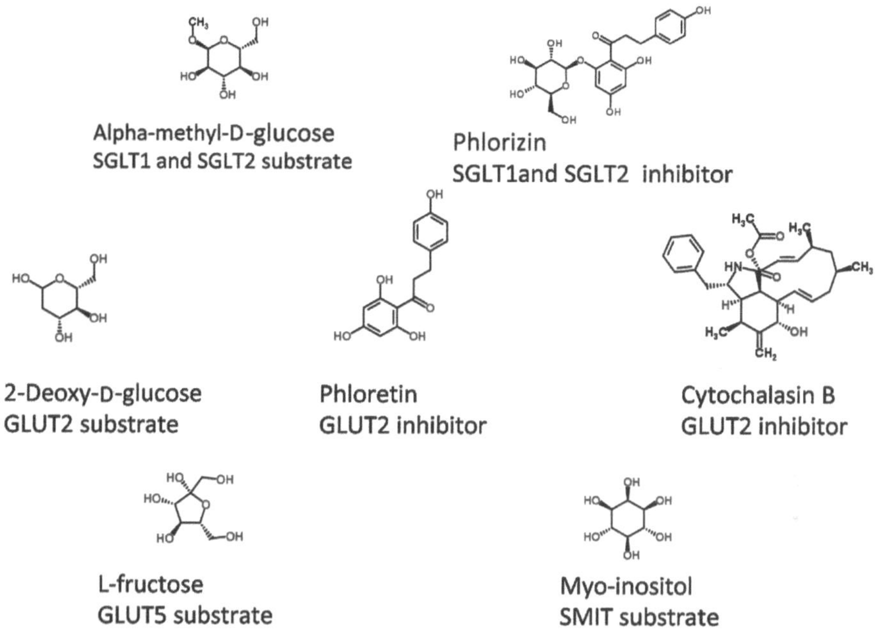

Figure 1 ***Chemical structure of substrates and inhibitors of transporters from the SLC2 and SLC5 families.*** For color version of this figure, the reader is referred to the online version of this book.

of SGLT1 in the brush border membrane of the epithelium. When not genetically expressed in null mice (Gorboulev et al., 2012), loss of glucose into the stool and diarrhea due to the osmotic load in the intestine occur. Similarly, mutations in the *sglt1* gene (Wright, Martin, & Turk, 2003) in humans cause the glucose–galactose malabsorption syndrome. The mutations can impede the proper function of the transporter itself, its turnover or sorting within the cell (Raja & Kinne, 2012; Wright, Loo, & Hirayama, 2011).

In the small intestine in vivo, there is also passive transfer of glucose when the concentration in the lumen is higher than in the blood. Due to the action of disaccharidases present on the surface of the microvilli, after a meal, local concentrations close to the transporter are estimated to be near or above 100 mM. Thus, a large concentration difference between intestinal lumen and blood exists, where the glucose concentration is only 5 mM. This passive absorption has been long overlooked. Work by Kellet suggests that these movements are facilitated by GLUT2, a member of the SLC2 family (Kellett, 2001; Kellett, Brot-Laroche, Mace, & Leturque, 2008).

At high glucose concentrations, GLUT2 is present and active in the brush border membrane. Facilitated, sodium-independent transport can be three to five times higher than the uptake by SGLT1. This hypothesis is, however, not undisputed (Wright et al., 2011). Normally, GLUT2 is found mainly in the basolateral membranes of the intestinal cells where it mediates the exit of D-glucose from the enterocytes (Uldry & Thorens, 2004).

GLUT proteins belong to the *slc2* gene family and are composed of 14 members, including the GLUT1 through the GLUT12 proteins and the proton–myo-inositol cotransporters (Uldry & Thorens, 2004). The transporters have a different affinity and specificity compared to SGLT1. Affinity for sugars is usually lower and translocation requires a free C1 hydroxyl group but no OH group at C2 on the sugar molecule. Thus, alpha-methyl-D-glucoside is not transported, whereas 2-deoxy-D-glucose is translocated (see Fig. 1).

Fructose (for chemical structure, see Fig. 1), the other component of ingested sucrose, is taken up in the intestine passively, because at any time the concentration in the blood is much lower than in the intestine. Fructose uptake is facilitated by GLUT5 in the brush border membrane, another member of the SLC2 family (Uldry & Thorens, 2004). Fructose can leave the cells by GLUT2 located in the basolateral membranes, which has a limited affinity for the sugar (Gould & Holman, 1993).

2.2. Carbohydrate Transport in the Kidney

2.2.1. *Glucose and Galactose*

In the kidney, clearance studies established that the renal reabsorption of glucose is a saturable process with a tubular transport maximum, Tm. Chinard, in turn, provided evidence that the removal of D-glucose from the urine is not due to an intracellular breakdown of the sugar by cell metabolism but rather represents transfer of the intact molecule from primary urine into the peritubular blood (Chinard, Taylor, Nolan, & Enn, 1957). Micropuncture and microperfusion studies have shown that the glucose transport characteristics of the early and late part of the proximal tubule are quite different (Barfuss & Schafer, 1981; Loeschke & Baumann, 1969; Loeschke, Baumann, Renschler, & Ullrich, 1969). The early proximal tubule exhibits low-affinity, high-capacity transport and the late proximal tubule is able to create large concentration differences at small transport rates. As first suggested by Scriver, this difference is due to the different expression of transporters in the two segments (Scriver, Chesney et al. 1976). In the early proximal tubule, SGLT2 is expressed which has a lower affinity to

D-glucose and does not transport galactose (Sabolic et al., 2012). In the late proximal tubule, SGLT1 is present that has a higher affinity and accepts both D-glucose and D-galactose (Balen et al., 2008). The two transporters also differ in their stoichiometry with regard to sodium, which explains the difference in their potential to lower the intratubular sugar concentration (Turner & Moran, 1982). In mice, the early proximal tubule (pars convoluta) extracts 78% of the filtered load; the late proximal tubule (pars recta) extracts another 15%. Mutations in the *sglt2* gene result in familiar renal glucosuria (Wright et al., 2003).

Glucose exit from the cells is facilitated mainly by GLUT2, as seen in the small intestine. A detailed microperfusion study of the substrate specificity of the basolateral GLUT2 transporter in the proximal tubule of rat kidney was performed by Ullrich (Ullrich & Papavassiliou, 1985). In later parts of the nephron, sugar uptake occurs from the capillaries, facilitated by GLUT2 located in the basolateral membranes (Grunewald & Kinne, 1988). In medullary and papillary cells, glucose is also used for the synthesis of the organic osmolyte sorbitol (Jans, Grunewald, & Kinne, 1989).

2.2.2. Myo-inositol

Other members of the SLC5 family are the sodium–myo-inositol transporters (SMIT) 1 and SMIT2, which also play important roles in renal function. SMIT1 maintains a high level of the organic osmolyte myo-inositol in cells exposed to hypertonicity in the renal medulla. SMIT1 is a sodium–solute cotransporter. It shows the following substrate specificity: myo-inositol (for chemical structure, see Fig. 1), scyllo-inositol > L-fucose > L-xylose > L-glucose, D-glucose, and alpha-methyl-D-glucopyranoside (Hager et al., 1995). At 100 mM NaCl, the affinity for myo-inositol is about 50 mM. At 0.5 mM myo-inositol, the K_m for sodium is about 76 mM. Two sodium ions are transported per molecule of myo-inositol. Phlorizin is an inhibitor with a voltage-dependent apparent K_i between 64 and 130 μM, much higher than for SGLT1 (Hager et al. 1995).

In rat kidneys, SMIT1 messenger RNA (mRNA) and protein are predominantly present in the medullary and cortical thick ascending limb of Henle's loop and, to a lesser extent, in the inner medullary-collecting duct (Yamauchi et al., 1995). SMIT1 protein is localized to the basolateral membranes of cells (Yamauchi et al., 1998).

In 2002, Coady et al. identified a second myo-inositol transporter, named SMIT2 (Coady, Wallendorff et al. 2002). Transport of myo-inositol by SMIT2 is sodium dependent, with a K_m of 13 mM, and inhibited by phlorizin, with a K_i of 76 μM. SMIT2 exhibits stereospecific cotransport of myo-inositol and both D-glucose and D-xylose but does not transport fucose. In addition, SMIT2 (but not SMIT1) transports D-chiro-inositol. In rabbit kidney, SMIT2 is mainly expressed in the renal cortex and present in the brush border membrane (Hammerman, Sacktor, & Daughaday, 1980), facilitating myo-inositol reabsorption.

2.3. Inhibitors of Sugar Transport

2.3.1. Aromatic Glucosides Such as Phlorizin

Phlorizin, a glucoside whose chemical structure is shown in Fig. 1, has played an important role in the elucidation of cellular and molecular mechanisms of sugar transport (Ehrenkranz, Lewis, Kahn, & Roth, 2005). It was isolated by French chemists in 1835 from the bark of the apple tree. In 1886, it was shown by von Mering to produce glucosuria (for references, see Ehrenkranz et al., 2005). Leaves of the Sweet Tea (*Lithocarpus polystachyus*) also contain significant amounts of the compound (Dong, Ning, Yu, Li, Lin, & Huang, 2007). Based on von Mering's observation, phlorizin became a tool for the study of renal function in humans. In the 1930s, using phlorizin to produce glucosuria in normal subjects, Homer Smith developed noninvasive clearance methods to measure glomerular filtration rate, tubular transport mechanisms and renal blood flow (Smith, 1951). In the 1960s, studies on the mechanism of action of phlorizin began to focus on the cellular and molecular effects. In 1962, Alvarado and Crane found that the phlorizin acts as a competitive inhibitor of active sugar transport in the intestine (Alvarado & Crane, 1962). In the following years, Diedrich investigated the relative potency of its congeners in detail (Diedrich, 1965).

Interaction of phlorizin with the SGLT transporter involves the sugar-binding site of the transporter and aglucone-binding sites (Koepsell, Fritzsch, Korn, & Madrala, 1990; Kumar, Tyagi, & Kinne, 2007). The affinity of the transporter for phlorizin is about 1000-fold higher than for glucose. Essential elements of the molecule for interaction with its binding pockets are hydrogen bonding via the hydroxyl groups at C (2), C (3), C (4), and C (6) of the sugar residue and at C (4) and C (6) of the aromatic ring A. Hydrophobic interactions are established via the pyranoside ring and the aromatic ring A. The dimension of the phlorizin-binding site on the sodium–D-glucose cotransporter has been estimated to be 17 × 10 × 7 Å (Wielert-Badt, Lin, Lorenz, Fritz, & Kinne, 2000). Similar dimensions have been derived from the homology modeling of hSGLT1 (Sala-Rabanal et al., 2012). Binding of phlorizin occurs at the exofacial side of the transporter. At the cytoplasmic face of the membrane, phlorizin does not inhibit glucose transport (Firnges, Lin, & Kinne, 2001). Derivatives of phlorizin are currently investigated for the use as antidiabetic drugs (for review see Kinne & Castaneda, 2011).

Phloretin (for chemical structure, see Fig. 1), the aglucone of phlorizin, inhibits SGLT transporter only partly, at much higher concentrations. Phloretin inhibits, however, GLUT2 effectively (Kellett, 2001). Phenylglucosides such as arbutin, phenylglucoside and helicin are transported by SGLT1. An OH group in the para position of the phenyl ring (in arbutin) facilitates transport, but an NCS group (in glucosylphenyl-isothiocyanate) prevents transport. In the ortho position, a CH_2OH group (in salicin) hinders interaction, but an aldehyde group (in helicin) permits the molecule to be transported (Lostao, Hirayama, Loo, & Wright, 1994).

2.3.2. *Alkylglucosides*

Alkylglucosides also interact with the transporter (Lin, Riedel, & Kinne, 1979). Optimum interaction requires a beta-anomer with a glycosidic bond. This places the alkyl chain into an equatorial position with regard to the D-glucose molecule. In addition, the alkyl chain has to be highly flexible (Kipp, Lin, & Kinne, 1996). Depending on their length, the alkyl chains interact with one or more hydrophobic sites at the carrier, in a slightly coiled conformation. Alkylglucosides are also transported, for example the omega-(18)F-fluoro-*n*-ethyl-beta-D-glucoside, used to locate SGLT1 transport activity in vivo, is transported by SGLT1 at a similar rate and with a similar affinity as alpha-methyl-D-glucoside. With an *n*-butyl or *n*-octyl alkyl chain, competitive inhibition of transport, but no translocation, is observed (de Groot et al., 2003).

2.3.3. *Cytochalasin B*

For the GLUT family of transporters, cytochalasin B, a mycotoxin, has played an essential role (for chemical structure, see Fig. 1). It inhibits facilitated diffusion of glucose in many cells. In erythrocytes, adipocytes, and chick embryo fibroblasts, the K_i for glucose uptake inhibition and the K_d for cytochalasin B binding range from 0.1 to 0.4 μM. Values for hepatocytes and liver membranes are an order of magnitude higher (Uldry & Thorens, 2004). GLUT2 is strongly inhibited by micromolar concentrations of cytochalasin B and phloretin, but only by high concentrations of phlorizin (Wheeler & Hinkle, 1985). Binding of cytochalasin B, which is membrane permeant, to the transporter occurs at the intracellular face (Carter, 1967; Estensen

& Plagemann, 1972; Mueckler & Makepeace, 2012; Thorens & Mueckler, 2010; Uldry & Thorens, 2004).

3. SODIUM–GLUCOSE SYMPORT AS THE BASIS FOR ACTIVE TRANSEPITHELIAL GLUCOSE TRANSPORT

The first step in the active sugar transport across epithelial cells of the small intestine and the renal proximal tubule involves sodium–D-glucose cotransport. Sodium and sugar are translocated together across the brush border membrane. Sodium moves down its electrochemical potential difference, whereas glucose is accumulated inside the cell. The development of this concept represents a major step in understanding a large number of transport phenomena in nature. Such transport processes are termed secondary active, because the driving forces for solute movement against a concentration difference are not derived from a direct coupling of the transport to an energy-yielding chemical reaction. In animal cells, the secondary active transport of organic or inorganic solutes is usually coupled to the sodium gradient, in bacteria, symport with protons predominates. In both instances, the ion providing the driving forces is subsequently removed from the cell by a primary active transport system such as the Na-K-ATPase or an H-ATPase.

3.1. Small Intestine

The concept for sodium–sugar cotransport was developed and proven between 1960 and 1975. As reviewed in 1973 (Kimmich, 1973), Riklis and Quastel observed in perfused guinea pig intestine that complete replacement of sodium by potassium abolished all active transport of D-glucose (Riklis & Quastel, 1958). Csaky and Thale (1960) confirmed the sodium dependence of active sugar transport in toad intestine and Clarkson and Rothstein (1960) noted the sodium dependence in rat intestine. In the same year, Crane suggested that the sugar movement across the membrane involves a carrier mechanism and depends, via osmotic coupling, on the activity of the sodium pump. In 1961, he summarized these ideas in the scheme shown in Fig. 2. A sodium–sugar–carrier complex is formed at the brush border membrane surface, which traverses the membrane to the cell interior. There occurs dissociation and glucose and sodium are released into the cell. Sodium is subsequently extruded from the cell by an energy-requiring (strophantidin inhibitable) process at adjacent membrane sites (Crane, Miller, & Bihler, 1961). In the

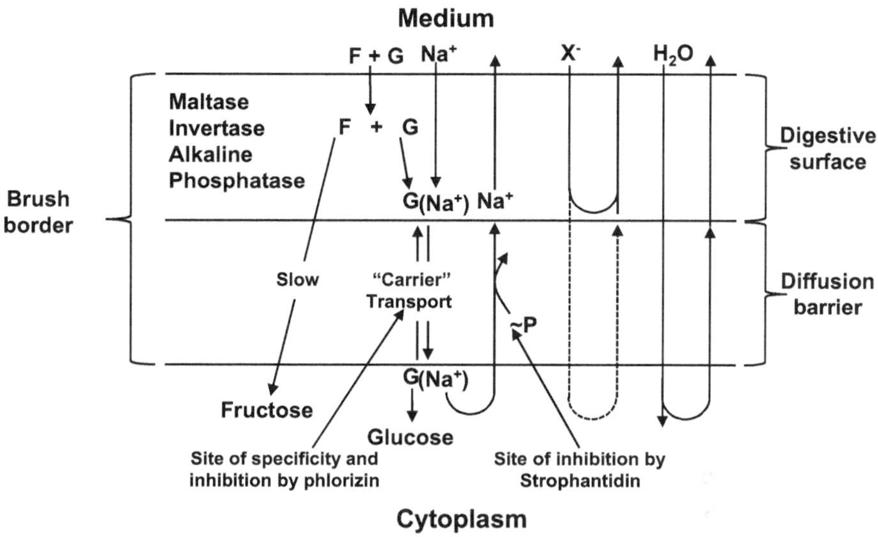

Figure 2 ***Original scheme of sodium–sugar cotransport in the small intestine as proposed by Crane.*** Note the "carrier transport" by which glucose and sodium are jointly moving across the brush border and the subsequent removal of sodium from the cytoplasm by a strophantidin-sensitive active sodium pump (Crane et al., 1961). *(Redrawn from Kimmich, G. A. (1973). Coupling between Na^+ and sugar transport in small intestine.* Biochimica et Biophysica Acta, *300(1), 31–78, copyright 1973, with permission from Elsevier)*

following years, evidence accumulated that the Na-dependent sugar accumulation and the (ouabain sensitive) active sodium removal are located at different sides of the epithelium. Previously, Ussing had developed a model for active sodium transport in frog skin. It consisted of passive entry of sodium, along a concentration gradient, at one cell side and an (ouabain sensitive) active extrusion by the sodium pump on the opposite cell side (Ussing & Zerahn, 1951). Experiments in the intestine suggested a similar asymmetry. Schultz and Curran (1970) incorporated this asymmetry in their model for active sugar absorption in the intestine, presented in Fig. 3. The sodium–glucose cotransporter is located in the luminal membrane and the Na-K-ATPase in the contraluminal membrane.

The model proposed by Schultz et al. also assumes that glucose leaves the cell passively, via a sodium-independent transporter along its concentration difference (Schultz & Curran, 1970). This transporter is GLUT2, which belongs to the large family of transporters mediating passive sugar uptake by the majority of cells (Thorens, Lodish, & Brown, 1990; Uldry & Thorens, 2004). Asymmetry of epithelial cells forms the basis for vectorial transepithelial transport of a number of solutes (for examples, see Kinne, 1991).

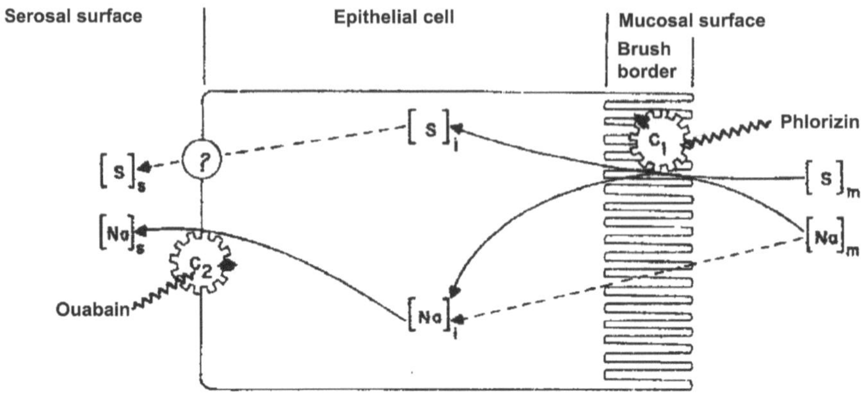

Figure 3 ***Interaction between active sodium transport and active glucose transport in rabbit ileum.*** Note the in-series arrangement of a sodium–sugar cotransporter (C1) in the brush border membrane and a sodium-independent transporter (?) at the basal-lateral membrane as well as the localization of the primary active sodium pump (C2) to the serosal cell side. C1 was later identified as SGLT1 and ? as GLUT2. *(From Schultz, S. G. & Curran, P. F. (1970). Coupled transport of sodium and organic solutes.* Physiological Reviews, *50(4), 637–718, copyright 1970, with permission from The American Physiological Society)*

In the same year, stereospecificity, sodium dependence and phlorizin inhibition of glucose transport were shown in the intestinal brush border vesicles (Hopfer, Nelson, Perotto, & Isselbacher, 1973). Subsequently, Murer and Hopfer (1974) demonstrated transient accumulation of glucose inside the vesicles, depending on the electrochemical difference of sodium across the membrane. Energetic coupling of sodium flux and solute flux is one of the hallmarks of cotransport systems.

3.2. Kidney

In the proximal tubule, sodium dependence of active glucose reabsorption was demonstrated in microperfusion experiments by Ullrich (Ullrich, Rumrich, & Klöss, 1974), indicating a mechanism similar to the small intestine. Earlier, Crane and Kleinzeller had observed that D-glucose uptake and accumulation in renal cells was sodium dependent (Kleinzeller, 1970; Krane & Crane, 1959). Sodium gradient-driven accumulation of D-glucose could also be demonstrated in renal brush border membrane vesicles (Heidrich, Kinne, Kinne-Saffran, & Hannig, 1972). Here, in addition, coupling of sodium flux to a glucose gradient could be shown, another proof for the operation of the cotransporter (Kinne, Murer, Kinne-Saffran, Thees, & Sachs, 1975). Beck and Sacktor analyzed the relative importance of the two components

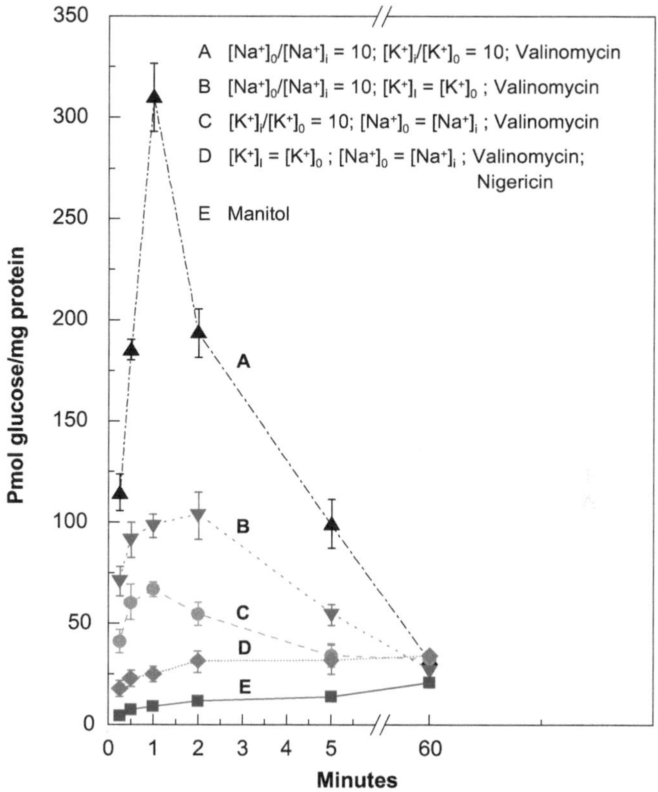

Figure 4 ***Demonstration of energetic coupling of glucose flux to the electrochemical potential difference of sodium across the membrane.*** Uptake studies in isolated renal brush border membrane vesicles are depicted. Note the intravesicular accumulation of glucose in the presence of a chemical gradient and electrical potential difference (generated by a potassium gradient in the presence of the potassium ionophore valinomycin). Similar studies were performed first by Hopfer and Murer (Hopfer, Nelson, Perotto, & Isselbacher, 1973; Murer & Hopfer, 1974). *(Redrawn from Beck, J. C & Sacktor, B. (1978). The sodium electrochemical potential-mediated uphill transport of D-glucose in renal brush border membrane vesicles.* The Journal of Biological Chemistry, *253 (15), 5531–5535, copyright 1978, with permission from the Journal of Biological Chemistry).* For color version of this figure, the reader is referred to the online version of this book.

of the electrochemical potential for sodium, the concentration difference and the electrical potential difference (see Fig. 4). They could demonstrate that the driving forces derived from a sodium concentration difference alone and from a potential difference alone are additive (Beck & Sacktor, 1978). Although, theoretically, the two driving forces are fully exchangeable, the driving forces exerted by the electrical potential differences were smaller than expected. Geck and Heinz (1976) showed earlier that depending on the theoretical model assumed for electrogenic cotransport, such differences indeed can exist.

Efficacy of driving forces also depends on the cotransport stoichiometry (Turner, 1985). SGLT2 has a stoichiometry of one sodium ion to one glucose molecule, whereas SGLT1 exhibits a stoichiometry of two sodium ions to one glucose molecule (Turner & Moran, 1982). The higher stoichiometry of SGLT1 allows accumulation of sugars inside the epithelial cells even at very low intraluminal concentrations.

3.3. Coupling to Sodium and Water Transport

Another sign of an interaction of two solutes with the same cotransporter molecule is that the transporter changes its properties when one of the cotransported solutes binds. Uptake studies in vivo and in vitro, in tissues and in vesicles, showed that sodium increases the affinity of the transporter for glucose suggesting conformational changes in the transporter molecule (for review, see Wright et al., 2011). These conformational changes will be discussed in the chapters below.

Sodium–glucose cotransport in the brush border membrane also implies a role of glucose in sodium and water transport across the epithelium. Indeed, Loo et al. determined that 260 molecules of water are translocated by SGLT1 for each sugar molecule transported. In human intestine, this would account for 5 l of water absorption per day (Loo, Zeuthen, Chnady, & Wright, 1996). GLUT2 also transports water together with the sugar, although to a much lower extent than SGLT1, and only 35 molecules of water are translocated per sugar molecule (Zeuthen, Zeuthen, & Macaulay, 2007).

4. CLONING OF TRANSPORTER cDNAs, SEQUENCE HOMOLOGIES AND CONSERVED AMINO ACIDS

4.1. SGLT and SMIT Transporters

In 1987, SGLT1 was cloned from rabbit intestinal epithelial cells (Hediger et al., 1987). The cloning procedure was quite unique at that time. A pool of mRNA was isolated and, after size fractionation and injection into oocytes, the expression of transport activity was measured. In the years before, several laboratories had tried to isolate the transporter from renal or intestinal membranes by biochemical methods, with frustratingly little success.

So far, six SGLT isoforms (SGLT1–6) have been cloned from various mammals and functionally characterized by the biochemical and electrophysiological methods, particularly after expression in *Xenopus laevis* oocytes (Hediger et al. 1987; Hediger and Rhoads, 1994; Wright, 1993). The transporters exhibit different substrate specificities, kinetic properties, and tissue expression profiles, indicating that each transporter plays a distinct role in maintaining body glucose homeostasis. The sequence homologies between human SGLT1 versus 2, 3, 4 and 5 isoforms are 60%, 70%, 56% and 51%, respectively (BLAST Run program: http://blast.ncbi.nlm.nih.gov). However, the similarity is higher (79%, 84%, 74% and 68% for hSGLT2, 3, 4 and 5, respectively). In contrast to the full-length sequence similarities, the last part of the C-terminus carries deletions of multiple amino acids (Fig. 5). To what extent these differences underlie the diverse functional properties observed in particular tissues remains to be established.

SGLT1 and the members of the SLC5 family share considerable homology among the proteins (21–70% amino acid identity to SGLT1) and contain several conserved sodium–solute symporter (SSS) family signatures that are characteristics of this family (GenBank accession numbers PS00456, PS00457 and PS50283).

The first SMIT, SMIT1, was cloned from MDCK cells (Kwon et al., 1992). In 2002, Coady et al. identified an orphan rat kidney gene as a second myo-inositol transporter, SMIT2 (Coady, Wallendorff et al. 2002). Furthermore, an SGLT1-like protein or xSGLT1L was cloned from *Xenopus*, which transports myo-inositol more efficiently than glucose

```
hSGLT1  544  ISLLTKPIPDVHLYRLCWSLRNSKEERIDLDAEEEN-------IQEGPKETIEIETQ-----VPEK-KKGIFRRAYDLFC  610
hSGLT2  544  VSLCTAPIPRKHLHRLVFSLRHSKEEREDLDADEQQG--SSLPVQNGCPESA-MEMN-----EPQAPAPSLFRQCLLWFC  615
hSGLT3  544  ISLLTKPIPDVHLYRLCWVLRNSTEERIDIDAEEKS-------QEETDDG---VEED-----YPEK-SRGCLKKAYDLFC  607
hSGLT4  548  VSLCTTPIPEEQLTRLTWWTRNCPLSELEKEAHESTPEISERPAGECPAGGGAAENSSLGQEQPEAPSRSWGKLLWSWFC  627
hSGLT5  531  GSLLTPPPQSVQIENLTWWTL----------------------AQDVPLGTKAGDGQ----------------------  565

                                          ----------------TM14----------------
hSGLT1  611  GLEQHGA---PKMTEEEEKAMKMKMTDTSEKPLWRTVLNVNGIILVTVAVFCHAYFA  664
hSGLT2  616  GMSRGGVGSPPPLTQEEAAAAARRLEDISEDPSWARVVNLNALLMMAVAVFLWGFYA  672
hSGLT3  608  GLQK--G---PKLTKEEEEALSKKLTDTSERPSWRTIVNINAILLLAVVVFIHGYYA  659
hSGLT4  628  GLSGTPEQA---LSPAEKAALEQKLTSIEEEPLWRHVCNINAVLLLAINIFLWGYFA  681
hSGLT5  566  ----TP---------------------QKHAFWARVCGFNAILLMCVNIFFYAYFA  596
```

Figure 5 ***Deduced amino acid sequences of the C-terminus of hSGLT1 to hSGLT5.*** Identical amino acids are shown in red, similar amino acids in blue, supposed transmembrane segments are indicated as yellow bars. The high sequence homology and similarity in the other parts of the molecules are discussed in the text. Note that the last part of the C-terminus carries deletions of multiple amino acids. Perhaps, such differences could underlie some characteristic features of functional differences among various isoforms of SGLT in particular tissues, for example the different interactions of SGLT1 and SGLT2 with inhibitors (Hummel et al., 2012). A phlorizin interaction site has been described in loop 13 of rbSGLT1 (Tyagi et al., 2011). See the color plate.

```
                                                                                      --------------TM2
Human SGLT1      1    MD-SSTWSPKTTAVTRPVETHELIRNAADISIIVIYFVVVMAVGLWAMFSTNRGTVGGFFLAGRSMVWWPIGASLFASNI  79
Xenopus SGLT-1   1    METSSQSSPQTTPGMEAFPKKS--LDTIDIVVLVLYFVFVLAVGLLSMCRTKRGTVKGYFLAGKDMAWWPVGASLFASNV  78
Human SMIT1      1    MR-A-------------------VLDTADIAIVALYFILVMCIGFFAMWKSNRSTVSGYFLAGRSMTWVTIGASLFVSNI  60
                      -----------      ----------TM3-----------                     ----------------TM4--
Human SGLT1      80   GSGHFVGLAGTGAASGIAIGGFEWNALVLVVVLGWLFVPIYIKAGVVTMPEYLRKRFGGQRIQVYLSLLSLLLYIFTKIS  159
Xenopus SGLT-1   79   GSGHFIGLAGSGAASGIAVTAYEWNGLFCVLALAWLFLPIYISAGVTTMPEYLQRRFGGKRIQIFLAILYLFIYIFTKIS  158
Human SMIT1      61   GSEHFIGLAGSGAASGFAVGAWEFNALLLQLLGWVFIPIYIRSGVYTMPEYLSKRFGGHRIQVYFAALSLILYIFTKLS  140
                      -----------       ------TM5----            --------TM6--------
Human SGLT1      160  ADIFSGAIFINLALGLNLYLAIFLLLAITALYTITGGLAAVIYTDTLQTVIMLVGSLILTGFAFHEVGGYDAFMEKYMKA  239
Xenopus SGLT-1   159  VDMYAGALFIQQALQWDLYVAVIGLLVITAIYTVAGGLAAVIYTDTLQTVIMIIGALILMAYSFIEIGGFEALQEKYFHA  238
Human SMIT1      141  VDLYSGALFIQESLGWNLYVSVILLIGMTALLTVTGGLVAVIYTDTLQALLMIIGALTLMIISIMEIGGFEEVKRRYMLA  220
                                                              ------------TM7----------    -------
Human SGLT1      240  IPTIVSD---GNTTFQEKC-YTPRADSFHIFRDPLTGDLPWPGFIFGMSILTLWYWCTDQVIVQRCLSAKNMSHVKGGCI  315
Xenopus SGLT-1   239  IPNTHSG--------NSTC-GIPREDAFHIFRDPVTSDLPWPGVLVGMTIPSLWYWCTDQVIVQRSLSAKNLSHAKAGSL  309
Human SMIT1      221  SPDVTSILLTYNLSNTNSCNVSPKKEALKMLRNPTDEDVPWPGFILGQTPASVWYWCADQVIVQRVLAAKNIAHAKGSTL  300
                      ------TM8------------                                         ---------------TM9--
Human SGLT1      316  LCGYLKLMPMFIMVMPGMISRILYTEKIACVVPSECEKYCGTKVGCTNIAYPTLVVELMPNGLRGLMLSVMLASLMSSLT  395
Xenopus SGLT-1   310  LAASLKVLPLFMMVLPGMISRVLFTDQVACADPELCKEICGNPSGCSDIAYPKMVIELLPTGLRGLMMSVMIAALMSSLT  389
Human SMIT1      301  MAGFLKLLPMFIIVVPGMISRILFTDDIACINPEHCMLVCGSRAGCSNIAYPRLVMKLVPVGLRGLMMAVMIAALMSDLD  380
                      ----------------        ------------TM10--------     -----------TM11-------
Human SGLT1      396  SIFNSASTLFTMDIYAKVRKRASEKELMIAGRLFILVLIGISIAWVPIVQSAQSGQLFDYIQSITSYLGPPIAAVFLLAI  475
Xenopus SGLT-1   390  SIFNSASTIFTMDLWRHIRPRSTEWELMIVGRVFVLVLVVVSILWIPLVQASQGGQLFIYIQSISSYLQPPVAVVFIAGC  469
Human SMIT1      381  SIFNSASTIFTLDVYKLIRKSASSRELMIVGRIFVAFMVVISIAWVPIIVEMQGGQMYLYIQEVADYLTPPVAALFLLAI  460
                      --     ---------TM12-----------                    ---------TM13---------
Human SGLT1      476  FWKRVNEPGAFWGLILGLLIGISRMITEFAYGTGSCMEPSNCPTIICGVHYLYFAIILFAISFITIVVISLLTKPIPDVH  555
Xenopus SGLT-1   470  FWKRTNEKGAFWGMTIGLVVGLIRMVLDFIYVAPQCDQPDTRPGVVKYVHYLYLSMILGLLTLLVVVAVSLWTEP-PSKQ  548
Human SMIT1      461  FWKRCNEQGAFYGGMAGFVLGAVRLILAFAYRAPECDQPDNRPGFIKDIHYMYVATGLFWVTGLITVIVSLLTPP-PTKE  539

Human SGLT1      556  LYRLC--WSLRN--------------SKEERIDLDAEEEN------IQEGPKETI-------------------------  588
Xenopus SGLT-1   549  MISRLTWFTRFDAEVEDPVETNHRP-----------AENGISVVEDISEEPHTTSTDAIYNDTT--------DDNPSSS-  608
Human SMIT1      540  QIRTTTFWSKKNLVVKENCSPKEEPYQMQEKSILRCSENNETINHIIPNGKSEDSIKGLQPEDVNLLVTCREEGNPVASL  619
                                                                                        -------
Human SGLT1      589  ---EIETQVP-------------EKKK-----GIFRRAYDLFCGLEQHGAPKMTEE---EEKAMKMKMTDTSEKPLWRTV  644
Xenopus SGLT-1   609  --------------------------------SLLKKTILWLCGMDSRKGDKHDQAPPAPLEPAEVLLYE---RPLLKQV  653
Human SMIT1      620  GHSEAETPVDAYSNGQAALMGEKERKKETDDGGRYWKFIDWFCGFKSKSLSKRSLRDLMEEEAVCLQMLEETRQV--KVI  697
                      ------TM14------------
Human SGLT1      645  LNVNGIILVTVAVFCHAYFA-  664
Xenopus SGLT-1   654  LNTAVILCMSAGVFLWAYFG-  673
Human SMIT1      698  LNIGLFAVCSLGIFMFVYFSL  718
```

Figure 6 ***Sequence comparison of SGLT1, SMIT1 and xSGLT1-like protein.*** Deduced amino acid sequences of the glucose-transporting SGLT1 are compared with the myo-inositol transporting SMIT1 and xSGLT1-like protein cloned from *Xenopus laevis*, which transports both substrates. Major differences are found in the region between aa 245 and aa 260 and in the C-terminus which constitutes the sugar translocating element in hSGLT1 (Wright et al., 2011). See the color plate.

(Nagata et al., 1999). Amino acid sequence, transport kinetics, and tissue expression pattern indicated that xSGLT1-like protein was more similar to SGLT1 than to SMIT1 (Wright, Hirsch, Loo, & Zampighi, 1997). Substrate specificity and absence of a Na^+ leak in xSGLT1L-like protein, however, resembled the properties of SMIT1. These findings suggested a close relationship between SGLT and SMIT transporter proteins. As shown in Fig. 6, the sequence homology between human SGLT1 and SMIT1 is 51%, with 70% sequence similarity. Similar to differences among SGLT isoforms, the C-termini of SGLT1, xSGLT1 and SMIT1 show remarkable differences in the amino acid composition with multiple deletions.

SGLT1 and SMIT1 differ in terms of their genomic DNAs. The coding region of SMIT mRNA resides within one exon, whereas SGLT1 is encoded in 15 exons (Berry et al., 1995; Turk, Martín, & Wright, 1994).

In addition to the expression of SGLT1 in intestine and kidney, mRNA for the transporter has been found in other epithelia such as salivary gland (Balen et al., 2008; Jurysta et al., 2012), human mammary gland (Obermeier, Hülseweh, Tinel, Kinne, & Kunz, 2000) and ruminal and omasal mucosa of lactating cows (Zhao, Okine, Cheeseman, Shirazi-Beechey, & Kennelly, 1998; Zhao, Okine, & Kennelly, 1999; Zhao, Zheng, Wall, & McFadden, 2005). Its physiological role in these epithelia is not quite clear (Zhao & Keating, 2007). SGLT1 might provide the carbohydrate components for sugars/glycoproteins produced by the glands and/or play a role in fluid secretion. Secretion rate varies largely and very rapid onsets of secretion are observed. To provide sufficient metabolic fuel for secretion, a high intracellular glucose level and high rate of glucose uptake might be crucial. In salivary glands, a role in glucose secretion has been postulated (Jurysta et al., 2012). SGLT1 was also found in the hepatic bile ducts, entero-endocrine cells and myo-enteric plexus in the small intestine (Balen et al., 2008). The transporter is transcribed in pyramidal cells of brain cortex and hippocampus, in Purkinje cells of cerebellum, and endothelium of the blood–brain barrier (Elfeber et al., 2004; O'Malley, Reimann, Simpson, & Gribble, 2006; Poppe et al., 1997). In the hypothalamus, SGLT1 acts as a glucose sensor (O'Malley et al., 2006).

SGLT2 is almost exclusively found in the kidney (Kanai, Lee, You, Brown, & Hediger, 1994; Sabolic et al., 2012). SGLT3 was originally cloned as a renal amino acid cotransporter in swine but later reclassified as a member of the SGLT family (Mackenzie, Panayotova-Heiermann, Loo, Lever, & Wright, 1994). It is also expressed in heart and skeletal muscle. SGLT3 can act as glucose sensor but might also transport sugar in humans. It was found to be increased in skeletal muscles by exercise in diabetic patients, increasing glycogen content of the muscle fibers (Castaneda-Sceppa & Castaneda, 2011). SGLT4 and 6 are expressed in a wide variety of tissues, whereas expression of SGLT5 is restricted to kidney (Wright, Loo, Hirayama, & Turk, 2004; Wright & Turk, 2004). SMIT1 is expressed in many tissues, such as kidney, skeletal muscle, lung, brain, pancreas, and heart but not in the ileal mucosa (Hediger et al., 1987; Kwon et al., 1992).

4.2. Isolation and Cloning of GLUT Transporter

GLUT transporters are expressed in almost every cell of the body (Uldry & Thorens, 2004). GLUT1 was isolated by biochemical methods from human erythrocytes, which have a high density of glucose transporters. Availability of the purified transporter protein allowed partial amino acid sequencing and the preparation of specific antibodies. Antibodies and sequence data were used for cloning of a human glucose transporter by screening an expression library prepared from the human hepatoma cell line, HepG2 (Mueckler et al., 1985). The cDNA of GLUT1 was then employed for cloning, by low-stringency screening, of GLUT2–5. All 12 GLUT members represent facilitative glucose transporters in humans (for review, see Thorens & Mueckler, 2010; Uldry & Thorens, 2004). Out of these 12 members, the sequence alignments of GLUT1–5 are shown in Fig. 7. The proteins are structurally conserved and related. Hydropathy plot analysis predicts a tertiary structure of 12 TM domains. Sequence comparisons of all 13 family members show that sequences are more conserved in putative TM regions

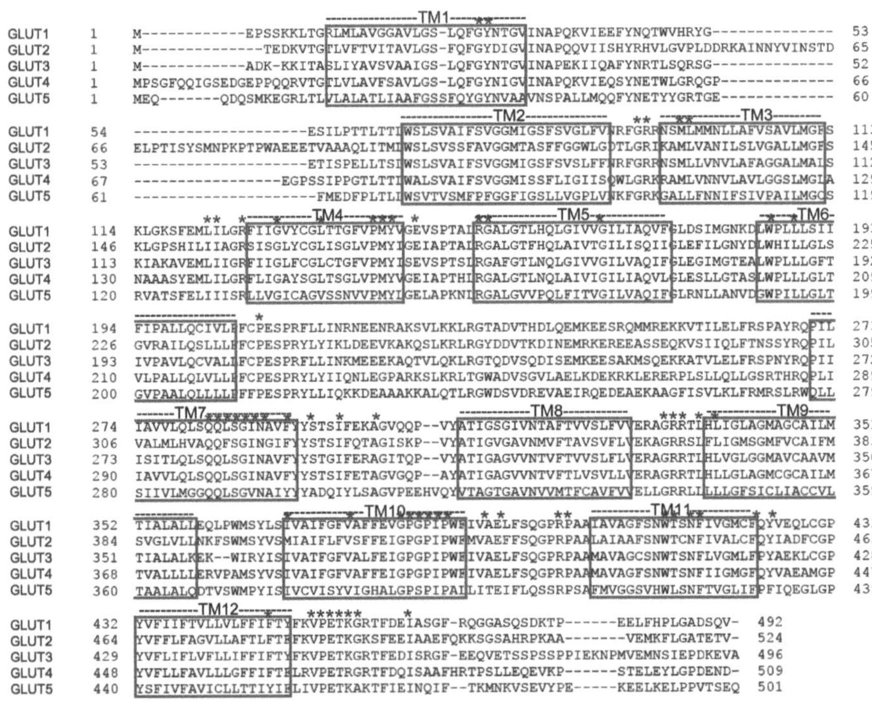

Figure 7 ***Sequence alignments of GLUT1–GLUT5.*** These proteins were found to be structurally conserved and related. Note that the sequences are more conserved in the putative transmembrane (TM) regions and more divergent in the loops between the TM and the C- and N-terminal regions. Highly conserved structures and sequences (denoted by red asterisk) are the characteristic sugar transporter signatures: PMY in TM4, QQLSGIN in TM7, GRR in loop 8, GPGPIP in TM10, and VPETKG in the C-terminal tail. For details and references, see text. See the color plate.

and more divergent in loops between the TM domains and in the C- and N-terminal regions (Fig. 7). Overall, the sequences among all members are 14–63% identical and 30–79% similar. Several highly conserved structures and sequences (denoted by red asterisk) are present; the characteristic sugar transporter signatures are PMY in TM4, QQLSGIN in TM7, GRR in loop 8, GPGPIP in TM10, and VPETKG in the C-terminal tail. In addition, there are 18 conserved glycine residues, 11 of them adjacent to at least one other conserved amino acid. The tryptophan in the GPGPIP/TW motif of TM10 (W388 of human GLUT1) has been shown to be critical for cytochalasin B and forskolin binding. The tryptophan conserved in all GLUTs in TM11 (W412) is essential for the GLUT1 transport activity and targeting to the plasma membrane.

5. TRANSPORTER DOMAINS, THEIR ARRANGEMENT IN THE MEMBRANE AND THEIR FUNCTION

Before crystal structures of transporters became available, information about the two-dimensional arrangement of transport molecules within the membrane was derived indirectly from the amino acid sequences and by comparison to other known transporters. Direct attempts to determine the intramembrane arrangement included insertion of sequences into the gene that carried a motif for *N*-glycosylation—suggesting an extracellular location of the protein segment (Hresko, Kruse, Strube, & Mueckler, 1994) or of a series of 6-histidines which could be recognized by a specific antibody (Lin, Kormanec, Homerová, & Kinne, 1999; Turk, Gasymov, Lanza, Horwitz, & Wright, 2006; Turk, Kerner, Lostao, & Wright, 1996). Reaction of the antibody with heterologously expressed transporters, before and after permeabilization of the cell membrane, indicated intra- or extracellular orientation of the segment. Sensitivity to digestion by trypsin beads (which during short exposure do not enter cells) or to epitope-specific antibodies was also employed (Lin et al., 1999). Accessibility from the outside of the membrane and proximity of nearest helix neighbors was also tested by combining cysteine scanning of the molecule with the analysis of the reactivity to membrane-permeant or non-permeant thio reagents of different size, charge and hydrophilicity. For review of the substituted cysteine accessibility method (SCAM), see Wright et al. (2011), Mueckler and Makepeace (2009), and Mueckler and Makepeace (2012). For SGLT1, the arrangement shown in Fig. 11 was derived from such studies. SGLTs are predicted to have 14–15 TM helices. The structural organization of most of the molecule is supported by glycosylation studies, antibody tagging, mass spectrometry and the more recent crystal structure of *Vibrio* SGLT (Faham et al., 2008; Wright & Turk, 2004; Zhao et al., 2005). There is, however, disagreement about the arrangement of the large loop between the TM helices 13 and 14 (loop 13).

5.1. Topology Studies of rbSGLT1 by Atomic Force Microscopy

In order to investigate the topology and function of extramembranous loops in intact cells, Puntheeranurak, in the laboratory of Hinterdorfer, recently developed new methods to apply single-molecule recognition atomic force microscopy (AFM) on living cells (Puntheeranurak, Neundlinger, Kinne, & Hinterdorfer, 2011).

In this approach, specific ligands are coupled to cantilevers of an atomic force microscope. Ligands can be epitope-specific antibodies, transport inhibitors or modified substrates, as illustrated in Fig. 8. Conditions of coupling reactions are chosen in such a way that only a small number of ligands are attached to the cantilever and only a defined chemical group of the ligand is involved. Immobilization on the cantilever guarantees that the ligand is presented to the cell from the extracellular space. With these specifically tagged cantilevers, force–distance cycles are performed, which are summarized in Fig. 9 (for a detailed description, see Puntheeranurak et al., 2011). The cantilever is brought in contact with the cells expressing the transporter and subsequently withdrawn from the cell surface. When the ligand has interacted with a component on the surface of the cell, the cantilever is deflected until an "unbinding force" is reached and the interaction is disrupted. This leads to characteristic patterns in force–distance diagrams (Puntheeranurak et al., 2011).

5.2. Three Surface Loops Form a Vestibule in rbSGLT1

Employing epitope-specific antibodies as ligands, the location of three extramembranous loops of rbSGLT1 (rabbit isoform) was investigated in intact cells. Three different antibodies were used: an antibody against aa 243–272 [loop 7, i.e. the loop between the TM helices 6 and 7], against aa 339–356 (loop 9), and against aa 606–630 (loop 13). Force–distance cycles were performed and the binding probability was determined. As shown in Fig. 10, all three antibodies are bound to the cells in a specific way, suggesting that all three loops are facing the extracellular medium. For hSGLT1, extracellular location of loop 13 was also confirmed in proteomic studies with isolated transporters, reconstituted in liposomes (Kumar, Tyagi, Arevalo, Miller,

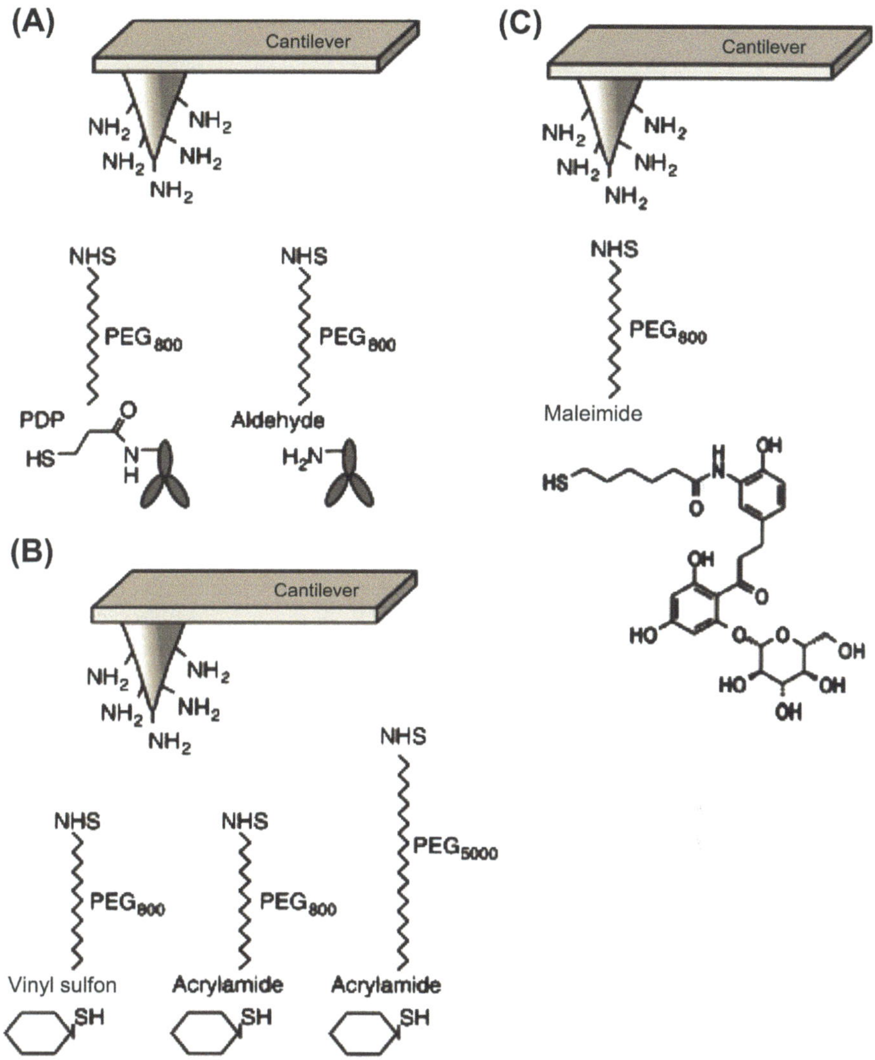

Figure 8 ***Linkage of ligands to AFM tips for studies on membrane topology.*** (A) Specific antibodies are covalently coupled to AFM tips via a heterobifunctional polyethylene glycol (PEG) derivative, PDP–PEG_{800}–NHS (left) and aldehyde–PEG_{800}–NHS (right). (B) Sugars are covalently linked to AFM tips via vinylsulfon–PEG_{800}–NHS (large end group) and acrylamide–PEG_{800}–NHS (small end group) or acrylamide–PEG_{5000}–NHS. (C) Maleimide–PEG_{800}–NHS was used to couple phlorizin to the AFM tips. The NHS ends of the PEG linkers are covalently bound to amines on the functionalized tip surface. *(From Puntheeranurak et al., 2006)*

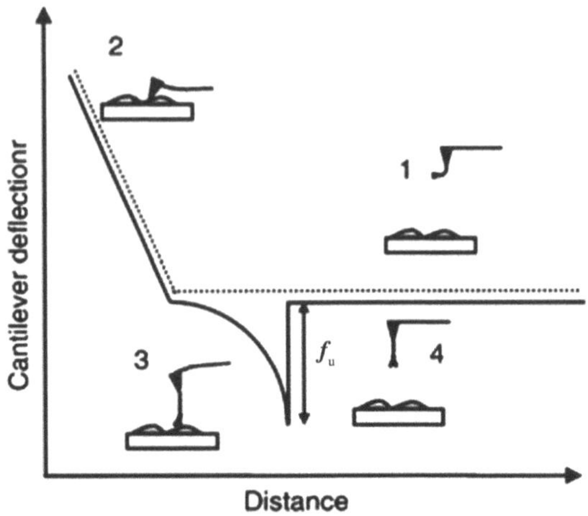

Figure 9 ***Schematic representation of an AFM force–distance cycle.*** The tip is moved toward the cell surface (dotted line, 1–2) and subsequently retracted (solid line) at a constant lateral position. During tip approach, the ligand interacts with a specific binding site on the cell surface, which leads to a force signal of distinct shape (3) during tip retraction (deflection). The force increases until dissociation occurs (4) at an unbinding force (f_u). *(Reprinted from Puntheeranurak et al., 2011)*

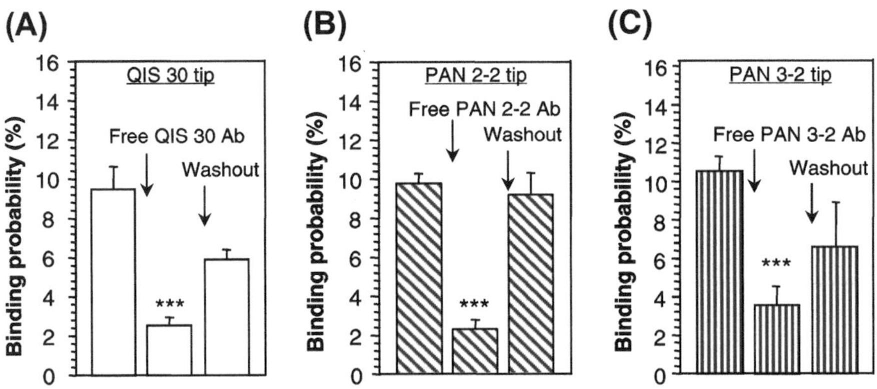

Figure 10 ***Recognition of extramembranous loops of SGLT1 on the surface of intact cells.*** AFM tips were tagged with epitope-specific antibodies (against subdomain I = PAN3-2 = loop 7, subdomain II = QUIS30 = loop 9, and subdomain III = PAN2-2 = loop 13) of rbSGLT1 (rabbit isoform). Quantitative comparison of binding probabilities of QIS30 (A), PAN2-2 (B), and PAN3-2 (C). Antibodies coupled to AFM tips were tested on G6D3 cells expressing rbSGLT1, in the absence or presence of free specific antibodies in the medium. *(Reprinted from Puntheeranurak, Kasch et al., 2007)*. For color version of this figure, the reader is referred to the online version of this book.

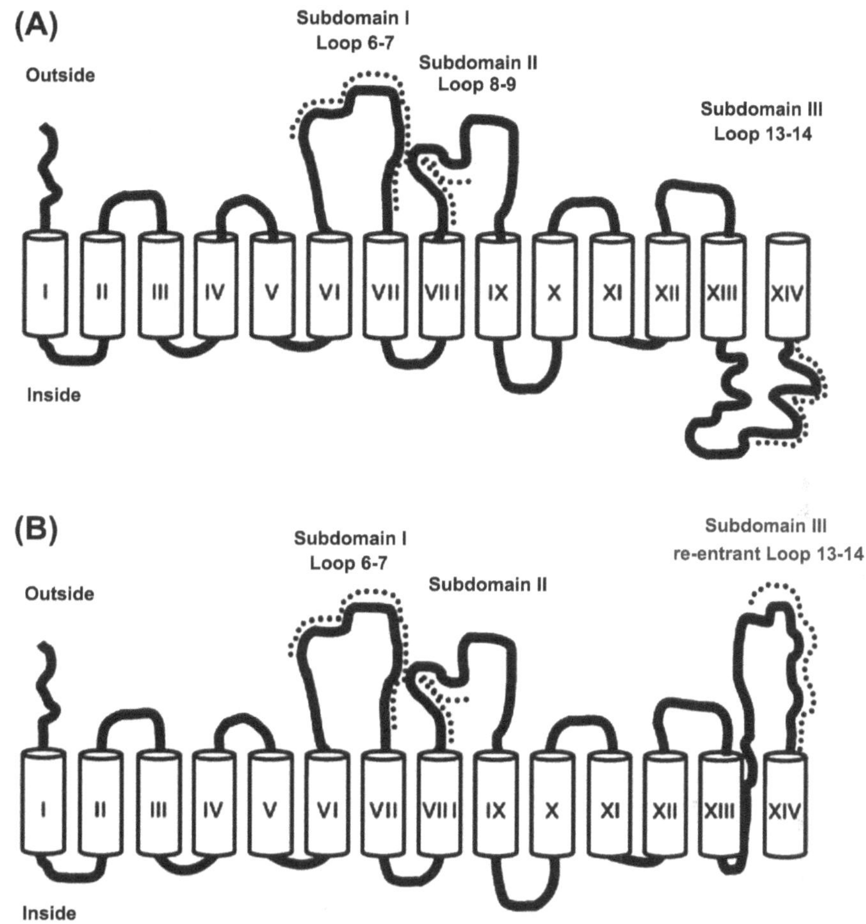

Figure 11 ***Membrane topology of SGLT1.*** (A) Secondary structure proposed of human SGLT1 by Turk, Kerner et al. (1996). The structure model shows 14 TM helices with both the NH_2 and COOH termini facing the extracellular compartment and loop 13 located completely in the cytoplasm. (B) Secondary structure modified according to the results obtained for rabbit SGLT1 with the antibody-tagged AFM tips in which the late part of loop 13 is exposed to the extracellular medium, as proposed earlier from 6-His scanning and trypsin digestion experiments (Lin et al., 1999). *(Reprinted from Puntheeranurak, Kasch et al., 2007).* For color version of this figure, the reader is referred to the online version of this book.

& Kinne, 2007). Therefore, the modified arrangement, shown in Fig. 11B, was proposed for the TM helices and extramembranous loops of SGLT1. It differs mainly at the C-terminus where an exposure of the late loop 13 to the extracellular medium is assumed.

In addition, a disulfide bond between C 255 (in loop 7) and C 608 (in loop 13) was detected by cysteine replacement mutagenesis and surface labeling studies with biotin (Puntheeranurak, Kasch, Xia, Hinterdorfer, & Kinne, 2007). This bond may bring the three extracellular loops closely together, creating a vestibule for the entry of glucose into the transporter.

An essential disulfide bond was also described for hSGLT1 between cysteine 255 (in loop 7) and cysteine 511 (in loop 9), suggesting a similar arrangement. During binding of the sodium and sugar, these regions undergo significant conformational changes (Gagnon, Frindel, & Lapointe, 2007).

5.3. Loop 13 of rbSGLT1 Contains Binding Sites for Transport Inhibitors

5.3.1. Phlorizin-Binding Site

Phlorizin inhibits sodium–sugar cotransport by binding to several binding sites. Its glucose residue interacts with the sugar-binding site and its aglucone with additional, yet undefined, binding sites (Frasch, Frohnert, Bode, Baumann, & Kinne, 1970; Glossmann & Neville, 1972; Koepsell et al., 1990; Kumar, Tyagi, & Kinne, 2007). In view of the hypothesis, that a vestibule is formed by the three extracellular loops 7, 9, and 13, the role of these loops in phlorizin binding was investigated. Mutagenesis studies, followed by the expression of rbSGLT1 in CHO cells and transport studies, showed that the replacement of amino acids in the region between 604 and 610 of loop 13 by bulkier hydrophilic amino acids resulted in markedly lower affinities for phlorizin. Reconstitution with alanine improved phlorizin affinity (Novakova, Homerova, Kinne, Kinne-Saffran, & Lin, 2001). These findings suggested that the amino acids 604–610 may be part of a complex formed when phlorizin binds to SGLT1. Direct evidence for binding of phlorizin to loop 13 was obtained in the tryptophan fluorescence studies on isolated loop 13. Intrinsic fluorescence was quenched significantly by phlorizin, with affinities about 10 times lower than observed for transport inhibition (Xia, Lin, & Kinne, 2003). Phloretin, but not D-glucose, had a similar effect. In further studies, the Trp molecules were introduced at positions indicated in Fig. 12 and fluorescence properties were investigated. In addition, truncated loop 13, photolabeled with the affinity label 3-azidophlorizin, was analyzed. The attachment site of the aromatic ring B of phlorizin was localized to Arg-602. From these data, a model

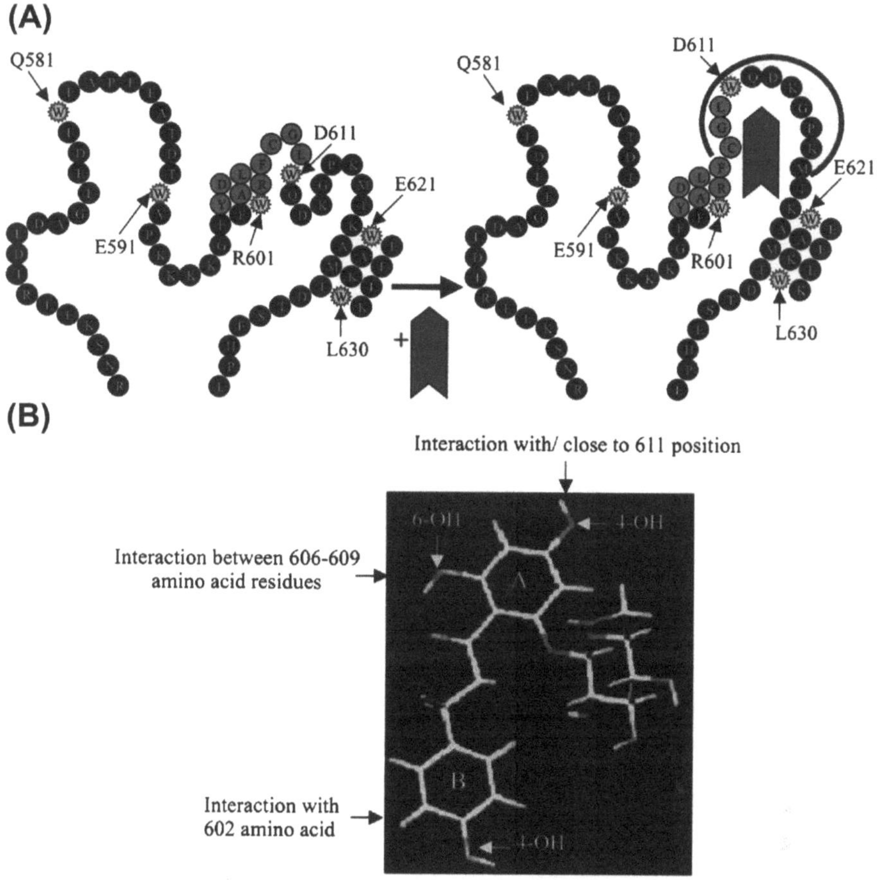

Figure 12 ***Model for the interaction of loop 13 of rbSGLT1 with phlorizin.*** (A) Two-dimensional model of truncated loop 13 of rbSGLT1 (amino acids 564–638). Left panel, shown is an arrangement of amino acids in the absence of phlorizin. α-Helical areas are based on a computer program; the arrangement of random coil takes into account the hydrophobicity of the tryptophan environment as predicted from the fluorescence maxima. Right panel, yellow represents the Trp mutations; green represents the presumed phlorizin-binding region; the red arrow indicates the proposed position of phlorizin in the binding region; and the blue arc highlights the main area of conformational changes. The exposure of the region around position 611 to a more hydrophilic environment is derived from the strong red shift in the tryptophan fluorescence maxima. (B) Three-dimensional conformation of phlorizin according to nuclear magnetic resonance studies (Wielert-Badt et al., 2000). The arrows show the interactions of aromatic ring A of phlorizin with the binding region of the loop. *(Reprinted from Raja et al., 2003).* See the color plate.

for interaction of phlorizin with loop 13 could be developed, which is shown in Fig. 12. Two short helices frame the binding site. The 4- and 6-OH groups of aromatic ring A of phlorizin (red arrow) interact with the region between amino acids 606 and 611. Aromatic ring B of the molecule interacts around amino acid 602. Phloretin binds to the same regions (Raja, Tyagi, & Kinne, 2003). The sugar residue of phlorizin remains free to associate with the sugar-binding site of the transporter.

These results, obtained with a truncated loop 13 dissociated from the transporter, could be confirmed after Tyagi et al. (2005) succeeded in expressing and purifying hSGLT1 from *Pichia pastoris* in an active form. Single Trp residues were introduced into a functional hSGLT1 mutant, initially devoid of Trps, at positions that previously had been postulated to be involved in the phlorizin binding. Mutant proteins were expressed in *P. pastoris*, purified, and reconstituted into liposomes. W602hSGLT1 and W609hSGLT1, with mutations in the putative aglucone-binding site of loop 13, exhibited normal sugar and phlorizin affinity. Phlorizin and phloretin, but not D-glucose, protected both mutants against collisional quenchers added to the outside medium (Kumar, Tyagi, & Kinne, 2007). These results support the conclusion that the aromatic amino acids in loop 13 contribute toward binding of the aglucone part of the phlorizin. However, species differences might exist (Gagnon et al., 2007).

5.3.2. Binding Sites for Alkylglucosides

Loop 13 also contains binding sites for alkylglucosides. A beta-configuration of the sugar moiety at C1, saturation of the side chain, and an alkyl chain length of six to eight carbon atoms optimizes interaction. C2 of the side chain interacts with loop 13 around residue 601, C3 and C4 interact with the region between residues 614 and 619, and C6–C8 interact with the region between residues 621 and 630 (Raja, Kipp, & Kinne, 2004). Thus, loop 13 of rbSGLT1 is a major binding domain for aglucone moieties of competitive D-glucose transport inhibitors.

5.4. Access to the Glucose Translocation Site in rbSGLT1

Carrier models for SGLT1 indicate that, in a translocation cycle, the sodium ions bind first to the transporter, only then the glucose can bind. This would mean that access of sugar to the translocation channel is controlled by sodium. AFM cantilevers, carrying glucose molecules, (see Figure 8) were used in single-molecule recognition studies to study the initial steps in glucose binding to the transporter. On rbSGLT1-expressing cells, binding events were only observed in the presence of sodium (Puntheeranurak, Wildling, Gruber, Kinne, & Hinterdorfer, 2006).

Furthermore, comparing inhibition of binding probability of glucose-tagged AFM cantilevers with that of sugar uptake revealed the following sequence of events in sugar recognition and selection for transport was proposed (Puntheeranurak, Wimmer et al., 2007). In the presence of sodium, sugars first reach a docking site in the upper part of the vestibule. Sugars that have a hydroxyl group (or *O*-methyl group) at C3 are then transferred to a second docking site, which requires a hydroxyl group in equatorial position at C2. Thereafter, sugars can be translocated across the membrane. However, more detailed studies at the molecular level are necessary to determine the number of interaction sites and their location, directly.

5.5. Membrane Topology of GLUT1

Figure 13 shows the TM arrangement of GLUT1 as proposed by Mueckler and Makepeace (2009). The figure is based on the extensive studies, in which all the 252 amino acid residues comprising the 12 predicted TM segments of human GLUT1 have been subjected to cysteine scanning mutagenesis and SCAM analysis. Forty-five of the 252 predicted TM residues noted were accessible to pCMBS (p-chloromercuribenzenesulfonic acid) from the external solvent (purple residues in Fig. 13). The data are mostly consistent with a basic model in which a central aqueous cavity, formed by eight inner helices, is stabilized by four outer helices. Inside this barrel, those amino acids are arranged (indicated in red), which are responsible for sugar binding and translocation. When glucose moves in the cavity, sugar–protein interactions form and break successively (Mueckler & Makepeace, 2009).

The location of the extramembranous loops has been confirmed by results from a variety of biochemical experiments, some of which were already available when the structure was proposed initially in 1985 (Mueckler et al., 1985).

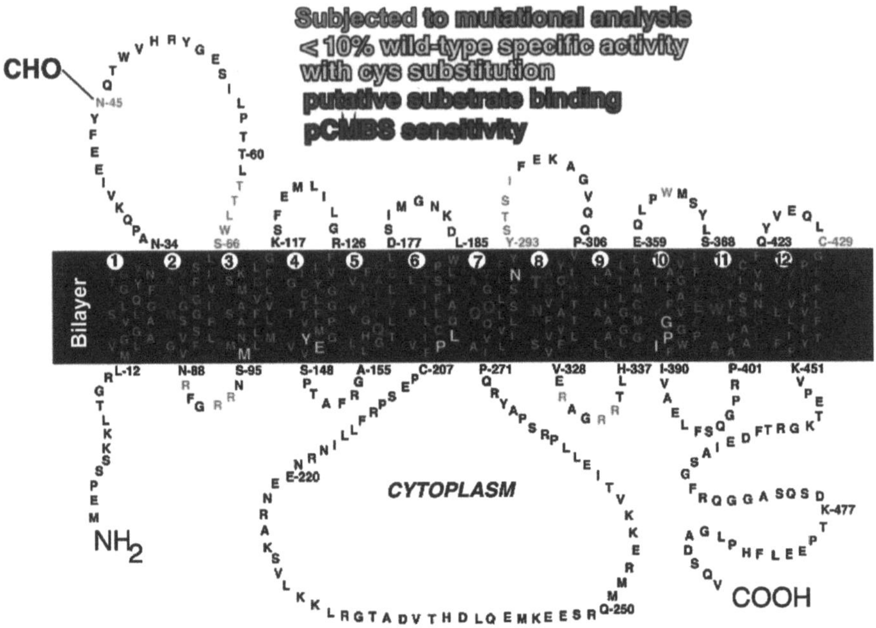

Figure 13 ***TM arrangement of GLUT1.*** Summary of mutagenesis data and SCAM analysis of GLUT1. The membrane is represented by a black rectangle, and the 12 TM segments are numbered consecutively from the N-terminus to the C-terminus. The termini of the TM segments are assigned as proposed in Mueckler et al. (1985). Amino acid residues that have been subjected to mutagenesis are colored green. Residues that represent sites of pCMBS sensitivity are colored purple. Residues corresponding to mutations that inhibit transport activity by >90% are colored yellow. Putative substrate-binding residues are colored red. *(Reprinted with permission from Mueckler, M. & Makepeace, C. (2009). Model of the exofacial substrate-binding site and helical folding of the human GLUT1 glucose transporter based on scanning mutagenesis.* Biochemistry, *48(25), 5934–5942. Copyright 2009, American Chemical Society. See text for a detailed discussion).* See the color plate.

6. CRYSTAL STRUCTURES AND HOMOLOGY MODELS OF MEMBERS OF SLC5 AND SLC2

6.1. The Crystal Structure of vSGLT

In 2008, the structure of SGLT from *Vibrio parahaemolyticus*, crystallized with galactose and sodium bound to it, was published (Faham et al., 2008). There were several major surprises. First of all, the 10 core helices (TM2 to TM11) are arranged in a topology repeat motif (see Fig. 14). A similar topology was found for the core structure of LeuT (TM1 to TM10), a

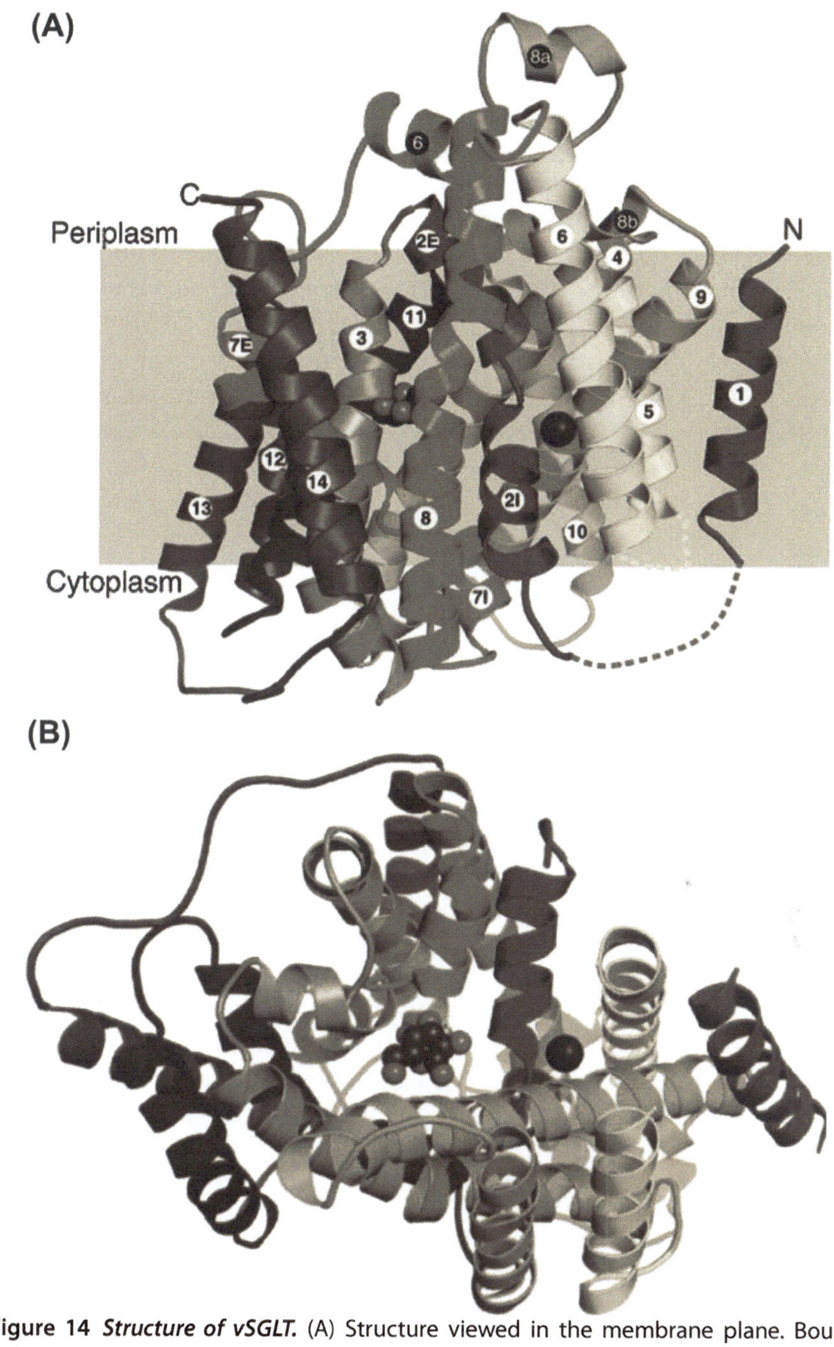

Figure 14 ***Structure of vSGLT.*** (A) Structure viewed in the membrane plane. Bound galactose is shown as black and red spheres for the C and O atoms. The proposed Na^+ ion is colored as a blue sphere. (B) Structure viewed from the intracellular side. The overall architecture has a maximal height and diameter of ~75 and ~55 Å. *(Reprinted from Faham et al. (2008). The crystal structure of a sodium galactose transporter reveals mechanistic insights into Na^+/sugar symport.* Science, *321(5890), 810–814, copyright 2008, with permission from AAAS).* See the color plate.

bacterial member of the neurotransmitter sodium symporter family. The two transporters have no significant (only 11.5%) similarity at the sequence level. The same core structure, the LeuT structure fold, is to date shared by transporters from four different gene families.

Second, two hydrophobic gates were identified that prevent the sugar from accessing the intra- and extracellular milieu. The extracellular gate is formed by a triad of hydrophobic residues (M73, Y87, and F424) (Fig. 15A). Directly above the hydrophobic residues, there are substantial interactions among TM11, TM3, TM2E, and TM7E with the loops from TM2–TM3, TM8–TM9, and TM10–TM11. These interactions stabilize a considerable protein mass, extending from the substrate-binding site to the extracellular surface. It further blocks access to the sugar-binding region from the extracellular milieu. On the intracellular side (Fig. 15B), Y263 from the discontinuous helix TM7E stacks with the pyranose ring. Together with the flanking residues Y262 and W264, a gate is established that prevents exit of the sugar to the large hydrophilic cavity contiguous with the intracellular compartment. Among members of the SSS family that bind sugars, the residues forming hydrophobic gates are highly conserved (Abramson et al., 2003; Sujatha & Balaji, 2004).

Several sodium cotransporters, which share the LeuT structure with vSGLT, have been crystallized in other intermediate states of the transport cycle. This allowed constructing homology models of the human SGLT1 and to predict, and experimentally test, the conformation of the inward-facing open and outward-open state. Such studies are beginning to "bridge the gap between structure and function" (Wright) (Sala-Rabanal et al., 2012). A current view of the transport cycle is shown in Fig. 16. This scheme depicts the movements of the two gates during the transport cycle as well as the opening and closing of hydrophilic cavities, to either the extracellular milieu (in the presence of sodium) or the intracellular fluid phase.

Figure 15 ***Galactose-binding site of vSGLT.*** (A) Hydrophobic gate residues (viewed from the extracellular side). The intracellular (Y263) and extracellular (M73, Y87, and F424) gates are shown as spheres and the galactose is shown as sticks. (B) The galactose-binding site [same view as in (A)], with the extracellular hydrophobic gate residues removed to view the galactose-binding site. Residues in the galactose-binding site are displayed as sticks colored by atom type. Hydrogen bonds are depicted as black dashed lines. *(Reprinted from Faham et al. (2008). The crystal structure of a sodium galactose transporter reveals mechanistic insights into Na^{+}/sugar symport.* Science, *321(5890), 810–814, copyright 2008, with permission from AAAS).* See the color plate.

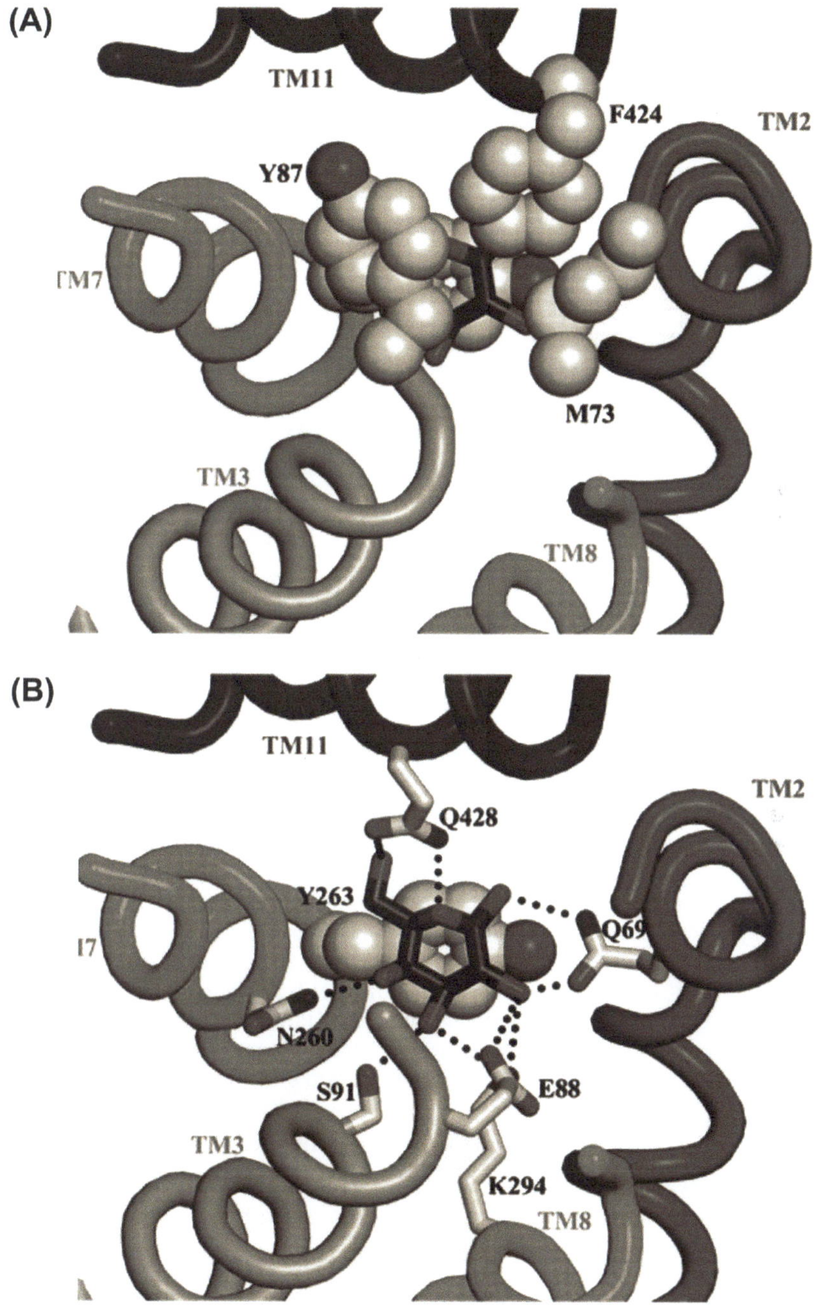
(A)
TM11
F424
TM2
Y87
TM7
M73
TM3
TM8
(B)
TM11
TM2
Q428
Y263
Q69
N260
S91
E88
TM3
K294
TM8

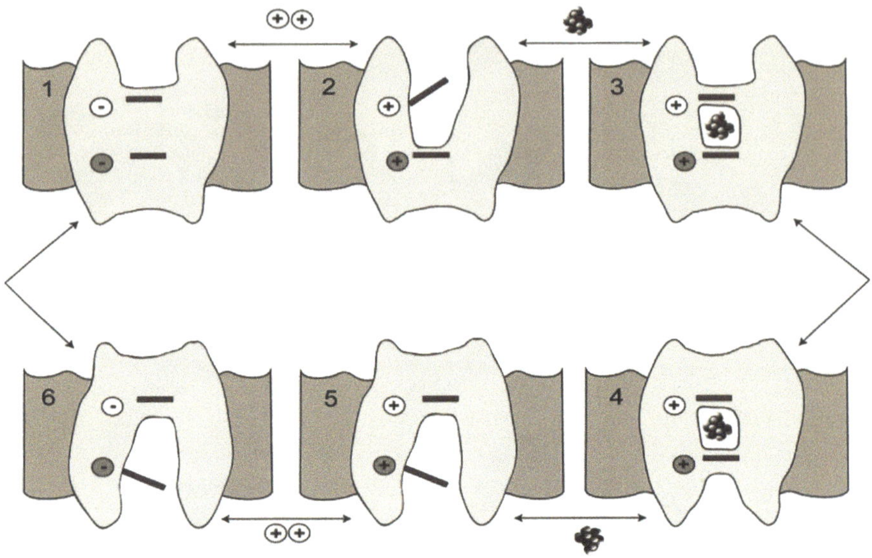

Figure 16 ***A six-state model of Na$^+$–glucose cotransporters (SGLTs).*** The model proposed by Sala-Rabanal et al. (2012) integrates the kinetic and structural data available to date. Na$^+$ binds first to the outside (C1) to open the outside gate [C2Na$_2$] (C2), permitting outside sugar to bind and be trapped in the binding site. This is followed by a conformational change from an outward occluded state [C3Na$_2$S] (C3) to an inward occluded state [C4Na$_2$S] (C4). On opening of the inward gate (C5→C6), the Na$^+$ ions and sugar are released into the cell interior. The transport cycle is completed by the change in conformation from the inward-facing ligand-free state (C6) to the outward-facing ligand-free state (C1). The structural states [C3N$_2$S] and [C4Na$_2$S] described here are different from the states C3 and C4 identified in previous kinetic studies (Parent, Supplisson, Loo, & Wright, 1992). The structural states are occluded, whereas the kinetic states C3 and C4 are open to the outer and inner membrane surfaces, respectively. The occluded states are intermediates: C3⇋[C3Na$_2$S]⇋[C4Na$_2$S]⇋C4. For simplicity, we have used C3 and C4 to represent [C3Na$_2$S] and [C4Na$_2$S]. *(Reprinted from Sala-Rabanal, M., Hirayama, B. A., Loo, D. D., Chaptal, V., Abramson, J., & Wright, E. M. (2012). Bridging the gap between structure and kinetics of human SGLT1.* American Journal of Physiology Cell Physiology, *302(9), C1293–C1305, copyright 2011, with permission from The American Physiological Society)* For color version of this figure, the reader is referred to the online version of this book.

Modeling of the phlorizin–carrier complex, using structural data (Sala-Rabanal et al. 2012), predicted an additional conformation of the carrier, in agreement with fluorescence studies on isolated hSGLT1 (Tyagi et al., 2005). The phlorizin-loaded carrier shows a highly condensed conformation. Thus, phlorizin inhibits transport not only by competing with sugars at the binding site but, in addition, transfers to the protein a rigidity that makes translocation of sugars impossible. Accordingly, in transport studies, phlorizin inhibits sugar influx as well as efflux (Kinne et al., 1975).

Homology modeling has also been used to delineate the potential structural changes of hSGLT1 caused by mutations leading to glucose–galactose malabsorption (Raja & Kinne, 2012). The modeling demonstrates how mutations of a single amino acid can destabilize and distort the structural elements of SGLT1 to yield proteins which cannot be processed properly by the cell.

6.2. Homology Model of GLUT1

GLUT proteins belong to the major facilitator superfamily that contains nearly 3600 members in all three kingdoms and represents the largest superfamily of membrane transporters. Although cloned as the first eukaryotic glucose transporter in 1985, attempts to obtain crystals suitable for the structural analysis have been unsuccessful, as of to date. However, the possibility to purify the transporter from erythrocytes and to incorporate it in liposomes made biochemical studies feasible that provide a substantial amount of data, which can be incorporated into the structural models.

The first concise GLUT1 structure was proposed in 2004 (Salas-Burgos, Iserovich, Zuniga, Vera, & Fischbarg, 2004). Two models were developed initially. One is based on the structure of *Escherichia coli* glycerol phosphate transporter (GLUT1A) and one on the structure of *E. coli* lac permease (GLUT1B) (PDB no. 1PW4; Huang, Lemieux, Song, Auer, & Wang, 2003) (PDB no. 1PV6; Abramson et al., 2003). The structure shown in Fig. 17 incorporates these biochemical results and the evidence from mutagenesis studies. It also gives insight on the molecular mechanism of protein flexibility, substrate migration, and binding of glucose and inhibitors like phloretin and cytochalasin B.

The central channel is mainly formed by helices 2, 4, 5, 7, 8, and 10. Residues crucial for pathogenicity (E247, K256, and R333) or transport (P387, W388) are near the segment of the long intracellular loop (loop 6–7) that restricts the transport pathway. Entrance at the extracellular end is funnel like, and at its bottom, the docking sites described below are located. Interestingly, the intracellular end of the channel is a horn-shaped pathway; it curves in a way that its exit is located almost to the side of the protein. It also expands into a cylindrical cavity ~12 Å in diameter. The position of the cytoplasmic orifice near the side of the protein may allow two monomer cytoplasmic cavities to join into a larger one for a dimer, as suggested by Carruthers and coworkers (Cloherty, Hamill, Levine, & Carruthers, 2001; Heard, Fidyk, & Carruthers, 2000).

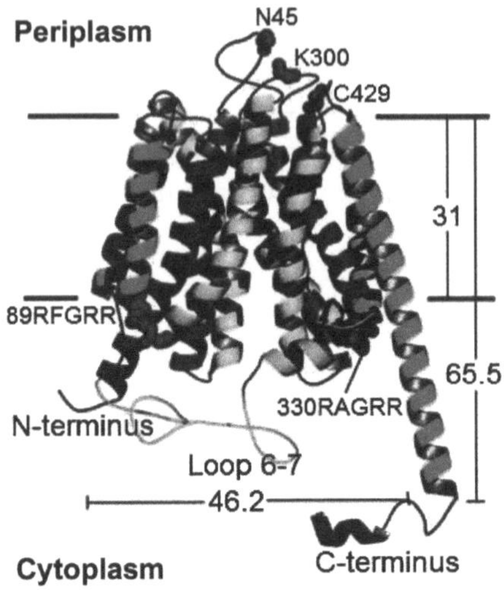

Figure 17 ***GLUT homology model.*** Homology model of GLUT1 as proposed by Salas-Burgos et al. (2004) Side view represents relative positions of the TM helices. Residues in red represent topology constraints involving N45, K300, and C429 for the extracellular side, and motifs 89RFGRR93 and 330RAGRR for the cytoplasmic side are highlighted. GLUT1 measures ~35.6 × 26.3 Å at the exofacial surface and 46.2 Å × 27.2 Å at the endofacial surface. The height is ~66 Å. *(Reprinted from Salas-Burgos et al. (2004). Predicting the three-dimensional structure of the human facilitative glucose transporter glut1 by a novel evolutionary homology strategy: insights on the molecular mechanism of substrate migration, and binding sites for glucose and inhibitory molecules.* Biophys J, *87(5), 2990–2999, copyright 2004, with permission from Elsevier).* See the color plate.

In the GLUT1 model, the internal segment of the transport pathway is lined by hydrophilic and hydrophobic amino acids. Among them, many have been shown to be crucial for transport activity—Q161, R126, Q279, Q282, N317, T321, W65, W388, W412, and V165 (see also Fig. 13). The presence of both hydrophilic and hydrophobic amino acids fits perfectly with the glucose molecule. OH groups render it hydrophilic and the pyranose ring offers hydrophobic interaction sites. The very well-conserved QLS motif in TM7 (Figs 7, 13, and 17) starting at Q279 determines the transport selectivity of GLUTs (Olsowski, Monden, Krause, & Keller, 2000; Seatter, De la Rue, Porter, & Gould, 1998). This site is at the very opening of the extracellular channel, which forms a bottleneck there. Salas-Burgos et al. hypothesized that, during translocation, glucose forms temporarily more than one hydrogen bond with the transporter. Formation of more hydrogen bonds stabilizes the substrate in position. Successive hydrogen bonding was also observed in molecular dynamics simulations based on a GLUT3 model (Dwyer, 2001). Scanning mutagenesis combined with the SCAM suggests that the cytochalasin B recruits the transporter to a conformational state in which the closure of inner helices occurs at the cytoplasmic face of the membrane. This occluded state, in which substrate is excluded from the endofacial binding site, provides a structural explanation for the competitive inhibition of substrate efflux by cytochalasin B (Mueckler & Makepeace, 2012).

Currently, the following model for sugar transport by GLUT1 is favored. A substrate-binding site (or two distinct sites) is alternately exposed at opposite faces of the membrane in two mutually exclusive transporter conformations (a kind of Patlak-Läuger gate; Läuger, 1979; Patlak, 1957). This alternating access model is supported by a large body of kinetic and biophysical data collected over the course of several decades (for review, see Mueckler & Makepeace, 2012). Crystal structures of the related transport proteins in the endofacial and exofacial conformation suggest that, during sugar translocation, inner TM helices alternatingly splay open to allow access to the substrate-binding site at one side of the membrane and pinch together at the other side to seal off aqueous cavities.

7. REGULATION OF TRANSPORTER ACTIVITY

7.1. Regulation of SGLT1

In view of the physiological importance of SGLT1, it is not surprising that a large number of experiments have been performed concerning the regulation of SGLT1-mediated transport activity. The studies comprise investigations in intact animals, isolated epithelia, transporters expressed heterologously in cell culture, purified membranes, and isolated transport proteins. They reveal a complex regulatory pattern which apparently differs in various species and experimental systems. In view of this complexity, only major mechanisms and sites of regulation are compiled in the following; every mechanism may contribute to a different extent to the final response of cells to a regulatory challenge.

7.1.1. Direct Modification of Transporters

Depending on the species, SGLT1 contains a varied number of phosphorylation sites for protein kinases A and C. hSGLT1 exhibits one potential PKA site and several PKC sites. PKA-mediated phosphorylation of hSGLT1, in vitro, changes the conformations of the empty carrier and the glucose–carrier complex (Subramanian, Glitz, Kipp, Kinne, & Castaneda,

2009). These changes probably cause the increase in transport affinity that is observed in the transport studies in CHO cells overexpressing rbSGLT1 (Castaneda-Sceppa, Subramanian, & Castaneda, 2010). A direct effect of PKA phosphorylation on the activity of SGLT1 has also been reported in rat small intestine, demonstrating the physiological relevance of PKA-mediated phosphorylation (Ishikawa, Eguchi, & Ishida, 1997). For rbSGLT, the group of Silverman observed that the stimulation of protein kinase C in Cos 7 cells increased the turnover rate of the transporter (the velocity of sugar translocation across the membrane) by a factor of 2, also suggesting a direct effect of phosphorylation on the properties of the transporter (Vayro & Silverman, 1999). Cellular uptake can also be increased without changing the intracellular distribution of the protein, if the transporters undergo an activation–inactivation cycle at the plasma membrane and/or in endosomal vesicles (Dieter et al., 2004; Khoursandi et al., 2004).

7.1.2. Regulation at the Cellular Level

SGLT1 transporters are present in an intracellular pool of endocytic vesicles and can be rapidly, and reversibly, incorporated into the plasma membrane. In CaCo-2 cells, about two third of the transporter protein resides in the endosomal compartment. Intracellular presence of SGLT1 has also been observed in human intestinal mucosa (Kipp, Khoursandi, Scharlau, & Kinne, 2003).

Protein kinase A stimulation increases the number of transporters in the (luminal) plasma membrane in rat small intestine and in CHO cells overexpressing rbSGLT1 (Subramanian et al., 2009). In oocytes, a stimulation of protein kinase C decreases the expression of transporter in the membrane (Hirsch, Loo, & Wright, 1996; Wright et al., 1997). In CHO cells, rbSGLT1 expression is increased after inhibition of protein kinase C, independent of the presence of phosphorylation sites on the transporter (Castaneda-Sceppa et al., 2010). In general, sorting to the plasma membrane is not dependent on the presence of "classical" phosphorylation sites but more on other, yet poorly defined, features in the structure of the transporters (Hirsch et al., 1996; Wright et al., 1997) or even on the expression system used (Ghezzi & Wright, 2012). RS1 (regulatory subunit), an intracellular protein which regulates SGLT1, controls release of SGLT1 from *trans*-Golgi network (Veyhl et al., 1993; Veyhl, Keller, Gorboulev, Vernaleken, & Koepsell, 2006). Increased transport activity, increased number of high-affinity phlorizin-binding sites and an increased amount of transporter protein are observed in rat jejunum after infusion with glucagon-like peptide 2. The translocation of SGLT1 from intracellular compartments to the plasma membrane requires the activation of PI3K (Cheeseman, 1997).

Changes in cellular sugar uptake via SGLT1 also can involve regulation at the transcriptional level (for a review, see Castaneda-Sceppa & Castaneda, 2011). The increase in intestinal transport activity in rats caused by a high-glucose diet is associated with an increase in SGLT1 mRNA, probably due to an increased transcription. For expression of the *sglt1* gene, the signal needs not to be a metabolizable or transportable substrate (Miyamoto et al., 1993). mRNA levels also increase in CHO cells overexpressing rbSGLT1 when protein kinase C activity is inhibited. In this response, intracellular signaling pathways (p38/MAPK, ERK/MAPK, JNK/MAPK, and PI3K/Akt/mTOR) are involved (Castaneda-Sceppa et al., 2010). In mice small intestine, SGLT1 mRNA abundance is critical for the level of total SGLT1 protein in the enterocytes, for the level of SGLT1 protein in brush border membranes and for SGLT1-mediated transport activity in the brush border membranes (Gorboulev et al., 2012). The intracellular protein RS1 is also involved in the long-term transcriptional regulation of SGLT1. The protein promotes transcription of SGLT1 during a D-glucose- and D-galactose-rich diet and increases the amount of active transporter in the brush border membrane. Modulation of nuclear location of RS1 via RNS-PKC is relevant for small intestinal D-glucose absorption during fasting or carbohydrate-poor nutrition (Filatova et al., 2009).

In LLC-PK1 cells, the expression of SGLT1 is transcriptionally regulated by cell confluence and D-glucose content in the medium (Amsler, 1994); again, nuclear location of RS1 seems to be important (Filatova et al., 2009). Protein kinase C activation appears to be involved in the dedifferentiation-induced decrease in SGLT1 mRNA in these cells (Shioda, Ohta, Isselbacher, & Rhoads, 1994).

7.2. Regulation of SGLT2

Recently, regulation of hSGLT2 was investigated in HEK 293T cells overexpressing transporters (Ghezzi & Wright, 2012). Br-cAMP increased sugar uptake and Na^+/glucose currents by 200% and protein kinase C activation augmented uptake and currents by 50%. In both cases, maximum rates of transport increased with no change in the glucose affinity. These effects were completely absent when a phosphorylation site of SGLT2, serine 624, was mutated to alanine. PKA, but not PKC, stimulation increased significantly membrane capacitance, suggesting a higher incorporation of endosomal vesicles containing hSGLT2 into the cell surface (Ghezzi & Wright, 2012).

Interestingly, insulin induced a 250% increase in Na^+/glucose transport mediated by the wild-type hSGLT2, but not the mutant S624A. The effect of insulin on hSGLT2 might suggest that in order to achieve complete recovery of glucose taken up after a meal, renal transport capacity also is increased (Ghezzi & Wright, 2012).

7.3. Regulation of SMIT1 and SMIT2

MDCK cells were the initial source for cloning SMIT1 and for a detailed study of its regulation (Handler & Kwon, 1997). Hypertonicity increases the transcription of cotransporter genes (Handler & Kwon, 1996). A tonicity-sensitive transcription factor, tonicity-responsive enhancer-binding protein (TonEBP), was identified (Garcia-Perez & Burg, 1991). In cells exposed to hypertonicity, transporter activity is further regulated by protein kinases A and C (Preston, Yamauchi, Kwon, & Handler, 1995). In intact rats, SMIT1 mRNA in the kidney medulla responds to the tonicity of the interstitium. Loading of rats with a high concentration of NaCl rapidly induces SMIT1 mRNA; furosemide, a diuretic, downregulates it. The most significant changes were found throughout the thick ascending loop of Henle, SMIT1 in the inner medullary-collecting duct is less subjected to acute regulation (Yamauchi et al., 1995).

When expressed in MDCK cells, SMIT2 transport activities also increase significantly in hypertonic media, but with a different time course and dose–response. The effect of osmolarity on the transfected SMIT2 is unrelated to TonEBP. Inhibition studies involving the hyperosmolarity-related MAPK suggest that p38 and JNK play a role in the induction of SMIT2 (Bissonnette, Lahjouji, Coady, & Lapointe, 2008).

7.4. Regulation of GLUT Transporter in the Intestine

In vivo, after ingestion of a carbohydrate-rich meal, GLUT2 molecules are inserted into the intestinal brush border membrane and facilitate sodium-independent glucose transport (Kellett, 2001). Insertion is very rapid and reversible ($t_{1/2}$ ~3.5 min). Apical GLUT2 also increases fructose uptake. The initial signal is depolarization of the apical membrane by the electrogenic transport of glucose and sodium through SGLT1. It stimulates Ca^{2+} entry via L-type channels Cav1.3. This causes contraction of the terminal web of the brush border, which is essential for the insertion of GLUT2 into the membrane. In addition, activation of the protein kinase C βII isoform is required (Kellett et al., 2008).

GLUT2 and GLUT5 expression is also regulated at the transcriptional level by the dietary sugar content. D-Galactose- and D-fructose-rich diets increase GLUT2 mRNA; GLUT5 mRNA levels are augmented only by D-fructose. For expression of the *glut2* gene, metabolism of the substrate in the liver may be necessary for signaling (Miyamoto et al., 1993). GLUT5 is, in addition, regulated by diurnal rhythm, substrate availability, and other factors (Douard, Cui, Soteropoulos, & Ferraris, 2008; Jones, Butler, & Brooks, 2011; Uldry & Thorens, 2004).

The reversible expression of sugar transporters in the plasma membrane by hormonal stimuli is also a hallmark of GLUT4, which mediates glucose uptake into skeletal muscle and fat cells (For reviews on the extensive literature, see Foley, Boguslavsky, & Klip, 2011; Stöckli, Fazakerley, & James, 2011; Thorens & Mueckler, 2010; Uldry & Thorens, 2004).

8. SYNOPSIS

"*It was the best of times, it was the worst of times, it was the age of wisdom, it was the age of foolishness* ..." (Dickens, 1859). In describing the different sugar transporters, one is reminded of this beginning of Charles Dickens' famous novel. In the case of SGLT, the age of wisdom might have arrived; in the case of GLUT, it might be still the worst of times, because crystallization has not yet been achieved. But the situation for GLUT is not so bad. Due to the availability of the protein in purified form, an enormous amount of biochemical and biophysical data are available which allows quite precise predictions on the structure and function.

Two types of transporters have evolved which catalyze the transfer of sugars across the cell membrane with high stereospecificity and efficiency. The affinities of the transporters as well as the selection of sugars have been adapted to the extra- and intracellular concentrations and to the metabolic reactions occurring inside the cell. Both achieve the transport by a basically similar model of alternating access—using the hydrophobic and hydrophilic residues for specific interactions of the sugars with the transporter proteins. Energy input for sodium cotransport occurs in a kind of push model of transepithelial transport: initial accumulation of sugar inside the cell and equilibration at the serosal cell side. In sodium-independent transfer, a facilitating gradient is available. For the regulation of transport activity, the whole array from direct modification of transporter molecules to changes in membrane sorting and transcription is used by both systems but is controlled, as would be expected, by different signals.

During the course of deciphering their properties, each transporter, at some time, has taken the lead and brought the field of membrane transport forward; experimental barriers were removed and new ideas emerged. Many concepts from previous investigators build the firm basis for conclusions reached today at the molecular level. As before, also the current interpretations have their limitations, but they create new questions which drive the quest of knowledge for an understanding of basic mechanisms in Biology. It is promising to see that, as previously in the case of membrane receptors and ion channels, sugar transporters also can now be approached by pharmacological tools. Thus, the efforts of many investigators over the last decades can be translated into the treatment of diseases (Kinne & Castaneda, 2011).

REFERENCES

Abramson, J., Smirnova, I., Kasho, V., Verner, G., Kaback, H. R., & Iwata, S. (2003). Structure and mechanism of the lactose permease of Escherichia coli. *Science, 301*(5633), 610–615.

Alvarado, F. C., & Crane, R. K. (1962). Phlorizin as a competitive inhibitor of the active transport of sugars by hamster small intestine in vitro. *Biochimica et Biophysica Acta, 56*, 170–172.

Amsler, K. (1994). Role of cell density/cell-cell contact, and growth state in expression of differentiated properties by the LLC-PK1 cell. *Journal of Cellular Physiology, 159*, 331–339.

Balen, D., Ljubojevic, M., Breljak, D., Brzica, H., Zlender, V., Koepsell, H., & Sabolic, I. (2008). Revised immunolocalization of the Na$^+$-D-glucose cotransporter SGLT1 in rat organs with an improved antibody. *American Journal of Physiology. Cell Physiology, 295*(2), C475–C489.

Barfuss, D. W., & Schafer, J. A. (1981). Differences in active and passive glucose transport along the proximal nephron. *American Journal of Physiology, 241*(3), F322–F332.

Beck, J. C., & Sacktor, B. (1978). The sodium electrochemical potential-mediated uphill transport of D-glucose in renal brush border membrane vesicles. *Journal of Biological Chemistry, 253*(15), 5531–5535.

Berry, G. T., Mallee, J. J., Kwon, H. M., Rim, J. S., Mulla, W. R., Muenke, M., & Spinner, N. B. (1995). The human osmoregulatory Na$^+$/myo-inositol cotransporter gene (SLC5A3): molecular cloning and localization to chromosome 21. *Genomics, 25*, 507–513.

Bissonnette, P., Lahjouji, K., Coady, M. J., & Lapointe, J. Y. (2008). Effects of hyperosmolarity on the Na$^+$-myo-inositol cotransporter SMIT2 stably transfected in the Madin-Darby canine kidney cell line. *American Journal of Physiology Cell Physiology, 295*(3), C791–C799.

Carter, S. B. (1967). Effects of cytochalasins on mammalian cells. *Nature, 213*(5073), 261–264.

Castaneda-Sceppa, C., & Castaneda, F. (2011). Sodium-dependent glucose transporter protein as a potential therapeutic target for improving glycemic control in diabetes. *Nutrition Reviews, 69*(12), 720–729.

Castaneda-Sceppa, C., Subramanian, S., & Castaneda, F. (2010). Protein kinase C mediated intracellular signaling pathways are involved in the regulation of sodium-dependent glucose co-transporter SGLT1 activity. *Journal of Cellular Biochemistry*, *109*(6), 1109–1117.

Cheeseman, C. (1997). Upregulation of SGLT-1 transport activity in rat jejunum induced by GLP-2 infusion in vivo. *American Journal of Physiology*, *273*(6 Pt 2), R1965–R1971.

Chinard, F. P., Taylor, W. R., Nolan, M. F., & Enn, T. (1957). Transport of glucose by the renal tubule cells of anesthetized dogs. *Science*, *125*, 736–737.

Clarkson, T. W., & Rothstein, A. (1960). Transport of monovalent cations by the isolated small intestine of the rat. *American journal of physiology*, *199*, 898–906.

Cloherty, E. K., Hamill, S., Levine, K., & Carruthers, A. (2001). Sugar transporter regulation by ATP and quaternary structure. *Blood Cells Molecules and Diseases*, *27*, 102–107.

Coady, M. J., Wallendorff, B., Gagnon, D. G., & Lapointe, J. Y. (2002). Identification of a novel Na^+/myo-inositol cotransporter. *Journal of Biological Chemistry*, *277*(38), 35219–35224.

Crane, R. K. (1960a). Intestinal absorption of sugars. *Physiological reviews*, *40*, 789–825.

Crane, R. K. (1960b). Studies on the mechanism of the intestinal absorption of sugars. III. Mutual inhibition, in vitro, between some actively transported sugars. *Biochimica et Biophysica Acta*, *45*, 477–482.

Crane, R. K., Miller, D., & Bihler, I. (1961). The restrictions on possible mechanisms of intestinal active transport of sugars. In A. Kleinzeller, & A. Kotyk (Eds.), *Membrane Transport and Metabolism* (pp. 429–449). London and New York: Academic Press.

Csaky, T. Z., & Thale, M. (1960). Effect of ionic environment on intestinal sugar transport. *Journal of Physiology*, *151*, 59–65.

de Groot, T. J., Veyhl, M., Terwinghe, C., Vanden Bempt, V., Dupont., P., Mortelmans, L., et al. (2003). Synthesis of 18F-fluoroalkyl-beta-D-glucosides and their evaluation as tracers for sodium-dependent glucose transporters. *Journal of Nuclear Medicine*, *44*(12), 1973–1981.

Dickens, C. (1859). A Tale of Two Cities. In C. Dickens (Ed.), *All The Year Round*. London: Chapman & Hall.

Diedrich, D. F. (1965). In vitro evaluation of relative inhibitory potency of phlorizin and its congeners. *American Journal of Physiology*, *209*(3), 621–626.

Dieter, M., Palmada, M., Rajamanickam, J., Aydin, A., Busjahn, A., Boehmer, C., et al. (2004). Regulation of glucose transporter SGLT1 by ubiquitin ligase Nedd4-2 and kinases SGK1, SGK3, and PKB. *Obesity Research*, *12*(5), 862–870.

Dong, H., Ning, Z., Yu, L., Li, L., Lin, L., & Huang, J. (2007). Preparative separation and identification of the flavonoid phlorhizin from the crude extract of Lithocarpus polystachyus Rehd. *Molecules*, *12*(3), 552–562.

Douard, V., Cui, X. L., Soteropoulos, P., & Ferraris, R. P. (2008). Dexamethasone sensitizes the neonatal intestine to fructose induction of intestinal fructose transporter (Slc2A5) function. *Endocrinology*, *149*(1), 409–423.

Dwyer, D. S. (2001). Model of the 3-D structure of the GLUT3 glucose transporter and molecular dynamics simulation of glucose transport. *Proteins*, *42*(4), 531–541.

Ehrenkranz, J. R.L., Lewis, N. G., Kahn, C. R., & Roth, J. (2005). Phlorizin: a review. *Diabetes/Metabolism Research and Reviews*, *21*(1), 31–38.

Elfeber, K., Köhler, A., Lutzenburg, M., Osswald, C., Galla, H. J., Witte, O. W., et al. (2004). Localization of the Na^+-D-glucose cotransporter SGLT1 in the blood-brain barrier. *Histochemistry and Cell Biology*, *121*(3), 201–207.

Estensen, R. D., & Plagemann, P. G. (1972). Cytochalasin B: inhibition of glucose and glucosamine transport. *Proceedings of the National Academy of Sciences of the United States of America*, *69*(6), 1430–1434.

Faham, S., Watanabe, A., Besserer, G. M., Cascio, D., Specht, A., Hirayama, B. A., et al. (2008). The crystal structure of a sodium galactose transporter reveals mechanistic insights into Na^+/sugar symport. *Science*, *321*(5890), 810–814.

Filatova, A., Leyerer, M., Gorboulev, V., Chintalapati, C., Reinders, Y., Mueller, T. D., et al. (2009). novel shuttling domain in a regulator (RSC1A1) of transporter SGLT1 steers cell cycle-dependent nuclear location. *Traffic, 10*, 1599–1618.
Firnges, M. A., Lin, J. T., & Kinne, R. K. (2001). Functional asymmetry of the sodium-D-glucose cotransporter expressed in yeast secretory vesicles. *The Journal of Membrane Biology, 179*(2), 143–153.
Foley, K., Boguslavsky, S., & Klip, A. (2011). Endocytosis, recycling, and regulated exocytosis of glucose transporter 4. *Biochemistry, 50*(15), 3048–3061.
Frasch, W., Frohnert, P. P., Bode, F., Baumann, K., & Kinne, R. (1970). Competitive inhibition of phlorizin binding by D-glucose and the influence of sodium: a study on isolated brush border membrane of rat kidney. *Pflügers Archiv European Journal of Physiology, 320*(3), 265–284.
Gagnon, D. G., Frindel, C., & Lapointe, J. Y. (2007). Voltage-clamp fluorometry in the local environment of the C255-C511 disulfide bridge of the Na^+/glucose cotransporter. *Biophysical Journal, 92*(7), 2403–2411.
Garcia-Perez, A., & Burg, M. B. (1991). Renal medullary organic osmolytes. *Physiological Reviews, 71*(4), 1081–1115.
Geck, P., & Heinz, E. (1976). Coupling in secondary transport. Effect of electrical potentials on the kinetics of ion-linked cotransport. *Biochimica et Biophysica Acta, 443*(1), 49–63.
Ghezzi, C., & Wright, E. M. (2012). Regulation of the human Na^+ dependent glucose cotransporter hSGLT2. *American journal of physiology. Cell physiology.*
Glossmann, H., & Neville, D. M. J. (1972). Phlorizin receptors in isolated kidney brush border membranes. *Journal of Biological Chemistry, 247*(23), 7779–7789.
Gorboulev, V., Schürmann, A., Vallon, V., Kipp, H., Jaschke, A., Klessen, D., et al. (2012). Na^+-D-glucose cotransporter SGLT1 is pivotal for intestinal glucose absorption and glucose-dependent incretin secretion. *Diabetes, 61*(1), 187–196.
Gould, G. W., & Holman, G. D. (1993). The glucose transporter family: structure, function and tissue-specific expression. *Biochemical Journal, 295*(Pt 2), 329–341.
Grunewald, R. W., & Kinne, R. K. (1988). Sugar transport in isolated rat kidney papillary collecting duct cells. *Pflügers Archiv European Journal of Physiology, 413*(1), 32–37.
Hager, K., Hazama, A., Kwon, H. M., Loo, D. D., Handler, J. S., & Wright, E. M. (1995). Kinetics and specificity of the renal Na^+/myo-inositol cotransporter expressed in Xenopus oocytes. *Journal of Membrane Biology, 143*, 103–113.
Hammerman, M. R., Sacktor, B., & Daughaday, W. H. (1980). Myo-inositol transport in renal brush border vesicles and its inhibition by D-glucose. *American Journal of Physiology, 239*, F113–F120.
Handler, J. S., & Kwon, H. M. (1996). Regulation of the myo-inositol and betaine cotransporters by tonicity. *Kidney International, 49*(6), 1682–1683.
Handler, J. S., & Kwon, H. M. (1997). Kidney cell survival in high tonicity. *Comparative Biochemistry and Physiology – Part A: Molecular & Integrative Physiology, 117*(3), 301–306.
Heard, K. S., Fidyk, N., & Carruthers, A. (2000). ATP-dependent substrate occlusion by the human erythrocyte sugar transporter. *Biochemistry, 39*, 3005–3014.
Hediger, M. A., Coady, M. J., Ikeda, T. S., & Wright, E. M. (1987). Expression cloning and cDNA sequencing of the Na/glucose co-transporter. *Nature, 330*, 379–381.
Hediger, M. A., & Rhoads, D. B. (1994). Molecular physiology of sodium-glucose cotransporters. *Physiological Reviews*, 993–1026.
Heidrich, H. G., Kinne, R., Kinne-Saffran, E., & Hannig, K. (1972). The polarity of the proximal tubule cell in rat kidney. Different surface charges for the brush-border microvilli and plasma membranes from the basal infoldings. *Journal of Cell Biology, 54*(2), 232–245.
Hirsch, J. R., Loo, D. D., & Wright, E. M. (1996). Regulation of Na^+/glucose cotransporter expression by protein kinases in Xenopus laevis oocytes. *Journal of Biological Chemistry, 271*, 14740–14746.

Hopfer, U., Nelson, K., Perotto, J., & Isselbacher, K. J. (1973). Glucose transport in isolated brush border membrane from rat small intestine. *Journal of Biological Chemistry*, *248*(1), 25–32.

Hresko, R. C., Kruse, M., Strube, M., & Mueckler, M. (1994). Topology of the Glut 1 glucose transporter deduced from glycosylation scanning mutagenesis. *Journal of Biological Chemistry*, *269*(32), 20482–20488.

Huang, Y., Lemieux, M. J., Song, J., Auer, M., & Wang, D. N. (2003). Structure and mechanism of the glycerol-3-phosphate transporter from *Escherichia coli*. *Science*, *301*(5633), 616–620.

Hummel, C. S., Lu, C., Liu, J., Ghezzi, C., Hirayama, B. A., Loo, D. D., et al. (2012). Structural selectivity of human SGLT inhibitors. *American Journal of Physiology. Cell Physiology*, *302*(2), C373–C382.

Ishikawa, Y., Eguchi, T., & Ishida, H. (1997). Mechanism of beta-adrenergic agonist-induced transmural transport of glucose in rat small intestine: regulation of phosphorylation of SGLT1 controls the function. *Biochimica et Biophysica Acta*, *1357*, 306–318.

Jans, A. W., Grunewald, R. W., & Kinne, R. K. (1989). Pathways for the synthesis of sorbitol from ^{13}C-labeled hexoses, pentose, and glycerol in renal papillary tissue. *Magnetic Resonance in Medicine*, *9*(3), 419–422.

Jones, H. F., Butler, R. N., & Brooks, D. A. (2011). Intestinal fructose transport and malabsorption in humans. *American Journal of Physiology. Gastrointestinal and Liver Physiology*, *300*(2), G202–G206.

Jurysta, C., Nicaise, C., Cetik, S., Louchami, K., Malaisse, W. J., & Sener, A. (2012). Glucose transport by acinar cells in rat parotid glands. *Cellular Physiology and Biochemistry*, *29*(3–4), 325–330.

Kanai, Y., Lee, W. S., You, G., Brown, D., & Hediger, M. A. (1994). The human kidney low affinity Na^+/glucose cotransporter SGLT2. Delineation of the major renal reabsorptive mechanism for D-glucose. *Journal of Clinical Investigation*, *93*(1), 397–404.

Kellett, G. L. (2001). The facilitated component of intestinal glucose absorption. *Journal of Physiology*, *531*(3), 585–595.

Kellett, G. L., Brot-Laroche, E., Mace, O. J., & Leturque, A. (2008). Sugar absorption in the intestine: the role of GLUT2. *Annual Review of Nutrition*, *28*, 35–54.

Khoursandi, S., Scharlau, D., Herter, P., Kuhnen, C., Martin, D., Kinne, R. K., et al. (2004). Different modes of sodium-D-glucose cotransporter-mediated d-glucose uptake regulation in Caco-2 cells. *American journal of physiology. Cell physiology*, *287*, C1041–C1047.

Kimmich, G. A. (1973). Coupling between Na^+ and sugar transport in small intestine. *Biochimica et Biophysica Acta*, *300*(1), 31–78.

Kinne, R., Murer, H., Kinne-Saffran, E., Thees, M., & Sachs, G. (1975). Sugar transport by renal plasma membrane vesicles. Characterization of the systems in the brush-border microvilli and basal-lateral plasma membranes. *Journal of Membrane Biology*, *21*(3–4), 375–395.

Kinne, R. K. (1991). Selectivity and direction: plasma membranes in renal transport. *American Journal of Physiology*, *260*(2 Pt 2), F153–F162.

Kinne, R. K., & Castaneda, F. (2011). SGLT inhibitors as new therapeutic tools in the treatment of diabetes. In M. Schwanstecher (Ed.), *Handbook of experimental pharmacology* (Vol. 203, pp. 105–126). Heidelberg: Springer Science & Business Media.

Kipp, H., Khoursandi, S., Scharlau, D., & Kinne, R. K. (2003). More than apical: distribution of SGLT1 in Caco-2 cells. *American Journal of Physiology. Cell Physiology*, *285*(4), C737–C749.

Kipp, H., Lin, J. T., & Kinne, R. K. (1996). Interactions of alkylglucosides with the renal sodium/D-glucose cotransporter. *Biochimica et Biophysica Acta*, *1282*(1), 125–130.

Kleinzeller, A. (1970). Active sugar transport in renal cortex cells: electrolyte requirement. *Biochimica et Biophysica Acta*, *211*, 277–292.

Koepsell, H., Fritzsch, G., Korn, K., & Madrala, A. (1990). Two substrate sites in the renal Na(+)-D-glucose cotransporter studied by model analysis of phlorizin binding and D-glucose transport measurements. *Journal of Membrane Biology, 114*(2), 113–132.
Krane, S. M., & Crane, R. K. (1959). The accumulation of D-galactose against a concentration gradient by slices of rabbit kidney cortex. *Journal of Biological Chemistry, 234*(2), 211–216.
Kumar, A., Tyagi, N. K., Arevalo, E., Miller, K. W., & Kinne, R. K. (2007). A proteomic study of sodium/D-glucose cotransporter 1 (SGLT1): topology of loop 13 and coverage of other functionally important domains. *Biochimica et Biophysica Acta, 1774*(8), 968–974.
Kumar, A., Tyagi, N. K., & Kinne, R. K. (2007). Ligand-mediated conformational changes and positioning of tryptophans in reconstituted human sodium/D-glucose cotransporter1 (hSGLT1) probed by tryptophan fluorescence. *Biophysical Chemistry, 127*(1–2), 69–77.
Kwon, H. M., Yamauchi, A., Uchida, S., Preston, A. S., Garcia-Perez, A., Burg, M. B., et al. (1992). Cloning of the cDNA for a Na^+/myo-inositol cotransporter, a hypertonicity stress protein. *Journal of Biological Chemistry, 267*, 6297–6301.
Läuger, P. (1979). A channel mechanism for electrogenic pumps. *Biochimica et Biophysica Acta, 552*, 143–161.
Lin, J., Kormanec, J., Homerová, D., & Kinne, R. K. (1999). Probing transmembrane topology of the high-affinity sodium/glucose cotransporter (SGLT1) with histidine-tagged mutants. *Journal of Membrane Biology, 170*(3), 243–252.
Lin, J. T., Riedel, S., & Kinne, R. (1979). The use of octyl beta-D-glucoside as detergent for hog kidney brush border membrane. *Biochimica et Biophysica Acta, 557*(1), 179–187.
Loeschke, K., & Baumann, K. (1969). Kinetic study of D-glucose reabsorption in the proximal convoluted tubule of rat kidney. *Pflügers Archiv European Journal of Physiology, 305*(2), 139–154.
Loeschke, K., Baumann, K., Renschler, H., & Ullrich, K. J. (1969). Differentiation of the active and passive components of D-glucose transport in the proximal tubule of rat kidney. *Pflügers Archiv European Journal of Physiology, 305*(2), 118–138.
Loo, D. F., Zeuthen, T., Chnady, G., & Wright, E. M. (1996). Cotransport of water by the Na/glucose cotransporter. *Proceedings of the National Academy of Sciences of the United States of America, 93*, 13367–13370.
Lostao, M. P., Hirayama, B. A., Loo, D. D., & Wright, E. M. (1994). Phenylglucosides and the Na^+/glucose cotransporter (SGLT1): analysis of interactions. *Journal of Membrane Biology, 142*(2), 161–170.
Mackenzie, B., Panayotova-Heiermann, M., Loo, D. D.F., Lever, J. E., & Wright, E. M. (1994). SAAT1 is a low affinity Na^+/glucose cotransporter and not an amino acid transporter: a reinterpretation. *Journal of Biological Chemistry, 269*, 22488–22491.
Miyamoto, K., Hase, K., Takagi, T., Fujii, T., Taketani, Y., Minami, H., et al. (1993). Differential responses of intestinal glucose transporter mRNA transcripts to levels of dietary sugars. *Biochemical Journal, 295*(Pt 1), 211–215.
Mueckler, M., Caruso, C., Baldwin, S. A., Panico, M., Blench, I., Morris, H. R., et al. (1985). Sequence and structure of a human glucose transporter. *Science, 229*(4717), 941–945.
Mueckler, M., & Makepeace, C. (2009). Model of the exofacial substrate-binding site and helical folding of the human Glut1 glucose transporter based on scanning mutagenesis. *Biochemistry, 48*(25), 5934–5942.
Mueckler, M., & Makepeace, C. (2012). Ligand-induced movements of inner transmembrane helices of Glut1 revealed by chemical cross-linking of di-cysteine mutants. *PLoS One*, 7(2), e31412.
Murer, H., & Hopfer, U. (1974). Demonstration of electrogenic Na^+-dependent D-glucose transport in intestinal brush border membranes. *Proceedings of the National Academy of Sciences of the United States of America, 71*(2), 484–488.
Nagata, K., Hori, N., Sato, K., Ohta, K., Tanaka, H., & Hiji, Y. (1999). Cloning and functional expression of an SGLT-1-like protein from the Xenopus laevis intestine. *American Journal of Physiology, 276*(5 Pt 1), G1251–G1259.

Novakova, R., Homerova, D., Kinne, R. K.H., Kinne-Saffran, E., & Lin, J. T. (2001). Identification of a region critically involved in the interaction of phlorizin with the rabbit sodium-D-glucose cotransporter SGLT1. *Journal of Membrane Biology*, *184*, 55–60.

O'Malley, D., Reimann, F., Simpson, A. K., & Gribble, F. M. (2006). Sodium-coupled glucose cotransporters contribute to hypothalamic glucose sensing. *Diabetes*, *55*(12), 3381–3386.

Obermeier, S., Hülseweh, B., Tinel, H., Kinne, R. K., & Kunz, C. (2000). Expression of glucose transporters in lactating human mammary gland epithelial cells. *European Journal of Nutrition*, *39*(5), 194–200.

Olsowski, A., Monden, I., Krause, G., & Keller, K. (2000). Cysteine scanning mutagenesis of helices 2 and 7 in GLUT1 identifies an exofacial cleft in both transmembrane segments. *Biochemistry*, *39*(10), 2469–2474.

Parent, L., Supplisson, S., Loo, D. D., & Wright, E. M. (1992). Electrogenic properties of the cloned Na^+/glucose cotransporter: II. A transport model under nonrapid equilibrium conditions. *Journal of Membrane Biology*, *125*, 63–79.

Patlak, C. S. (1957). Contributions to the theory of active transport II. The gate type noncarrier mechanism and generalizations concerning tracer flow, efficiency, and measurement of energy expenditure. *Bulletin of Mathematical Biophysics*, *19*, 209–235.

Poppe, R., Karbach, U., Gambaryan, S., Wiesinger, H., Lutzenburg, M., Kraemer, M., et al. (1997). Expression of the Na^+-D-glucose cotransporter SGLT1 in neurons. *Journal of Neurochemistry*, *69*(1), 84–94.

Preston, A. S., Yamauchi, A., Kwon, H. M., & Handler, J. S. (1995). Activators of protein kinase A and of protein kinase C inhibit MDCK cell myo-inositol and betaine uptake. *Journal of the American Society of Nephrology*, *6*(6), 1559–1564.

Puntheeranurak, T., Kasch, M., Xia, X., Hinterdorfer, P., & Kinne, R. K. (2007). Three surface subdomains form the vestibule of the Na^+/glucose cotransporter SGLT1. *Journal of Biological Chemistry*, *282*(35), 25222–25230.

Puntheeranurak, T., Neundlinger, I., Kinne, R. K., & Hinterdorfer, P. (2011). Single-molecule recognition force spectroscopy of transmembrane transporters on living cells. *Nature Protocols*, *6*(9), 1443–1452.

Puntheeranurak, T., Wildling, L., Gruber, H. J., Kinne, R. K., & Hinterdorfer, P. (2006). Ligands on the string: single-molecule AFM studies on the interaction of antibodies and substrates with the Na^+-glucose co-transporter SGLT1 in living cells. *Journal of Cell Science*, *119*(Pt 14), 2960–72960.

Puntheeranurak, T., Wimmer, B., Castaneda, F., Gruber, H. J., Hinterdorfer, P., & Kinne, R. K. (2007). Substrate specificity of sugar transport by rabbit SGLT1: single-molecule atomic force microscopy versus transport studies. *Biochemistry*, *46*(10), 2797–2804.

Raja, M., & Kinne, R. K. (2012). Structural insights into genetic variants of Na(+)/glucose cotransporter SGLT1 causing glucose-galactose malabsorption: vSGLT as a model structure. *Cell Biochemistry and Biophysics*, *63*(2), 151–158.

Raja, M. M., Kipp, H., & Kinne, R. K. (2004). C-terminus loop 13 of Na^+ glucose cotransporter SGLT1 contains a binding site for alkyl glucosides. *Biochemistry*, *43*(34), 10944–10951.

Raja, M. M., Tyagi, N. K., & Kinne, R. K. (2003). Phlorizin recognition in a C-terminal fragment of SGLT1 studied by tryptophan scanning and affinity labeling. *Journal of Biological Chemistry*, *278*(49), 49154–49163.

Riklis, E., & Quastel, J. H. (1958). Effects of cations on sugar absorption by isolated surviving guinea pig intestine. *Canadian Journal of Biochemistry and Physiology*, *36*, 347–362.

Sabolic, I., Vrhovac, I., Eror, D. B., Gerasimova, M., Rose, M., Breljak, D., et al. (2012). Expression of Na^+-D-glucose cotransporter SGLT2 in rodents is kidney-specific and exhibits sex and species differences. *American Journal of Physiology. Cell Physiology*, *302*(8), C1174–C1188.

Sala-Rabanal, M., Hirayama, B. A., Loo, D. D., Chaptal, V., Abramson, J., & Wright, E. M. (2012). Bridging the gap between structure and kinetics of human SGLT1. *American Journal of Physiology. Cell Physiology*, *302*(9), C1293–C1305.

Salas-Burgos, A., Iserovich, P., Zuniga, F., Vera, J. C., & Fischbarg, J. (2004). Predicting the three-dimensional structure of the human facilitative glucose transporter glut1 by a novel evolutionary homology strategy: insights on the molecular mechanism of substrate migration, and binding sites for glucose and inhibitory molecules. *Biophysical Journal*, *87*(5), 2990–2999.

Schultz, S. G., & Curran, P. F. (1970). Coupled transport of sodium and organic solutes. *Physiological Reviews*, *50*(4), 637–718.

Scriver, C. R., Chesney, R. W., & McInnes, R. R. (1976). Genetic aspects of renal tubular transport: diversity and topology of carrier. *Kidney International*, *9*, 149–171.

Seatter, M. J., De la Rue, S. A., Porter, L. M., & Gould, G. W. (1998). QLS motif in transmembrane helixVII of the glucose transporter family interacts with the C-1 position of D-glucose and is involved in substrate selection at the exofacial binding site. *Biochemistry*, *37*(5), 1322–1326.

Shioda, T., Ohta, T., Isselbacher, K. J., & Rhoads, D. B. (1994). Confluence dependent expression of the Na^+/glucose cotransporter (SGLT1) in LLC-PK1 cells: role of protein kinase C activation and ongoing transcription. *Proceedings of the National Academy of Sciences of the United States of America*, *91*(25), 11919–11923.

Smith, H. W. (1951). *The Kidney: Structure and Function in Health and Disease*. New York: Oxford Univ Press.

Stöckli, J., Fazakerley, D. J., & James, D. E. (2011). GLUT4 exocytosis. *Journal of Cell Science*, *124*(Pt 24), 4147–4159.

Subramanian, S., Glitz, P., Kipp, H., Kinne, R. K., & Castaneda, F. (2009). Protein kinase-A affects sorting and conformation of the sodium-dependent glucose co-transporter SGLT1. *Journal of Cellular Biochemistry*, *106*(3), 444–452.

Sujatha, M. S., & Balaji, P. V. (2004). Identification of common structural features of binding sites in galactose-specific proteins. *Proteins*, *55*(1), 44–65.

Thorens, B., Lodish, H. F., & Brown, D. (1990). Differential localization of two glucose transporter isoforms in rat kidney: Glut 2 in S1 and S2 Glut1 in S3. *American Journal of Physiology*, *259*, C286–C294.

Thorens, B., & Mueckler, M. (2010). Glucose transporters in the 21st century. *American Journal of Physiology. Endocrinology and Metabolism*, *298*(2), E141–E145.

Turk, E., Gasymov, O. K., Lanza, S., Horwitz, J., & Wright, E. M. (2006). A reinvestigation of the secondary structure of functionally active vSGLT, the Vibrio sodium/galactose cotransporter. *Biochemistry*, *45*, 1470–1479.

Turk, E., Kerner, C. J., Lostao, M. P., & Wright, E. M. (1996). Membrane topology of the human Na^+/glucose cotransporter SGLT1. *Journal of Biological Chemistry*, *271*(4), 1925–1934.

Turk, E., Martín, M. G., & Wright, E. M. (1994). Structure of the human Na^+/glucose cotransporter gene SGLT1. *Journal of Biological Chemistry*, *269*, 15204–15209.

Turner, R. J. (1985). Stoichiometry of cotransport systems. *Annals of the New York Academy of Sciences*, *456*, 10–25.

Turner, R. J., & Moran, A. (1982). Heterogeneity of sodium-dependent D-glucose transport sites along the proximal tubule: evidence from vesicle studies. *American Journal of Physiology*, *242*, F406–F414. Renal Fluid Electrolyte Physiol. 11.

Tyagi, N. K., Goyal, P., Kumar, A., Pandey, D., Siess, W., & Kinne, R. K. (2005). High-yield functional expression of human sodium/D-glucose cotransporter1 in pichia pastoris and characterization of ligand-induced conformational changes as studied by tryptophan fluorescence. *Biochemistry*, *44*, 15514–15524.

Tyagi, N. K., Puntheeranurak, T., Raja, M., Kumar, A., Wimmer, B., Neundlinger, I., et al. (2011). A biophysical glance at the outer surface of the membrane transporter SGLT1. *Biochimica et Biophysica Acta*, *1808*(1), 1–18.

Uldry, M., & Thorens, B. (2004). The SLC2 family of facilitated hexose and polyol transporters. The ABC of solute carriers, *Pflügers Archiv European Journal of Physiology*, *447*, 480–489.

Ullrich, K. J., & Papavassiliou, F. (1985). Contraluminal transport of hexoses in the proximal convolution of the rat kidney in situ. *Pflügers Archiv European Journal of Physiology*, *404*(2), 150–156.

Ullrich, K. J., Rumrich, G., & Klöss, S. (1974). Specificity and sodium dependence of the active sugar transport in the proximal convolution of the rat kidney. *Pflügers Archiv European Journal of Physiology*, *351*(1), 35–48.

Ussing, H. H., & Zerahn, K. (1951). Active transport of sodium as the source of electric current in the short-circuited isolated frog skin. *Acta Physiologica Scandinavica*, *23*(2–3), 110–127.

Vayro, S., & Silverman, M. (1999). PKC regulates turnover rate of rabbit intestinal Na^+-glucose transporter expressed in COS-7 cells. *American Journal of Physiology*, *276*(5 Pt 1), C1053–C1060.

Veyhl, M., Spangenberg, J., Puschel, B., Poppe, R., Dekel, C., Fritzsch, G., et al. (1993). Cloning of a membrane-associated protein which modifies activity and properties of the Na^+-D-glucose cotransporter. *Journal of Biological Chemistry*, *268*, 25041–25053.

Veyhl, M., Keller., T., Gorboulev, V., Vernaleken, A., & Koepsell, H. (2006). RS1(RSC1A1) regulates the exocytotic pathway of Na^+-D-glucose cotransporter SGLT1. *American Journal of Physiology. Renal Physiology*, *291*, F1213–F1223.

Wheeler, T. J., & Hinkle, P. C. (1985). The glucose transporter of mammalian cells. *Annual Review of Physiology*, *47*, 503–517.

Wielert-Badt, S., Lin, J. T., Lorenz, M., Fritz, S., & Kinne, R. K. (2000). Probing the conformation of the sugar transport inhibitor phlorizin by 2D-NMR, molecular dynamics studies, and pharmacophore analysis. *Journal of Medicinal Chemistry*, *43*(9), 1692–1698.

Wright, E. M., Hirsch, J. R., Loo, D. D., & Zampighi, G. A. (1997). Regulation of Na^+/glucose cotransporters. *Journal of Experimental Biology*, *200*(Pt 2), 287–293.

Wright, E. M., Loo, D. D., Hirayama, B. A., & Turk, E. (2004). Surprising versatility of Na^+-glucose cotransporters: SLC5. *Physiology*, *19*(370–376), 370.

Wright, E. M., Loo, D. F., & Hirayama, B. A. (2011). Biology of human sodium glucose cotransporters. *Physiological Reviews*, *91*, 733–794.

Wright, E. M., Martin, M. D., & Turk, E. (2003). Intestinal absorption in health and disease—sugars. *Best Practice & Research Clinical Gastroenterology*, *17*(6), 943–956.

Wright, E. M. (1993). The intestinal Na/glucose cotransporter. *Annual Reviews of Physiology*, *55*, 575–589.

Wright, E. M., & Turk, E. (2004). The sodium/glucose cotransport family SLC5. *Pflügers Archiv European Journal of Physiology*, *447*(5), 510–518.

Xia, X., Lin, J. T., & Kinne, R. K. (2003). Binding of phlorizin to the isolated C-terminal extramembranous loop of the Na^+/glucose cotransporter assessed by intrinsic tryptophan fluorescence. *Biochemistry*, *42*(20), 6115–6120.

Yamauchi, A., Miyai, A., Shimada, S., Minami, Y., Tohyama, M., Imai, E., & Ueda, N. (1995). Localization and rapid regulation of Na^+/myo-inositol cotransporter in rat kidney. *Journal of Clinical Investigation*, *96*, 1195–1201.

Yamauchi, A., Sugiura, T., Kitamura., H., Matsuoka, Y., Imai, E., & Hori, M. (1998). Expression of the Na^+/myo-inositol cotransporter in the juxtaglomerular region. *Kidney International Supplement*, *67*, S183–S185.

Zeuthen, T., Zeuthen, E., & Macaulay, N. (2007). Water transport by GLUT2 expressed in Xenopus laevis oocytes. *Journal of Physiology*, *579*(Pt 2), 345–361.

Zhao, F. -Q., Okine, E. R., Cheeseman, C., Shirazi-Beechey, S. P., & Kennelly, J. J. (1998). Gene expression of Na^+/glucose cotransporter in lactating bovine gastrointestinal tract. *Journal of Animal Science*, *76*, 2921–2929.

Zhao, F. -Q., Okine, E. R., & Kennelly, J. J. (1999). Glucose transporter gene expression in bovine mammary gland. *Journal of Animal Science*, 77, 2517–2522.
Zhao, F. Q., & Keating, A. F. (2007). Expression and regulation of glucose transporters in the bovine mammary gland. *Journal of Dairy Science*, *90*(Suppl. 1), E76–E86.
Zhao, F. Q., Zheng, Y. C., Wall, E. H., & McFadden, T. B. (2005). Cloning and expression of bovine sodium/glucose cotransporters. *Journal of Dairy Science*, *88*(1), 182–194.

CHAPTER THREE

SLC4A Transporters

Inyeong Choi*
Department of Physiology, Emory University, Atlanta, Georgia, USA
*Corresponding author: E-mail: ichoi@emory.edu

Contents

Abstract

SLC4A gene family proteins include bicarbonate transporters that move HCO_3^- across the plasma membrane and regulate intracellular pH and transepithelial movement of acid–base equivalents. These transporters are Cl/HCO_3 exchangers, electrogenic Na/HCO_3 cotransporters, electroneutral Na/HCO_3 cotransporters, and Na^+-driven Cl/HCO_3 exchanger.

Current Topics in Membranes, Volume 70
ISSN 1063-5823
http://dx.doi.org/10.1016/B978-0-12-394316-3.00003-X

Studies of the bicarbonate transporters in vitro and in vivo have demonstrated their physiological importance for acid–base homeostasis at the cellular and systemic levels. Recent advances in structure/function analysis have also provided valuable information on domains or motifs critical for regulation, ion translocation, and protein topology. This chapter focuses on the molecular mechanisms of ion transport along with associated structural aspects from mutagenesis of particular residues and from chimeric constructs. Structure/function studies have helped to understand the mechanism by which ion substrates are moved via the transporters. This chapter also describes some insights into the structure of SLC4A1 (AE1) and SLC4A4 (NBCe1) transporters. Finally, as some SLC4A transporters exist in concert with other proteins in the cells, the structural features associated with protein–protein interactions are briefly discussed.

1. INTRODUCTION

In most cells, the internal ionic environment is regulated by membrane proteins that move ions across the cell membranes. The SLC4A bicarbonate transporters are a subset of these proteins that are specialized to move HCO_3^-. HCO_3^- buffers H^+, and therefore, HCO_3^- transport is equivalent to H^+ transport in the opposite direction from a pH standpoint. The primary function of the bicarbonate transporters is to regulate intracellular H^+. Some bicarbonate transporters move HCO_3^- into the cytosol and raise intracellular pH (acid extruders), whereas others remove HCO_3^- from the cytosol and lower intracellular pH (acid loaders). The activity of acid extruders and loaders affects steady-state pH in the cells. Cells also utilize these pH regulators to transfer large amounts of Na^+ and HCO_3^- from one body compartment to another to maintain extracellular and total body acid–base balance. The bicarbonate transporters are also widely distributed in many epithelial tissues where they play a role in transepithelial absorption and secretion of HCO_3^-. This transepithelial HCO_3^- movement contributes to tissue-specific physiological processes such as reabsorbing HCO_3^- or excreting acid into urine (kidney), neutralizing intestinal gastric acid (pancreatic ducts), or maintaining the cerebrospinal fluid pH (choroid plexus).

Many studies have been done to determine the functional and physiological significance of the SLC4A bicarbonate transporters in the body and to understand their pathological involvement in acid–base disorders. Cellular signals responsible for transport regulation have also been examined. In addition, the members of SLC4A transporters exhibit high homology in amino acid sequence, and this homology has enabled researchers to investigate domains or residues important for distinct molecular and functional properties of ion movement. Such efforts have produced valuable results to advance our knowledge of the underlying molecular mechanisms of ion transport.

Furthermore, similar structural features among bicarbonate transporters have facilitated our understanding of transporter structure and the macromolecular complexes with other proteins. This chapter provides some insight into ion transport properties, structure, and protein–protein interactions of SLC4A bicarbonate transporters. This chapter focuses on anion exchangers AE1–3 and Na/HCO_3 cotransporters NBCe1 and NBCn1 because less has been done with other Na^+-coupled transporters.

2. SLC4A TRANSPORTERS

The SLC4A transporters (Table 1) are Cl/HCO_3 exchangers, Na/HCO_3 cotransporters, Na^+-driven Cl/HCO_3 exchanger, and Na/borate transporter. These are distinct from the SLC26A anion exchangers that

Table 1 Human SLC4A transporters

Human gene	Protein	Function	pH_i regulation	Human chromosome
SLC4A1	AE1 Band 3	Cl^-/HCO_3^- exchange	Acid loader	17q21–q22
SLC4A2	AE2	Cl^-/HCO_3^- exchange	Acid loader	7q35–36
SLC4A3	AE3	Cl^-/HCO_3^- exchange	Acid loader	2q36
SLC4A4	NBCe1 NBC1	Na^+/HCO_3^- cotransport (electrogenic)	Acid extruder (except NBCe1-A)	4q21
SLC4A5	NBCe2 NBC4	Na^+/HCO_3^- cotransport (electrogenic)	Acid extruder	2p13
SLC4A6	–			–
SLC4A7	NBCn1 NBC3	Na^+/HCO_3^- cotransport	Acid extruder	3p22
SLC4A8	NDCBE	$Na^+/Cl^-/2HCO_3^-$ cotransport/exchange*	Acid extruder	12q13
SLC4A9	AE4	Cl^-/HCO_3^- exchange	Acid loader	5q31
SLC4A10	NBCn2 NBCE	Na^+/HCO_3^- cotransport	Acid extruder	2q23-q24
SLC4A11	NaBC BTR1	$Na^+/B(OH)_4^-$, OH^- cotransport	Acid extruder for OH^- transport	20p12

*NDCBE cotransports external $1Na^+$ and $2HCO_3^-$ in exchange for internal Cl^-.

transport a variety of monovalent and divalent anions (Sindic, Chang, Mount, & Romero, 2007). Some SLC26A proteins transport HCO_3^- in exchange for Cl^- or other anions.

2.1. Cl/HCO_3 Exchangers (AEs)

2.1.1. AE1 (SLC4A1), AE2 (SLC4A2), AE3 (SLC4A3)

The three Cl/HCO_3 exchangers (AE1–3) primarily move external HCO_3^- in exchange for internal Cl^- (Alper, 2009; Bonar & Casey, 2008). AE1 is the first cloned HCO_3^- transporter (Kopito & Lodish, 1985). The human AE1 contains 911 amino acids with the predicted structure of 13 transmembrane (TM) (The TM topology of AE1 is defined according to the 13-TM model (Bonar & Casey, 2008), in which 13 TMs and one unusual membrane-spanning segment (called extended structure) between TM11 and TM12 are present. The topology of NBCe1 is defined according to the 14-TM model (Zhu et al., 2010), in which 14 TMs are present.) segments and cytoplasmic N- and C-terminal domains (Zhu, Lee, & Casey, 2003). The large N-terminal domain (~400 amino acids) plays a regulatory role, while the transmembrane domain is responsible for anion exchange activity. The N-terminal domain interacts with other proteins such as ankyrin, glycolytic enzymes, and hemoglobin, whereas the C-terminal domain interacts with carbonic anhydrase II (CA II) (Vince & Reithmeier, 1998). AE1 exists as homodimers (Wang, Sarabia, Reithmeier, & Kuhlbrandt, 1994). AE1 is the most abundant membrane protein in red blood cells and plays a key role in CO_2 delivery from tissues to the lungs. AE1 is also localized to the basolateral membrane of type-A intercalated cells in the kidney collecting ducts (van Adelsberg, Edwards, & Al Awqati, 1993). The renal kAE1 lacks the first 65 amino acids because of an alternate promoter site. AE1 also transports other anions such as SO_4^{2-} at a very low rate (Milanick & Gunn, 1982).

AE2 was originally isolated from the kidney (Alper, Kopito, Libresco, & Lodish, 1988). AE2 is ~300 amino acids larger than AE1 and exists as multiple variants because of alternative promoter sites. Similar to AE1, the N- and C-terminal domains of AE2 play regulatory roles in its pH dependence, as well as response to NH_4^+ and hypertonicity (Chernova et al., 2003; Stewart, Chernova, Shmukler, Wilhelm, & Alper, 2002), while the transmembrane domains are responsible for exchanger function (Stewart, Kurschat, Vaughan-Jones, Shmukler, & Alper, 2007). AE2 is widely distributed in the body, particularly in epithelial cells including the thick ascending limb and cortical collecting duct intercalated cells in the kidney (Brosius et al., 1995), and contributes to transepithelial movement of HCO_3^- and salt.

Human AE3 comprised the cardiac variant (cAE3) and the brain variant (bAE3) with different N-terminal domains (Kopito et al., 1989). Both variants are found in many different tissues in addition to the heart and brain. AE3 has a lower exchange activity than AE1 and AE2 because its membrane expression is weak (Fujinaga, Loiselle, & Casey, 2003).

2.1.2. AE4 (SLC4A9)

Human AE4 contains 983 amino acids and has three alternative splice variants according to the NCBI database. AE4 is more closely related to Na/HCO_3 transporters than AEs in amino acid sequence. AE4 is functionally characterized as a Cl/HCO_3 exchanger (Tsuganezawa et al., 2001; Ko, Luo et al., 2002), but it is also suggested to be an electroneutral NBC without Cl/HCO_3 exchange activity (Parker et al., 2002). AE4 is localized to renal collecting duct type-A intercalated cells, but its membrane polarity appears to be species-specific. It is localized to the basolateral membrane of the intercalated cells in rats and mice, but to the apical and lateral membranes in rabbits (Ko, Luo et al., 2002).

2.2. Na^+-Coupled Bicarbonate Transporters (NCBTs)

2.2.1. NBCe1 (SLC4A4), NBCe2 (SLC4A5)

Romero and Boron cloned the electrogenic NBCe1 from the salamander kidney (Romero, Hediger, Boulpaep, & Boron, 1997). On the basis of predicted topology and similarity to AE1, NBCe1 (1035 amino acids) was proposed to have a cytoplasmic N-terminal domain, 12–14 TM segments, and a cytoplasmic C-terminal domain. The most widely distributed variant is NBCe1-B containing the unique N-terminal 85 amino acids, which replace the first 41 amino acids of the renal variant NBCe1-A (Abuladze et al., 1998; Choi, Romero, Khandoudi, Bril, & Boron, 1999). The brain-specific NBCe1-C is identical to NBCe1-B except for the C-terminal 61 amino acids (Bevensee, Schmitt, Choi, Romero, & Boron, 2000). NBCe1-A heterologously expressed in *Xenopus* oocytes produces an electrogenic HCO_3^- current with a $1Na^+$ and $2HCO_3^-$ stoichiometry (Ducoudret, Diakov, Muller-Berger, Romero, & Frömter, 2001; Grichtchenko et al., 2001; Sciortino & Romero, 1999), mediating net Ducoudret, Diakov, Muller-Berger, Romero, & Frömter, 2001 influx. In contrast, the transporter in the proximal tubules moves $1Na^+$ and $2HCO_3^-$ across the basolateral membrane, mediating net HCO_3^- efflux. Studies show that the same transport molecule presumably changes its stoichiometry depending on the cellular environment and parameters such as intracellular Ca^{2+} levels (Muller-Berger, Ducoudret,

Diakov, & Frömter, 2001) or phosphorylation activated by cAMP (Gross et al., 2003). Similar to AEs, NBCe1 forms a homodimer composed of two individually functional subunits (Gill & Boron, 2006a; Kao et al., 2008).

NBCe2 is the second type of the electrogenic Na/HCO_3 cotransporter (Pushkin et al., 2000; Virkki, Wilson, Vaughan-Jones, & Boron, 2002). NBCe2 is strongly expressed in hepatocytes and bile duct cholangiocytes in the liver and renal pelvis uroepithelial cells in the kidney (Abuladze et al., 2004). It is also found in the renal collecting ducts, choroid plexus epithelia (Bouzinova et al., 2005; Damkier, Nielsen, & Praetorius, 2007), and sarcolemmal membranes in skeletal muscles (Kristensen, Kristensen, & Juel, 2004). NBCe2 produces a HCO_3^--induced current with the $1Na^+$:$2HCO_3^-$ stoichiometry when expressed in *Xenopus* oocytes (Virkki et al., 2002). However, it operates with the $1Na^+$:$3HCO_3^-$ stoichiometry when expressed in the renal proximal tubule cell line (Sassani et al., 2002) and in intact cells prepared from choroid plexus epithelia (Millar & Brown, 2008).

2.2.2. NBCn1 (SLC4A7), NBCn2 (SLC4A10)

The electroneutral NBCn1 was isolated from rat vascular smooth muscle (Choi, Aalkjaer, Boulpaep, & Boron, 2000) and human skeletal muscle (Pushkin et al., 1999). Human NBCn1 contains 1214 amino acids with a predicted structure similar to that of NBCe1. NBCn1 exists as multiple variants because of N- and C-terminal splice events. The transporter is widely distributed in the body (Boedtkjer, Praetorius, Fuchtbauer, & Aalkjaer, 2008). NBCn1 is expressed in vascular smooth muscle cells and endothelial cells and modulates artery tone and blood pressure control (Boedtkjer et al., 2011). The variant containing the N-terminal cassette of 123 amino acids (cassette II) is predominantly found in the cardiovascular system (Cooper et al., 2006). Among SLC4A bicarbonate transporters, NBCn1 is least sensitive to the anion channel/transporter blocker 4,4′-diisothiocyanatostilbene disulfonate (DIDS) (Choi et al., 2000). In addition, NBCn1 has HCO_3^--independent, Na^+ channel-like activity (Cooper et al., 2005).

NBCn2 is the second cloned electroneutral Na/HCO_3 cotransporter (Wang, Yano, Nagashima, & Seino, 2000). This transporter is predominantly expressed in neurons and choroid plexus epithelia (Damkier et al., 2007). NBCn2 operates with the $1Na^+$:$1HCO_3^-$ stoichiometry with Cl^- self-exchange activity (Parker, Musa-Aziz et al., 2008). Other studies show that NBCn2 directly transports Cl^- in exchange for Na/HCO_3 cotransport and functions as a Na^+-driven Cl/HCO_3 exchanger NCBE (Damkier, Aalkjaer, & Praetorius, 2010; Giffard, Lee, Ouyang, Murphy, & Monyer, 2003).

2.2.3. *NDCBE1 (SLC4A8)*

NDCBE1 is the Na^+-driven Cl/HCO_3 exchanger (Grichtchenko et al., 2001; Parker, Bouyer, Daly, & Boron, 2008). This transporter is predominantly expressed in neurons (Chen, Haddad, & Boron, 2008; Sinning et al., 2011), where Na^+-driven Cl/HCO_3 exchange has been known to primarily govern HCO_3^--dependent acid extrusion (Schwiening & Boron, 1994). NDCBE moves external $1Na^+$ and $2HCO_3^-$ in exchange for internal Cl^-.

2.3. Sodium Borate (NaBC) Transporter

NaBC1 (SLC4A11) transports Na^+ and borate (Park, Li, Shcheynikov, Zeng, & Muallem, 2004). This transporter has substantial sequence homology to SLC4A bicarbonate transporters and was assigned as SLC4A11 before its function was characterized. NaBC1 operates with the stoichiometry of $1Na^+$ and at least $2B(OH)_4^-$ (Park et al., 2004) The transporter moves Na^+ and OH^- (H^+) in the absence of borate.

3. ION TRANSPORT PROPERTIES

3.1. Ion Dependence and Selectivity

3.1.1. *AEs*

AEs are permissive to many anions in addition to HCO_3^- and Cl^-. For example, AE1 has the capacity to move several monovalent and divalent anions in either direction or via anion self exchange (Knauf & Pal, 2003). AE2 also mediates the exchange of anions including NO_3^-, Br^-, and I^- with low affinity when expressed in *Xenopus* oocytes (Humphreys, Jiang, Chernova, & Alper, 1994). The amino acid residue best characterized for anion selectivity is Glu^{681} (TM8) in erythrocyte AE1 and the corresponding Glu site in AE2. Chemical modification of Glu^{681} to an alcohol leads to electrogenic SO_4^{2-}/Cl^- exchange and electroneutral SO_4^{2-}/SO_4^{2-} self exchange (Jennings, 1995; Jennings & Al Rhaiyel, 1988). Similar changes are also observed from the site-directed mutagenesis of the corresponding Glu in cloned AE1 and AE2 (Chernova et al., 1997; Sekler, Lo, & Kopito, 1995), indicating that the role of Glu^{681} is conserved among AEs. The underlying mechanism involves an interaction of sulfate with H^+ that is donated by the Glu residue lining the pore. Therefore, Glu^{681} contributes to anion translocation, as discussed below. Other residues in AE are identified to serve as the anion selectivity sites in addition to Glu^{681} of TM8. By scanning cysteine accessibility mutagenesis (SCAM) of AE1, Casey and his colleagues (Zhu & Casey, 2004) found that the region of Ser[852]–Leu[857] in the extracellular loop between the last two TMs serves as an anion selectivity filter. Whether the residues Ser[852]–Leu[857] are mechanistically associated with Glu[681] for function is unclear.

While it is generally accepted that the transmembrane domain plays an essential role in anion selectivity, the cytoplasmic domains may also contribute to maintaining HCO_3^- selectivity. Deletion and missense mutants without the C-terminal putative CAII-binding site lose Cl^-/HCO_3^- exchange while retaining Cl^-/Cl^- self-exchange (Dahl et al., 2003).

3.1.2. NCBTs

All NCBTs are dependent upon Na^+ (Amlal, Wang, Burnham, & Soleimani, 1998; Sciortino & Romero, 1999). The current–voltage relationship for electrogenic HCO_3^- transport shows a positive and linear slope with outward currents at positive voltages. The current is highly selective to HCO_3^- and is not elicited by other anions (Fig. 1). The charge-conserved substitution of Glu[555] with an Asp (i.e. D555E) produces dose-dependent Cl^- currents in the absence of HCO_3^- (Yang, Kim et al., 2009). Nonetheless, D555E produces a HCO_3^- current when both Cl^- and HCO_3^- are present. Anion selectivity experiments reveal that D555E is broadly permissive to other anions including NO_3^-, SCN^-, I^-, and Br^-, whereas wild-type NBCe1 shows a high selectivity to HCO_3^-. D555 may act to distinguish HCO_3^- from other trigonal planar polyanions such as NO_3^-, as well as provide steric hindrance for small anions

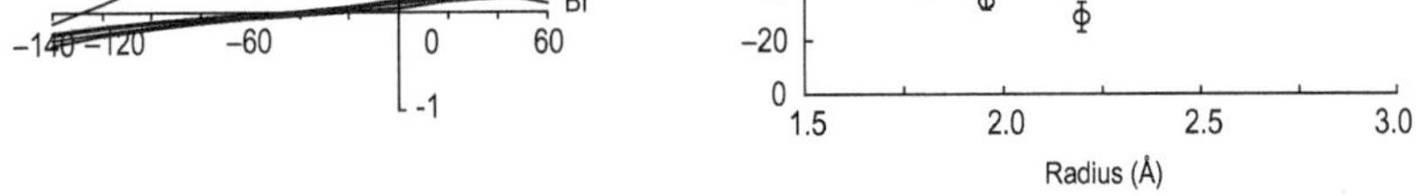

Figure 1 Anion selectivity of NBCe1. A) NBCe1 expressed in *Xenopus* oocytes produces electrogenic outward currents in response to HCO_3^-, but not other anions. Recording was done while oocytes were exposed to different anions (25 mM for each). B) Zero current voltage (E_{rev}) versus anion radius. Note that HCO_3^- has a large effective radius, but produces a current. The molecular or effective radius (in Å) is 1.81 Cl^-, 1.89 NO_3^-, 1.96 Br^-, 2.13 SCN^-, 2.20 I^- and 2.8 HCO_3^-. $n = 6$ for each.

and SCN^-. The mechanism appears to involve a salt bridge between Asp^{555} and adjacent charge residues in TM5 (unpublished observation by Soojung Lee). NBCe1 may also transport OH^- at high extracellular pH (Amlal et al., 1998).

The apparent affinity of NBCe1-A for Na^+ is about 30 mM, and independent of different voltages and external $[HCO_3^-]$ (Sciortino & Romero, 1999). Similar affinity values were also observed for all three NBCe1 variants, -A, -B, and -C (apparent K_m of 21–36 mM) (McAlear & Bevensee, 2006). The apparent K_m of NBCe1 for HCO_3^- is 4–6.5 mM (Grichtchenko, Romero, & Boron, 2000; McAlear & Bevensee, 2006).

3.2. Ion Translocation

3.2.1. AEs

In addition to anion dependence and selectivity, Glu^{681} plays a role in ion translocation. Chemical modification or site-directed mutagenesis of this residue alters anion exchange kinetics (Chernova et al., 1997; Sekler et al., 1995), indicating that it is located in the anion translocation pathway. An SCAM study on TM8 (Tang, Kovacs, Sterling, & Casey, 1999) also identified residues that are predicted to line the translocation pore adjacent to Glu^{681}. These are Ala^{666}, Ser^{667}, Leu^{669}, Leu^{673}, Leu^{677}, and Leu^{680} located in one face of α-helical TM8. Furthermore, cysteine mutants at Ile^{684} and Ile^{688} in the intracellular loop are also inhibited by sulfhydryl reagents and thus may be parts of the pore. Other TMs have been implicated as part of anion translocation. Lys^{539} in TM5 is the site where the anion channel/transporter blocker 4,4′-diisothiocyanostilbene-2,2′-disulfonate (DIDS) binds (Bartel, Hans, & Passow, 1989). Mutations of His^{703}, Arg^{731}, His^{735}, His^{816}, and His^{834} located in the region of TM10–13 result in inhibition of Cl^- flux (Muller-Berger, Karbach, Konig et al., 1995). The His residues, particularly His^{735}, are proposed to interact with Glu^{681} and form an access channel (Muller-Berger, Karbach, Kang et al., 1995). Several other TMs from 9 to 13 are required for transport function or contribute to ion translocation. Cysteine substitutions G714C, S725C, S731C, S762C, G790C, and F806C all reduced transporter activity to <40% of wild-type AE1 (Fujinaga, Tang, & Casey, 1999). The authors also found that the sulfhydryl reagent decreases the activity of AE constructs S852C (TM12) and A858C (TM13).

3.2.2. NCBTs

Regarding ion translocation of NBCe1, site-directed mutagenesis of amino acid residues in NBCe1 has been performed to identify the structural requirements for cotransport function (Abuladze et al., 2005). Many charged and polar residues, such as Glu, Asp and Arg, scattered throughout all the TMs were found critical for transport function. Nonetheless, while these residues contribute to charge interaction with ion substrates, it is unclear whether they are responsible for ion translocation or structural components of the protein. Furthermore, noncharged or nonpolar residues may play a more important role in ion translocation. An SCAM analysis of residues in TM8 identified Leu^{750} among others involved in ion translocation (McAlear & Bevensee, 2006). The sulfhydryl reagent *p*CMBS accessibility of L750C is strongly influenced by Na^+ and HCO_3^-, DIDS binding, and membrane potential. Kurtz and his colleagues used a similar approach to evaluate amino acid residues in TM1 and found that T442C, A435C, and A428C have substantial sensitivity to the methylsulfonate reagents (Zhu et al., 2009). The α-helical structure of TM1 suggests that these residues would be located on the same face of the TM and serve as binding sites for ion substrates or lining the pore. The functional significance of TM1 is also evident for S427L, which has a defect in either the voltage sensor or Na^+-binding site and fails to mediate Na/HCO_3 efflux (Dinour et al., 2004). An SCAM analysis of residues in TM10-14 reveals that the C-terminal TMs are tightly folded with limited access to the extracellular aqueous environment (Zhu et al., 2010). D894C (in a loop between TM12 and TM13) and L937C (in TM14) appear to display a severe decrease in NBC function. Figure 2 summarizes amino acid residues that have been identified as residues critical for ion translocation via NBCe1. These residues are located in TM1, TM5, TM7, TM8, and TM14.

While structure/function studies continue to investigate the transmembrane domain for ion translocation, the cytoplasmic domain may also play

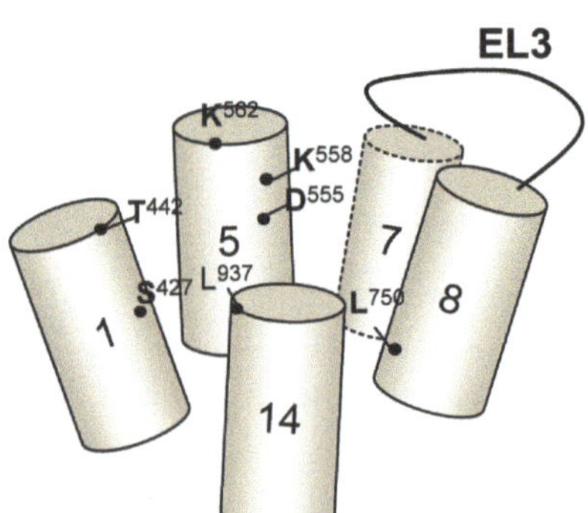

Figure 2 Amino acid residues critical for ion translocation through NBCe1. The residues are located in TM1, TM5, TM7, TM8, and TM14. Identification and characterization of these residues are described in the text.

a role. The naturally occurring mutant R298S is located in the cytoplasmic N-terminal domain of NBCe1 and has reduced transport activity. By modeling NBCe1 onto the crystal structure of the AE1 N-terminal domain, Romero and his colleagues hypothesized that Arg^{298} interacts with Glu^{91} or Glu^{295} via hydrogen bonds or charge–charge interactions (Chang, Dipiero, Sonnichsen, & Romero, 2008). The double mutation E91R/R298E rescues transport activity, indicating that the N-terminal domain contributes to HCO_3^- permeation.

3.3. Mechanisms of Ion Transport

The following two different mechanistic models are considered for SLC4A transporters: a sequential (or ping-pong) mechanism and a simultaneous mechanism (Fig. 3). In the sequential mechanism, one ion substrate binds to the transporter, causes a conformational change, and is released at the *trans* side of the membrane. Then, a second ion substrate binds to the transporter, regenerates the original conformation, and is released on the original *cis* side of the membrane. Therefore, the sequential mechanism involves a series of association–dissociation steps and conformational changes. However, the simultaneous mechanism proposes that the transporter forms a ternary

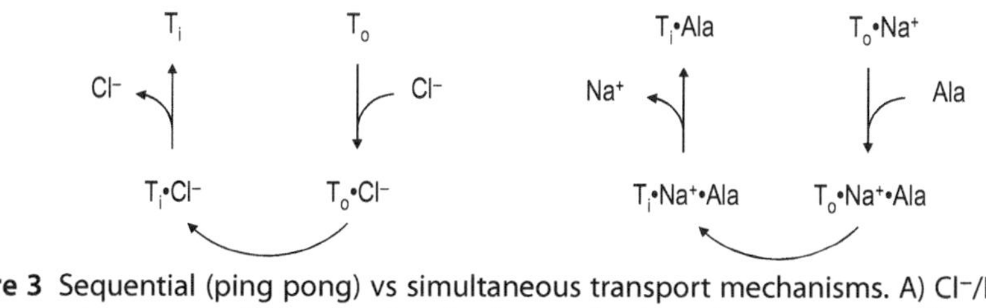

Figure 3 Sequential (ping pong) vs simultaneous transport mechanisms. A) Cl^-/HCO_3 exchange via a sequential mechanism. In this mechanism, the transporter (T) accepts only one ion substrate at a time at each membrane surface. T_i and T_o represent the transporter in the inward- and outward-facing states, respectively. B) Na^+/alanine cotransport via a simultaneous mechanism. Both Na^+ and alanine (Ala) bind to the transporter before translocation occurs. The tertiary complex $T{\cdot}Na^+{\cdot}Ala$ undergoes a conformational change. Releasing the two substrates on the opposite side of the membrane causes the transporter to be returned to the original state.

complex with substrates before translocation occurs. The first and second ion substrates bind to the transporter and form a tertiary complex. This causes a conformational transition that makes the bound substrates face the *trans* side of the membrane. The two substrates are subsequently released.

The sequential mechanism has been proposed to account for the exchange kinetics of Cl^- and HCO_3^- by AEs (Frohlich & Gunn, 1986). H_2-DIDS can bind preferentially to the outward-facing state (Jennings, Whitlock, & Shinde, 1998). The exchange rates for HCO_3^- or Cl^- are different depending upon the inward facing or outward facing state (Knauf, Law, Leung, Gehret, & Perez, 2002). In addition, for the E699Q mutation of mouse AE1 (corresponding to E681Q in human AE1), elevating intracellular $[SO_4^{2-}]$ increases the apparent K_m for extracellular Cl^- and SO_4^{2-} (Chernova, Stewart, Barry, Jennings, & Alper, 2008). All these data are consistent with the sequential kinetics.

It is unclear if the mechanism used by NCBTs is simultaneous or sequential. The simultaneous mechanism can be distinguished from the sequential mechanism experimentally by studying the relationship of K_m and V_{max} as shown for the Na/alanine transporter (Jauch & Lauger, 1986). The simultaneous mechanism exhibits a variable K_m/V_{max} ratio for the variable substrate at different concentrations of the fixed substrate, whereas the sequential mechanism shows a constant K_m/V_{max} ratio. The apparent K_m and V_{max} of NBCe1 for Na^+ are unaffected at 10 and 33 mM $[HCO_3^-]$ (Sciortino & Romero, 1999), implying that NBCe1-mediated transport may occur via a sequential mechanism although additional experiments are needed to confirm this. The ability of D555E to produce a Cl^- current (Yang, Kim et al., 2009) provides some insights into the binding order. The mutant transporter favorably selects HCO_3^- when HCO_3^- and Cl^- are present, but this HCO_3^- selection requires Na^+. In the absence of Na^+, D555E fails to select HCO_3^- and instead selects Cl^-. This indicates that HCO_3^- binding requires the precondition of Na^+ binding from the perspective of the ion binding order.

3.4. Electrogenicity and Stoichiometry

The electrogenicity of a given ion transporter determines whether it is affected by the membrane voltage in addition to the ion gradient across the cell membrane. Among SLC4A transporters, NBCe1 and NBCe2 are electrogenic despite their sequence similarity with other transporters. The molecular basis of electrogenicity is poorly understood. Studies on chimeric transporters constructed from NBCe1 and NBCn1 reveal that the NBCe1 electrogenicity is conferred by the conserved transmembrane domain, while the cytoplasmic N and C termini and the large third extracellular loop (EL3) between TM5 and TM6 are not critical (Choi, Yang, & Boron, 2007). Furthermore, electrogenicity is mediated via interactions between TM1–5 and TM6–14 because swapping these two regions with NBCe1 and NBCn1 results in electroneutral Na/HCO_3 transporters. The EL4 (32 amino acids) between TM7 and TM8 has been found to be critical for electrogenicity (Chen, Liu, & Boron, 2011). Chimeric transporters with NBCe1–EL4 exhibit electrogenic properties, while those with NBCn1–EL4 exhibit electroneutral properties. EL4 may directly constitute a part of the ion-binding vestibule or coordinate with TMs to modulate ion binding.

Related to the issue of electrogenicity is the ion transport stoichiometry. The Na^+:HCO_3^- stoichiometry of NBCe1 is particularly essential for HCO_3 reabsorption in the proximal tubules of the kidney, where the basolateral NBCe1-A moves $1Na^+$ and $3HCO_3^-$ into the interstitium (Soleimani, Grassl, & Aronson, 1987). If the stoichiometry were $1Na^+$ and $2HCO_3^-$, then the transporter would instead move ions in the opposite direction (i.e. into the cell) because the equilibrium potential for $1Na^+/2HCO_3^-$ transport is estimated to be more negative than the resting potential, typically −70 mV (Boron & Boulpaep, 1983). Indeed, NBCe1 in astrocytes has a $1Na^+$:$2HCO_3^-$ stoichiometry and transports ions into the cells (Bevensee, Apkon, & Boron, 1997; Bevensee, Weed, & Boron, 1997). Similarly, NBCe1 in pancreatic ducts has a $1Na^+$:$2HCO_3^-$ stoichiometry and transports ions into the cells (Ishiguro, Steward, Lindsay, & Case, 1996; Zhao, Star, & Muallem, 1994). Given the different N-terminal amino acids between NBCe1-A and NBCe1-B/C, it was initially expected that the stoichiometry would be determined by the N-terminal domain of the transporter. However, NBCe1-A expressed in *Xenopus* oocytes mediates Na/HCO_3 influx with a $1Na^+$:$2HCO_3^-$ stoichiometry (Ducoudret et al., 2001; Grichtchenko et al., 2001; Sciortino & Romero, 1999). The stoichiometry changes from $1Na^+$:$2HCO_3^-$ to $1Na^+$:$3HCO_3^-$ when the cytosolic Ca^{2+} concentration increases (Muller-Berger et al., 2001). Both NBCe1-A and -B show a $1Na^+$:$3HCO_3^-$ stoichiometry when expressed in a proximal tubule cell line and a $1Na^+$:$2HCO_3^-$ stoichiometry when expressed in a mouse collecting tubule cell line (Gross et al., 2001). These findings provide a strong argument that the same transport molecule can express different stoichiometries, depending on the cellular environment. Gross and his colleagues demonstrate that the stoichiometry can be changed by a cAMP-dependent protein kinase A phosphorylation site (Ser^{982} in NBCe1-A and Ser^{1026} in NBCe1-B) in the C-terminal domain (Gross et al., 2003).

3.5. Pharmacological Inhibition of HCO_3^- Transport

All SLC4A HCO_3^- transporters except NBCn1 are inhibited by stilbene derivatives such as DIDS, 4-acetamide-4′-isothiocyanatostilbene-2,2′-disulfonic acid (SITS) and 4,4′-dinitrostilbene-2,2′ disulfonate (DNDS). These derivatives are broad inhibitors of anion transporters/channels (Cabantchik, Knauf, & Rothstein, 1978). DIDS binds to the pore entry of anion channels/transporters and occludes the pore vestibule (Tombola, Del Giudice, Papini, & Zoratti, 2000). The inhibition constant K_i of AE1 in erythrocytes for DIDS is as low as 0.04 μM (Knauf, Law, & Hahn, 1995), while the values of cloned AEs are much larger. For NBCe1, the inhibition constant is 36 μM (Liu, Williams, Sumpter, & Bevensee, 2007). DIDS reversibly and covalently binds to the first lysine (K) in the SKLIK motif in TM 5 of AE1 (Bartel et al., 1989). NBCe1 has the homologous KKMIK motif, and mutation analysis of this motif shows the second K (i.e. K^{559}) with the most profound effect on the reversible DIDS binding (Lu & Boron, 2007). However, in the absence of DIDS, the mutation at K^{559} produces a HCO_3^- conductance similar to those of other mutations in the KKMIK motif. Thus, K^{559} does not appear to serve directly as a HCO_3^- translocation site. Other NCBTs contain similar DIDS motifs in the corresponding region of TM5, but their role in DIDS sensitivity has not been tested.

In addition to DIDS, fluorescent oxonol dyes such as WW-781, diBA(3)C4 and diBA(5)C4 are potent inhibitors for AEs (Alper et al., 1988). The inhibition potency of these oxonol dyes is higher than that of DIDS and varies among AE isoforms. The oxonol dyes diBA(3)C4 and diBA(5)C4 also inhibit NBCe1 (Liu et al., 2007). The oxonol dyes and DIDS compete with each other, and therefore appear to share at least part of the same external binding site. It is suggested that SLC4A transporters may contain a relatively large inner vestibule, in which a number of large anions bind and block Cl^- permeation. Other chemicals that inhibit AEs and/or NBCs include niflumic acid (Cousin & Motais, 1976; Liu et al., 2007), benzamil and tenidap (Ducoudret et al., 2001; Lu & Boron, 2007), and an N-cyanosulphonamide inhibitor S8059 (Ch'en, Villafuerte, Swietach, Cobden, & Vaughan-Jones, 2008).

3.6. HCO_3^- vs CO_3^- Transport

The possibility that HCO_3 transporters would carry CO_3^{2-} is based on an unexpected observation by Grichtchenko and Chesler (1994). They monitored extracellular pH in the brain slices and found a pH decrease that was caused by electrogenic Na/HCO_3 influx via NBCe1 into the astrocytes.

They anticipated that inhibiting extracellular carbonic anhydrase (CA) would reduce the supply of extracellular HCO_3^- and thus blunt the pH decrease. However, CA inhibition had the opposite effect and stimulated the pH decrease. They hypothesized that NBCe1 transports CO_3^{2-} instead of HCO_3^-. At physiological pH (i.e. pH 7.4 in the extracellular space), HCO_3^- is in the millimolar range, while CO_3^{2-} is in the micromolar range and H^+ is in the nanomolar range. If CO_3^{2-} is consumed via NBCe1, the reaction $HCO_3^- \rightarrow CO_3^{2-} + H^+$ would produce a large increase in $[H^+]$. However, $[HCO_3^-]$ would remain relatively unchanged as it is in the millimolar range. From the reaction $CO_2 + H_2O \leftrightarrow HCO_3^- + H^+$, one can expect that an increase in $[H^+]$ makes the reaction run leftward. Therefore, inhibiting extracellular carbonic anhydrase would blunt this leftward reaction and consequently enhance a decrease in extracellular pH. Boron and his colleagues (Grichtchenko & Boron, 2002; Lee, Grichtchenko, & Boron, 2011) tested this model by coexpressing NBCe1 and GPI-linked CAIV in *Xenopus* oocytes and measuring surface pH (pH_s). They found that stimulating NBCe1 caused pHs to change as predicted from CO_3^{2-} transport and inhibiting CAIV augmented all pH_S changes. These results are consistent with CO_3^{2-} transport rather than HCO_3^- by NBCe1.

3.7. Channel-Like Activity

Historically, ion transporters have been considered mechanistically and structurally different from ion channels (Hille, 2001). Ion transporters move ions slowly and exhibit distinct substrate coupling (i.e. stoichiometry) and kinetics, whereas ion channels move ions rapidly ($>10^6$ ions per second) and exhibit gating and permeation. A typical view is that ion transporters structurally undergo rather marked conformational changes between inward-facing and outward-facing states, whereas ion channels undergo more subtle gating changes that open and close the pore. However, this distinction between transporters and channels became increasingly unclear as some membrane proteins share functional characteristics of both groups. Neurotransmitter transporters, including serotonin transporters, norepinephrine transporters, glutamate transporters and dopamine transporters, have imbedded channel properties (DeFelice & Blakely, 1996; Torres & Amara, 2007). Similarly, NBCn1 has robust Na^+ channel-like activity ($Na^+ \gg K^+ = Cs^+ = NMDG^+$) (Choi et al., 2000; Cooper et al., 2005). This activity can cause a positive shift in the membrane potential, and raise intracellular Na^+ when NBCn1 is expressed in HEK 293 cells or *Xenopus* oocytes. The channel-like activity is not coupled to Na/HCO_3 cotransport.

The domain responsible for the channel-like activity of NBCn1 is the region between TM6 and TM14 (Choi et al., 2007). NBCn1 is not the only HCO_3^- transporter exhibiting this activity, other SLC4A proteins also have similar activities. Erythrocyte AE1 mediates an anion flux that is stimulated by cell hyperpolarization, implicating a conductive pathway (Freedman & Novak, 1997). Trout AE1 expressed in *Xenopus* oocytes induces Cl^- currents (Fievet, Gabillat, Borgese, & Motais, 1995). The electroneutral NDAE1 in *Drosophila* has a small inward current (Romero et al., 2000). NBCe1 variants display a Na^+-independent HCO_3^- current (McAlear, Liu, Williams, McNicholas-Bevensee, & Bevensee, 2006). Furthermore, the Cl/HCO_3 exchangers Slc26a3 and Slc26a6 produce large NO_3^- and SCN^- currents (Ko, Shcheynikov et al., 2002; Shcheynikov et al., 2008).

The channel-like activity is more than just an oddity of the transporter because it affects cellular membrane potential and controls entry of ions at least in heterologous expression systems. For example, the Na^+ channel activity of NBCn1 lowers an electrochemical Na^+ gradient across the cell membrane. The lowered Na^+ gradient may alter activities of other Na^+-conducting ion channels if such channels and NBCn1 are present together. The significance of this effect has yet to be explored.

4. STRUCTURE OF SLC4A TRANSPORTERS

AE1 is predominantly composed of dimers although both monomers and tetramers are known to exist (Reithmeier, Chan, & Popov, 1996). A structural analysis by electron microscopy and three-dimensional image reconstruction at 20 Å resolution reveal the dimeric membrane domain in the U-shaped structure (Dolder, Walz, Hefti, & Engel, 1993; Wang et al., 1994). The two monomers are in contact at the lower part (called the basal domain) embedded in the membrane, while forming a pair of protrusions at the upper part that faces the cytosol. The protrusions form the sides of a canyon, which surrounds a wide space and then converges into a depression in the basal domain. This depression may represent the pore entry or the opening to a transporter vestibule located at the dimer interface.

In support of the above model, the three-dimensional structure of the AE1 membrane domain shows V-shaped densities near the center of the dimer at 7.5 Å resolution (Hirai, Hamasaki, Yamaguchi, & Ikeda, 2011; Yamaguchi et al., 2010). Interestingly, similar V-shaped densities exist in the previously reported structure of a prokaryotic chloride channel (ClC) protein. The ClC protein contains 14 TMs with two internal repeats of 7 TMs inserted into the membrane in the transverse orientation (Dutzler, Campbell, Cadene, Chait, & MacKinnon, 2002). A V-shaped pair of helical repeats (TM1 +2 and TM8 +9) in AE1 are similar to corresponding repeats in the ClC. This similarity in the density and projection maps coincides with a sequence identity of 20%. Nonetheless, there are difficulties with this model. All the TMs cannot be clearly accounted for at 7.5 Å. The beginning of TM1 predicted from NMR studies (Chambers, Bloomberg, Ring, & Tanner, 1999) does not align with the beginning of that for ClC.

The cytoplasmic N-terminal domain of NBCe1-A has been crystallized at 3 Å (Gill & Boron, 2006b). The domain forms two monomers related by a 2-fold axis. The dimer architecture is similar to that observed for the N-terminal crystals of AE1 (Zhang, Kiyatkin, Bolin, & Low, 2000).

5. INTERACTION WITH OTHER PROTEINS

Similar to many other membrane proteins, some SLC4A transporters exist in concert with other proteins and form macromolecular complexes. It would be worth discussing the structural features associated with protein–protein interactions, which mainly occur in the cytoplasmic N- and C-terminal domains of the transporters.

5.1. AEs

The N-terminal domain in AE1 has an ankyrin-binding motif in residues 175–185 (Chang & Low, 2003). The AE1/ankyrin interaction tethers the cytoskeletal protein spectrin to the microdomains near the plasma membrane. Tyrosine phosphorylation promotes dissociation of AE1 from the spectrin-actin skeleton and may enable erythrocytes to undergo adaptive changes in response to stimuli such as malaria parasite invasion, cell shrinkage, and aging (Ferru et al., 2011). AE1 also binds to glycolytic enzymes, hemoglobin and protein 4.1 and 4.2.

The cytoplasmic C-terminal domain of AE1 interacts with CAII (McMurtrie et al., 2004; Vince & Reithmeier, 2000). The consensus sequence for the CAII binding site is DADD, where the first Asp makes an electrostatic interaction with His or Lys residues in the N-terminal region of CAII. Other AEs and different CA isoforms also interact with each other (Casey, Sly, Shah, & Alvarez, 2009). The AE1/CAII interaction significantly increases the rate of HCO_3^- transport (Sterling, Reithmeier, & Casey, 2001). Cl^-/HCO_3^- exchange via AE1 in erythrocytes is a rate-limiting step for CO_2 delivery from tissues to the lungs (Wieth, Andersen, Brahm, Bjerrum,

& Borders, 1982). Maximizing the HCO_3^- transport rate helps fully utilize the CO_2-transport capacity of the blood.

5.2. NCBTs

IP_3 receptor binding protein released with IP_3 (IRBIT) is a signaling protein that is released from IP_3 receptors and regulates downstream target molecules. IRBIT binds to the N terminus of NBCe1-B, but not NBCe1-A, and increases HCO_3^- transport activity (Shirakabe et al., 2006). IRBIT-mediated NBCe1-B stimulation has been demonstrated in pancreatic ducts (Yang, Shcheynikov et al., 2009a). The N-terminal 87 amino acids of NBCe1-B is an autoinhibitory domain that, when deleted, stimulates transport activity (McAlear et al., 2006). Deleting the N-terminal residues 2–16 of NBCe1-B abolishes the IRBIT-stimulated effect while maintaining autoinhibition (Lee, Boron, & Parker, 2012). Thus, autoinhibitory and IRBIT-binding determinants are not identical.

Some SLC4A transporters contain a motif in their C terminus to which PDZ proteins bind (PDZ stands for Post synaptic density protein PSD95, Drosophila disc large tumor suppressor Dlg1, and Zonula occludens-1 protein zo-1). PDZ proteins are proteins that interact with a variety of membrane and cytosolic proteins, modulate activities of their binding partners, associate the partners with cytoskeletal protein complexes near the membrane, and recruit signaling proteins (Kim & Sheng, 2004). NBCn1 contains the PDZ-binding motif (amino acid sequence ETSL) and interacts with the 56 kDa subunit of H-ATPase (Pushkin et al., 2003). The two proteins colocalize to the collecting duct intercalated cells of the kidney. NBCn1 also interacts with ezrin-binding protein 50 (EBP50; *alias* Na/H exchanger regulatory factor 1 NHERF-1) in the apical membrane of the pancreatic ducts and salivary glands (Park et al., 2002). The NBCn1/EBP50 interaction helps scavenge luminal HCO_3^- at rest but facilitates Cl^--mediated HCO_3^- release during stimulation. NBCn1 also interacts with the postsynaptic density protein PSD-95 at synapses in the hippocampus (Park et al., 2010). The NBCn1/PSD-95 interaction appears to increase channel-like activity of the transporter. NBCn1 also interacts with harmonin to form a protein complex in the ear (Reiners et al., 2005). Other SLC4A bicarbonate transporters capable of interacting with PDZ proteins include NBCn2-C, which contains ETCL at its C terminus and interacts with the cytoskeleton and EBP5 (Lee, Ouyang, & Giffard, 2006).

Similar to AEs, NCBTs are also reported to interact with CAII isoforms. For example, the cytoplasmic C-terminal domain of NBCe1 contains $D^{986}NDD$ for the CA II binding and the interaction enhances transport function (Becker & Deitmer, 2007; Pushkin et al., 2004). In contrast, other studies reveal no interaction between NBCe1 and CA II (Lu et al., 2006) or normal transport activity of a mutant lacking the CAII-binding domain (Yamada, Horita, Suzuki, Fujita, & Seki, 2011).

6. CONCLUSION

SLC4A transporters regulate intracellular pH in a variety of cells. Almost all cells in the body possess at least one of these transporters, and changes in pH influence activities of other physiologically important proteins. For example, pH alters ion permeation, sensitivity, and response to agonists of ion channel receptors in the nervous system. Slc4a-knockout mice exhibit aberrant neuronal activities that are caused by acid–base disturbance. In addition, the bicarbonate transporters are essential for transepithelial movement of acid–base equivalents in epithelial tissues such as the kidney where the transporters contribute to regulating plasma pH. Studying the molecular mechanism of how ions are selected and carried via the transporters helps better understand acid–base homeostasis at the cellular and whole-body levels. Studying the transporter structure is valuable for drug designs for transporter-related pathophysiology. Future molecular studies of SLC4A bicarbonate transporters will certainly advance our understanding of transporter physiology.

ACKNOWLEDGMENTS

I thank Drs Mark Bevensee and Christian Aalkjaer for their valuable input, and Dr. Mark Bevensee for contributing some material. This work was supported by the NIH GM078502.

REFERENCES

Abuladze, N., Azimov, R., Newman, D., Liu, W., Tatishchev, S., Pushkin, A., et al. (2005). Critical amino acid residues involved in the electrogenic sodium bicarbonate cotransporter kNBC1-mediated transport. *Journal of Physiology*, *565*(15), 717–730.

Abuladze, N., Lee, I., Newman, D., Hwang, J., Boorer, K., Pushkin, A., et al. (1998). Molecular cloning, chromosomal localization, tissue distribution, and functional expression of the human pancreatic sodium bicarbonate cotransporter. *Journal of Biological Chemistry*, *273*(28), 17689–17695.

Abuladze, N., Pushkin, A., Tatishchev, S., Newman, D., Sassani, P., & Kurtz, I. (2004). Expression and localization of rat NBC4c in liver and renal uroepithelium. *American Journal of Physiology Cell Physiology*, *287*(3), C781–C789.

Alper, S. L. (2009). Molecular physiology and genetics of Na^+-independent SLC4 anion exchangers. *Journal of Experimental Biology*, *212*(11), 1672–1683.

Alper, S. L., Chernova, M. N., Williams, J., Zasloff, M., Law, F. Y., & Knauf, P. A. (1988). Differential inhibition of AE1 and AE2 anion exchangers by oxonol dyes and by novel polyaminosterol analogs of the shark antibiotic, squalamine. *Biochemistry and Cell Biology*, *76*(5), 799–806.

Alper, S. L., Kopito, R. R., Libresco, S. M., & Lodish, H. F. (1988). Cloning and characterization of a murine band 3-related cDNA from kidney and from a lymphoid cell line. *Journal of Biological Chemistry, 263*(32), 17092–17099.

Amlal, H., Wang, Z., Burnham, C., & Soleimani, M. (1998). Functional characterization of a cloned human kidney Na^+:HCO_3^- cotransporter. *Journal of Biological Chemistry, 273*(27), 16810–16815.

Bartel, D., Hans, H., & Passow, H. (1989). Identification by site directed mutagenesis of Lys-558 as the covalent attachment site of dihydro DIDS in the mouse erythroid band 3 protein. *Biochimica et Biophysica Acta, 985*(3), 355–358.

Becker, H. M., & Deitmer, J. W. (2007). Carbonic anhydrase II increases the activity of the human electrogenic Na^+/HCO_3^- cotransporter. *Journal of Biological Chemistry, 282*(18), 13508–13521.

Bevensee, M. O., Apkon, M., & Boron, W. F. (1997). Intracellular pH regulation in cultured astrocytes from rat hippocampus. II. Electrogenic Na/HCO_3 cotransport. *Journal of General Physiology, 110*(4), 467–483.

Bevensee, M. O., Schmitt, B. M., Choi, I., Romero, M. F., & Boron, W. F. (2000). An electrogenic Na/HCO_3 cotransporter (NBC) with a novel C terminus, cloned from rat brain. *American Journal of Physiology Cell Physiology, 278*(6), C1200–C1211.

Bevensee, M. O., Weed, R. A., & Boron, W. F. (1997). Intracellular pH regulation in cultured astrocytes from rat hippocampus. I. Role of HCO_3^-. *Journal of General Physiology, 110*(4), 453–465.

Boedtkjer, E., Praetorius, J., Fuchtbauer, E. M., & Aalkjaer, C. (2008). Antibody-independent localization of the electroneutral Na^+-HCO_3 cotransporter NBCn1 (Slc4a7) in mice. *American Journal of Physiology Cell Physiology, 294*(2), C591–C603.

Boedtkjer, E., Praetorius, J., Matchkov, V.V., Stankevicius, E., Mogensen, S., Fuchtbauer, A. C., et al. (2011). Disruption of Na^+, HCO_3^- cotransporter NBCn1 (slc4a7) inhibits NO-mediated vasorelaxation, smooth muscle Ca^{2+} sensitivity, and hypertension development in mice. *Circulation, 124*(17), 1819–1829.

Bonar, P. T., & Casey, J. R. (2008). Plasma membrane Cl^-/HCO_3^- exchangers: structure, mechanism and physiology. *Channels (Austin), 2*(5), 337–345.

Boron, W. F., & Boulpaep, E. L. (1983). Intracellular pH regulation in the renal proximal tubule of the salamander: basolateral HCO_3^- transport. *Journal of General Physiology, 81*(1), 53–94.

Bouzinova, E.V., Praetorius, J., Virkki, L.V., Nielsen, S., Boron, W. F., & Aalkjaer, C. (2005). Na^+-dependent HCO_3^- uptake into the rat choroid plexus epithelium is partially DIDS sensitive. *American Journal of Physiology Cell Physiology, 289*(6), C1448–C1456.

Brosius, F. C., Nguyen, K., Stuart-Tilley, A. K., Haller, C., Briggs, J. P., & Alper, S. L. (1995). Regional and segmental localization of AE2 anion exchanger mRNA and protein in rat kidney. *American Journal of Physiology, 269*(4 Pt 2), F461–F468.

Cabantchik, Z. I., Knauf, P. A., & Rothstein, A. (1978). The anion transport system of the red blood cell. The role of membrane protein evaluated by the use of 'probes'. *Biochimica et Biophysica Acta, 515*(3), 239–302.

Casey, J. R., Sly, W. S., Shah, G. N., & Alvarez, B.V. (2009). Bicarbonate homeostasis in excitable tissues: role of AE3 Cl^-/HCO_3^- exchanger and carbonic anhydrase XIV interaction. *American Journal of Physiology Cell Physiology, 297*(5), C1091–C1102.

Ch'en, F. F., Villafuerte, F. C., Swietach, P., Cobden, P. M., & Vaughan-Jones, R. D. (2008). S0859, an N-cyanosulphonamide inhibitor of sodium-bicarbonate cotransport in the heart. *British Journal of Pharmacology, 153*(5), 972–982.

Chambers, E. J., Bloomberg, G. B., Ring, S. M., & Tanner, M. J. (1999). Structural studies on the effects of the deletion in the red cell anion exchanger (band 3, AE1) associated with South East Asian ovalocytosis. *Journal of Molecular Biology, 285*(3), 1289–1307.

Chang, M. H., Dipiero, J., Sonnichsen, F. D., & Romero, M. F. (2008). Entry to "Formula Tunnel" revealed by SLC4A4 human mutation and structural model. *Journal of Biological Chemistry*, *283*(26), 18402–18410.

Chang, S. H., & Low, P. S. (2003). Identification of a critical ankyrin-binding loop on the cytoplasmic domain of erythrocyte membrane band 3 by crystal structure analysis and site-directed mutagenesis. *Journal of Biological Chemistry*, *278*(9), 6879–6884.

Chen, L. M., Haddad, G. G., & Boron, W. F. (2008). Effects of chronic continuous hypoxia on the expression of SLC4A8 (NDCBE) in neonatal versus adult mouse brain. *Brain Research*, *1238*, 85–92.

Chen, L. M., Liu, Y., & Boron, W. F. (2011). Role of an extracellular loop in determining the stoichiometry of Na^+-HCO_3^- cotransporters. *Journal of Physiology*, *589*(Pt 4), 877–890.

Chernova, M. N., Jiang, L., Crest, M., Hand, M., Vandorpe, D. H., Strange, K., et al. (1997). Electrogenic sulfate/chloride exchange in *Xenopus* oocytes mediated by murine AE1 E699Q. *Journal of General Physiology*, *109*(3), 345–360.

Chernova, M. N., Stewart, A. K., Barry, P. N., Jennings, M. L., & Alper, S. L. (2008). Mouse Ae1 E699Q mediates $SO_4^{2-}{}_i$/aniono exchange with $[SO_4^{2-}]_i$-dependent reversal of wild-type pHo sensitivity. *American Journal of Physiology Cell Physiology*, *295*(2), C302–C312.

Chernova, M. N., Stewart, A. K., Jiang, L., Friedman, D. J., Kunes, Y. Z., & Alper, S. L. (2003). Structure-function relationships of AE2 regulation by Ca_i^{2+}-sensitive stimulators NH_4^+ and hypertonicity. *American Journal of Physiology. Cell Physiology*, *284*(5), C1235–C1246.

Choi, I., Aalkjaer, C., Boulpaep, E. L., & Boron, W. F. (2000). An electroneutral sodium/bicarbonate cotransporter NBCn1 and associated sodium channel. *Nature*, *405*(6786), 571–575.

Choi, I., Romero, M. F., Khandoudi, N., Bril, A., & Boron, W. F. (1999). Cloning and characterization of a human electrogenic Na^+-HCO_3^- cotransporter isoform (hhNBC). *American Journal of Physiology*, *276*(3 Pt 1), C576–C584.

Choi, I., Yang, H. S., & Boron, W. F. (2007). The electrogenicity of the sodium/bicarbonate cotransporter NBCe1 requires interactions among transmembrane segments of the transporter. *Journal of Physiology (London)*, *578*(Pt 1), 131–142.

Cooper, D. S., Lee, H. J., Yang, H. S., Kippen, J., Yun, C. C., & Choi, I. (2006). The electroneutral sodium/bicarbonate cotransporter containing an amino terminal 123-amino-acid cassette is expressed predominantly in the heart. *Journal of Biomedical Science*, *13*(4), 593–595.

Cooper, D. S., Saxena, N. C., Yang, H. S., Lee, H. J., Moring, A. G., Lee, A., et al. (2005). Molecular and functional characterization of the electroneutral Na/HCO_3 cotransporter NBCn1 in rat hippocampal neurons. *Journal of Biological Chemistry*, *280*(18), 17823–17830.

Cousin, J. L., & Motais, R. (1976). The role of carbonic anhydrase inhibitors on anion permeability into ox red blood cells. *Journal of Physiology*, *256*(1), 61–80.

Dahl, N. K., Jiang, L., Chernova, M. N., Stuart-Tilley, A. K., Shmukler, B. E., & Alper, S. L. (2003). Deficient HCO_3^- transport in an AE1 mutant with normal Cl^- transport can be rescued by carbonic anhydrase II presented on an adjacent AE1 protomer. *Journal of Biological Chemistry*, *278*(45), 44949–44958.

Damkier, H. H., Aalkjaer, C., & Praetorius, J. (2010). Na^+-dependent HCO_3^- import by the slc4a10 gene product involves Cl^- export. *Journal of Biological Chemistry*, *285*(35), 26998–27007.

Damkier, H. H., Nielsen, S., & Praetorius, J. (2007). Molecular expression of SLC4-derived Na^+-dependent anion transporters in selected human tissues. *American Journal of Physiology - Regulatory, Integrative and Comparative Physiology*, *293*(5), R2136–R2146.

DeFelice, L. J., & Blakely, R. D. (1996). Pore models for transporters? *Biophysical Journal*, *70*(2), 579–580.

Dinour, D., Chang, M. H., Satoh, J., Smith, B. L., Angle, N., Knecht, A., et al. (2004). A novel missense mutation in the sodium bicarbonate cotransporter (NBCe1/SLC4A4) causes proximal tubular acidosis and glaucoma through ion transport defects. *Journal of Biological Chemistry*, *279*(50), 52238–52246.

Dolder, M., Walz, T., Hefti, A., & Engel, A. (1993). Human erythrocyte band 3. Solubilization and reconstitution into two-dimensional crystals. *Journal of Molecular Biology, 231*(1), 119–132.

Ducoudret, O., Diakov, A., Muller-Berger, S., Romero, M. F., & Frömter, E. (2001). The renal Na-HCO_3^-cotransporter expressed in *Xenopus laevis* oocytes: inhibition by tenidap and benzamil and effect of temperature on transport rate and stoichiometry. *Pflügers Archiv European Journal of Physiology, 442*(5), 709–717.

Dutzler, R., Campbell, E. B., Cadene, M., Chait, B.T., & MacKinnon, R. (2002). X-ray structure of a ClC chloride channel at 3.0 A° reveals the molecular basis of anion selectivity. *Nature, 415*(6869), 287–294.

Ferru, E., Giger, K., Pantaleo, A., Campanella, E., Grey, J., Ritchie, K., et al. (2011). Regulation of membrane-cytoskeletal interactions by tyrosine phosphorylation of erythrocyte band 3. *Blood, 117*(22), 5998–6006.

Fievet, B., Gabillat, N., Borgese, F., & Motais, R. (1995). Expression of band 3 anion exchanger induces chloride current and taurine transport: structure-function analysis. *EMBO Journal, 14*(21), 5158–5169.

Freedman, J. C., & Novak, T. S. (1997). Electrodiffusion, barrier, and gating analysis of DIDS-insensitive chloride conductance in human red blood cells treated with valinomycin or gramicidin. *Journal of General Physiology, 109*(2), 201–216.

Frohlich, O., & Gunn, R. B. (1986). Erythrocyte anion transport: the kinetics of a single-site obligatory system. *Biochimica et Biophysica Acta, 864*(2), 169–194.

Fujinaga, J., Loiselle, F. B., & Casey, J. R. (2003). Transport activity of chimaeric AE2-AE3 chloride/bicarbonate anion exchange proteins. *Biochemical Journal, 371*(Pt 3), 687–696.

Fujinaga, J., Tang, X. B., & Casey, J. R. (1999). Topology of the membrane domain of human erythrocyte anion exchange protein, AE1. *Journal of Biological Chemistry, 274*(10), 6626–6633.

Giffard, R. G., Lee, Y. S., Ouyang, Y. B., Murphy, S. L., & Monyer, H. (2003). Two variants of the rat brain sodium-driven chloride bicarbonate exchanger (NCBE): developmental expression and addition of a PDZ motif. *European Journal of Neuroscience, 18*(11), 2935–2945.

Gill, H. S., & Boron, W. F. (2006a). Expression and purification of the cytoplasmic N-terminal domain of the Na/HCO_3 cotransporter NBCe1-A: structural insights from a generalized approach. *Protein Expression and Purification, 49*(2), 228–234.

Gill, H. S., & Boron, W. F. (2006b). Preliminary X-ray diffraction analysis of the cytoplasmic N-terminal domain of the Na/HCO3 cotransporter NBCe1-A. *Acta Crystallographica Section F: Structural Biology and Crystallization Communications, 62*(Pt 6), 534–537.

Grichtchenko, & Boron. (2002). Evidence for CO_3^- transport by NBCe1, based on surface-pH measurements in voltage-clamped *Xenopus* oocytes co-expressing NBCe1 and CAIV: evidence for CO_3^- transport. *FASEB Journal, 16,* A795.

Grichtchenko, I. I., Choi, I., Zhong, X., Bray-Ward, P., Russell, J. M., & Boron, W. F. (2001). Cloning, characterization, and chromosomal mapping of a human electroneutral Na^+-driven Cl-HCO_3 exchanger. *Journal of Biological Chemistry, 276*(11), 8358–8363.

Grichtchenko, I. I., Romero, M. F., & Boron, W. F. (2000). Extracellular HCO_3^- dependence of electrogenic Na/HCO_3 cotransporters cloned from salamander and rat kidney. *Journal of General Physiology, 115*(5), 533–545.

Gross, E., Hawkins, K, Abuladze, N, Pushkin, A, Cotton, C.U., Hopfer, U & Kurtz, I. (2001). The stoichiometry of the electrogenic sodium bicarbonate cotransporter NBC1 is cell-type dependent.. *Journal of Physiology, 531*(Pt 3), 597–603.

Gross, E, Fedotoff, O., Pushkin, A., Abuladze, N., Newman, D., & Kurtz, I. (2003). Phosphorylation-induced modulation of pNBC1 function: distinct roles for the amino- and carboxy-termini. *Journal of Physiology, 549*(Pt 3), 673–682.

Hille, B. (2001). *Ion channels of Exitable Membranes (chapter 11: Elementary properties of pores).* Massachusetts: Sinauer Associates, Inc.

Hirai, T., Hamasaki, N., Yamaguchi, T., & Ikeda, Y. (2011). Topology models of anion exchanger 1 that incorporate the anti-parallel V-shaped motifs found in the EM structure. *Biochemistry and Cell Biology*, *89*(2), 148–156.

Humphreys, B. D., Jiang, L., Chernova, M. N., & Alper, S. L. (1994). Functional characterization and regulation by pH of murine AE2 anion exchanger expressed in *Xenopus* oocytes. *American Journal of Physiology*, *267*(5 Pt 1), C1295–C1307.

Ishiguro, H., Steward, M. C., Lindsay, A. R.G., & Case, R. M. (1996). Accumulation of intracellular HCO_3^- by Na^+-HCO_3^- cotransport in interlobular ducts from the guinea-pig pancreas. *Journal of Physiology*, *495*(Pt 1), 169–178.

Jauch, P., & Lauger, P. (1986). Electrogenic properties of the sodium-alanine cotransporter in pancreatic acinar cells: II. Comparison with transport models. *Journal of Membrane Biology*, *94*(2), 117–127.

Jennings, M. L. (1995). Rapid electrogenic sulfate-chloride exchange mediated by chemically modified band 3 in human erythrocytes. *Journal of General Physiology*, *105*(1), 21–47.

Jennings, M. L., & Al Rhaiyel, S. (1988). Modification of a Carboxyl Group that appears to cross the permeability barrier inthe red blood cell anion transporter. *Journal of General Physiology*, *92*(2), 161–178.

Jennings, M. L., Whitlock, J., & Shinde, A. (1998). Pre-steady state transport by erythrocyte band 3 protein: uphill countertransport induced by the impermeant inhibitor H_2DIDS. *Biochemistry and Cell Biology*, *76*(5), 807–813.

Kao, L., Sassani, P., Azimov, R., Pushkin, A., Abuladze, N., Peti-Peterdi, J., et al. (2008). Oligomeric structure and minimal functional unit of the electrogenic sodium bicarbonate cotransporter NBCe1-A. *Journal of Biological Chemistry*, *283*(39), 26782–26794.

Kim, E., & Sheng, M. (2004). PDZ domain proteins of synapses. *Nature Reviews Neuroscience*, *5*(10), 771–781.

Knauf, P.A., Law, F.Y., & Hahn, K. (1995). An oxonol dye is the most potent known inhibitor of band 3-mediated anion exchange. *American Journal of Physiology*, *269*(4 Pt 1), C1073–C1077.

Knauf, P. A., Law, F. Y., Leung, T. W., Gehret, A. U., & Perez, M. L. (2002). Substrate-dependent reversal of anion transport site orientation in the human red blood cell anion-exchange protein, AE1. *Proceedings of the National Academy of Sciences of the United States of America*, *99*(16), 10861–10864.

Knauf, P. A., & Pal, P. (2003). *Band 3-mediated transport in red cell membrane transport in Health and Disease*. Berlin: Springer.

Ko, S. B., Luo, X., Hager, H., Rojek, A., Choi, J. Y., Licht, C., et al. (2002). AE4 is a DIDS-sensitive Cl^-/HCO_3^- exchanger in the basolateral membrane of the renal CCD and the SMG duct. *American Journal of Physiology Cell Physiology*, *283*(4), C1206–C1218.

Kopito, R. R., Lee, B. S., Simmons, D. M., Lindsey, A. E., Morgans, C. W., & Schneider, K. (1989). Regulation of intracellular pH by a neuronal homolog of the erythrocyte anion exchanger. *Cell*, *59*(5), 927–937.

Kopito, R. R., & Lodish, H. F. (1985). Primary structure and transmembrane orientation of the murine anion exchange protein. *Nature*, *316*(6025), 234–238.

Ko, S. B., Shcheynikov, N., Choi, J. Y., Luo, X., Ishibashi, K., Thomas, P. J., et al. (2002). A molecular mechanism for aberrant CFTR-dependent HCO_3^- transport in cystic fibrosis. *EMBO Journal*, *21*(21), 5662–5672.

Kristensen, J. M., Kristensen, M., & Juel, C. (2004). Expression of Na^+/HCO_3^- co-transporter proteins (NBCs) in rat and human skeletal muscle. *Acta Physiologica Scandinavica*, *182*(1), 69–76.

Lee, S. K., Boron, W. F., & Parker, M. D. (2012). Relief of autoinhibition of the electrogenic Na-HCO_3 cotransporter NBCe1-B: role of IRBIT *vs.* amino-terminal truncation. *AAmerican Journal of Physiology Cell Physiology*, *302*(3), C518–C526.

Lee, S. K., Grichtchenko, I. I., & Boron, W. F. (2011). Distinguishing HCO_3^- from CO_3^{2-} transport by NBCe1-A. *FASEB Journal*, *25*, 656.9.

Lee, Y. S., Ouyang, Y. B., & Giffard, R. G. (2006). Regulation of the rat brain Na+ -driven Cl^-/HCO_3^- exchanger involves protein kinase A and a multiprotein signaling complex. *FEBS Letters*, *580*(20), 4865–4871.

Liu, X., Williams, J. B., Sumpter, B. R., & Bevensee, M. O. (2007). Inhibition of the Na/bicarbonate cotransporter NBCe1-A by diBAC oxonol dyes relative to niflumic acid and a stilbene. *Journal of Membrane Biology*, *215*(2–3), 195–204.

Lu, J., & Boron, W. F. (2007). Reversible and irreversible interactions of DIDS with the human electrogenic Na/HCO_3 cotransporter NBCe1-A: role of lysines in the KKMIK motif of TM5. *American Journal of Physiology Cell Physiology*, *292*(5), C1787–C1798.

Lu, J., Daly, C. M., Parker, M. D., Gill, H. S., Piermarini, P. M., Pelletier, M. F., et al. (2006). Effect of human carbonic anhydrase II on the activity of the human electrogenic Na/HCO_3 cotransporter NBCe1-A in Xenopus oocytes. *Journal of Biological Chemistry*, *281*(28), 19241–19250.

McAlear, S. D., & Bevensee, M. O. (2006). A cysteine-scanning mutagenesis study of transmembrane domain 8 of the electrogenic Sodium/Bicarbonate cotransporter NBCe1. *Journal of Biological Chemistry*, *281*(43), 32417–32427.

McAlear, S. D., Liu, X., Williams, J. B., McNicholas-Bevensee, C. M., & Bevensee, M. O. (2006). Electrogenic Na/HCO_3 cotransporter (NBCe1) variants expressed in Xenopus oocytes: functional comparison and roles of the amino and carboxy termini. *Journal of General Physiology*, *127*(6), 639–658.

McMurtrie, H. L., Cleary, H. J., Alvarez, B.V., Loiselle, F. B., Sterling, D., Morgan, P. E., et al. (2004). The bicarbonate transport metabolon. *Journal of Enzyme Inhibition and Medicinal Chemistry*, *19*(3), 231–236.

Milanick, M. A., & Gunn, R. B. (1982). Proton-sulfate co-transport: mechanism of H^+ and sulfate addition to the chloride transporter of human red blood cells. *Journal of General Physiology*, *79*(1), 87–113.

Millar, I. D., & Brown, P. D. (2008). NBCe2 exhibits a 3 HCO_3^-:1 Na^+ stoichiometry in mouse choroid plexus epithelial cells. *Biochemical and Biophysical Research*, *373*(4), 550–554.

Muller-Berger, S., Ducoudret, O., Diakov, A., & Frömter, E. (2001). The renal Na-HCO_3 cotransporter expressed in *Xenopus laevis* oocytes: change in stoichiometry in response to elevation of cytosolic Ca^{2+} concentration. *Pflügers Archiv European Journal of Physiology*, *442*(5), 718–728.

Muller-Berger, S., Karbach, D., Kang, D., Aranibar, N., Wood, P. G., Ruterjans, H., et al. (1995). Roles of histidine 752 and glutamate 699 in the pH dependence of mouse band 3 protein-mediated anion transport. *Biochemistry*, *34*(29), 9325–9332.

Muller-Berger, S., Karbach, D., Konig, J., Lepke, S., Wood, P. G., Appelhans, H., et al. (1995). Inhibition of mouse erythroid band 3-mediated chloride transport by site-directed mutagenesis of histidine residues and its reversal by second site mutation of Lys 558, the locus of covalent H2DIDS binding. *Biochemistry*, *34*(29), 9315–9324.

Parker, M. D., Bouyer, P., Daly, C. M., & Boron, W. F. (2008). Cloning and characterization of novel human SLC4A8 gene products encoding Na^+-driven Cl^-/HCO_3^- exchanger variants NDCBE-A, -C, and -D. *Physiological Genomics*, *34*(3), 265–276.

Parker, M. D., Musa-Aziz, R., Rojas, j. d., Choi, I., Daly, C. M., & Boron, W. F. (2008). Characterization of human SLC4A10 as an electroneutral Na/HCO_3 cotransporter (NBCn2) with Cl^- self-exchange activity. *Journal of Biological Chemistry*, *283*(19), 12777–12788.

Park, M., Ko, S. B.H., Davidson, N., Muallem, G., Thomas, P.J., Pushkin, A., et al. (2002). The cystic fibrosis transmembrane conductance regulator interacts with and regulates the activity of the HCO salvage transporter human Na^+-HCO_3 cotransporter isoform 3. *Journal of Biological Chemistry*, *277*(52), 50503–50509.

Park, M., Li, Q., Shcheynikov, N., Zeng, W., & Muallem, S. (2004). NaBC1 is a ubiquitous electrogenic Na^+ -coupled borate transporter essential for cellular boron homeostasis and cell growth and proliferation. *Molecular Cell, 16*(3), 331–341.

Park, H. J., Rajbhandari, I., Yang, H. S., Lee, S., Cucoranu, D., Cooper, D. S., et al. (2010). Neuronal expression of sodium/bicarbonate cotransporter NBCn1 (SLC4A7) and its response to chronic metabolic acidosis. *American Journal of Physiology Cell Physiology, 298*(5), C1018–C1028.

Pushkin, A., Abuladze, N., Gross, E., Newman, D., Tatishchev, S., Lee, I., et al. (2004). Molecular mechanism of kNBC1-carbonic anhydrase II interaction in proximal tubule cells. *Journal of Physiology, 559*(Pt 1), 55–65.

Pushkin, A., Abuladze, N., Lee, I., Newman, D., Hwang, J., & Kurtz, I. (1999). Cloning, tissue distribution, genomic organization, and functional characterization of NBC3, a new member of the sodium bicarbonate cotransporter family. *Journal of Biological Chemistry, 274*(23), 16569–16575.

Pushkin, A., Abuladze, N., Newman, D., Lee, I., Xu, G., & Kurtz, I. (2000). Cloning, characterization and chromosomal assignment of NBC4, a new member of the sodium bicarbonate cotransporter family. *Biochimica et Biophysica Acta, 1493*(1–2), 215–218.

Pushkin, A., Abuladze, N., Newman, D., Muronets, V., Sassani, P., Tatishchev, S., et al. (2003). The COOH termini of NBC3 and the 56-kDa H^+-ATPase subunit are PDZ motifs involved in their interaction. *American Journal of Physiology Cell Physiology, 284*(3), C667–C673.

Reiners, J., van Wijk, E., Marker, T., Zimmermann, U., Jurgens, K., te, B. H., et al. (2005). Scaffold protein harmonin (USH1C) provides molecular links between Usher syndrome type 1 and type 2. *Human Molecular Genetics, 14*(24), 3933–3943.

Reithmeier, R. A., Chan, S. L., & Popov, M. (1996). Structure of the erythrocyte Band 3 anion exchanger. In W. N. Konings, H. R. Kaback, & J. S. Lolkema (Eds.), *Handbook of biological physics* (pp. 281–309). , Amsterdam: Elsevier Science.

Romero, M. F., Hediger, M. A., Boulpaep, E. L., & Boron, W. F. (1997). Expression cloning and characterization of a renal electrogenic Na^+/HCO_3^- cotransporter. *Nature, 387*(6631), 409–413.

Romero, M. F., Henry, D., Nelson, S., Harte, P. J., Dillon, A. K., & Sciortino, C. M. (2000). Cloning and characterization of a Na^+-driven anion exchanger (NDAE1). A new bicarbonate transporter. *Journal of Biological Chemistry, 275*(32), 24552–24559.

Sassani, P., Pushkin, A., Gross, E., Gomer, A., Abuladze, N., Dukkipati, R., et al. (2002). Functional characterization of NBC4: a new electrogenic sodium- bicarbonate cotransporter. *American Journal of Physiology Cell Physiology, 282*(2), C408–C416.

Schwiening, C. J., & Boron, W. F. (1994). Regulation of intracellular pH in pyramidal neurons from the rat hippocampus by Na^+-dependent Cl^--HCO_3^- exchange. *Journal of Physiology, 475*(1), 59–67.

Sciortino, C. M., & Romero, M. F. (1999). Cation and voltage dependence of rat kidney electrogenic Na^+-HCO_3^- cotransporter, rkNBC, expressed in oocytes. *American Journal of Physiology, 277*(4 Pt 2), F611–F623.

Sekler, I., Lo, R. S., & Kopito, R. R. (1995). A conserved glutamate is responsible for ion selectivity and pH dependence of the mammalian anion exchangers AE1 and AE2. *Journal of Biological Chemistry, 270*(48), 28751–28758.

Shcheynikov, N., Yang, D., Wang, Y., Zeng, W., Karniski, L. P., So, I., et al. (2008). The Slc26a4 transporter functions as an electroneutral $Cl^-/I^-/HCO_3^-$ exchanger: role of Slc26a4 and Slc26a6 in I^- and HCO_3^- secretion and in regulation of CFTR in the parotid duct. *Journal of Physiology, 586*(16), 3813–3824.

Shirakabe, K., Priori, G., Yamada, H., Ando, H., Horita, S., Fujita, T., et al. (2006). IRBIT, an inositol 1,4,5-trisphosphate receptor-binding protein, specifically binds to and activates pancreas-type Na^+/HCO_3^- cotransporter 1 (pNBC1). *Proceedings of the National Academy of Sciences of the United States of America, 103*(25), 9542–9547.

Sindic, A., Chang, M. H., Mount, D. B., & Romero, M. F. (2007). Renal physiology of SLC26 anion exchangers. *Current Opinion in Nephrology and Hypertension, 16*(5), 484–490.
Sinning, A., Liebmann, L., Kougioumtzes, A., Westermann, M., Bruehl, C., & Hubner, C. A. (2011). Synaptic glutamate release is modulated by the Na^+-driven Cl^-/HCO_3^- exchanger Slc4a8. *Journal of Neuroscience, 31*(20), 7300–7311.
Soleimani, M., Grassl, S. M., & Aronson, P. S. (1987). Stoichiometry of Na^+-HCO_3^- cotransport in basolateral membrane vesicles isolated from rabbit renal cortex. *Journal of Clinical Investigation, 79*, 1276–1280.
Sterling, D., Reithmeier, R. A., & Casey, J. R. (2001). A transport metabolon: functional interaction of carbonic anhydrase II and chloride/bicarbonate exchangers. *Journal of Biological Chemistry, 276*(51), 47886–47894.
Stewart, A. K., Chernova, M. N., Shmukler, B. E., Wilhelm, S., & Alper, S. L. (2002). Regulation of AE2-mediated Cl^- transport by intracellular or by extracellular pH requires highly conserved amino acid residues of the AE2 NH_2-terminal cytoplasmic domain. *Journal of General Physiology, 120*(5), 707–722.
Stewart, A. K., Kurschat, C. E., Vaughan-Jones, R. D., Shmukler, B. E., & Alper, S. L. (2007). Acute regulation of mouse AE2 anion exchanger requires isoform-specific amino acid residues from most of the transmembrane domain. *Journal of Physiology, 584*(Pt 1), 59–73.
Tang, X. B., Kovacs, M., Sterling, D., & Casey, J. R. (1999). Identification of residues lining the translocation pore of human AE1, plasma membrane anion exchange protein. *Journal of Biological Chemistry, 274*(6), 3557–3564.
Tombola, F., Del Giudice, G., Papini, E., & Zoratti, M. (2000). Blockers of VacA provide insights into the structure of the pore. *Biophysical Journal, 79*(2), 863–873.
Torres, G. E., & Amara, S. G. (2007). Glutamate and monoamine transporters: new visions of form and function. *Current Opinion In Neurobiology, 17*(3), 304–312.
Tsuganezawa, H., Kobayashi, K., Iyori, M., Araki, T., Koizumi, A., Watanabe, S. I., et al. (2001). A new member of the HCO_3^- transporter superfamily is an apical anion exchanger of beta-intercalated cells in the kidney. *Journal of Biological Chemistry, 276*(16), 8180–8189.
van Adelsberg, J. S., Edwards, J. C., & Al Awqati, Q. (1993). The apical Cl/HCO_3 exchanger of ß intercalated cells. *Journal of Biological Chemistry, 268*(15), 11283–11289.
Vince, J. W., & Reithmeier, R. A. (1998). Carbonic anhydrase II binds to the carboxyl terminus of human band 3, the erythrocyte Cl^-/HCO_3^- exchanger. *Journal of Biological Chemistry, 273*(43), 28430–28437.
Vince, J. W., & Reithmeier, R. A. (2000). Identification of the carbonic anhydrase II binding site in the Cl^-/HCO_3^- anion exchanger AE1. *Biochemistry, 39*(18), 5527–5533.
Virkki, L. V., Wilson, D. A., Vaughan-Jones, R. D., & Boron, W. F. (2002). Functional characterization of human NBC4 as an electrogenic Na^+-HCO_3^- cotransporter (NBCe2). *American Journal of Physiology Cell Physiology, 282*(6), C1278–C1289.
Wang, D. N., Sarabia, V. E., Reithmeier, R. A., & Kuhlbrandt, W. (1994). Three-dimensional map of the dimeric membrane domain of the human erythrocyte anion exchanger, Band 3. *EMBO Journal, 13*(14), 3230–3235.
Wang, C. Z., Yano, H., Nagashima, K., & Seino, S. (2000). The Na^+-driven Cl^-/HCO_3^- exchanger: cloning, tissue distribution, and functional characterization. *Journal of Biological Chemistry, 275*(45), 35486–35490.
Wieth, J. O., Andersen, O. S., Brahm, J., Bjerrum, P. J., & Borders, C. L., Jr. (1982). Chloride–bicarbonate exchange in red blood cells: physiology of transport and chemical modification of binding sites. *Philosophical Transactions of the Royal Society of London B Biological Sciences, 299*(1097), 383–399.
Yamada, H., Horita, S., Suzuki, M., Fujita, T., & Seki, G. (2011). Functional role of a putative carbonic anhydrase II-binding domain in the electrogenic Na^+-HCO_3^- cotransporter NBCe1 expressed in Xenopus oocytes. *Channels (Austin), 5*(2), 106–109.

Yamaguchi, T., Ikeda, Y., Abe, Y., Kuma, H., Kang, D., Hamasaki, N., et al. (2010). Structure of the membrane domain of human erythrocyte anion exchanger 1 revealed by electron crystallography. *Journal of Molecular Biology, 397*(1), 179–189.

Yang, D., Shcheynikov, N., Zeng, W., Ohana, E., So, I., Ando, H., et al. (2009). IRBIT coordinates epithelial fluid and HCO_3^- secretion by stimulating the transporters pNBC1 and CFTR in the murine pancreatic duct. *Journal of Clinical Investigation, 119*(1), 193–202.

Yang, H. S., Kim, E., Lee, S., Park, H. J., Cooper, D. S., Rajbhandari, I., et al. (2009). Mutation of aspartate 555 of the sodium/bicarbonate transporter SLC4A4/NBCe1 induces chloride transport. *Journal of Biological Chemistry, 284*(23), 15970–15979.

Zhang, D., Kiyatkin, A., Bolin, J. T., & Low, P. S. (2000). Crystallographic structure and functional interpretation of the cytoplasmic domain of erythrocyte membrane band 3. *Blood, 96*(6), 2925–2933.

Zhao, H., Star, R. A., & Muallem, S. (1994). Membrane localization of H^+ and HCO_3^- transporters in the rat pancreatic duct. *Journal of General Physiology, 104*(1), 57–85.

Zhu, Q., Azimov, R., Kao, L., Newman, D., Liu, W., Abuladze, N., et al. (2009). NBCe1-A transmembrane segment 1 lines the ion translocation pathway. *Journal of Biological Chemistry, 284*(13), 8918–8929.

Zhu, Q., & Casey, J. R. (2004). The substrate anion selectivity filter in the human erythrocyte Cl^-/HCO_3^- exchange protein, AE1. *Journal of Biological Chemistry, 279*(22), 23565–23573.

Zhu, Q., Kao, L., Azimov, R., Abuladze, N., Newman, D., Pushkin, A., et al. (2010). Structural and functional characterization of the C-terminal transmembrane region of NBCe1-A. *Journal of Biological Chemistry, 285*(48), 37178–37187.

Zhu, Q., Lee, D. W., & Casey, J. R. (2003). Novel topology in C-terminal region of the human plasma membrane anion exchanger, AE1. *Journal of Biological Chemistry, 278*(5), 3112–3120.

CHAPTER FOUR

The SLC10 Carrier Family: Transport Functions and Molecular Structure

Barbara Döring*, Thomas Lütteke†, Joachim Geyer*,[1], Ernst Petzinger*

*SLC10 family research group, Institute of Pharmacology and Toxicology, Justus Liebig University Giessen, Biomedical Research Center (BFS), Giessen, Germany
†Institute of Veterinary Physiology and Biochemistry, Justus Liebig University Giessen, Giessen, Germany
[1]Corresponding author: E-mail: joachim.m.geyer@vetmed.uni-giessen.de

Contents

Current Topics in Membranes, Volume 70
ISSN 1063-5823
http://dx.doi.org/10.1016/B978-0-12-394316-3.00004-1

Abstract

The SLC10 family represents seven genes containing 1–12 exons that encode proteins in humans with sequence lengths of 348–477 amino acids. Although termed solute carriers (SLCs), only three out of seven (i.e. SLC10A1, SLC10A2, and SLC10A6) show sodium-dependent uptake of organic substrates across the cell membrane. These include the uptake of bile salts, sulfated steroids, sulfated thyroidal hormones, and certain statin drugs by SLC10A1 (Na^+-taurocholate cotransporting polypeptide (NTCP)), the uptake of bile salts by SLC10A2 (apical sodium-dependent bile acid transporter (ASBT)), and uptake of sulfated steroids and sulfated taurolithocholate by SLC10A6 (sodium-dependent organic anion transporter (SOAT)). The other members of the family are orphan carriers not all localized in the cell membrane. The name "bile acid transporter family" arose because the first two SLC10 members (NTCP and ASBT) are carriers for bile salts that establish their enterohepatic circulation. In recent years, information has been obtained on their 2D and 3D membrane topology, structure–transport relationships, and on the ligand and sodium-binding sites. For SLC10A2, the putative 3D morphology was deduced from the crystal structure of a bacterial SLC10A2 analog, $ASBT_{NM}$. This information was used in this chapter to calculate the putative 3D structure of NTCP. This review provides first an introduction to recent knowledge about bile acid synthesis and newly found bile acid hormonal functions, and then describes step-by-step each individual member of the family in terms of expression, localization, substrate pattern, as well as protein topology with emphasis on the three functional SLC10 carrier members.

1. INTRODUCTION

The solute carrier family 10 (SLC10) is also referred to as the family of Na^+-dependent bile acid transporters because the first two members, i.e. SLC10A1 and SLC10A2, are typical bile salt carriers in the liver and gut, respectively. The name "bile acid" transporters is misleading because at physiological pH in blood and in the lumen of the ileum, bile acids are in the anionic form and actually are bile salts. Regarding transport, we will use the term bile salt (BS). Furthermore, the family name needs to be reconsidered because only these two members of the SLC10 family transport BSs, whereas the others do not (see below).

1.1. Synthesis of Bile Acids

In humans, primary bile acids (BAs) such as cholic acid (CA) and chenodeoxycholic acid (CDCA) are formed in the liver from cholesterol by means of the activity of a cascade of at least 16 enzymes that catalyze 17 reactions to produce a conjugated BA. The synthesis takes place under physiological conditions in pericentral hepatocytes (Baumgartner, Baier, Birke, & Farthmann, 2003; Twisk et al., 1995). Enzymes for BA synthesis are located in the endoplasmic reticulum, cytosol, mitochondria, and peroxisomes (Russell, 2003; Russell & Setchell, 1992), through which BA precursors shuttle. The majority of enzymes belong to the family of cytochrome P450 (CYP) monooxygenases that perform hydroxylation reactions. Primary BAs are either di- or trihydroxylated at the sterol moiety with each hydroxyl group in the alpha position. In humans, normally only the following three carbon atoms in the steroid nucleus are hydroxylated: C_3, C_7, and/or C_{12}. Altered hydroxylation patterns occur in other species, during development, and in states of disease (Hofmann & Hagey, 2008). Two pathways, the neutral pathway and the acidic pathway, exist in the liver for their synthesis (Russell, 2009). This is depicted in Fig. 1.

Carbon$_{24}$ (C_{24}) primary BAs (Fig. 2) are extensively conjugated by the amidation reactions with amino acids (aa) glycine or taurine by the conjugating enzyme bile acid-CoA amino acid N-acyltransferase (BAAT). In humans, glycine conjugation is prevailing, whereas in rats, primary BAs are conjugated with taurine. Substrates of BAAT are coenzyme A-activated primary BAs that are activated by means of the enzyme bile acid-CoA ligase (BACL, syn. SLC27A5). Using these enzymes, the liver completely reconjugates BAs that are returned from the gut after bacterial deamidation. Genetic mutations of BAAT and BACL with defective BA amidation have been detected (Carlton et al., 2003). Such patients only synthesize nonamidated BAs in the liver that cannot be transported by the liver canalicular bile salt export pump (BSEP) (ABCB11) into bile (Noe, Stieger, & Meier, 2002). These patients suffer from malabsorption of fat and fat-soluble vitamins from the gut, show coagulopathies without jaundice, and in several cases, develop neonatal cholestasis with growth failure (Chong et al., 2012; Heubi, Setchell, & Bove, 2007). Several other diseases are connected with

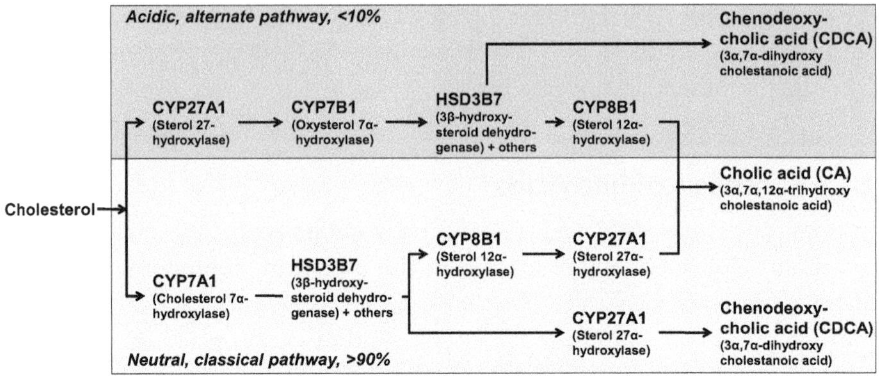

Figure 1 ***Acidic and neutral bile acid synthesis pathway.*** In the prevailing neutral pathway, the rate-limiting step of synthesis is the α-hydroxylation of cholesterol by CYP7A1 in the endoplasmic reticulum at position C_7. This reaction is followed by the isomerization of the original 3β-hydroxy group in cholesterol to the 3α-hydroxy group in primary bile acids. The enzymatic step of this reaction is complex and involves enzymes HSD3B7, AKR1D1, and AKR1C4. After the ring hydroxylations at C_3 and C_7 or at C_3, C_7, and C_{12} are completed, the C_{27} side chain of cholesterol is shortened by β-oxidation to yield the C_{24} primary bile acids cholic acid (CA) and chenodeoxycholic acid (CDCA), respectively. In the acidic pathway which accounts for approximately 10% of human bile acid synthesis (Duane & Javitt, 1999), the side chain at position C_{27} is first hydroxylated by mitochondrial CYP27A1 to yield 27-hydroxycholesterol, which forms the respective cholestenoic acid that contains an acidic C_{27} carboxy end group. This reaction is immediately followed by hydroxylation on C_7 by CYP7B1 to yield 3β,7α-dihydroxy-5-cholestenoic acid. As in the neutral pathway, the 3β-hydroxy group is isomerized to the corresponding 3α-hydroxy group. By side-chain modification, the acidic pathway yields as end product CDCA. Disorders of the bile acid synthesis pathways should manifest in jaundice or steatorrhea, in both or even in neurologic disorders (Clayton, 2011; Heubi et al., 2007; Sundaram, Bove, Lovell, & Sokol, 2008).

altered bile acid synthesis because of enzyme defects in their synthesis cascade (for review, see Heubi et al., 2007).

1.2. Functions of Bile Acids

1.2.1. Bile Acids in Nutrition and Liver Function

BAs are known for their essential activities for nutrition in general, and are particularly important for absorption of fat and lipophilic vitamins in the small intestine. (Forsgren, 1969; Maldonado-Valderrama, Wilde, Macierzanka, & Mackie, 2011; Schmidt et al., 2010; Weber, 1983). BAs are unavoidable as major driving forces for bile acid-dependent bile formation by liver parenchymal cells (Klaassen & Watkins, 1984; Strange, 1984). At the canalicular cell membrane of hepatocytes, each secreted molecule of BA is accompanied

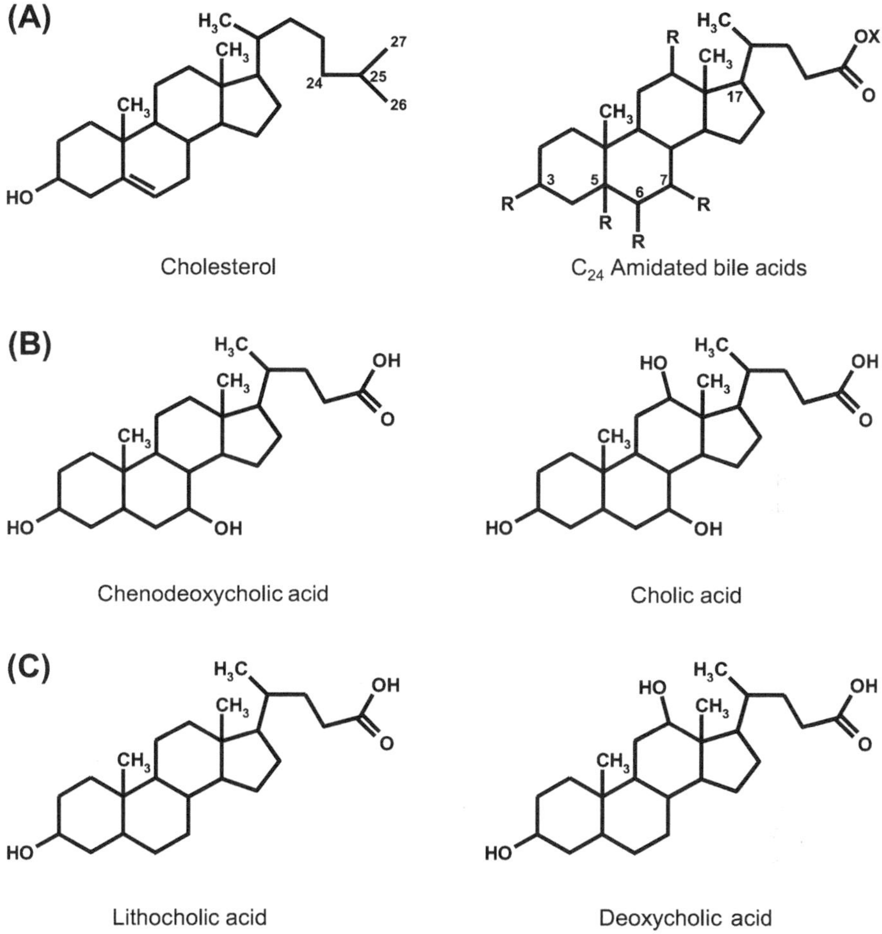

Figure 2 ***Molecular structure of cholesterol and bile acids.*** (A) Structure of cholesterol and general scheme of a C_{24} amidated bile acid. R = hydroxy group, X = hydrogen or, in amidated bile acids, glycine or taurine. (B) Primary bile acids originating from human liver. (C) Secondary bile acids originating from bacterial degradation in human gut.

by the secretion of a fluid volume that quantitatively accounts for about half of the daily formed canalicular bile volume in humans, i.e. approximately 200 ml of 423 ml total canalicular bile volume. That is about one-third of the 500–600 ml of 24 h total bile volume in humans (Boyer, 1980; Boyer & Bloomer, 1974). BAs have been recognized to function as signaling molecules regulating BA synthesis and BS transporter expression in the liver. For example, BAs serve as endogenous ligands and cellular regulators in diverse metabolic pathways via the farnesoid (FXR) and liver X nuclear receptor (LXR) (Hylemon et al., 2009; Makishima et al., 1999; Redinger, 2003). This is also probably applicable via tumor suppressor protein p53 (Kim & Lee, 2011). This includes regulation of their own biosynthesis by repressing CYP7A1 expression (Chiang, Kimmel, Weinberger, & Stroup, 2000; Eloranta & Kullak-Ublick, 2008) and of their own biliary secretion via BSEP timely together with an increase of phospholipid secretion by multidrug resistance protein MDR3 (ABCB4) (Ananthanarayanan, Balasubramanian, Makishima, Mangelsdorf, & Suchy, 2001; Huang et al., 2003). BAs also regulate glucose and lipid metabolism by binding and activating FXR (Stayrook et al., 2005; Trauner, Claudel, Fickert, Moustafa, & Wagner, 2010; Wang, Chen, Moore, & Huang, 2008) and by activating the liver insulin Akt/protein kinase B signaling pathway (Hylemon et al., 2009). The effects of activated FXR are mediated in the cell nucleus of liver cells by binding to the gene encoding small heterodimer partner (SHP) that contains an FXR responsive element in its promoter region. Several other target genes regulated by BA activated FXR have been described (Gardmo, Tamburro, Modica, & Moschetta, 2011; Li & Chiang, 2012; Modica, Gadaleta, & Moschetta, 2010) and have revealed that FXR is a master regulator of BA-mediated cellular effects.

1.2.2. Bile Acids as Hormones

Numerous other functions of BAs, some resembling hormones, have been observed in the cardiovascular system (Khurana, Raufman, & Pallone, 2011), on blood glucose and energy expenditure (Chen, Lou, Meng, & Huang, 2011; Keitel & Häussinger, 2012; Thomas et al., 2009), and also on gallbladder functions (Keitel et al., 2009). These effects are mediated by G-protein-coupled cell membrane receptors (GPCR) such as the bile acid receptor M-Bar, syn. GB37, syn. TGR5, syn. Gpbar-1 (Fiorucci, Mencarelli, Palladino, & Cipriani, 2009; Li et al., 2011). In human enterocytes, BAs stimulate the synthesis of the fibroblast growth factor (FGF)-19 (corresponding to FGF-15 of mouse), a peptide hormone that is delivered to the hepatocyte via portal blood. FGF-15/19, upon binding to its liver receptor FGFR4, represses the CYP7A1 gene by the JNK1/2 signaling pathway (Hylemon et al., 2009). Via TGR5, taurochenodeoxycholic acid (TCDCA) promotes monocyte differentiation and induces a dendritic cell phenotype with suppressed IL-12 secretion in response to LPS or LPS–interferon-γ stimulation (Ichikawa et al., 2012). This effect might offer a new therapeutic strategy against chronic inflammatory processes, particularly in the gut (Ichikawa et al., 2012). BAs also induce hepatocyte cell polarity via a cAMP-dependent protein kinase Epac and the MEK-LKB1 pathway (Fu, Wakabayashi, Lippincott-Schwartz, & Arias, 2011).

1.2.3. Bile Acids as Therapeutics

In the first half of the twentieth century, the only therapeutic use of BAs was the administration of dehydrocholic acid (DHCA), which is an artificial 3,7,12-trioxo-cholic acid, in order to achieve choleresis in patients with liver disease (Hofmann & Hagey, 2008). This treatment is now largely abandoned. Nowadays, the BA ursodeoxycholic acid (UDCA) is approved as a drug for the treatment of primary biliary cirrhosis (Calmus & Poupon, 1991; Dilger et al., 2012; EASL Clinical Practice Guidelines, 2009; Paumgartner, 2010; Poupon, Poupon, & Balkau, 1994), although potent effects have been questioned (Gluud & Christensen, 2002). This BA originates from bear bile and is present in minor amounts in human bile. UDCA and also CDCA were previously recommended for the nonsurgical dissolution of cholesterol gallstones (Bachrach & Hofmann, 1982a, 1982b; Hofmann, 1984; Leuschner, Leuschner, Sieratzki, Kurtz, & Hübner, 1985). Today, this recommendation has changed as only a small group of patients (about 15%) benefit from UDCA treatment (Di Ciaula, Wang, Wang, Bonfrate, & Portincasa, 2010; Portincasa, Di Ciaula, Wang, Moschetta, & Wang, 2009; Venneman et al., 2006), and CDCA treatment is hepatotoxic. A modified UDCA, 24-*nor*UDCA, is capable of hypercholeresis (Hofmann et al., 2005) and thus regarded of beneficial effects in the elimination of hepatotoxic compounds during the process of intrahepatic cholestasis induced by drugs (Beuers, 2006) or toxic BAs (Stapelbroek, van Erpecum, Klomp, & Houwen, 2010; Trauner, Fickert, Halilbasic, & Moustafa, 2008). In certain cases of genetic-based intrahepatic cholestasis such as progressive familial intrahepatic cholestasis type 3 (PFIC3), but less successful in PFIC type 2, both UDCA bile acids are of therapeutic value (Stapelbroek et al., 2010). However, transporter(s) for hepatocellular uptake of either UDCA or 24-*nor*UDCA were not resolved (König, Klatt, Dilger, & Fromm, 2012).

Recent progress in modern BA perspectives has lead to the development of numerous medicinal chemistry programs for the development of lead structures for BA derivatives. These are developed for different therapeutic targets, mainly with efficacy in the treatment of metabolic disorders and certain cancers (Sharma, Long, & Gilmer, 2011). Neurosteroids, namely intermediates of BAs of the two BA synthesis pathways (C_{24} and C_{27} intermediates) (Fig. 1), have been detected in the cerebrospinal fluid with partial activity to activate LXR (Ogundare et al., 2010). Because LXR activation has been found to reduce the symptoms of Alzheimer disease (Fan, Donkin, & Wellington, 2009) and because UDCA and TUDCA appear to exhibit neuroprotection in vitro that may be relevant for Huntington's and Alzheimer disease (Sharma, Long et al., 2011), a new approach for BAs as neuroprotectants is feasible.

1.2.4. Bile Acids as Pathological Factors

Some BAs, particularly unconjugated di- and monohydroxy ones such as the secondary BAs litho- and deoxycholic acid (Fig. 2), are assumed to play a role in the genesis and/or progression of colon cancer (Bernstein et al., 2011; Chaplin, 1998; Cheng & Raufman, 2005; Hill 1991; Nagengast, Grubben, & van Munster, 1995). In addition, BAs may be involved in the development of esophageal cancer (Stamp, 2003, 2006). BAs can provoke acute biliary pancreatitis upon BA regurgitation at the *papilla vateri* (*papilla duodeni major*), the common outlet of bile and pancreatic juice into the duodenum, and may cause pancreatic disorders (Wan et al., 2012). Interestingly, the pancreas is the only nonhepatic tissue with expression of a functional BS transporter Ntcp in the rat (Kim et al., 2002). The sodium-dependent uptake of taurocholate into rat pancreatic acini causes cell death because of the generation of Ca^{2+}-associated inflammation signals (Kim et al., 2002).

1.3. From Early Transport Studies to Bile Salt Carriers

The BA pool of 2–4 g in humans circulates up to 10 times per day, and therefore requires a daily BS uptake of 30 g by human intestine enterocytes and liver parenchymal cells. This circulation is long known as the enterohepatic circulation (EHC) of bile and bile components. EHC was historically published by Schiff (1870) on dogs and with radiolabeled CA by Lindstedt (1957) on humans (reviewed by Hislop, 1970).

Early in vivo and in vitro transport studies in the precloning era on BS transport on dogs (Glasinovic, Dumont, Duval, & Erlinger, 1975) with perfused rat liver (Reichen & Paumgartner, 1976), isolated rat hepatocytes (Anwer & Hegner, 1978; Schwarz, Burr, Schwenk, Pfaff, & Greim, 1975), as well as with membrane vesicles (Inoue, Kinne, Tran, & Arias, 1982; Suchy, Bucuvalas, Goodrich, Moyer, & Blitzer, 1986), revealed a common BS transporter for hepatic uptake and discriminated between the sodium-dependent versus the sodium-independent uptake of the primary BSs taurocholate (TC), cholate (C), and chenodeoxycholate (CDC). These studies revealed K_m and V_{max} data for saturable BS transport in the period between 1979 and 1990 albeit via nonidentified transporters (reviewed by Frimmer & Ziegler, 1988; Petzinger, 1994). The sodium-dependent uptake of the BSs TC, glycocholate (GC), C, and CDC in the liver later on turned out to occur by the cloned NTCP protein (SLC10A1). The reclamation of BAs in intestine includes mainly active transport and passive nonionic diffusion (Dietschy, 1968). Active transport of conjugated BSs was located mainly in terminal ileum, whereas passive mechanisms can occur along the whole intestine. This was verified in vivo in different animal models such as rat, mouse, guinea pig, hamster, dog, chicken, and human (Dietschy, 1968; Glasser, Weiner, & Lack, 1965). Furthermore, in vitro studies using everted gut sacs and ileal brush-border membrane vesicles identified this transport as sodium-dependent secondary active uptake system (Barnard & Ghishan, 1987; Dietschy, 1968; Glasser et al., 1965; Lücke, Stange, Kinne, & Murer, 1978).

The mentioned EHC affords BS carriers in the gut and the liver. This carrier gap was closed when step-by-step liver NTCP (SLC10A1), intestinal and biliary tract ASBT (SLC10A2), liver BSEP (ABCB11), and intestinal and liver OST α/β (SLC51A1/A2) were cloned and their function as BS transporters in health and disease was described (Alrefai & Gill, 2007; Dawson, Lan, & Rao, 2009; Kosters & Karpen, 2008; Meier & Stieger, 2002). The sodium-independent BS uptake system of the precloning era was identified to belong to the family SLC21 syn. SLCO of OATP transporters (organic anion transporting polypeptides). This important family of at least 10 transporters in human (Hagenbuch & Meier, 2004) facilitates the transport of several endogenous compounds and even more of xenobiotics (Bossuyt, Müller, Hagenbuch, & Meier, 1996). This was recently reviewed comprehensively (Klaassen & Aleksunes, 2010; Koenen, Kroemer, Grube & Meyer zu Schwabedissen, 2011; König, 2011; Niemi, Pasanen, & Neuvonen, 2011) and is not further considered in this chapter. A recent summary on BS EHC is given by Hofmann, 2009 and is depicted in Fig. 3.

2. THE SLC10 FAMILY

A previous review entitled "The sodium bile salt cotransport family SLC10" (Hagenbuch & Dawson, 2004) listed, in addition to the two cloned members SLC10A1 (NTCP) and SLC10A2 (ASBT), three further putative orphan members in humans (SLC10A3, A4, and A5), which were added to the SLC10 family by EST data homologies. Our review entitled "The solute carrier family SLC10: more than a family of bile acid transporters regarding function and phylogenetic relationships" published in 2006 (Geyer, Wilke, & Petzinger, 2006) and was embedded in a general topic about "Drug transporters in pharmacokinetics" (Petzinger & Geyer, 2006). The new gene

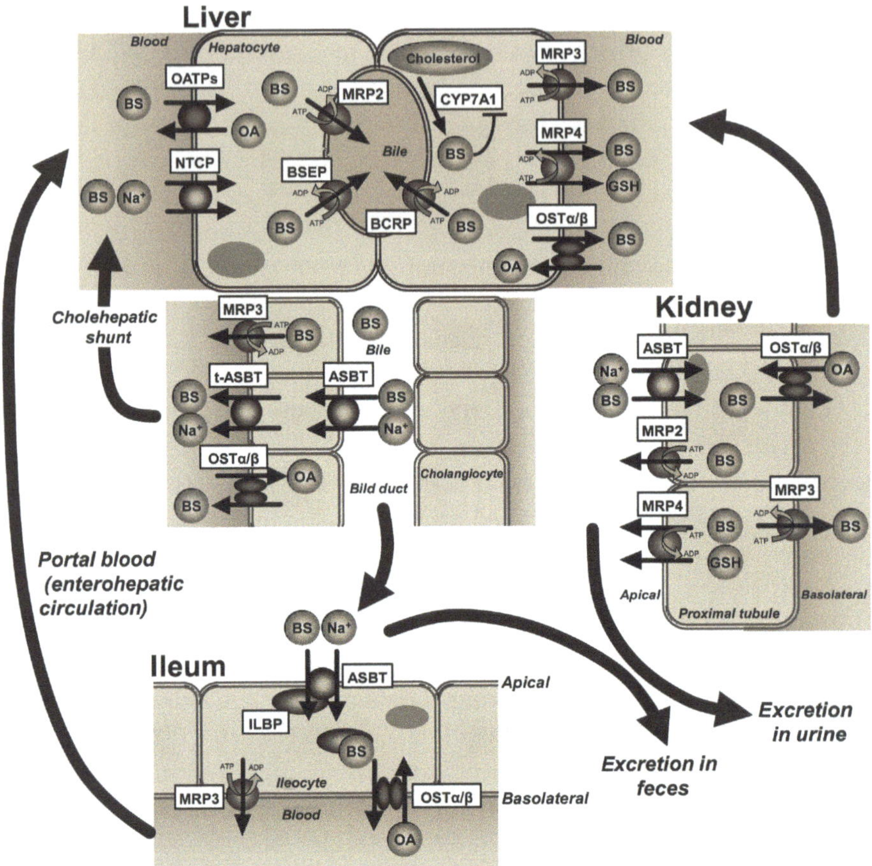

Figure 3 ***Enterohepatic circulation of bile salts and transporters for bile salts in liver, gut, and kidney.*** Freshly synthesized liver bile salts are delivered to bile via ATP-dependent canalicular transporters BSEP, syn. bile salt export pump (ABCB11) (major excretion pathway) and MRP2, multidrug resistance associated protein (ABCC2) (minor excretion pathway). Under cholestatic circumstances, significant efflux across the basolateral membrane into blood is achieved by MRP3 (ABCC3), MRP4 (ABCC4), and the dimeric half-transporters OSTα/β (SLC51A1/A2). In the bile ductules, bile salts are subjected to an intrahepatic recirculation (cholehepatic shunt) by means of ASBT (SLC10A2), truncated ASBT (t-ASBT), OSTα/β, and MRP3. Amidated bile salts delivered into gut are almost completely reabsorbed by small gut ileal enterocytes via ASBT where they bind to the ileal lipid binding protein (ILBP), shuttled to the basolateral membrane and released by MRP3 and OSTα/β into the portal blood circulation. From portal blood, reuptake of bile salts into hepatocytes occurs via NTCP at high and via SLCO family members (OATPs) at lower affinities. In kidney, glomerular filtrated bile salts are reabsorbed from urine via ASBT followed by efflux into the systemic circulation by OSTα/β and MRP3. Under cholestatic conditions MRP2 and MRP4, partially supported by glutathione (GSH), can act as efflux mechanisms at apical tubule membranes. The combination of the different transporter pathways serves for an almost complete reabsorption and conservation of bile salts. BS, bile salt; OA, organic counter anion. See the color plate.

members SLC10A3, A4, A5, A6, and A7, with the exception of A3, were first cloned in our lab between 2003 and 2005 (Fernandes et al., 2007; Geyer, Godoy, & Petzinger, 2004; Geyer et al., 2007, 2008; Godoy et al., 2007). The proteins were tentatively named P3, P4, P5, P6, and P7 and their phylogenetic relationship with many other SLC10 genes from mammalian, bird, fish, and amphibian species of the SLC10 family was described previously (Geyer et al., 2006). Here, we review for the first time recent data about functional and structural aspects of all members of the SLC10 transporter family.

3. THE LIVER BILE SALT TRANSPORTER SLC10A1 (NTCP)

This review on human NTCP is more selective than encyclopedic and focuses on biochemical and structural data, transport properties, and transported substrates. Exceptionally, data from rat Ntcp is addressed if it was not available for human NTCP.

3.1. Cloning of SLC10A1

Animal Slc10a1 Ntcps (small letter nomenclature) were cloned in the 1990s from rat (Hagenbuch, Lübbert, Stieger, & Meier, 1990; Hagenbuch, Stieger, Foguet, Lübbert, & Meier, 1991), mouse (Cattori, Eckhardt, & Hagenbuch, 1999), and rabbit (Kramer et al., 1999) and human (capital letter nomenclature SLC10A1/NTCP) (Hagenbuch & Meier, 1994) and include members from many other species (Geyer et al., 2006). The human, rat, and mouse SLC10A1/Slc10a1 genes span 21.4, 13.6, and 12.5 kb and map to chromosomes 14q24, 6q24, and 12D1, respectively. The human SLC10A1 gene comprises five exons, and contains an open reading frame of 1047 bp coding for a 349-aa protein (Craddock et al., 1998). The rat and mouse Ntcp proteins have a slightly longer C terminus resulting in proteins with 362-aa length. In mouse, two Ntcp isoforms were isolated. Besides a 1579-bp transcript (Ntcp1), coding for the 362-aa protein, an alternatively spliced isoform (Ntcp2) with 1411 bp, coding for a 317-aa protein missing the C-terminal 45 aa, could be found. This shorter isoform could also be detected in human liver samples and was 50-fold lower expressed than Ntcp1, the major isoform in mouse liver (Cattori et al., 1999; Cui, Gunewardena et al., 2012). Ntcp2 transports sodium-dependent TC, GC, and C but TC with lower K_m and V_{max}, and probably represents a high-affinity and low-capacity transport system (Cattori et al., 1999). Transcription of the human SLC10A1 gene is stimulated by dexamethasone via a glucocorticoid responsive element (Eloranta, Jung, & Kullak-Ublick, 2006) in a similar manner as for the human SLC10A2/ASBT gene (Jung, Fantin, Scheurer, Fried, & Kullak-Ublick, 2004). It was assumed that glucocorticoids may coordinately regulate the major BS uptake systems in liver and intestine.

The calculated molecular mass of the protein is approximately 38 kD and corresponds to 349 aa (Hagenbuch & Meier, 1994). The sequence identity of human NTCP protein with rat Ntcp is >73% but with dog, >85%. Phylogenetic approaches showed that the SLC10A1 gene belongs to Clade I of the SLC10 family that comprises the members SLC10A1, A2, A4, and A6 (Geyer et al., 2006). Within this clade, SLC10A1 is more closely related to SLC10A4 although the genomic structures are quite different (SLC10A4 exhibits only three exons) probably because of a very early gene separation. The human NTCP protein shares 29.7% identity and 53.7% similarity with P4 protein. This is much less than with ASBT, for which identity is 34.9% but similarity is 62.8% (Geyer et al., 2006).

3.2. Expression of NTCP

There is only one kind of sodium-dependent BS uptake carrier in the hepatocyte. This was concluded from over 95% inhibition of TC uptake by specific Ntcp antisense oligonucleotides in rat liver cRNA-expressing *Xenopus laevis* oocytes (Hagenbuch & Meier, 1996). Only one functional expression of rat Slc10a1 outside the liver occurs in acini of the rat pancreas (Kim et al., 2002).

In human liver, the NTCP cDNA hybridized by Northern blot with a 1.8-kb transcript, a shorter transcript of 1.1 kb is present in placenta that might be nonfunctional (Craddock et al., 1998). Transcripts were absent in kidney and ileum. In late-term placenta and in mononucleated trophoblast (MT) cells, only low-level expression of NTCP was detected by nested RT-PCR. If TC uptake was measured in membrane vesicles from human term placental trophoblast (Marin, Serrano, el-Mir, Eleno, & Boyd, 1990) and in isolated human MT, no significant initial uptake was observed (Ugele, St-Pierre, Pihusch, Bahn, & Hantschmann, 2003). Organic anion transport carriers of the SLC10 and SLCO families become implemented in placenta and the fetal liver only during the late gestational period of the rat fetus (St-Pierre et al., 2004). In fetal rat liver, Ntcp mRNA is detected from day 18–21 of gestation along with Ntcp protein (Boyer et al., 1993; Gao, St-Pierre, Stieger, & Meier, 2004; Hardikar, Ananthanarayanan, & Suchy, 1995) and sodium-dependent BS transport in basolateral membrane vesicles (Suchy et al., 1986). Around birth and during the first postnatal day, the mRNA level of rat Ntcp rises rapidly and only one 1.7-kb type transcript occurs during the next 4 weeks such as in adult rats (Hardikar et al., 1995). A similar ontogenic mRNA expression pattern was observed for mouse Ntcp1 transcript, whereas Ntcp2 was very lowly expressed during pre- and postnatal liver development (Cui, Aleksunes et al., 2012; Cui, Gunewardena et al., 2012). The sharp expression onset is not reflected by a concomitant gain of TC transport capacity around birth. Instead, BS transport capacity slowly increases in neonatal rat hepatocytes apparently because of slow onset of Ntcp glycosylation (Hardikar et al., 1995). Transport studies with isolated hepatocytes from 5-day-old rats reached about 50% of uptake in hepatocytes from 3-months-old rats (Petzinger & Frimmer, 1980; Ziegler, Petzinger, Grundmann, & Frimmer, 1979). In dedifferentiated AS30D rat hepatoma cells, BS transport was absent (Kroker, Anwer, & Hegner, 1978) and decreased in isolated rat hepatocytes upon tumorigenesis by gavage treatment with diethylnitrosamine over 16 weeks (Ziegler, Petzinger, & Frimmer, 1980) or by feeding a carcinogen-rich diet (Blumrich, Pack, Oesch, Petzinger, & Steinberg, 1994). In hepatocytes isolated from such fed rats, strongly reduced V_{max} transport maxima were measured without a change of the kinetic affinity constant K_m. The human hepatoblastoma-derived HepG2 cell line lacks SLC10A1 expression and after injection of HepG2 mRNA into *X. laevis* oocytes, sodium-dependent TC transport was not observed (Boyer et al., 1993; Kullak-Ublick, Beuers, & Paumgartner, 1996). In contrast, hepatocellular carcinomas (HCCs) excised from human liver tissue expressed SLC10A1 mRNA at 50% level of normal liver cells (Kullak-Ublick et al., 1997). Injection of their cRNA into *X. laevis* oocytes revealed sodium-dependent TC uptake, indicating conservation of TC uptake in HCC tumor tissue albeit at lower levels. Thus, expression and localization of the NTCP protein depends on the grade of differentiation of the liver and of liver-derived tumor cells.

Dedifferentiation toward a fetal type of liver-cell transporters occurs in cultures of rat primary hepatocytes on collagen-coated flasks and concomitantly BS uptake decreased considerably within 24 h and got almost lost within 2 days (Föllmann, Petzinger, & Kinne, 1990; Petzinger & Frimmer, 1988). Cultured rat hepatocytes significantly downregulate mRNA expression of Ntcp and reduce maximal transport rates (Liang, Hagenbuch, Stieger, & Meier, 1993). However, in simple monolayer cell cultures with rat hepatocytes, the basolateral rat Ntcp uptake carrier disappears and the expression of canalicular transporters such as Mdr1a, Mdr2, and Bsep remains very constant over 9 days (Borlak & Klutcka, 2004). Changes of the culture conditions to three-dimensional sandwich cultures of human hepatocytes, however, stabilize transport properties for BS transport significantly (reviewed by Kotani et al., 2011; Marion, Perry, St Claire, & Brouwer, 2012; Swift, Pfeifer, & Brouwer, 2010).

3.3. Transport Functions of NTCP

3.3.1. Sodium Cotransport

All sodium-dependent human BS transporters of the SLC10 family (NTCP and ASBT), so-far studied, are electrogenic (Weinman, 1997, Weinman, Carruth, & Dawson, 1998). It is believed that two joined transport processes occur via Ntcp in basolateral rat liver membrane vesicles: BS transport and simultaneous cotransport with sodium ions (Inoue et al., 1982; Ruifrok & Meijer, 1982). The transport stoichiometry is >1, likely 2 Na^+ plus 1 (anionic) TC molecule (Bear, Shaffer, & Davison, 1987; Lidofsky, Fitz, Weisiger, & Scharschmidt, 1993). Such TC transport mode generates a short depolarizing impulse on the cell membrane due to the entry of surplus positive Na^+ charges (Lidofsky et al., 1993; Wehner, 1993). A small inward current flow of about 2–8 pA was expected (Lidofsky et al., 1993; reviewed by Weinman, 1997). The 2:1 stoichiometry was affirmed for rat Ntcp in expressing *X. laevis* oocytes by a calculated Hill coefficient of 1.9 for Na^+/TC uptake (Hagenbuch & Meier, 1996) and highest coupling probability for 2 Na^+ with one molecule BS in COS cells (Craddock et al., 1998). This was indirectly proven when it was shown that the transport of a fluorescent BS derivative with a single negative charge via rat Ntcp was electrogenic (as was TC itself) but another fluorescent derivative, CGamF (cholylglycylamidofluorescein), with a net negative charge of −2, lacked electrogenic transport due to charge compensation likely by the two positive sodium ions (Grüne, Meng, & Weinman, 1996). The dislocation of positive charges on cultured rat hepatocytes exposed to 50–100 μM TC caused membrane depolarization of about 10 mV and lasted even after complete withdrawal of the BS for a time period of at least 5 min in cultured rat hepatocytes (Wehner, 1993). The degree of depolarization amounts to about 25% of the −30 to −40 mV resting membrane potential of hepatocytes. The effect was interpreted that the BS additionally increased a long-lasting sodium ion conductance even in the absence of its own transport (Wehner, 1993). Whether this membrane potential change triggers effects on, e.g. potential-dependent ion channels or receptors, is, however, unknown. It may contribute to the newly described

BA-dependent signal cascades generated by BA-sensitive membrane receptors in- and outside the liver as described in Section 1.2.

The transport direction of NTCP is from outside to inside the cell. It is unknown whether the system also operates in the reversal direction under artificial ion- and/or substrate-gradient conditions, as it was demonstrated, e.g. for the sodium-dependent glucose transporter SGLT1 (Wright, Loo, & Hirayama, 2011). No electrophysiological transport studies with excised membrane patches to our knowledge exist that would confirm reverse transport.

Yet unresolved is the question about the amino acid binding partners for the sodium ions in the NTCP/Ntcp protein. Positively charged ions most likely interact with negatively charged amino acids such as aspartate (Asp) and glutamate (Glu) and two candidates for a putative sodium sensor were claimed (Zahner, Eckhardt, & Petzinger, 2003; Geyer et al., 2006). Figure 4 shows a 9-transmembrane domain (TMD) model of human NTCP adopted from crystal structure of the bacterial homolog of human ASBT from *Neisseria meningitidis* ($ASBT_{NM}$) (Hu, Iwata, Cameron, & Drew, 2011). TC transport of $ASBT_{NM}$ is coupled to transport of two sodium ions similarly as NTCP/Ntcp and ASBT/Asbt. Amino acids responsible for sodium binding in $ASBT_{NM}$ are highly conserved between members of the SLC10 family and possibly also mediate sodium binding of human NTCP. E257 was found to be involved in sodium binding of rat Ntcp (Zahner et al., 2003) and also participated in sodium binding in the depicted model.

3.3.2. Substrate Transport

BSs are the most specific organic substrates of human NTCP. But despite this preference, rat Ntcp was regarded as a multispecific BS transporter (Meier, 1995; Meier, Eckhardt, Schroeder, Hagenbuch, & Stieger, 1997; Petzinger et al., 1996; Zimmerli, Valantinas, & Meier, 1989). Unlike human ASBT (Craddock et al., 1998), the human liver NTCP interacts in addition with some sulfated steroids, thyroidal hormones, xenobiotics, and drug–BA conjugates (Table 1).

3.3.2.1. Bile Salts

Transport of TC in different experimental systems by rat, mouse, and human NTCP/Ntcp was overviewed recently (Stieger, 2011). There is a plethora of literature addressing BS transport by rat Ntcp in various experimental models (Meier, 1995). Here, the focus is on BS transport via

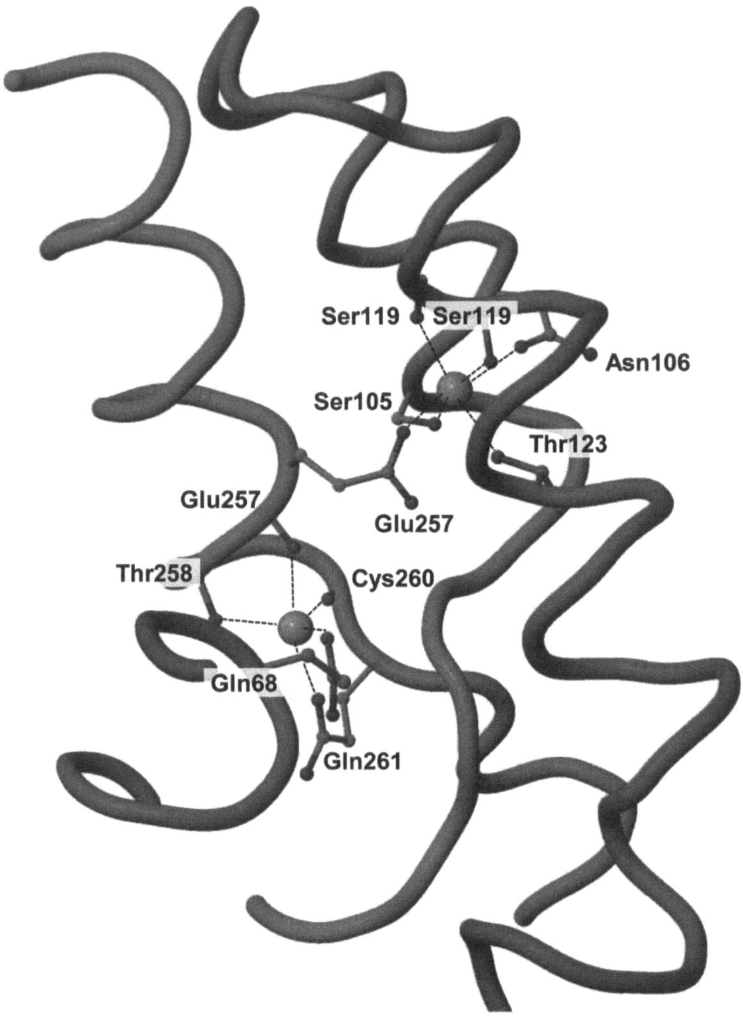

Figure 4 ***Amino acids involved in sodium binding of human NTCP.*** Ribbon representation of human NTCP structure deduced from $ASBT_{NM}$ crystal structure. Most interactions between the proteins and the sodium ions are established via charged or polar atoms in the amino acid side chains. In case of Ser119 and Glu257, oxygen atoms of the protein backbone in addition to the side chain atoms are involved in binding (reason for depicting these amino acids twice), while Thr258 and Cys260 interact with a sodium ion exclusively via backbone oxygen atoms. The illustrated structure is a homology model that was calculated with Yasara (Krieger et al., 2009) and evaluated using the MolProbity Server (Chen et al., 2010). The model is based on Protein Data Bank (PDB) entry 3zuy of $ASBT_{NM}$ (Hu et al., 2011) and a multiple sequence alignment of $ASBT_{NM}$ and human SLC10A1-SLC10A7. The alignment was generated with ClustalW (Larkin et al., 2007). The first 17 amino acids of NTCP correspond to TMD 1 of $ASBT_{NM}$, resulting in an intracellular N-terminus. However, the N-terminus of NTCP is known to be extracellular (Hagenbuch & Meier, 1994; Hagenbuch, 1997; Hallén, Mareninova, Branden, & Sachs, 2002). Therefore, these residues were excluded from the model. Sodium ions are depicted as purple spheres. Interaction of amino acid residues with sodium ions are marked with discontinuous lines. See the color plate.

Table 1 Substrate transport via human NTCP in human isolated hepatocytes, transfected cells, and *Xenopus laevis* oocytes

Substrate	K_m (μM)	V_{max}	Inhibitor	Experimental model	References
[^{3}H]taurocholate	46	0.51*	Cyclosporine	Human hepatocytes	Azer & Stacey, 1993
	62	300†	KCl, ouabain, vecuronium	Human hepatocytes	Sandker et al., 1994
	14.6	0.19*	Pregnanolone sulfates	Human hepatocytes	Abu-Hayyeh et al., 2010
		5.4‡	Choline	Fresh human hepatocytes	Shitara et al., 2003
	2–8	42.6†	Choline	Cryopreserved human hepatocytes	Shitara et al., 2003
	6.3	n.d.	13 bile salts, bumetanide, and BSP	*Xenopus laevis* oocytes	Hagenbuch & Meier, 1994
	10	0.3*		COS cells	Craddock et al., 1998
	7.9	0.049*	C, GC, CDC, DC, UDC, BSP, several drugs	HeLa cells	Kim et al., 1999
	8.4	n.d.	Simvastatin, fluvastatin	CHO cells	Greupink et al., 2011
	8.5	n.d.	MTSET, MTSES	HEK293 cells	Hallén et al., 2000
	13.2	0.38*		HEK293-FlpIn cells	Pan et al., 2011
	7.5	0.41*		HeLa cells	Ho et al., 2004
	34	1.25*	TC, TCDC, C, GC, BSP	Basolateral membrane vesicles	Novak, Ryckman, & Suchy, 1989
[^{3}H]cholate	34	0.23*	Cyclosporine A	Human hepatocytes	Azer & Stacey, 1993
	12	0.14*		HeLa cells	Ho et al., 2004
[^{14}C]glycocholate	101	1.92*	Cyclosporine A	Human hepatocytes	Azer & Stacey, 1993
	21	n.d.	Di- and trihydroxy bile salts, choline chloride	HEK293 cells (at 22 °C)	Bonge et al., 2000
[^{3}H]17α-ethinylestradiol-3-O-sulfate	2.6	0.018*	BSP, bumetanide, TC	HEK293 cells	Han et al., 2010
[^{3}H]estrone-3-sulfate	60	0.11*		COS cells	Craddock et al., 1998
	27	0.068*		HeLa cells	Ho et al., 2004

Continued

Table 1 Substrate transport via human NTCP in human isolated hepatocytes, transfected cells, and *Xenopus laevis* oocytes—*cont'd*

Substrate	K_m (μM)	V_{max}	Inhibitor	Experimental model	References
[^{3}H]rosuvastatin	65	74*	Cyclosporine	Human hepatocytes	Ho et al., 2006
	97	0.45*		HEK293-FlpIn cells	Pan et al., 2011
[^{14}C]pitavastatin	15	16¶		*Xenopus laevis* oocytes	Fujino et al., 2005
Fluvastatin (analyzed by LC-MS/MS)	250	1340§	TC, rifampicin	CHO cells	Greupink et al., 2011
Atorvastatin (analyzed by LC-MS/MS)	40	256**		*Xenopus laevis* oocytes	Choi et al., 2011
Pitavastatin (analyzed by LC-MS/MS)	37	224**		*Xenopus laevis* oocytes	Choi et al., 2011
Rosuvastatin (analyzed by LC-MS/MS)	26	96**		*Xenopus laevis* oocytes	Choi et al., 2011
[^{125}J]T3S/[^{125}J]T4S (sulfated jodothyronines)	Δ 30 min uptake at 1 μM substrate concentration		TC, DHEAS, E_1S		Visser et al., 2010
[^{3}H]taurocholate–chlorambucil conjugate	11	11**		*Xenopus laevis* oocytes	Kullak-Ublick et al., 1997
[^{3}H]taurocholate	6	30**			

n.d. not determined.

BSP, bromosulfophthalein; C, cholate; DHEAS, dehydroepiandrosterone sulfate; E_1S, estrone-3-sulfate; GC, glycocholate; TC, taurocholate; TCD, taurochenodeoxycholate.

*nmol/mg protein/min.

†pmol/10^6 cells/min.

‡μl/10^6 cells/min (uptake clearance).

§ng/mg protein/min.

¶pmol/oocyte/h.

**fmol/oocyte/min.

the human NTCP (Table 1). All kinetic studies with strong evidence of specific sodium-dependent BS uptake were performed with radiolabeled TC. Its transport was inhibited in the heterologous *X. laevis* expression model by 12 other BSs except taurodehydrocholate (TDHC) (Hagenbuch & Meier, 1994). In the human hepatocyte cell model, the apparent kinetic parameters of affinity (K_m) and capacity (V_{max}) for TC uptake differed from data of rat Ntcp. A direct comparison of both species showed that in human hepatocytes, the K_m was 2–3 times higher (Sandker et al., 1994) or quite similar (Azer & Stacey, 1993) but V_{max} was consistently much lower than for TC transport in rat hepatocytes (e.g. V_{max} 300 pmol/min/10^6 human cells versus 1800 pmol/min/10^6 rat cells), reflecting less-efficient uptake of the BS in human hepatocytes (Azer & Stacey, 1993; Sandker et al., 1994). In NTCP transfected cells such as HeLa cell, COS cells as well as in *X. laevis* oocytes K_m for TC transport was, however, always lower (Table 1) than in human hepatocytes (Hagenbuch & Meier, 1994; Kim et al., 1999), reflecting higher affinity in transfected cell systems. The V_{max} values and lesser the K_m parameters highly depend on the quality of the cells, the expression efficacy, the artificial buffer systems, the energy status, incubation temperature, as well as on the calculation methods and cell system limitations. In particular, transport studies at room temperature show reduced V_{max} compared with studies on 37 °C likely due to an inappropriate sodium gradient. In a study with cryopreserved human hepatocytes that were stored over several months and thawed before transport measurements, the cells exhibited inconsistent kinetic data with respect to individual donors. The K_m values were 2–8 μM and the uptake capacity was reduced to 42% of that of freshly prepared nonfrozen cells (Shitara et al., 2003).

BSs of apparently any molecular structure, except the artificial choleretic BSs (T)DHC (Hagenbuch & Meier, 1994) and probably sulfated BSs (Craddock et al., 1998) are substrates of NTCP. Whether NTCP indeed transports sulfated BSs is not clear, although sulfated and nonsulfated BSs inhibit the uptake of each other competitively in isolated rat hepatocytes (Bartholomew & Billing, 1983). In the isolated perfused rat liver, a C_3-sulfonated CDC showed only 20% of the influx of its mother compound CDC (Gärtner, Goeser, Stiehl, Raedsch, & Wolkoff, 1990). Sulfation of BAs on C_3, which is chemically a sulfonation reaction, is regarded as detoxification system of BAs; sulfated BSs are predominantly eliminated into urine and contribute only marginally to the EHC (Alnouti, 2009). The spectrum of transported BSs was analyzed particularly with rat (Hata et al., 2003) and rabbit Ntcp (Kramer et al., 1999) in transfected cells. A comparably complete study on human NTCP is missing. In a review from 1997 transport of TC, GC, C, taurochenodeoxycholate (TCDC), TUDC, estrone-3-sulfate (E_1S), and bromosulfophthalein (BSP) in human NTCP-expressing *X. laevis* oocytes was reported, however, kinetic data were not given (Meier et al., 1997). The rat and human sodium-dependent BS carriers transport all major physiological BSs but surprisingly nonamidated BSs are transported very slowly although these BSs return from gut to liver (Meier et al., 1997; Hata et al., 2003). This might reflect also the transport competition for nonamidated BSs by Oatps since <50% of cholate uptake into rat hepatocytes is sodium-dependent (Meier, 1995). On the other hand, the taurine-conjugated monohydroxy BS taurolithocholate (TLC) was a very strong inhibitor of TC uptake in cells transfected with rabbit Ntcp (Kramer et al., 1999) and inhibition concentrations and K_i values of the more hydrophobic dihydroxy BSs are lower than that of trihydroxy BSs (Bonge, Hallén, Fryklund, & Sjötröm, 2000). This reveals that an additional parameter for the interference of BSs with the Ntcp protein either as a substrate or as an inhibitor is the BSs lipophilicity. This conclusion fits well with the pharmacophore model for BS interactions on the rabbit Slc10a2 protein Asbt (Baringhaus, Matter, Stengelin, & Kramer, 1999) and on the human NTCP (Greupink et al., 2012), where three of five substrate–protein interacting sites are hydrophobic.

3.3.2.2. Steroid Hormones

In isolated human hepatocytes, the uptake of [^{3}H]estradiol-17β-glucuronide was not sodium-dependent, indicating uptake by the OATP1B1 carrier (Greupink, Dillen, Monshouwer, Huisman, & Russel, 2011). Long known is the sodium-independent uptake of E_1S by OATPs, in particular, by OATP1B1 in human liver (Tamai et al., 2000). However, sodium-dependent transport of E_1S is mediated by human NTCP after transfection into COS cells (Craddock et al., 1998). The K_m of 60 μM is higher than K_m of 27 μM with rat Ntcp in the same cell system (Schroeder et al., 1998) and is about sixfold higher than for TC uptake (Craddock et al., 1998). The only other sulfated estrogen steroid transported by NTCP is the 3-O-sulfate of ethinylestradiol (ethinyl-E_2-S) (Han et al., 2010) (Table 1). Whether other sulfated steroids such as dehydroepiandrosterone sulfate (DHEAS) or pregnenolone sulfate (PREGS) are transported by human NTCP was not directly proven hitherto. Recently, competitive inhibition of TC uptake by two differently sulfated pregnenolone derivatives in human hepatocytes and in NTCP-expressing *X. laevis* oocytes was described that might suggest relevant uptake for steroids that cause hypercholanemia in normal as well as pregnancy-induced cholestasis (Abu-Hayyeh et al., 2010). In a report without full kinetic data, it was shown that human HepG2 cells transfected with rat Ntcp-GFP (green fluorescent protein) exhibited sodium-dependent uptake of DHEAS and E_1S (Kullak-Ublick et al., 2000). However, transport of E_1S and likely also of other sulfated steroids in the liver is more a combination of NTCP- and OATP1B1-mediated transport. When both carriers were co-expressed in *X. laevis* oocytes, E_1S uptake was solely dependent on OATP1B1. However, since both carriers transport E_1S, this discrimination is doubtable as well as the conclusion that OATP1B1 is the prevailing uptake carrier of sulfated estrone in human liver. The release of sulfated steroids as well as of BAs from the hepatocytes into blood, however, is achieved by the facilitated heterodimeric diffusion carrier OST α/β (SLC51A1/A2) (Ballatori et al., 2009) (Fig. 3).

3.3.2.3. Thyroid Hormones

Human and rat NTCP/Ntcp transport sodium-dependent thyroid gland hormones. In *X. laevis* oocytes injected with rat Slc10a1 cRNA, iodothyronines T3/T4 were weakly transported, whereas uptake of their sulfated counterparts was strongly increased (Friesema et al., 1999). This was confirmed by transfection of human SLC10A1 in HEK293 cells while the other human SLC10 family members SLC10A2, 3, 5, 6, 7, and rat Slc10a4 did not transport sulfated or nonsulfated thyroid gland hormones (Visser et al., 2010). Because SLC10A3, 4, 5, and 7 are orphan carriers without any reported transport activity, lack of iodothyronine transport by these SLC10 proteins could also simply reflect the inability of the proteins to transport substrates at all, rather than that iodothyronines are inadequate substrates. Iodothyronines are, however, transported in addition to NTCP/Ntcp by the monocarboxylate transporter MCT8 (SLC16A2) (Friesema et al., 2008; Karapanou & Papadimitriou, 2011) and by members of the OATP/SLCO family (Pizzagalli et al., 2002; Sugiyama et al., 2003).

3.4. Pharmacological Importance of NTCP: Transport of Xenobiotics

Drugs which interfere with BS transporters at the canalicular membrane, e.g. with ABCB11 (BSEP), cause intrahepatic cholestasis (Padda, Sanchez, Akhtar, & Boyer, 2011; Trauner, Meier, & Boyer, 1998). There is much less information whether drug interference with sinusoidal BS uptake transporters, e.g. NTCP or OATPs, contributes to cholestasis, but during cholestasis manifestation, basolateral NTCPs and OATPs are downregulated through the FXR (for review, see Pauli-Magnus & Meier, 2006; Trauner et al., 1998). In an experimental cell model co-transfected with the BSEP and NTCP, cholestasis-inducing drugs inhibited both proteins (Mita et al., 2006). A large number of drugs inhibited TC uptake via human NTCP in transfected HeLa cells (Kim et al., 1999) but only a few drugs were considered to be true transportates of NTCP. In order to document xenobiotic transport, either direct transport kinetics with radiolabeled drugs were performed in solitarily NTCP-expressing systems or conclusions on transport were drawn from inhibition studies where the transport of radiolabeled TC was competitively or at least dose-dependently inhibited by a (nonlabeled) drug. Such studies were mostly performed with rat hepatocytes (Zimmerli et al., 1989) and were critically reviewed (Meier et al., 1997) since this kind of experiments could be misleading. Therefore, the arguments in favor of drug transport via NTCP/Ntcp that are based on inhibition studies of BS uptake are only indicative but not conclusive. This concerns BSP, propranolol, progesterone, cyclosporine, ketoconazole, bumetanide, and furosemide, which dose-dependently inhibited TC uptake into HeLa cells transfected with human SLC10A1 (Kim et al., 1999). However, neither propranolol (Kim et al., 1999) is transported by human NTCP, nor is bumetanide (Horz, Honscha, & Petzinger, 1996) and the MDR1 substrate cyclosporine A (Schroeder et al., 1998) transported by rat Ntcp.

3.4.1. Statins

A drug group that exhibits pharmacokinetic behavior strongly dependent on membrane transporters of the SLCO (OATP) and SLC10 families is the statins (Table 1). These HMG-CoA reductase inhibitors are used to inhibit cholesterol biosynthesis and to lower cholesterol level in blood. Their sodium-independent uptake into human liver is mediated primarily by OATP1B1 (König, 2011) and in extrahepatic human tissues by OATP2B1 (Grube et al., 2006; Ho et al., 2006; Niessen et al., 2009). However, OATP1B1 protein is strongly polymorphic and the in vivo disposition of statins varies significantly (Igel et al., 2006; Niemi et al., 2011). Statins with direct evidence for being transported by human NTCP are rosuvastatin, pitavastatin, atorvastatin, and possibly fluvastatin (Choi et al., 2011; Fujino, Saito, Ogawa, & Kojima, 2005; Greupink et al., 2011; Ho, Leake, Roberts, Lee, & Kim, 2004; Ho et al., 2006) (Table 1). Surprisingly, rosuvastatin transport was not found in rat Ntcp transfected HeLa cells (Ho et al., 2006) nor was simvastatin transport observed in human NTCP-expressing

X. laevis oocytes (Choi et al., 2011). Obviously, not every member of the statin drug family is transported by NTCP/Ntcp. BS and statin transport has recently been studied also in naturally occurring genetic NTCP variants (Table 2). Regarding rosuvastatin, about 35% of overall uptake is due to NTCP (Ho et al., 2006) and surprisingly, even the genetic mutant NTCP*2 (C800T → Ser267Phe) which lacks almost completely TC transport (Ho et al., 2004) performed rosuvastatin and atorvastatin transport even better than wild-type NTCP (Choi et al., 2011) (Table 2). If NTCP was co-transfected with OATP1B1 in *X. laevis* oocytes, the cellular accumulation of rosuvastatin, pitavastatin, and atorvastatin was higher than after transfection of each single transporter alone. However, the internal clearance CL_{int} (ratio of V_{max}/K_m) of these statins in *X. laevis* oocytes was larger via OATP1B1 than via NTCP, and OATP1B1 showed higher affinity for the statins than NTCP (Choi et al., 2011). Although under physiological conditions, both systems operate in combination for statin uptake in the liver, there might be dominance by OATP1B1, in particular, at lower statin concentrations.

3.4.2. Diagnostics

Diagnostic xenobiotics, which are used since a long time to assess hepatic clearance functions such as indocyanine green (ICG) and BSP, are considered to be substrates of OATP/Oatp transporters, which means a prevailing sodium-independent uptake by the liver (Meier et al., 1997). Indeed BSP was the founding substrate for the cloning of rat Oatp1a1 (Jacquemin, Hagenbuch, Stieger, Wolkoff, & Meier, 1994). Recently, sodium-dependent transport of BSP (Hata et al., 2003) and ICG (de Graaf et al., 2011) by NTCP/Ntcp was reported. In transfected HeLa cells, the K_m of sodium-dependent BSP uptake by rat Ntcp was 3.7 μM which was the same as for rat Oatp1a1 (Hata et al., 2003). ICG uptake by NTCP could not be assessed kinetically, but was also sodium-dependent in NTCP-transfected CHO cells (de Graaf et al., 2011). The results are controversial since ICG is also transported by liver-specific OATP1B3 (de Graaf et al., 2011). Nonetheless, ICG inhibited competitively the transport of E_1S by OATP1B1 and OATP1B3 and of TC by NTCP in the transfected cells (de Graaf et al. 2011).

3.4.3. Cyclic Peptides

Several authors have claimed that the green death cap mushroom toxins phalloidin (Münter, Mayer, & Faulstich, 1986; Petzinger, Ziegler, & Frimmer, 1979; Petzinger & Frimmer, 1980) and α-amanitin (Gundala et al.,

Table 2 Transport of substrates via NTCP and polymorphic NTCP variants

dbSNP accession number			**Not assigned**		**rs61745930**		**rs2296651**		**rs72547507**		**rs72547506**	
Nucleotide substitution			**G190A**		**T668C**		**C800T**		**T836C**		**A940G**	
Amino acid substitution	**Wild type**		**Ala64Thr**		**Ile223Thr**		**Ser267Phe**		**Ile279Thr**		**Lys314Glu**	
	CL_{int}	K_m	CL_{int}	K_m	CL_{int}	K_m	CL_{int}	K_m	CL_{int}	K_m	CL_{int}	K_m
TC*	55	7.5	—	—	15.5	6.8	0.63	35	15.2	13	14	10
TC†	28.6	13	19	46	—	—	3.35	109	—	—	—	—
C*	11	12	—	—	1.7	21	0.39	49	2.5	17	1.8	17
E_1S*	2.5	27	—	—	0.53	19	2.4	29	1.4	30	0.8	74
Rosuvastatin†	4.7	97	n.d.	n.d.	—	—	15.2	141	—	—	—	—

$CL_{int} = V_{max}/K_m$ (μl/mg protein/min); K_m = μM.
C, cholate; E_1S, estrone-3-sulfate; TC, taurocholate; n.d., not determined; —, SNP not found in the tested cohorts.
*HeLa cells (Ho et al., 2004).
†HEK293-FlpIn cells (Pan et al., 2011).

2004; Kröncke et al., 1986), both bicyclic peptides, are substrates of rat Ntcp. However, this could not be verified in *X. laevis* oocytes expressing rat Ntcp (Meier-Abt, Faulstich, & Hagenbuch, 2004). With respect to phalloidin, specific liver uptake by the human OATP1B1 and rat oatp1b2 was measured (Meier-Abt et al., 2004), which was confirmed in HEK293 cells expressing solitarily OATP1B1 that transported phalloidin with K_m of 39 μM (Fehrenbach, Cui, Faulstich, & Keppler, 2003). The toxin is released into bile by rat multidrug-resistance-associated protein Mrp2 (Abcc2) (Gavrilova, Geyer, & Petzinger, 2007). With respect to α-amanitin, the evidence for Ntcp-mediated transport is weak and indirect, i.e. HepG2 cells transfected with rat Ntcp were more sensitive for the toxic effects of α-amanitin (Gundala et al., 2004). However, kinetic studies with [^{3}H]-labeled α-amanitin have directly shown that human OATP1B3 transported the toxin into MDCKII cells with K_m of 3.7 μM, whereas the transfected human OATP1B1 and OATP2B1 lacked any transport activity for this cyclopeptide (Letschert, Faulstich, Keller, & Keppler, 2006). Unfortunately, α-amanitin transport by NTCP-expressing cells was not tested.

3.4.4. Trojan Horses

An exceptional position is set by drug–bile acid conjugates, which were developed in the 1990s by Werner Kramer and Günther Wess at the former Hoechst pharmaceutical company in Frankfurt, Germany, and were tested in cooperation (Kramer, Wess et al., 1992; Kramer, Wess, Enhsen et al. 1994; Petzinger et al., 1995). The concept of liver-specific drug targeting by the covalent attachment of a BA to a drug in order to gain uptake of the drug via BS carriers was recently named the Trojan horse concept (Kramer, 2011; Marin, Romero, Vallejo, Perez, & Briz, 2005). With nearly all the BA conjugates of HMG-CoA reductase inhibitors, peptides, or oligonucleotides from that era, an increase in the affinity of the drug to liver cells and an interaction with the BS uptake transporter Ntcp was observed (Petzinger, Wickboldt, Pagels, Starke, & Kramer, 1999). In some cases such as the BA-HMG CoA reductase inhibitors (Petzinger et al., 1995) and the BA-oligonucleotide conjugates (Starke et al., 2001), the attachment of one or two BAs to the drug enhanced significantly biliary excretion or even permitted the drug to reach bile. The canalicular excretory system responsible for biliary elimination of the BA–oligonucleotide conjugate was rat Mrp2 (Lischka et al., 2003). With a single BA–drug conjugate, i.e. the TC–chlorambucil conjugate, transport via the human NTCP was directly shown after expression of SLC10A1 in *X. laevis* oocytes (Kullak-Ublick et al., 1997). The conjugate showed less affinity and only one-third of the maximal transport velocity when compared with uncoupled TC (Table 1). However, no measurement of the uptake of chlorambucil alone was performed for comparison. Conjugates of BAs with cytotoxic metals (e.g. cis-platin) named Bamets were generated to induce inhibition of tumor growth in the liver and biliary tract and were substrates of NTCP (Briz et al., 2002). Finally, chemical modification of glycocholic acid (GCA) was achieved by conjugation with polyamine molecules. These compounds inhibited sodium-dependent TC uptake via rat Ntcp in *X. laevis* oocytes, but direct transport by the carrier protein was not tested (Vicens, Medarde et al., 2007).

3.5. Naturally Occurring NTCP Polymorphisms

In humans, several single-nucleotide polymorphisms (SNPs) of SLC10A1 were described and kinetic studies with taurocholate, E_1S, or rosuvastatin were performed (Choi et al., 2011; Ho et al., 2004; Pan et al., 2011) (Table 2). Four nonsynonymous SNPs were found in the American population (Ho et al., 2004) and two in Korean, Chinese, and Vietnamese populations (Pan et al., 2011). The NTCP variants showed ethnic distribution; for example, the variant NTCP*3 (Ile223Thr) was only found in African-Americans, whereas NTCP*2 (Ser267Phe) occurred only in Chinese Americans (Ho et al., 2004). With the latter variant, residual uptake of TC was 2% of control uptake (Ho et al., 2004). No other variant showed such a massive loss of BS transport function but on the other hand, rosuvastatin uptake was increased in this variant (Pan et al., 2011). Most variants showed markedly decreased V_{max} rates but also the K_m values differed significantly. Therefore, the internal substrate clearance CL_{int} is given in Table 2 in addition to K_m values. This revealed that in variant NTCP*2, in addition to TC, also C uptake was markedly decreased, whereas E_1S uptake was not changed at all (Ho et al., 2004) (Table 2).

3.6. Membrane Insertion Topology

The human and rat NTCP/Ntcp protein bears five putative N-linked glycosylation sites but only the N-terminal located N-glycosylation sites (Asn5, Asn11) were demonstrated to be glycosylated and, therefore, located extracellular (Hagenbuch, 1997; Hallén, Mareninova, Branden, & Sachs, 2002). The native form of the rat Ntcp protein in the cell membrane shows a molecular weight of about 51 kDa in Western blots (Ananthanarayanan, Ng, Boyer, & Suchy, 1994; Stieger et al., 1994), whereas the molecular weight estimation for the deglycosylated form is 33 kDa. For the human NTCP, comparable efficiency of glycosylation was found: in HeLa cells, the molecular weight of cell surface expressed NTCP was about 50 kDa, which was reduced to 37 kDa upon complete enzymatic deglycosylation (Ho et al., 2004). Furthermore, C terminus of NTCP/Ntcp could be localized intracellular by immunostaining and glycosylation experiments (Ananthanarayanan et al., 1994; Hagenbuch & Meier, 1994; Hallén, Mareninova et al., 2002).

The initial suggested protein structure based on Kyte-Doolittle hydrophobicity plots for rat Ntcp (Hagenbuch, Stieger, Foguet, Lübbert, & Meier, 1991) and human NTCP (Hagenbuch & Meier, 1994) showed 7 TMDs with the N- and C-termini in opposite location to the cell membrane as demonstrated experimentally. Other algorithms for the predictions of the membrane topology of NTCP predicted either 8 or 9 TMDs (Hallén, Björquist, Ostlund-Lindqvist, & Sachs, 2002). In order to define the domain structure, a combined approach of insertion scanning with glycosylation flag constructs, glycosylation site mutagenesis, and alanine insertion was applied that suggested a protein topology with 9-TMD segments (Hallén, Mareninova et al., 2002). A reanalysis of NTCP topology with cell transfections of complete NTCP transporter constructs labeled by GFP/YFP in HEK293 cells and tryptic digestion of the membrane-inserted proteins reincarnated the 7-TMD model of human NTCP (Mareninova et al., 2005). Recently, the crystal structure of a bacterial $ASBT_{NM}$ was published that showed a 10-TMD structure when crystallized with bound Na^+ and TC (Hu et al., 2011).

Shortly before the cloning of NTCP/Ntcp, the determination of its functional molecular mass by the radiation-inactivation method revealed a 170-kDa transporter cluster that maintained sodium-dependent TC transport in basolateral plasma membrane vesicles from rat liver (Elsner & Ziegler, 1989). Very recently, a minimal dimeric architecture of the human NTCP with a molecular weight of 100–150 kDa was suggested after chemical cross-linking of the NTCP protein in expressing U2OS cells and of >250 kDa after the treatment of rat liver plasma membranes with cross-linking agents (Bijsmans, Bouwmeester, Geyer, Faber, & van de Graaf, 2012). Dimerization could be observed by coimmunoprecipitation of differently tagged NTCP proteins of varying glycosylation status as well as by fluorescence resonance energy transfer (FRET) analysis of diverse fluorescence-labeled NTCP proteins in the plasma membrane. This dimerization seems not to be essential for transport activity as heterodimers of native NTCP with a nonfunctional

NTCP (NTCP Glu257Asn) mutant or with the steroid sulfate transporter SOAT (SLC10A6) did not influence transport activity. This suggests BS transport by each individual functional NTCP subunit. Coexpression of native NTCP with proteins rarely sorted to the plasma membrane (e.g. SLC10A4, NTCP Tyr307X) leads to impaired BS transport due to irregular NTCP trafficking and retained residence in the endoplasmic reticulum. In vivo consequences could be considered in humans having heterozygous gene alterations of NTCP (see above). This could lead approximately to 50% transport reduction after dimerization with nonfunctional proteins, or even higher transport restriction by dimerization with proteins not sorted appropriately to the plasma membrane (Bijsmans et al., 2012). Finally, it was shown by coimmunoprecipitation and colocalization in U2OS cells that heterodimers were only formed with SLC10A4 and SLC10A6 but not with SLC10A3, SLC10A5, and SLC10A7 (Bijsmans et al., 2012).

3.7. 3D Homology Model of Human NTCP

As mentioned above, the crystal structure of $ASBT_{NM}$ showed 10 TMDs. This bacterial protein exhibited sodium-dependent TC uptake in transformed *Escherichia coli* with $K_m = 52\ \mu M$ and $V_{max} = 0.45$ nmol/mg protein/min, which means transport at lower affinity but comparable capacity of human ASBT ($K_m = 11.3\ \mu M$ and $V_{max} = 0.31$ nmol/mg protein/min) (Hu et al., 2011) but with affinities and capacities close to TC transport of human NTCP in human hepatocytes (Table 1). For this reason, we attempted here to calculate an assumed 3D homology model for human NTCP based on the crystal structure information of $ASBT_{NM}$ (Figs 4 and 5). Because the first TMD in $ASBT_{NM}$ is not recognized by topology prediction programs in other functional SLC10 carrier proteins (ASBT/Asbt, NTCP/Ntcp, SOAT/Soat) and experimental verification showed extracellular localization of their N-terminus (Banerjee & Swaan, 2006; Geyer et al., 2007; Hallén, Branden, Dawson, & Sachs, 1999; Zhang et al., 2004), it was claimed that the 9-TMD model could picture their protein structures at best (Hu et al., 2011). Therefore, the calculated homology model of human NTCP consists of 9 TMDs (Figs 4 and 5).

On the basis of this NTCP model, we re-evaluated a synthetically engineered amino acid substitution and a naturally occurring polymorphism with impaired transport activity. Cys96 of rat Ntcp was claimed not to be directly involved in TC transport as the Cys96Ala mutant was fully transport active. However, substitution of Cys96 by tryptophan leads to loss of complete transport (Zahner et al., 2003). As shown in Fig. 5A, Cys96

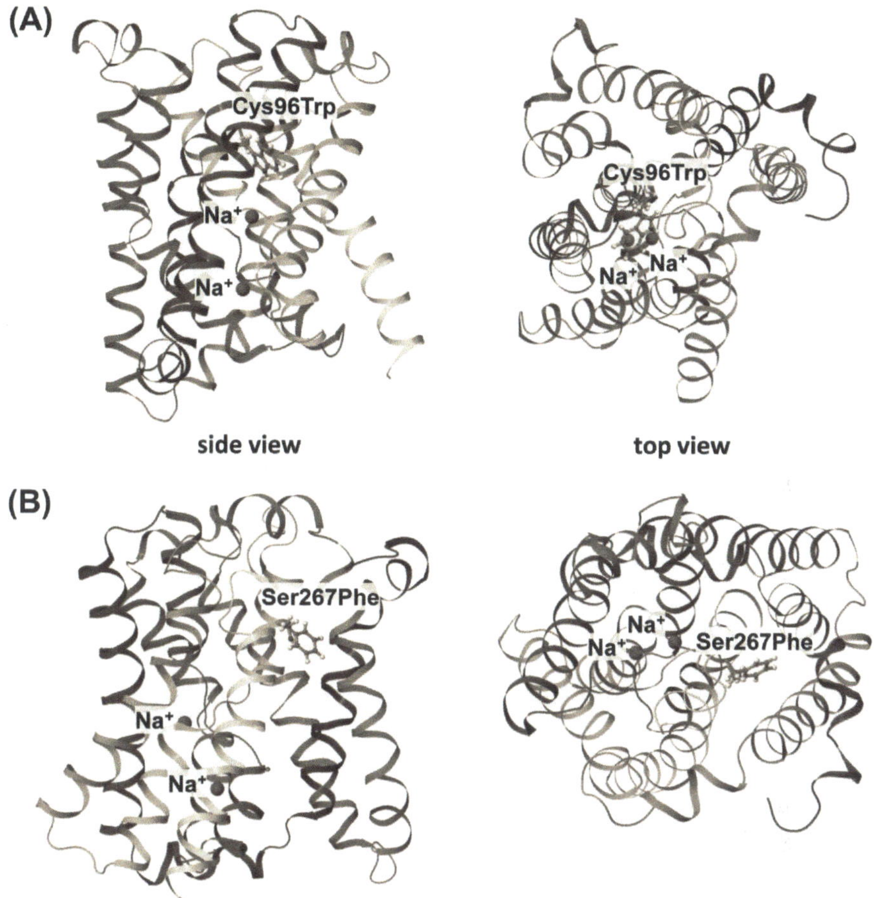

Figure 5 ***Illustration of a 9 TMD-model of human NTCP with amino acid substitutions leading to impaired bile salt transport activity.*** TMD model of human NTCP was calculated according to the crystal structure of $ASBT_{NM}$. The homology models were generated as described in the legend to Fig. 4, with the only difference that the mutations described below were integrated into the alignment. Transmembrane helices are shown in gray, sodium ions are depicted as purple spheres, amino acids of wild-type NTCP are shown in light blue, substituted amino acids in orange. (A) Presentation of amino acid substitution Cys96Trp in top and side view. Cys96 seems to be located in the proposed sodium ion pathway. Substitution into the bulky tryptophan residue possibly blocks sodium ion binding and prohibits bile salt transport. (B) Presentation of the naturally occurring SNP Ser267Phe in top and side view. Ser267 seems to be located in the proposed bile salt pathway. Substitution of serine into phenylalanine blocks the thought binding cavity or transport path of bile salts followed by transport inhibition. See the color plate.

directly lies in the possible permeation pathway of Na^+. Insertion of bulky tryptophan at this position seems to close this pathway for sodium ions, which could explain the functional inactivation of this mutant. A further interesting naturally occurring SNP is Ser267Phe. As mentioned above, transport activity of TC and C by this mutant is nearly abolished, whereas normal transport activity occurred for the steroid sulfate E_1S. A gain of function was observed for the statin rosuvastatin (Ho et al., 2004; Pan et al., 2011). Ser267 lies in a region thought to be the substrate-binding cavity and transport pathway in NTCP (Fig. 5B). The bulkier phenylalanine extends into this cleft and could possibly inhibit BS binding or transport. Unfortunately, no information can be drawn for binding/transport of the steroid sulfate and the statin, which could give an explanation for the preserved transport function of these two substrates in contrast to the abolished transport of BSs.

4. THE ILEAL BILE SALT TRANSPORTER ASBT (SLC10A2)

The apical sodium-dependent bile acid transporter, ASBT/Asbt (also called ileal apical sodium bile acid cotransporter, IBAT/Ibat), represents the second member of the SLC10 family (gene SLC10A2/Slc10a2). This chapter gives a short overview over its physiological function and the current insights into structure and transport function.

4.1. Cloning of ASBT

The first Slc10a2 carrier was identified through expression cloning of a hamster ileal cDNA library in 1994 (Wong, Oelkers, Craddock, & Dawson, 1994). In the next years, identification of human (Wong, Oelkers, & Dawson, 1995), rat (Shneider et al., 1995), rabbit (Kramer et al., 1999), and mouse ASBT/Asbt (Saeki et al., 1999) was carried out. The human, rat, and mouse genes have a length of 22.8, 22.2, and 19.6 kb and are located on chromosome 13q33, 16q12, and 8A1, respectively. All three genes comprise six exons with an open reading frame of 1047 bp and 348 aa (Geyer et al., 2006). Further orthologs could be identified on experimental level and predictions from the genome projects in progress. Some examples are Asbt from vertebrates such as chimpanzee, orangutan, dog, cow, chicken, zebrafish, horse, elephant, panda, swine, zebra finch, and frog, which have 56–99.5% protein identity to each other (unpublished own results). In 2011, a bacterial homolog from *N. meningitidis*, $ASBT_{NM}$, was reported with 26% identity and 54%

similarity to human ASBT (Hu et al., 2011). Comparable sequence identities were calculated for the other members of the SLC10 family that question the classification of this bacterial protein as ASBT homolog. Furthermore, phylogenetic relationship does not clearly attribute this protein as ASBT ortholog. In fact, the $ASBT_{NM}$ branch is localized between SLC10A1–6 and SLC10A7 and should be named $Slc10_{NM}$ (Lionarons, Boyer, & Cai, 2012). Recently, two ASBT orthologs from the sea lamprey and little skate were cloned representing early stages in the vertebrate lineage and sharing 58% and 64% amino acid identity to human ASBT, respectively (Lionarons et al., 2012). No direct ASBT ortholog could be identified in late nonvertebrate sea squirt (*Ciona intestinalis*), indicating that an enterohepatic cycling of BSs seems to have emerged at the early beginning of vertebrate evolution. The most closely related sea squirt analog was localized at the common branch of SLC10A6 and SLC10A2 (Lionarons et al., 2012), probably representing the earlier assumed common ancestor gene of these two carriers (Geyer et al., 2006).

4.2. Expression and Physiological Function of ASBT

ASBT is predominantly expressed in the apical brush-border membrane of ileal enterocytes where it works as reclamation system for BSs in order to maintain their EHC in cooperation with liver NTCP (Fig. 3). ASBT/Asbt of rat (Shneider et al., 1995), mouse (Dawson et al., 2003), hamster (Wong et al., 1994), rabbit (Kramer, Wess, Bewersdorf et al., 1997) and human (Balesaria et al., 2008; Duane, Hartich, Bartman, & Ho, 2000; Meier et al., 2007) are predominantly expressed in enterocytes of terminal distal ileum. Rat Asbt expression on mRNA and protein level increases *post partum* especially after weaning as well as its glycosylation rate in the early weeks of life (Shneider et al., 1995). Also, mice Asbt mRNA increases especially at postnatal day 20 (Cui, Aleksunes et al., 2012). Human ASBT mRNA could further be detected in the duodenum (Balesaria et al., 2008; Hruz et al., 2006; Meier et al., 2007). Lamprey and skate Asbt also showed highest Asbt expression in distal intestine and kidney (lamprey), indicating a similar role of Asbt in these evolutionary old species as in mammals (Lionarons et al., 2012). In mammalian species, ASBT/Asbt is additionally expressed in proximal tubule cells of kidney where it maintains the reabsorption of glomerular filtrated BSs (Christie, Dawson, Thevananther, & Shneider, 1996; Craddock et al., 1998) (Fig. 3). Furthermore, ASBT/Asbt was detected in human biliary epithelium (Chignard et al., 2001) and in the apical membrane of rat cholangiocytes, i.e. in epithelial cells of the hepatic bile ductules and bile duct (Alpini, Glaser, Rodgers et al., 1997; Lazaridis et al., 1997). In addition to native rat Asbt, a truncated Asbt (t-Asbt) was located in ileum, kidney, and basolateral membrane of cholangiocytes (Fig. 3). By alternative splicing of exon 2, the open reading frame is shifted and a premature stop codon leads to generation of a protein with only 154 aa and a molecular weight of 19 kDa. This variant showed even higher expression in ileum and cholangiocytes than native Asbt. In contrast to native Asbt, t-Asbt performed efflux of BSs in *X. laevis* oocytes instead of uptake (Lazaridis et al., 2000). The function of t-Asbt is not clearly understood. Probably native and t-Asbt mediate together an alternative route for BSs between bile duct and liver (*cholehepatic shunt*) (Alpini et al., 2005) (Fig. 3). Cholehepatic shunting of BSs allow the recovery of BSs during extrahepatic cholestasis. Furthermore, this uptake activates various regulating pathways for secretion, proliferation, and survival of cholangiocytes (Alpini, Glaser, Robertson et al. 1997; Alpini, Glaser, Rodgers et al., 1997; Alpini et al., 2005; Chignard et al., 2001, 2003; Xia, Francis, Glaser, Alpini, & Lesage, 2006).

4.3. Transport Function of ASBT

ASBT/Asbt exhibits a narrow and more specific substrate spectrum than NTCP/Ntcp. Transportates are unconjugated as well as glycine- and taurin-conjugated BSs with higher affinity for the dihydroxy BSs, CDC and deoxycholate (DC), than for the trihydroxy BS, TC (Craddock et al., 1998; Kramer et al., 1999; Wong et al., 1994). This transport is sodium-dependent and electrogenic with a Na^+:BS stoichiometry of 2:1 as verified by sodium activation kinetic and electrophysiology experiments (Craddock et al., 1998; Sun, Balasubramaniyan, Chen, Shahid, & Suchy, 2006; Weinman et al., 1998). In addition, the ASBT orthologs from skate, lamprey, and *N. meningitidis* show a transport activity for BSs (Hu et al., 2011; Lionarons et al., 2012). More "ancient" 5α-C_{27} and 5β-C_{27} BSs such as 5α-petromyzonol sulfate, 5α-cyprinol sulfate, scymnol sulfate, or lamprey hepatic BSs are also substrates of ASBT/Asbt, indicating that ASBT retained its primary function during evolution (Lionarons et al., 2012). In contrast to NTCP/Ntcp, no transport of xenobiotics and steroid sulfates was detected in phylogenetic old and new ASBT/Asbt orthologs (Craddock et al., 1998; Ho et al., 2006; Lionarons et al., 2012). The substrate specificity of ASBT/Asbt is in line with its proposed function as recovery system for BSs especially from gut and also kidney (Fig. 3). This was verified in Asbt knockout mice, which showed a 10- to 20-fold higher loss of BSs followed by a reduced BS pool compared to wild-type mice (Dawson et al., 2003). On the other hand, in a recent publication, two developmental nonsteroidal compounds were reported as weak ASBT substrates and might be the first xenobiotic ASBT transportates identified so far (Kolhatkar, Diao, Acharya, Mackerell, & Polli, 2012).

4.4. Pharmacological Importance of ASBT

ASBT is an interesting target for pharmacological manipulation: firstly, ASBT serves as prodrug target, used to increase the oral bioavailability and drug permeability of therapeutics poorly absorbed in gut (Balakrishnan & Polli, 2006; Kramer, 2011; Kramer & Wess, 1996). Once absorbed, prodrugs are converted into the active form and act in its desired function. This concept was tested for peptides (e.g. renin inhibitory peptides), HIV protease inhibitors (Kagedahl et al., 1997; Kim et al., 1993; Kramer, Wess, Enhsen et al., 1997; Kramer, Wess, Neckermann et al., 1994; Swaan, Hillgren, Szoka, & Oie, 1997), gabapentin (Rais, Fletcher, & Polli, 2011), niacin, and ketoprofen (Zheng & Polli, 2010). Best results are reached for the antiviral drug acyclovir, whose oral bioavailability was doubled as prodrug (Tolle-Sander, Lentz, Maeda, Coop, & Polli, 2004). To improve this strategy, a better understanding of the structural needs for substrate recognition is required. Therefore, several BA analogs with diverse numbers, positions, and altered stereochemistry of hydroxyl groups (Kramer et al. 1999) and also BA conjugates (Kolhatkar et al., 2012; Rais, Acharya, Mackerell, & Polli, 2010; Zheng, Pan, Acharya, Swaan, & Polli, 2010) were investigated for their transport and inhibition characteristics on ASBT. These resulted in the identification of position C_3 (Baringhaus et al., 1999; Kramer et al., 1999; Kolhatkar & Polli, 2012) and C_{24} (Swaan, Hillgren et al., 1997) as preferred coupling position for a drug.

Second, ASBT is a target for cholesterol-lowering therapies to prevent atherosclerosis (Balakrishnan & Polli, 2006; Dawson, 2011; Geyer et al., 2006; Izzat, Deshazer, & Loose-Mitchell, 2000; Kramer & Glombik, 2006). This is founded on the negative feedback mechanism of BAs on the expression of hepatic cholesterol 7α-hydroxylase CYP7A1 (Section 1.2.1), which represents the rate-limiting step in BA synthesis (Chiang et al., 2000). Blockage of BS reabsorption in terminal ileum leads to reduced BA pools, which triggers de novo BA synthesis from plasma cholesterol resulting in reduced plasma cholesterol levels (Fig. 3). This blockage is achieved by bile acid reabsorption inhibitors (BARIs) that are highly specific, nonabsorbable ASBT inhibitors of diverse chemical structure. These BARIs reduced cholesterol levels in animal studies with rat, mouse, hamster, rabbit, minipig, guinea pig, and monkey and, therefore, were found a suitable method for treating hypercholesterolemia (Bhat et al., 2003; Hara et al., 1997; Higaki et al., 1998; Huang et al., 2005; Huff et al., 2002; Ichihashi et al., 1998; Iwasaki et al., 1995; Kitayama et al., 2006; Kurata et al., 2004; Lewis, Brieaddy, & Root, 1995; Li et al., 2004; Root et al., 2002; Takashima et al., 1994; Telford et al., 2003 ; Tremont et al., 2005; Vicens, Macias et al., 2007; West, Ramjiganesh, Roy, Keller, & Fernandez, 2002; West, Zern, Butteiger, Keller, & Fernandez, 2003). To increase inhibitory potency of BARIs, several quantitative structure–activity relationship (QSAR) analyses were developed, which describe the required molecular properties of an inhibitor to interact with its target molecule (Baringhaus et al., 1999; Gonzáles, Acharya, Mackerell, & Polli, 2009; Rais, Acharya, Tririya, Mackerell, & Polli, 2010; Swaan, Szoka, & Oie, 1997; Sharma, Singh, Pilania, Sarbhai, & Prabhakar, 2011; Zheng, Ekins, Raufman, & Polli, 2009; Zheng et al., 2010). The intention of BARI applications has changed since ASBT inhibitors are beneficial for treatment of chronic idiopathic constipation (Chey, Camilleri, Chang, Rikner, & Graffner, 2011; Simren, Bajor, Gillberg, Rudling, & Abrahamsson, 2011; Wong et al., 2011), diabetes mellitus type 2, hypertriglyceridemia (Chen et al., 2011, 2012; Lundåsen et al., 2012) and necrotizing enterocolitis (Halpern et al., 2010). For a more comprehensive overview over pharmacological importance of ASBT/Asbt, different reviews were published in the recent years (Balakrishnan & Polli, 2006; Dawson, 2011; Geyer et al., 2006; Kramer, 2011; Kramer & Glombik, 2006).

4.5. Naturally Occurring ASBT Polymorphisms

In various studies, naturally occurring SNPs of ASBT were identified which are suggested to be associated with different diseases (Table 3). In patients with primary bile acid malabsorption, which is ongoing with congenital diarrhea, steatorrhea, and disruption of EHC of BSs, nonfunctional ASBT mutations were identified. These included a 3-bp substitution at donor-splice site of exon 3 followed by alternative splicing, and the amino acid substitutions Leu243Pro and Thr262Met, all leading to a dysfunctional transporter (Oelkers, Kirby, Heubi, & Dawson, 1997). However, only rare incidences of idiopathic bile salt malabsorption seem to be associated with polymorphisms/mutations in ASBT gene indicating additional causes for this disease (Montagnani, Love, Rossel, Dawson, & Qvist, 2001; Montagnani et al., 2006). Reduced ASBT expression was also found in patients with hypertriglyceridemia, gallstone disease, and Crohn's disease (Bergheim et al., 2006; Duane et al., 2000; Holzer et al.,

Table 3 Nonsynonymous SNPs in the coding region of human ASBT and characterization of taurocholate transport

dbSNP accession number	Nucleotide substitution	Amino acid substitution	TCA uptake (% of control uptake)	K_m = μM	V_{max} = pmol/mg protein/min	CL_{int} = μl/mg protein/min (x-fold reduction to control)	Total/cell surface expression (x-fold of control)
rs71640247[††]	G197A	Trp66X	n.d.	n.d.	n.d.	—	n.d.
rs55971546[*,†,††]	G292A	Val98Ile[*,†]	40%	4.5[§§]	2.5[§§]	0.55 (5)	~1.4 / 1.0
			↔	15.0[¶¶]	↔	—	↔
rs143238545[*]	G431A	Cys144Tyr[*]	10%	2.8[§§]	1[§§]	0.36 (8)	~1.4 / 0.75
rs60380298[*,†,**,††]	G475A	Val159Ile[*,†]	80%	1.5[§§]	2.3[§§]	1.5 (0.5)	~1.2 / 0.70
				12.0[¶¶]	↑	—	↔
rs188096[‡,§,¶,††]	G511T	Ala171Ser[*,†,§]	80%	4.2[§§]	4.3[§§]	1 (3)	~1.3 / 0.80
			↔	26.0[¶¶]	↑	—	↔
			↔	n.d.	n.d.	—	↔
not assigned[†]	646insG	246X[†]	∅	—	—	—	n.d.
rs121917848[*,††,§]	T728C	Leu243Pro[*,§]	∅	—	—	—	~1.1 / 1.0
			∅	—	—	—	↔
rs72547505[*,§,††]	C785T	Thr262Met[*,§]	∅	—	—	—	↔
			∅	—	—	—	↔
not assigned[*]	A790G	Met264Val[*]	∅	—	—	—	~1.2 / 1.5
rs56398830[*,††,‡‡]	C868T	Pro290Ser[*,‡‡]	∅	—	—	—	~1.5 / 1.4
			∅	—	—	—	↔
rs71640248[††]	T886C	Phe296Leu	n.d.	n.d.	n.d.	—	n.d.
rs61966074[††]	T910C	Phe304Leu	n.d.	n.d.	n.d.	—	n.d.

n.d., not determined.
[*]Ho et al., 2011.
[†]Love et al., 2001.
[‡]Montagnani et al., 2006.
[§]Oelkers et al., 1997.
[¶]Pan et al., 2011.
[**]Renner, Harsch, Schaeffeler, Schwab et al., 2009.
[††]Renner, Harsch, Schaeffeler, Winter et al., 2009.
[‡‡]Wong et al., 1995.
[§§]Reference wild-type ASBT K_m = 1.7 μM, V_{max} = 4.7 pmol/mg protein/min, CL_{int} = 2.76 μl/mg protein/min, transiently transfected HeLa cells.
[¶¶]Reference wild-type ASBT K_m = 16 μM, V_{max} = not deterimined (n.d.), transiently transfected COS cells.

2008; Jung et al., 2004) as was also caused by a SLC10A2 haplotype block that includes 10 genetic variations (Renner, Harsch, Schaeffeler, Schwab et al., 2009). A patient with Morbus Crohn carried the ASBT loss of function mutation Pro290Ser (Wong et al., 1995) and other ASBT alterations were linked to increased susceptibility for cholelithiasis (Renner, Harsch, Schaeffeler, Winter et al., 2009) and risk of colorectal adenomas (Wang et al., 2001). On the other hand, these relationships could not be verified by a meta-analysis and case–control study (Grünhage et al., 2008; Tönjes et al., 2011). Further naturally occurring SNPs with (Val98Ile, Cys144Tyr) and without (Ala171Ser) diminished transport activity could also not be associated to a specific disorder (Ho et al., 2011; Love et al., 2001; Oelkers et al., 1997). Because of the ambiguous results, the impact of decreased ASBT expression or function for the mentioned disorders is currently not clear.

4.6. Membrane Insertion Topology Model of ASBT

The ASBT/Asbt protein represents a glycoprotein as confirmed by N-deglycosylation studies of protein preparations from native tissues and different in vitro models (Banerjee, Ray, Chang, & Swaan, 2005; Bonge et al., 2000; Chignard et al., 2001; Dawson et al., 2005; Duane et al., 2000; Shneider et al., 1995; Wong et al., 1995; Zhang et al., 2004). Of two potential N-glycosylation sites (Asn10, Asn328), Asn10 of human ASBT N terminus is utilized, which implies extracellular localization of the N-terminal domain (Dawson & Oelkers, 1995; Zhang et al., 2004). As the C terminus was demonstrated to be localized intracellular (Banerjee & Swaan, 2006; Hallén et al., 1999), the transmembrane structure of ASBT/Asbt likely consists of an odd number of transmembrane helices. Hydrophobicity analyses predicted protein structure of 7–9 TMDs (Zhang et al., 2004). Questionable regions lie between amino acid position 69–120 (one or two TMD) and 255–280 (no or one TMD) (Banerjee & Swaan, 2006). To distinguish between the 7- and 9-TMD models, various approaches have been used. These include membrane insertion scanning, indicating a 9-TMD model (Hallén et al., 1999) as well as N-glycosylation scanning (Zhang et al., 2004) and dual label epitope insertion (Banerjee & Swaan, 2006) that support a transmembrane organization with 7 TMDs. An ASBT antibody raised against amino acids 104–117 recognized ASBT and reduced transport activity suggesting extracellular localization of the epitope which fits to the 7-TMD model (Montagnani et al., 2009). However, other studies supported the 9-TMD model (Hallén, Fryklund, & Sachs, 2000; Hallén,

Mareninova et al., 2002). Therefore, Zhang et al. (2004) discussed the possibility of a substrate-sensitive re-entrant loop which could explain the dynamic coexistence of 7- and 9-TMD models. Furthermore, a homology model of ASBT was generated with bacteriorhodopsin as scaffold, sharing topological (7 TMDs) but not mechanistic similarity (Zhang et al., 2004). Recently, the crystal structure of ASBT from *N. meningitides* ($ASBT_{NM}$) was reported, which allowed additional insights into the organization of ASBT TMDs (Hu et al., 2011). $ASBT_{NM}$ shares 26% identity to human ASBT and transports TC in a sodium-dependent manner. It was suggested that $ASBT_{NM}$ could be an appropriate functional template model for vertebrate ASBT/Asbt. $ASBT_{NM}$ crystallization was done in presence of the substrate, TC, and the cosubstrate, sodium. X-ray diffraction showed a hydrophobic inward facing cavity which was surrounded by TMDs (Fig. 6). In contrast to predicted 7- or 9-TMD models of vertebrate ASBT/Asbt and NTCP/Ntcp with an N_{ex}/C_{in} localization (see also Section 3.6), 10 TMDs with N_{in}/C_{in} localization were observed for $ASBT_{NM}$ (Hu et al., 2011). As mentioned in Section 3.7, the first TMD of $ASBT_{NM}$

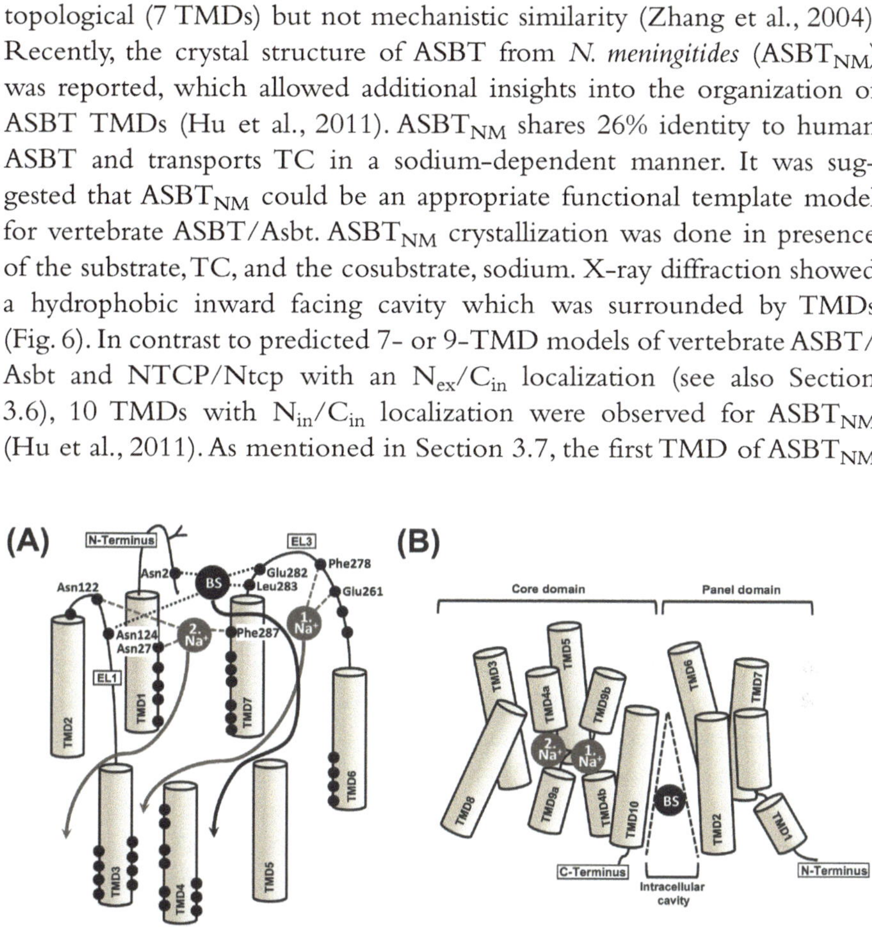

Figure 6 ***Transmembrane domain topology of human ASBT and $ASBT_{NM}$.*** (A) ASBT 7-TMD model and putative ASBT transport cycle (adopted from da Silva et al., 2011). Amino acid residues lining the transport pathway are designated as black dots, amino acids believed to be involved in sodium ion (Na^+) or bile salt (BS) binding are individually named. Binding of sodium ions lead to conformational changes of the ASBT protein followed by creation of the bile salt binding pocket and bile salt binding. Permeations of sodium ions (gray arrows) and bile salt (black arrows) follow the proposed pathways. (B) Ten-TMD architecture of ASBT of *Neisseria meningitides* ($ASBT_{NM}$) adopted from the inward facing crystal structure with bound two Na^+ and one bile salt (BS). Bile salt binding cavity lies between the core and the panel domain (Hu et al., 2011). BS, bile salt; EL, extracellular loop; TMD, transmembrane domain.

is likely not recognized in vertebrate SLC10 carriers, which favor a 9-TMD structure of ASBT/Asbt.

Furthermore, there is evidence that, similar to NTCP (Bijsmans et al., 2012), ASBT/Asbt aggregate to homo- and heterodimeric complexes. Western Blot analysis of ASBT/Asbt in tissues and in vitro transfected cell systems revealed molecular weights of 45–48 kDa (glycosylated ASBT/Asbt) as well as higher molecular bands of about 92–99 kDa which are believed to be dimeric forms of ASBT/Asbt (Bonge et al., 2000; Christie et al., 1996; Hallén et al., 1999; Kramer et al., 1993; Kramer, Wess, Bewersdorf et al., 1997; Kramer et al., 2001; Montagnani et al., 2009; Stravitz et al., 1997; Sun et al., 1998). In intact rabbit ileal tissues, the functional transporting unit of Na^{+}/bile acid cotransport was suggested to be composed of heterodimeric complexes of several transporter subunits with ileal lipid-binding protein ILBP (Kramer et al., 1993, 1998, 2001; Kramer, Girbig, Gutjahr, & Kowalewski, 1995; Kramer, Wess, Bewersdorf et al., 1997) (Fig. 3).

4.7. Proposed Substrate Interaction Sites and Transport Mode of ASBT

As ASBT is a pharmacological target for cholesterol-lowering therapies, strong effort was undertaken to identify functional important amino acids and domains in the protein. Photoaffinity labeling of rabbit Asbt with 7-azo derivatives of BAs identified amino acids 67–56 in C-terminal domain (TMD 7 and C terminus in 7-TMD model) as interaction site between substrate and transporter (Kramer, Nicol et al., 1992; Kramer et al. 2001). The working group of Peter W. Swaan in Baltimore used cysteine scanning mutagenesis and site-directed alkylation technique to identify solvent-accessible domains of ASBT. An ASBT mutant (Cys270Ala) was generated that was insensitive to thiol modification by methanethiosulfonate (MTS) reagents on cysteine residues but showed only a slight change in transport capacity and affinity to the substrate TC (Banerjee et al., 2005; Hallén et al., 2000). On the basis of the previously predicted 7-TMD model (Banerjee & Swaan, 2006; Zhang et al., 2004), this ASBT Cys270Ala mutant was used as scaffold to introduce substituted cysteine mutations on nearly all amino acid positions in TMD 1 (da Silva, Hussainzada, Khantwal, Polli, & Swaan, 2011), TMD 3 (Hussainzada, da Silva, & Swaan, 2009), TMD 4 (Khantwal & Swaan, 2008), TMD 6 (Hussainzada, Khandewal, & Swaan, 2008), TMD 7 (Gonzáles, Hussainzada, Swaan, Mackerell, & Polli, 2012; Hussainzada, Banerjee, & Swaan, 2006), EL 1 (Hussainzada, da Silva, Zhang,

& Swaan, 2008), and EL 3 (Banerjee, Hussainzada, Khandelwal, & Swaan, 2008). All mutants were preincubated with membrane impermeant MTS reagents in the presence or absence of sodium ions and/or ASBT substrates followed by measurements of TC transport. The different MTS preincubation conditions provoked conformational changes of sensitive binding regions for sodium and/or substrate and should result in different accessibilities of introduced cysteines to MTS reagents. According to the results and the additional information of a generated homology model of human ASBT (Zhang et al., 2004), a putative ASBT transport cycle was proposed (da Silva et al., 2011) (Fig. 6A). The first step includes the binding and stabilizing of the first sodium ion to the EL 3 through electrostatic interaction with the negatively charged sodium sensor Glu261 (Banerjee et al., 2008; Sun et al., 2006) and formation of cation-π forces with the aromatic ring of Phe278 (Banerjee et al., 2008). The resulting conformational change drives the sodium permeation and facilitates the binding of the second sodium ion to Asn122 in EL 1. Stabilization of this sodium occurs through amino-aromatic and cation-π interactions with Phe287 as well as with the amide proton in Asn27 (Sun et al., 2006). By means of the initiated structural change, a binding pocket is created in which BS can interact with amino acids Glu282, Leu293, Asn2 (Zhang et al., 2004), and possibly Asp124. Subsequent, permeation of sodium and BS follows the translocation pathway formed by amino acids in TMD 1 (da Silva et al., 2011), TMD 6 (Hussainzada, Khandewal et al., 2008), and TMD 7 (Gonzáles et al. 2012; Hussainzada et al., 2006) as well as in TMD 3 (Hussainzada et al., 2009) and TMD 4 (Khantwal & Swaan, 2008) as intracellular release route. Homology modeling of human ASBT showed an additional ligand-binding domain which is localized within a cleft area between EL 3 and N terminus. Val21 and Asn27 as well as Asn10 and Ala11 are forming stable interactions with COOH group and 3α- or 7α-OH group of BSs, respectively (Zhang et al., 2004).

On the other hand, the crystal structure of $ASBT_{NM}$ proposed a transport mode on the basis of 10 TMDs (presumably resembling nine in vertebrate) (Fig. 6B) (Hu et al., 2011). Between a panel-like domain formed by TMD 1/2 and 6/7 and a core domain formed by TMD 3–5 and 8–10, an inward open substrate-binding cavity is formed which extended halfway through the protein. Comparing the inward-facing $ASBT_{NM}$ structure with an outward-facing model suggested the movement of the panel domain during the transport process. This movement is proposed to be triggered through sodium ion binding at the two sodium ion binding sites identified in the core domain followed by conformational changes (Hu et al., 2011). The identification of two sodium-binding sites is consistent with the established transport stoichiometry of two sodium ions per BS for ASBT/Asbt and NTCP/Ntcp (Craddock et al., 1998; Hagenbuch & Meier, 1996; Weinman, 1997; Weinman et al., 1998) (Sections 3.3.1 and 4.3). The amino acids directly involved in sodium binding are highly conserved in all functionally characterized members of the SLC10 family (NTCP, ASBT, SOAT), reflecting their important role in the transport mode (Hu et al., 2011; Lionarons et al., 2012). On the other hand, interaction of $ASBT_{NM}$ with BS occurs in the intracellular binding cleft at amino acids localized in TMD 9 (Asn265) and TMD 10 (His294, Asn295, Asn265) (Fig. 6), corresponding to TMD 8 and 9 in mammal members, respectively (Hu et al., 2011).

5. NOVEL MEMBERS OF THE SLC10 FAMILY

5.1. The Steroid Sulfate Carrier SLC10A6 (SOAT)

The SLC10A6 transcript was identified for the first time in our group in 2003 from rat adrenal gland mRNA. This transcript consist of six exons and is coding for a 370-aa protein that was named sodium-dependent organic anion transporter SOAT (Geyer et al., 2004). By functional transport studies in *X. laevis* oocytes, we identified E_1S and DHEAS as substrates of this novel carrier. In agreement with the transport mode of NTCP and ASBT, the uptake of E_1S and DHEAS by SOAT was strictly sodium-dependent (Geyer et al., 2004). More systematic analyses of its functions and expression were performed with the human SOAT protein (Geyer et al., 2007). Human SOAT showed a predominant expression in testes and additional significant expression in pancreas, placenta, and mammary gland. So far, no SOAT/Soat transcript variants could be identified. For functional characterization, a stably transfected SOAT-HEK293 cell line was generated which showed sodium-dependent transport activity for DHEAS, E_1S, and PREGS with apparent K_m values of 28.7, 12.0, and 11.3 μM, respectively. In contrast, free steroids such as estrone and DHEA, glucuronidated steroids such as estradiol-3β-D-glucuronide, BSs such as C, CDC, DC, or lithocholate (LC) as well as the heart glycosides digoxin and ouabain, all with steroid structure, were not transported by SOAT, indicating that SOAT transport is quite specific for sulfated steroid hormones. In contrast, most BSs are not transported by SOAT, although they are quite good inhibitors. This inhibition was most pronounced by the sulfated BAs taurolithocholate-3-sulfate

(TLCS), glycolithocholate-3-sulfate, and lithocholate-3-sulfate, which almost completely blocked the transport of 2.5 μM DHEAS at 10-fold molar excess. Furthermore, organosulfates such as 4-sulfooxymethylpyrene (4-SMP) and α-naphthylsulfate as well as BSP proved to be potent SOAT inhibitors. By direct transport experiments, it was shown that 2-SMP, 4-SMP, and TLCS are transported substrates by SOAT in a sodium-dependent manner, thus, indicating that the substrate spectrum is not limited to sulfated steroid hormones alone (Geyer et al., 2007). In contrast, the sulfated thyroid hormones, sT3 and sT4, were not transported by SOAT although they are substrates for NTCP (Visser et al., 2010).

The human SOAT protein consists of 377 aa and exhibits higher sequence identity/similarity of 42%/70% to the human ASBT as to the human NTCP (33%/63%). SOAT has a calculated molecular mass of 41.2 kDa but appeared after immunoprecipitation and SDS–PAGE separation with an additional band at 46 kDa that disappeared after deglycosylation by PNGase F treatment. This finding is consistent with N-glycosylation of at least one of the potential Asn4, Asn14, or Asn157 N-glycosylation sites. Similar to the NTCP and ASBT, several membrane topology prediction programs calculated 7, 8, or 9 TMDs with a different orientation of the C- and N-terminal domains. By the use of an SOAT-specific antibody directed against the 2–17 N-terminal amino acids, the SOAT protein could be readily detected by immunofluorescence even when the cells were kept under nonpermeabilizing conditions indicating that the N terminus has an outward extracellular orientation. In contrast, an FLAG epitope added to the SOAT C terminus was only accessible by an anti-FLAG antibody if the cells were fixed and permeabilized, demonstrating an intracellular orientation of the C-terminal end. On the basis of these experiments, an 8-TMD topology was excluded but the calculated 7- and 9-TMD models were not further verified (Geyer et al., 2007).

5.2. The Orphan Carrier SLC10A3

The SLC10A3 gene was identified in 1988 by Alcalay & Toniolo. It consists of one exon coding for a 477-aa and 473-aa protein in human and rodents (mouse, rat), respectively. Owing to sequence identity of approximately 20% to the other SLC10 members, SLC10A3 was retrospectively included into the SLC10 family (Hagenbuch & Dawson, 2004). SLC10A3 transcripts were identified in human placenta, T lymphocytes, mammary carcinoma cells, neuroblastoma cells, chorioncarcinoma cells, and many other sites (Alcalay & Toniolo, 1988), but more systematic expression analyses were not performed in any species so far. Furthermore, the functional role of the SLC10A3 protein still remains completely unresolved.

5.3. The Neuronal Orphan Carrier SLC10A4

The rat Slc10a4 transcript was for the first time cloned in our group in 2004 (GenBank accession number AY825923 from December 6th 2004). Later on, the sequence of the human SLC10A4 transcript was published (Geyer et al., 2006; Splinter, Lazaridis, Dawson, & LaRusso, 2006). The SLC10A4/Slc10a4 genes of different species consist of three exons which encode for the 437-aa SLC10A4 protein in man, rat, and mouse. This protein shows the highest sequence homology (30% identity, 54% similarity) to NTCP. Systematic expression analyses in man, rat, and mouse revealed highest SLC10A4/Slc10a4 expression levels in the brain of all species. Furthermore, significant expression levels were detected in rat adrenal gland, mouse eye, and human small intestine (Geyer et al., 2008). Expression analyses at the protein level then revealed expression of the SLC10A4 protein in vesicles of cholinergic and monoaminergic neurons of the central and peripheral nervous system, so that a function for neurotransmitter release and/or uptake rather than for BS transport was considered (Burger et al., 2011; Geyer et al., 2008). Indeed transport experiments performed in transfected HEK293, CHO, and U2OS cells or *X. laevis* oocytes showed no transport function for BSs and steroid sulfates as the typical substrates of the other SLC10 carriers (Bijsmans et al., 2012; Geyer et al., 2008; Splinter et al., 2006; Visser et al., 2010). Currently, no transport function is known so that SLC10A4 still remains an orphan carrier.

5.4. The Liver and Kidney Orphan Carrier SLC10A5

The SLC10A5 transcripts from man, rat, and mouse were cloned and analyzed in our group between 2004 and 2007 (Fernandes et al., 2007). The human SLC10A5 protein consists of 438 aa and is coded from only one coding exon of the SLC10A5 gene. SLC10A5 has the highest sequence identity of 34% to the SLC10A3 protein, whereas sequence homology to NTCP, ASBT, and SOAT is quite low, at around 22%. After expression in HEK293 or U2SO cells, the SLC10A5 protein was localized to the Golgi network and could not be detected at the plasma membrane (Bijsmans et al., 2012). By contrast, in *X. laevis* oocytes, the SLC10A5 protein was sorted into the plasma membrane but showed similar as SLC10A4 no transport activity for BSs or steroid sulfates (Fernandes et al., 2007). SLC10A5 mRNA expression was systematically analyzed for human, rat, and mouse by real-time PCR and consistently showed a predominant expression in the liver and kidney. By further in situ hybridization experiments in the rat, Slc10a5 mRNA expression could be directed to hepatocytes in the liver and to epithelial cells of the proximal tubules in the kidney (Fernandes et al., 2007).

5.5. The Discrete Orphan Carrier SLC10A7

Among the SLC10 carriers, SLC10A7 phylogenetically has a somewhat discrete position. It was primarily cloned from human, rat, mouse, and frog between 2004 and 2007 (Godoy et al., 2007) and was subsequently assigned to the SLC10 carrier family by the Hugo Gene Nomenclature Committee despite its low sequence identity of <15% to the SLC10A1–A6 members. Although the tissue expression, membrane topology, and transport function of SLC10A7 were intensely investigated in our laboratory over the past years, its functional properties could not completely be elucidated. SLC10A7 is widely expressed in most of the tissues analyzed in rat, mouse, and frog including gastrointestinal tract, liver, kidney, heart, brain, lung, spleen, adrenal gland, and ovary. In man, SLC10A7 showed the highest expression levels in testis and liver (Godoy et al., 2007). The SLC10A7 gene is by far the most complex SLC10 gene and consists of 12 coding exons from which at least 10 different transcript variants and at least eight protein isoforms with different size are expressed (Godoy et al., 2007).

Apart from several eukaryotic SLC10A7 sequences, a plenty of SLC10A7 homologs were identified in prokaryotes (Godoy et al., 2007), which share <20% sequence identity to the eukaryotic proteins. For functional transport measurements, the human SLC10A7 protein was expressed in *X. laevis* oocytes but showed no transport activity for the typical SLC10 substrates TC, C, CDC, E_1S, DHEAS, and PREGS. However, disruption of a SLC10A7-homologous gene in *Candida albicans* cells yielded a hypersensitive phenotype against high extracellular calcium concentrations and, therefore, the *C. albicans* SLC10A7 homolog was referred to as RCH1 (regulator of the Ca^{2+} homoeostasis 1) (Jiang et al., 2012). In the RCH1 mutants, the Ca^{2+} elimination was comparable to the wild type, so that RCH1 was considered to be a negative regulator of cellular calcium influx. However, whether vertebrate SLC10A7 homologs of RCH1 also are involved in the regulation of calcium homeostasis is currently unknown.

6. CONCLUSION

The SLC10 family is a rather heterologous gene family with respect to gene structure, expression sites, and substrate patterns. So far, observed substrate transport has always been sodium-dependent. However, the functions of some SLC10 members are not resolved. To elucidate this problem, research will need to be carried out on yet unknown substrates or on protein functions eventually different from transport. A future approach to uncover the function of the orphan members will be the use of knockout mice that were up to now not investigated. Finally, the SLC10 family is not restricted to transporters for bile salts. Already suggested structure–function relationships of the carrier members and of their membrane topology allow conclusions on the proposed transport mechanistics.

ABBREVIATIONS

aa amino acid
AKR aldo-keto reductase family
ASBT apical sodium-dependent bile acid transporter
BA bile acid
BAAT bile acid-CoA amino acid N-acyltransferase
BACL bile acid-CoA ligase
BARI bile acid reabsorption inhibitor
BS bile salt
BSEP bile salt export pump
BSP bromosulfophthalein
C cholate
CA cholic acid
CDC chenodeoxycholate
CDCA chenodeoxycholic acid
CGamF cholylglycylamido fluorescein
CL_{int} internal substrate clearance
CYP cytochrome P450
DC deoxycholate
DHCA dehydrocholic acid
DHEA(S) dehydroepiandrosterone (sulfate)
EHC enterohepatic circulation
EL extracellular loop
E_1S estrone-3-sulfate

Ethinyl-E2-S ethinylestradiol 3-O-sulfate
FGF fibroblast growth factor
FRET fluorescence resonance energy transfer
FXR farnesoid X receptor
GC glycocholate
GCA glycocholic acid
GFP green fluorescence protein
GPCR G-protein coupled cell membrane receptor
HCC hepatocellular carcinoma
HSD hydroxysteroid dehydrogenase
IBAT ileal bile acid transporter
ICG indocyanine green
IL interleukin
ILBP ileal lipid binding protein
JNK c-Jun N-terminal kinase
LC lithocholate
LPS lipopolysaccharide
LXR liver X receptor
MCT monocarboxylate transporter
MDR multidrug resistance protein
MRP multidrug resistance associated protein
MT mononucleated trophoblast
MTS methanethiosulfonate
NTCP Na+-taurocholate cotransporting polypeptide
OA organic anion
OATP organic anion transporting polypeptide
OST organic solute transporter
PDB protein data base
PFIC progressive familial intrahepatic cholestasis
PREGS pregnenolone 3-sulfate
QSAR quantitative structure–activity relationship
SGLT sodium-dependent glucose transporter
SHP small heterodimer partner
SLC solute carrier
SMP sulfooxymethylpyrene
SNP single-nucleotide polymorphisms
SOAT sodium-dependent organic anion transporter
t-Asbt truncated Asbt
TC taurocholate
TCDC taurochenodeoxycholate
TCDCA taurochenodeoxycholic acid
TDHC taurodehydrocholate
TLC(S) taurolithocholate (3-sulfate)
TMD transmembrane domain
(T)UDCA (tauro)ursodeoxycholic acid
YFP yellow fluorescence protein

REFERENCES

Abu-Hayyeh, S., Martinez-Becerra, P., Sheikh Abdul Kadir, S. H., Selden, C., Romero, M. R., Rees, M., et al. (2010). Inhibition of Na^+-taurocholate Co-transporting polypeptide-mediated bile acid transport by cholestatic sulfated progesterone metabolites. *Journal of Biological Chemistry*, *285*(22), 16504–16512.

Alcalay, M., & Toniolo, D. (1988). CpG islands of the X chromosome are gene associated. *Nucleic Acids Research*, *16*(20), 9527–9543.

Alnouti, Y. (2009). Bile Acid sulfation: a pathway of bile acid elimination and detoxification. *Toxicological Sciences*, *108*(2), 225–246.

Alpini, G., Glaser, S., Baiocchi, L., Francis, H., Xia, X., & Lesage, G. (2005). Secretin activation of the apical Na^+-dependent bile acid transporter is associated with cholehepatic shunting in rats. *Hepatology*, *41*(5), 1037–1045.

Alpini, G., Glaser, S., Robertson, W., Phinizy, J. L., Rodgers, R. E., Caligiuri, A., et al. (1997). Bile acids stimulate proliferative and secretory events in large but not small cholangiocytes. *American Journal of Physiology*, *273*(2 Pt 1), G518–G529.

Alpini, G., Glaser, S. S., Rodgers, R., Phinizy, J. L., Robertson, W. E., Lasater, J., et al. (1997). Functional expression of the apical Na^+-dependent bile acid transporter in large but not small rat cholangiocytes. *Gastroenterology*, *113*(5), 1734–1740.

Alrefai, W. A., & Gill, R. K. (2007). Bile acid transporters: structure, function, regulation and pathophysiological implications. *Pharmaceutical Research*, *24*(10), 1803–1823.

Ananthanarayanan, M., Balasubramanian, N., Makishima, M., Mangelsdorf, D. J., & Suchy, F. J. (2001). Human bile salt export pump promoter is transactivated by the farnesoid X receptor/bile acid receptor. *Journal of Biological Chemistry*, *276*(31), 28857–28865.

Ananthanarayanan, M., Ng, O. C., Boyer, J. L., & Suchy, F. J. (1994). Characterization of cloned rat liver Na(+)-bile acid cotransporter using peptide and fusion protein antibodies. *American Journal of Physiology*, *267*(4 Pt 1), G637–G643.

Anwer, M. S., & Hegner, D. (1978). Effect of Na on bile acid uptake by isolated rat hepatocytes. Evidence for a heterogeneous system. *Hoppe-Seyler's Zeitschrift für physiologische Chemie*, *359*(2), 181–192.

Azer, S. A., & Stacey, N. H. (1993). Differential effects of cyclosporin A on the transport of bile acids by human hepatocytes. *Biochemical Pharmacology*, *46*(5), 813–819.

Bachrach, W. H., & Hofmann, A. F. (1982a). Ursodeoxycholic acid in the treatment of cholesterol cholelithiasis. Part I. *Digestive Diseases and Sciences*, *27*(8), 737–761.

Bachrach, W. H., & Hofmann, A. F. (1982b). Ursodeoxycholic acid in the treatment of cholesterol cholelithiasis. Part II. *Digestive Diseases and Sciences*, *27*(9), 833–856.

Balakrishnan, A., & Polli, J. E. (2006). Apical sodium dependent bile acid transporter (ASBT, SLC10A2): a potential prodrug target. *Molecular Pharmacology*, *3*(3), 223–230.

Balesaria, S., Pell, R. J., Abbott, L. J., Tasleem, A., Chavele, K. M., Barley, N. F., et al. (2008). Exploring possible mechanisms for primary bile acid malabsorption: evidence for different regulation of ileal bile acid transporter transcripts in chronic diarrhoea. *European Journal of Gastroenterology & Hepatology*, *20*(5), 413–422.

Ballatori, N., Li, N., Fang, F., Boyer, J. L., Christian, W. V., & Hammond, C. L. (2009). OST alpha-OST beta: a key membrane transporter of bile acids and conjugated steroids. *Frontiers in Bioscience*, *14*, 2829–2844.

Banerjee, A., Hussainzada, N., Khandelwal, A., & Swaan, P. W. (2008). Electrostatic and potential cation-pi forces may guide the interaction of extracellular loop III with Na^+ and bile acids for human apical Na^+-dependent bile acid transporter. *Biochemical Journal*, *410*(2), 391–400.

Banerjee, A., Ray, A., Chang, C., & Swaan, P. W. (2005). Site-directed mutagenesis and use of bile acid-MTS conjugates to probe the role of cysteines in the human apical sodium-dependent bile acid transporter (SLC10A2). *Biochemistry*, *44*(24), 8908–8917.

Banerjee, A., & Swaan, P. W. (2006). Membrane topology of human ASBT (SLC10A2) determined by dual label epitope insertion scanning mutagenesis. New evidence for seven transmembrane domains. *Biochemistry, 45*(3), 943–953.

Baringhaus, K. H., Matter, H., Stengelin, S., & Kramer, W. (1999). Substrate specificity of the ileal and the hepatic Na(+)/bile acid cotransporters of the rabbit. II. A reliable 3D QSAR pharmacophore model for the ileal Na(+)/bile acid cotransporter. *Journal of Lipid Research, 40*(12), 2158–2168.

Barnard, J. A., & Ghishan, F. K. (1987). Taurocholate transport by human ileal brush border membrane vesicles. *Gastroenterology, 93*(5), 925–933.

Bartholomew, T. C., & Billing, B. H. (1983). The effect of 3-sulphation and taurine conjugation on the uptake of chenodeoxycholic acid by rat hepatocytes. *Biochimica et Biophysica Acta, 754*(1), 101–109.

Baumgartner, U., Baier, P., Birke, O., & Farthmann, E. H. (2003). The 3alpha-hydroxysteroid-dehydrogenase-mRNA and -protein are more prevalent in pericentral than in periportal hepatocytes. *European Journal of Gastroenterology & Hepatology, 15*(5), 509–513.

Bear, C., Shaffer, E. A., & Davison, J. S. (1987). The electrogenic effect of sodium taurocholate on rat hepatocyte couplets. *Proceedings of the Western Pharmacology Society, 30*, 109–111.

Bergheim, I., Harsch, S., Mueller, O., Schimmel, S., Fritz, P., & Stange, E. F. (2006). Apical sodium bile acid transporter and ileal lipid binding protein in gallstone carriers. *Journal of Lipid Research, 47*(1), 42–50.

Bernstein, C., Holubec, H., Bhattacharyya, A. K., Nguyen, H., Payne, C. M., Zaitlin, B., et al. (2011). Carcinogenicity of deoxycholate, a secondary bile acid. *Archives of Toxicology, 85*(8), 863–871.

Beuers, U. (2006). Drug insight: mechanisms and sites of action of ursodeoxycholic acid in cholestasis. *Nature Clinical Practice Gastroenterology & Hepatology, 3*(6), 318–328.

Bhat, B. G., Rapp, S. R., Beaudry, J. A., Napawan, N., Butteiger, D. N., Hall, K. A., et al. (2003). Inhibition of ileal bile acid transport and reduced atherosclerosis in apoE-/- mice by SC-435. *Journal of Lipid Research, 44*(9), 1614–1621.

Bijsmans, I. T., Bouwmeester, R. A., Geyer, J., Faber, K. N., & van de Graaf, S. F. (2012). Homo- and hetero-dimeric architecture of the human liver Na^+-dependent taurocholate co-transporting protein. *Biochemical Journal, 441*(3), 1007–1015.

Blumrich, M., Pack, R., Oesch, F., Petzinger, E., & Steinberg, P. (1994). Deficiency of bile acid transport and synthesis in oval cells from carcinogen-fed rats. *Hepatology, 19*(3), 722–727.

Bonge, H., Hallén, S., Fryklund, J., & Sjöström, J. E. (2000). Cytostar-T scintillating microplate assay for measurement of sodium-dependent bile acid uptake in transfected HEK-293 cells. *Analytical Biochemistry, 282*(1), 94–101.

Borlak, J., & Klutcka, T. (2004). Expression of basolateral and canalicular transporters in rat liver and cultures of primary hepatocytes. *Xenobiotica, 34*(11–12), 935–947.

Bossuyt, X., Müller, M., Hagenbuch, B., & Meier, P. J. (1996). Polyspecific drug and steroid clearance by an organic anion transporter of mammalian liver. *The Journal of Pharmacology and Experimental Therapeutics, 276*(3), 891–896.

Boyer, J. L. (1980). New concepts of mechanisms of hepatocyte bile formation. *Physiological Reviews, 60*(2), 303–326.

Boyer, J. L., & Bloomer, J. R. (1974). Canalicular bile secretion in man. Studies utilizing the biliary clearance of (14C)mannitol. *Journal of Clinical Investigation, 54*(4), 773–781.

Boyer, J. L., Hagenbuch, B., Ananthanarayanan, M., Suchy, F., Stieger, B., & Meier, P. J. (1993). Phylogenic and ontogenic expression of hepatocellular bile acid transport. *Proceedings of the National Academy of Sciences of the United States of America, 90*(2), 435–438.

Briz, O., Serrano, M. A., Rebollo, N., Hagenbuch, B., Meier, P. J., Koepsell, H., et al. (2002). Carriers involved in targeting the cytostatic bile acid-cisplatin derivatives cis-diammine-chloro-cholylglycinate-platinum(II) and cis-diammine-bisursodeoxycholate-platinum(II) toward liver cells. *Molecular Pharmacology, 61*(4), 853–860.

Burger, S., Döring, B., Hardt, M., Beuerlein, K., Gerstberger, R., & Geyer, J. (2011). Co-expression studies of the orphan carrier protein Slc10a4 and the vesicular carriers VAChT and VMAT2 in the rat central and peripheral nervous system. *Neuroscience, 193*, 109–121.

Calmus, Y., & Poupon, R. (1991). Ursodeoxycholic acid (UDCA) in the treatment of chronic cholestatic diseases. *Biochimie, 73*(10), 1335–1338.

Carlton, V. E., Harris, B. Z., Puffenberger, E. G., Batta, A. K., Knisely, A. S., Robinson, D. L., et al. (2003). Complex inheritance of familial hypercholanemia with associated mutations in TJP2 and BAAT. *Nature Genetics, 34*(1), 91–96.

Cattori, V., Eckhardt, U., & Hagenbuch, B. (1999). Molecular cloning and functional characterization of two alternatively spliced Ntcp isoforms from mouse liver1. *Biochimica et Biophysica Acta, 1445*(1), 154–159.

Chaplin, M. F. (1998). Bile acids, fibre and colon cancer: the story unfolds. *Journal of the Royal Society of Health, 118*(1), 53–61.

Chen, V. B., Arendall, W. B., III, Headd, J. J., Keedy, D. A., Immormino, R. M., Kapral, G. J., et al. (2010). MolProbity: all-atom structure validation for macromolecular crystallography. *Acta Crystallographica Section D: Biological Crystallography, 66*(Pt 1), 12–21.

Cheng, K., & Raufman, J. P. (2005). Bile acid-induced proliferation of a human colon cancer cell line is mediated by transactivation of epidermal growth factor receptors. *Biochemical Pharmacology, 70*(7), 1035–1047.

Chen, X., Lou, G., Meng, Z., & Huang, W. (2011). TGR5: a novel target for weight maintenance and glucose metabolism. *Experimental Diabetes Research, 2011*, 853501.

Chen, L., Yao, X., Young, A., McNulty, J., Anderson, D., Liu, Y., et al. (2012). Inhibition of apical sodium-dependent bile acid transporter as a novel treatment for diabetes. *American Journal of Physiology. Endocrinology and Metabolism, 302*(1), E68–E76.

Chey, W. D., Camilleri, M., Chang, L., Rikner, L., & Graffner, H. (2011). A randomized placebo-controlled phase IIb trial of a3309, a bile acid transporter inhibitor, for chronic idiopathic constipation. *American Journal of Gastroenterology, 106*(10), 1803–1812.

Chiang, J. Y., Kimmel, R., Weinberger, C., & Stroup, D. (2000). Farnesoid X receptor responds to bile acids and represses cholesterol 7alpha-hydroxylase gene (CYP7A1) transcription. *Journal of Biological Chemistry, 275*(15), 10918–10924.

Chignard, N., Mergey, M., Veissiere, D., Parc, R., Capeau, J., Poupon, R., et al. (2001). Bile acid transport and regulating functions in the human biliary epithelium. *Hepatology, 33*(3), 496–503.

Chignard, N., Mergey, M., Veissiere, D., Poupon, R., Capeau, J., Parc, R., et al. (2003). Bile salts potentiate adenylyl cyclase activity and cAMP-regulated secretion in human gallbladder epithelium. *American Journal of Physiology – Gastrointestinal and Liver Physiology, 284*(2), G205–G212.

Choi, M. K., Shin, H. J., Choi, Y. L., Deng, J. W., Shin, J. G., & Song, I. S. (2011). Differential effect of genetic variants of Na(+)-taurocholate co-transporting polypeptide (NTCP) and organic anion-transporting polypeptide 1B1 (OATP1B1) on the uptake of HMG-CoA reductase inhibitors. *Xenobiotica, 41*(1), 24–34.

Chong, C. P., Mills, P. B., McClean, P., Gissen, P., Bruce, C., Stahlschmidt, J., et al. (2012). Bile acid-CoA ligase deficiency–a new inborn error of bile acid metabolism. *Journal of Inherited Metabolic Disease, 35*(3), 521–530.

Christie, D. M., Dawson, P. A., Thevananther, S., & Shneider, B. L. (1996). Comparative analysis of the ontogeny of a sodium-dependent bile acid transporter in rat kidney and ileum. *American Journal of Physiology, 271*(2 Pt 1), G377–G385.

Clayton, P. T. (2011). Disorders of bile acid synthesis. *Journal of Inherited Metabolic Disease, 34*(3), 593–604.

Craddock, A. L., Love, M. W., Daniel, R. W., Kirby, L. C., Walters, H. C., Wong, M. H., et al. (1998). Expression and transport properties of the human ileal and renal sodium-dependent bile acid transporter. *American Journal of Physiology, 274*(1 Pt 1), G157–G169.

Cui, J. Y., Aleksunes, L. M., Tanaka, Y., Fu, Z. D., Guo, Y., Guo, G. L., et al. (2012). Bile acids via FXR initiate the expression of major transporters involved in the enterohepatic circulation of bile acids in newborn mice. *American Journal of Physiology – Gastrointestinal and Liver Physiology*, *302*(9), G979–G996.

Cui, J. Y., Gunewardena, S. S., Yoo, B., Liu, J., Renaud, H. J., Lu, H., et al. (2012). RNA-Seq reveals different mRNA abundance of transporters and their alternative transcript isoforms during liver development. *Toxicological Sciences*, *127*(2), 592–608.

da Silva, T. C., Hussainzada, N., Khantwal, C. M., Polli, J. E., & Swaan, P. W. (2011). Transmembrane helix 1 contributes to substrate translocation and protein stability of bile acid transporter SLC10A2. *Journal of Biological Chemistry*, *286*(31), 27322–27332.

Dawson, P. A. (2011). Role of the intestinal bile acid transporters in bile acid and drug disposition. *Handbook of Experimental Pharmacology*, *201*, 169–203.

Dawson, P. A., Haywood, J., Craddock, A. L., Wilson, M., Tietjen, M., Kluckman, K., et al. (2003). Targeted deletion of the ileal bile acid transporter eliminates enterohepatic cycling of bile acids in mice. *Journal of Biological Chemistry*, *278*(36), 33920–33927.

Dawson, P. A., Hubbert, M., Haywood, J., Craddock, A. L., Zerangue, N., Christian, W. V., et al. (2005). The heteromeric organic solute transporter alpha-beta, Ostalpha-Ostbeta, is an ileal basolateral bile acid transporter. *Journal of Biological Chemistry*, *280*(8), 6960–6968.

Dawson, P. A., Lan, T., & Rao, A. (2009). Bile acid transporters. *Journal of Lipid Research*, *50*(12), 2340–2357.

Dawson, P. A., & Oelkers, P. (1995). Bile acid transporters. *Current Opinion in Lipidology*, *6*(2), 109–114.

de Graaf, W., Häusler, S., Heger, M., van Ginhoven, T. M., van Capellen, G., Bennink, R. J., et al. (2011). Transporters involved in the hepatic uptake of (99m)Tc-mebrofenin and indocyanine green. *Journal of Hepatology*, *54*(4), 738–745.

Di Ciaula, A., Wang, D. Q., Wang, H. H., Bonfrate, L., & Portincasa, P. (2010). Targets for current pharmacologic therapy in cholesterol gallstone disease. *Gastroenterology Clinics of North America*, *39*(2), 245–264.

Dietschy, J. M. (1968). Mechanisms for the intestinal absorption of bile acids. *Journal of Lipid Research*, *9*(3), 297–309.

Dilger, K., Hohenester, S., Winkler-Budenhofer, U., Bastiaansen, B. A., Schaap, F. G., Rust, C., et al. (2012). Effect of ursodeoxycholic acid on bile acid profiles and intestinal detoxification machinery in primary biliary cirrhosis and health. *Journal of Hepatology*, *57*(1), 133–140.

Duane, W. C., Hartich, L. A., Bartman, A. E., & Ho, S. B. (2000). Diminished gene expression of ileal apical sodium bile acid transporter explains impaired absorption of bile acid in patients with hypertriglyceridemia. *Journal of Lipid Research*, *41*(9), 1384–1389.

Duane, W. C., & Javitt, N. B. (1999). 27-hydroxycholesterol: production rates in normal human subjects. *Journal of Lipid Research*, *40*(7), 1194–1199.

EASL Clinical Practice Guidelines: management of cholestatic liver diseases. (2009). *Journal of Hepatology*, *51*(2), 237–267.

Eloranta, J. J., Jung, D., & Kullak-Ublick, G. A. (2006). The human Na^+-taurocholate cotransporting polypeptide gene is activated by glucocorticoid receptor and peroxisome proliferator-activated receptor-gamma coactivator-1alpha, and suppressed by bile acids via a small heterodimer partner-dependent mechanism. *Molecular Endocrinology*, *20*(1), 65–79.

Eloranta, J. J., & Kullak-Ublick, G. A. (2008). The role of FXR in disorders of bile acid homeostasis. *Physiology (Bethesda.)*, *23*, 286–295.

Elsner, R., & Ziegler, K. (1989). Determination of the apparent functional molecular mass of the hepatocellular sodium-dependent taurocholate transporter by radiation inactivation. *Biochimica et Biophysica Acta*, *983*(1), 113–117.

Fan, J., Donkin, J., & Wellington, C. (2009). Greasing the wheels of Abeta clearance in Alzheimer's disease: the role of lipids and apolipoprotein E. *Biofactors*, *35*(3), 239–248.

Fehrenbach, T., Cui, Y., Faulstich, H., & Keppler, D. (2003). Characterization of the transport of the bicyclic peptide phalloidin by human hepatic transport proteins. *Naunyn-Schmiedeberg's Archives of Pharmacology*, *368*(5), 415–420.

Fernandes, C. F., Godoy, J. R., Döring, B., Cavalcanti, M. C., Bergmann, M., Petzinger, E., et al. (2007). The novel putative bile acid transporter SLC10A5 is highly expressed in liver and kidney. *Biochemical and Biophysical Research Communications*, *361*(1), 26–32.

Fiorucci, S., Mencarelli, A., Palladino, G., & Cipriani, S. (2009). Bile-acid-activated receptors: targeting TGR5 and farnesoid-X-receptor in lipid and glucose disorders. *Trends in Pharmacological Sciences*, *30*(11), 570–580.

Föllmann, W., Petzinger, E., & Kinne, R. K. (1990). Alterations of bile acid and bumetanide uptake during culturing of rat hepatocytes. *American journal of physiology*, *258*(4 Pt 1), C700–C712.

Forsgren, L. (1969). Studies on the intestinal absorption of labelled fat-soluble vitamins (A, D, E, and K) via the thoracic-duct lymph in the absence of bile in man. *Acta Chirurgica Scandinavica Supplement*, *399*, 1–29.

Friesema, E. C., Docter, R., Moerings, E. P., Stieger, B., Hagenbuch, B., Meier, P. J., et al. (1999). Identification of thyroid hormone transporters. *Biochemical and Biophysical Research Communications*, *254*(2), 497–501.

Friesema, E. C., Jansen, J., Jachtenberg, J. W., Visser, W. E., Kester, M. H., & Visser, T. J. (2008). Effective cellular uptake and efflux of thyroid hormone by human monocarboxylate transporter 10. *Molecular Endocrinology*, *22*(6), 1357–1369.

Frimmer, M., & Ziegler, K. (1988). The transport of bile acids in liver cells. *Biochimica et Biophysica Acta*, *947*(1), 75–99.

Fujino, H., Saito, T., Ogawa, S., & Kojima, J. (2005). Transporter-mediated influx and efflux mechanisms of pitavastatin, a new inhibitor of HMG-CoA reductase. *Journal of Pharmacy and Pharmacology*, *57*(10), 1305–1311.

Fu, D., Wakabayashi, Y., Lippincott-Schwartz, J., & Arias, I. M. (2011). Bile acid stimulates hepatocyte polarization through a cAMP-Epac-MEK-LKB1-AMPK pathway. *Proceedings of the National Academy of Sciences of the United States of America*, *108*(4), 1403–1408.

Gao, B., St-Pierre, M. V., Stieger, B., & Meier, P. J. (2004). Differential expression of bile salt and organic anion transporters in developing rat liver. *Journal of Hepatology*, *41*(2), 201–208.

Gardmo, C., Tamburro, A., Modica, S., & Moschetta, A. (2011). Proteomics for the discovery of nuclear bile acid receptor FXR targets. *Biochimica et Biophysica Acta*, *1812*(8), 836–841.

Gärtner, U., Goeser, T., Stiehl, A., Raedsch, R., & Wolkoff, A. W. (1990). Transport of chenodeoxycholic acid and its 3-alpha- and 7-alpha-sulfates by isolated perfused rat liver. *Hepatology*, *12*(4 Pt 1), 738–742.

Gavrilova, O., Geyer, J., & Petzinger, E. (2007). In vivo relevance of Mrp2-mediated biliary excretion of the Amanita mushroom toxin demethylphalloin. *Biochimica et Biophysica Acta*, *1768*(9), 2070–2077.

Geyer, J., Döring, B., Meerkamp, K., Ugele, B., Bakhiya, N., Fernandes, C. F., et al. (2007). Cloning and functional characterization of human sodium-dependent organic anion transporter (SLC10A6). *Journal of Biological Chemistry*, *282*(27), 19728–19741.

Geyer, J., Fernandes, C. F., Döring, B., Burger, S., Godoy, J. R., Rafalzik, S., et al. (2008). Cloning and molecular characterization of the orphan carrier protein Slc10a4: expression in cholinergic neurons of the rat central nervous system. *Neuroscience*, *152*(4), 990–1005.

Geyer, J., Godoy, J. R., & Petzinger, E. (2004). Identification of a sodium-dependent organic anion transporter from rat adrenal gland. *Biochemical and Biophysical Research Communications*, *316*(2), 300–306.

Geyer, J., Wilke, T., & Petzinger, E. (2006). The solute carrier family SLC10: more than a family of bile acid transporters regarding function and phylogenetic relationships. *Naunyn-Schmiedeberg's Archives of Pharmacology*, *372*, 413–431.

Glasinovic, J. C., Dumont, M., Duval, M., & Erlinger, S. (1975). Hepatocellular uptake of bile acids in the dog: evidence for a common carrier-mediated transport system. An indicator dilution study. *Gastroenterology*, *69*(4), 973–981.

Glasser, J. E., Weiner, I. M., & Lack, L. (1965). Comparative physiology of intestinal taurocholate transport. *American Journal of Physiology*, *208*, 359–362.

Gluud, C., & Christensen, E. (2002). Ursodeoxycholic acid for primary biliary cirrhosis. *Cochrane Database of Systematic Reviews* (1). CD000551.

Godoy, J. R., Fernandes, C., Döring, B., Beuerlein, K., Petzinger, E., & Geyer, J. (2007). Molecular and phylogenetic characterization of a novel putative membrane transporter (SLC10A7), conserved in vertebrates and bacteria. *European Journal of Cell Biology*, *86*(8), 445–460.

González, P. M., Acharya, C., Mackerell, A. D., Jr., & Polli, J. E. (2009). Inhibition requirements of the human apical sodium-dependent bile acid transporter (hASBT) using aminopiperidine conjugates of glutamyl-bile acids. *Pharmaceutical Research*, *26*(7), 1665–1678.

González, P. M., Hussainzada, N., Swaan, P. W., Mackerell, A. D., Jr., & Polli, J. E. (2012). Putative irreversible inhibitors of the human sodium-dependent bile acid transporter (hASBT; SLC10A2) support the role of transmembrane domain 7 in substrate binding/translocation. *Pharmaceutical Research*, *29*(7), 1821–1831.

Greupink, R., Dillen, L., Monshouwer, M., Huisman, M. T., & Russel, F. G. (2011). Interaction of fluvastatin with the liver-specific Na^+-dependent taurocholate cotransporting polypeptide (NTCP). *European Journal of Pharmaceutical Sciences*, *44*(4), 487–496.

Greupink, R., Nabuurs, S., Zarzycka, B., Verweij, V., Monshouwer, M., Huisman, M., et al. (2012). In silico identification of potential cholestasis-inducing agents via modeling of Na^+-dependent taurocholate cotransporting polypeptide (NTCP) substrate specificity. *Toxicological Sciences*10.1093/toxsci/kfs188.

Grube, M., Köck, K., Oswald, S., Draber, K., Meissner, K., Eckel, L., et al. (2006). Organic anion transporting polypeptide 2B1 is a high-affinity transporter for atorvastatin and is expressed in the human heart. *Clinical Pharmacology & Therapeutics*, *80*(6), 607–620.

Grüne, S., Meng, X. J., & Weinman, S. A. (1996). cAMP stimulates fluorescent bile acid uptake into hepatocytes by membrane hyperpolarization. *American Journal of Physiology*, *270*(2 Pt 1), G339–G346.

Grünhage, F., Jungck, M., Lamberti, C., Keppeler, H., Becker, U., Schulte-Witte, H., et al. (2008). Effects of common haplotypes of the ileal sodium dependent bile acid transporter gene on the development of sporadic and familial colorectal cancer: a case control study. *BMC Medical Genetics*, *9*, 70.

Gundala, S., Wells, L. D., Milliano, M. T., Talkad, V., Luxon, B. A., & Neuschwander-Tetri, B. A. (2004). The hepatocellular bile acid transporter Ntcp facilitates uptake of the lethal mushroom toxin alpha-amanitin. *Archives of Toxicology*, *78*(2), 68–73.

Hagenbuch, B. (1997). Molecular properties of hepatic uptake systems for bile acids and organic anions. *Journal of Membrane Biology*, *160*(1), 1–8.

Hagenbuch, B., & Dawson, P. (2004). The sodium bile salt cotransport family SLC10. *Pflügers Archiv European Journal of Physiology*, *447*(5), 566–570.

Hagenbuch, B., Lübbert, H., Stieger, B., & Meier, P. J. (1990). Expression of the hepatocyte Na^+/bile acid cotransporter in Xenopus laevis oocytes. *Journal of Biological Chemistry*, *265*(10), 5357–5360.

Hagenbuch, B., & Meier, P. J. (1994). Molecular cloning, chromosomal localization, and functional characterization of a human liver Na^+/bile acid cotransporter. *Journal of Clinical Investigation*, *93*(3), 1326–1331.

Hagenbuch, B., & Meier, P. J. (1996). Sinusoidal (basolateral) bile salt uptake systems of hepatocytes. *Seminars in Liver Disease*, *16*(2), 129–136.

Hagenbuch, B., & Meier, P. J. (2004). Organic anion transporting polypeptides of the OATP/SLC21 family: phylogenetic classification as OATP/SLCO superfamily, new nomenclature and molecular/functional properties. *Pflügers Archiv European Journal of Physiology.*, *447*(5), 653–665.

Hagenbuch, B., Stieger, B., Foguet, M., Lübbert, H., & Meier, P. J. (1991). Functional expression cloning and characterization of the hepatocyte Na^+/bile acid cotransport system. *Proceedings of the National Academy of Sciences of the United States of America*, *88*(23), 10629–10633.

Hallén, S., Björquist, A., Ostlund-Lindqvist, A. M., & Sachs, G. (2002). Identification of a region of the ileal-type sodium/bile acid cotransporter interacting with a competitive bile acid transport inhibitor. *Biochemistry*, *41*(50), 14916–14924.

Hallén, S., Branden, M., Dawson, P. A., & Sachs, G. (1999). Membrane insertion scanning of the human ileal sodium/bile acid co-transporter. *Biochemistry*, *38*(35), 11379–11388.

Hallén, S., Fryklund, J., & Sachs, G. (2000). Inhibition of the human sodium/bile acid cotransporters by side-specific methanethiosulfonate sulfhydryl reagents: substrate-controlled accessibility of site of inactivation. *Biochemistry*, *39*(22), 6743–6750.

Hallén, S., Mareninova, O., Branden, M., & Sachs, G. (2002). Organization of the membrane domain of the human liver sodium/bile acid cotransporter. *Biochemistry*, *41*(23), 7253–7266.

Halpern, M. D., Weitkamp, J. H., Mount Patrick, S. K., Dobrenen, H. J., Khailova, L., et al. (2010). Apical sodium-dependent bile acid transporter upregulation is associated with necrotizing enterocolitis. *American Journal of Physiology – Gastrointestinal and Liver Physiology*, *299*(3), G623–G631.

Han, Y. H., Busler, D., Hong, Y., Tian, Y., Chen, C., & Rodrigues, A. D. (2010). Transporter studies with the 3-O-sulfate conjugate of 17alpha-ethinylestradiol: assessment of human liver drug transporters. *Drug Metabolism and Disposition*, *38*(7), 1072–1082.

Hara, S., Higaki, J., Higashino, K., Iwai, M., Takasu, N., Miyata, K., et al. (1997). S-8921, an ileal Na^+/bile acid cotransporter inhibitor decreases serum cholesterol in hamsters. *Life Sciences*, *60*(24), L-70.

Hardikar, W., Ananthanarayanan, M., & Suchy, F. J. (1995). Differential ontogenic regulation of basolateral and canalicular bile acid transport proteins in rat liver. *Journal of Biological Chemistry*, *270*(35), 20841–20846.

Hata, S., Wang, P., Eftychiou, N., Ananthanarayanan, M., Batta, A., Salen, G., et al. (2003). Substrate specificities of rat oatp1 and ntcp: implications for hepatic organic anion uptake. *American Journal of Physiology – Gastrointestinal and Liver Physiology*, *285*(5), G829–G839.

Heubi, J. E., Setchell, K. D., & Bove, K. E. (2007). Inborn errors of bile acid metabolism. *Seminars in Liver Disease*, *27*(3), 282–294.

Higaki, J., Hara, S., Takasu, N., Tonda, K., Miyata, K., Shike, T., et al. (1998). Inhibition of ileal Na^+/bile acid cotransporter by S-8921 reduces serum cholesterol and prevents atherosclerosis in rabbits. *Arteriosclerosis, Thrombosis, and Vascular Biology*, *18*(8), 1304–1311.

Hill, M. J. (1991). Bile acids and colorectal cancer: hypothesis. *European Journal of Cancer Prevention*, *2*(Suppl. 1), 69–74.

Hislop, I. G. (1970). The absorption and entero-hepatic circulation of bile salts. An historical review. *The Medical journal of Australia*, *1*(24), 1223–1226.

Hofmann, A. F. (1984). Medical treatment of cholesterol gallstones by bile desaturating agents. *Hepatology*, *4*(Suppl. 5), 199S–208S.

Hofmann, A. F. (2009). The enterohepatic circulation of bile acids in mammals: form and functions. *Frontiers in Bioscience*, *14*, 2584–2598.

Hofmann, A. F., & Hagey, L. R. (2008). Bile acids: chemistry, pathochemistry, biology, pathobiology, and therapeutics. *Cellular and Molecular Life Sciences*, *65*(16), 2461–2483.

Hofmann, A. F., Zakko, S. F., Lira, M., Clerici, C., Hagey, L. R., Lambert, K. K., et al. (2005). Novel biotransformation and physiological properties of norursodeoxycholic acid in humans. *Hepatology*, *42*(6), 1391–1398.

Ho, R. H., Leake, B. F., Roberts, R. L., Lee, W., & Kim, R. B. (2004). Ethnicity-dependent polymorphism in Na^+-taurocholate cotransporting polypeptide (SLC10A1) reveals a domain critical for bile acid substrate recognition. *Journal of Biological Chemistry, 279*(8), 7213–7222.

Ho, R. H., Tirona, R. G., Leake, B. F., Glaeser, H., Lee, W., Lemke, C. J., et al. (2006). Drug and bile acid transporters in rosuvastatin hepatic uptake: function, expression, and pharmacogenetics. *Gastroenterology, 130*(6), 1793–1806.

Ho, R. H., Leake, B. F., Urquhart, B. L., Gregor, J. C., Dawson, P. A., & Kim, R. B. (2011). Functional characterization of genetic variants in the apical sodium-dependent bile acid transporter (ASBT; SLC10A2). *Journal of Gastroenterology and Hepatology, 26*(12), 1740–1748.

Holzer, A., Harsch, S., Renner, O., Strohmeyer, A., Schimmel, S., Wehkamp, J., et al. (2008). Diminished expression of apical sodium-dependent bile acid transporter in gallstone disease is independent of ileal inflammation. *Digestion, 78*(1), 52–59.

Horz, J. A., Honscha, W., & Petzinger, E. (1996). Bumetanide is not transported by the Ntcp or by the oatp: evidence for a third organic anion transporter in rat liver cells. *Biochimica et Biophysica Acta, 1300*(2), 114–118.

Hruz, P., Zimmermann, C., Gutmann, H., Degen, L., Beuers, U., Terracciano, L., et al. (2006). Adaptive regulation of the ileal apical sodium dependent bile acid transporter (ASBT) in patients with obstructive cholestasis. *Gut, 55*(3), 395–402.

Huang, H. C., Tremont, S. J., Lee, L. F., Keller, B. T., Carpenter, A. J., Wang, C. C., et al. (2005). Discovery of potent, nonsystemic apical sodium-codependent bile acid transporter inhibitors (Part 2). *Journal of Medicinal Chemistry., 48*(18), 5853–5868.

Huang, L., Zhao, A., Lew, J. L., Zhang, T., Hrywna, Y., Thompson, J. R., et al. (2003). Farnesoid X receptor activates transcription of the phospholipid pump MDR3. *Journal of Biological Chemistry, 278*(51), 51085–51090.

Huff, M. W., Telford, D. E., Edwards, J. Y., Burnett, J. R., Barrett, P. H., Rapp, S. R., et al. (2002). Inhibition of the apical sodium-dependent bile acid transporter reduces LDL cholesterol and apoB by enhanced plasma clearance of LDL apoB. *Arteriosclerosis, Thrombosis, and Vascular Biology, 22*(11), 1884–1891.

Hu, N. J., Iwata, S., Cameron, A. D., & Drew, D. (2011). Crystal structure of a bacterial homologue of the bile acid sodium symporter ASBT. *Nature, 478*(7369), 408–411.

Hussainzada, N., da Silva, T. C., & Swaan, P. W. (2009). The cytosolic half of helix III forms the substrate exit route during permeation events of the sodium/bile acid cotransporter ASBT. *Biochemistry, 48*(36), 8528–8539.

Hussainzada, N., da Silva, T. C., Zhang, E. Y., & Swaan, P. W. (2008). Conserved aspartic acid residues lining the extracellular loop 1 of sodium-coupled bile acid transporter ASBT Interact with Na^+ and 7alpha-OH moieties on the ligand cholestane skeleton. *Journal of Biological Chemistry, 283*(30), 20653–20663.

Hussainzada, N., Banerjee, A., & Swaan, P. W. (2006). Transmembrane domain VII of the human apical sodium-dependent bile acid transporter ASBT (SLC10A2) lines the substrate translocation pathway. *Molecular Pharmacology, 70*, 1565–1574.

Hussainzada, N., Khandewal, A., & Swaan, P. W. (2008). Conformational flexibility of helix VI is essential for substrate permeation of the human apical sodium-dependent bile acid transporter. *Molecular Pharmacology, 73*(2), 305–313.

Hylemon, P. B., Zhou, H., Pandak, W. M., Ren, S., Gil, G., & Dent, P. (2009). Bile acids as regulatory molecules. *Journal of Lipid Research, 50*(8), 1509–1520.

Ichihashi, T., Izawa, M., Miyata, K., Mizui, T., Hirano, K., & Takagishi, Y. (1998). Mechanism of hypocholesterolemic action of S-8921 in rats: S-8921 inhibits ileal bile acid absorption. *Journal of Pharmacology and Experimental Therapeutics, 284*(1), 43–50.

Ichikawa, R., Takayama, T., Yoneno, K., Kamada, N., Kitazume, M. T., Higuchi, H., et al. (2012). Bile acids induce monocyte differentiation toward interleukin-12 hypo-producing dendritic cells via a TGR5-dependent pathway. *Immunology, 136*(2), 153–162.

Igel, M., Arnold, K. A., Niemi, M., Hofmann, U., Schwab, M., Lutjohann, D., et al. (2006). Impact of the SLCO1B1 polymorphism on the pharmacokinetics and lipid-lowering efficacy of multiple-dose pravastatin. *Clinical Pharmacology & Therapeutics*, *79*(5), 419–426.

Inoue, M., Kinne, R., Tran, T., & Arias, I. M. (1982). Taurocholate transport by rat liver sinusoidal membrane vesicles: evidence of sodium cotransport. *Hepatology*, *2*(5), 572–579.

Iwasaki, T., Kondo, K., Nishitani, T., Kuroda, T., Hirakoso, K., Ohtani, A., et al. (1995). Arylnaphthalene lignans as novel series of hypolipidemic agents raising high-density lipoprotein level. *Chemical & Pharmaceutical Bulletin (Tokyo)*, *43*(10), 1701–1705.

Izzat, N. N., Deshazer, M. E., & Loose-Mitchell, D. S. (2000). New molecular targets for cholesterol-lowering therapy. *Journal of Pharmacology and Experimental Therapeutics*, *293*(2), 315–320.

Jacquemin, E., Hagenbuch, B., Stieger, B., Wolkoff, A. W., & Meier, P. J. (1994). Expression cloning of a rat liver Na(+)-independent organic anion transporter. *Proceedings of the National Academy of Sciences of the United States of America*, *91*(1), 133–137.

Jiang, L., Alber, J., Wang, J., Du, W., Yang, X., Li, X., et al. (2012). The Candida albicans plasma membrane protein Rch1p, a member of the vertebrate SLC10 carrier family, is a novel regulator of cytosolic Ca^{2+} homoeostasis. *Biochemical Journal*, *444*(3), 497–502.

Jung, D., Fantin, A. C., Scheurer, U., Fried, M., & Kullak-Ublick, G. A. (2004). Human ileal bile acid transporter gene ASBT (SLC10A2) is transactivated by the glucocorticoid receptor. *Gut*, *53*(1), 78–84.

Kagedahl, M., Swaan, P. W., Redemann, C. T., Tang, M., Craik, C. S., Szoka, F. C., Jr., et al. (1997). Use of the intestinal bile acid transporter for the uptake of cholic acid conjugates with HIV-1 protease inhibitory activity. *Pharmaceutical Research*, *14*(2), 176–180.

Karapanou, O., & Papadimitriou, A. (2011). Thyroid hormone transporters in the human. *Hormones (Athens)*, *10*(4), 270–279.

Keitel, V., Cupisti, K., Ullmer, C., Knoefel, W. T., Kubitz, R., & Häussinger, D. (2009). The membrane-bound bile acid receptor TGR5 is localized in the epithelium of human gallbladders. *Hepatology*, *50*(3), 861–870.

Keitel, V., & Häussinger, D. (2012). Perspective: TGR5 (Gpbar-1) in liver physiology and disease. *Clinics and Research in Hepatology and Gastroenterology*10.1016/j.clinre.2012.03.008.

Khantwal, C. M., & Swaan, P. W. (2008). Cytosolic half of transmembrane domain IV of the human bile acid transporter hASBT (SLC10A2) forms part of the substrate translocation pathway. *Biochemistry*, *47*(12), 3606–3614.

Khurana, S., Raufman, J. P., & Pallone, T. L. (2011). Bile acids regulate cardiovascular function. *Clinical and Translational Science*, *4*(3), 210–218.

Kim, D. C., Harrison, A. W., Ruwart, M. J., Wilkinson, K. F., Fisher, J. F., Hidalgo, I. J., et al. (1993). Evaluation of the bile acid transporter in enhancing intestinal permeability to renin-inhibitory peptides. *Journal of Drug Targeting*, *1*(4), 347–359.

Kim, J. Y., Kim, K. H., Lee, J. A., Namkung, W., Sun, A. Q., Ananthanarayanan, M., et al. (2002). Transporter-mediated bile acid uptake causes Ca^{2+}-dependent cell death in rat pancreatic acinar cells. *Gastroenterology*, *122*(7), 1941–1953.

Kim, R. B., Leake, B., Cvetkovic, M., Roden, M. M., Nadeau, J., Walubo, A., et al. (1999). Modulation by drugs of human hepatic sodium-dependent bile acid transporter (sodium taurocholate cotransporting polypeptide) activity. *Journal of Pharmacology and Experimental Therapeutics*, *291*(3), 1204–1209.

Kim, D. H., & Lee, J. W. (2011). Tumor suppressor p53 regulates bile acid homeostasis via small heterodimer partner. *Proceedings of the National Academy of Sciences of the United States of America*, *108*(30), 12266–12270.

Kitayama, K., Nakai, D., Kono, K., van der Hoop, A. G., Kurata, H., de Wit, et al. (2006). Novel non-systemic inhibitor of ileal apical Na+-dependent bile acid transporter reduces serum cholesterol levels in hamsters and monkeys. *European Journal of Pharmacology*, *539*(1–2), 89–98.

Klaassen, C. D., & Aleksunes, L. M. (2010). Xenobiotic, bile acid, and cholesterol transporters: function and regulation. *Pharmacological Reviews, 62*(1), 1–96.
Klaassen, C. D., & Watkins, J. B., III (1984). Mechanisms of bile formation, hepatic uptake, and biliary excretion. *Pharmacological Reviews, 36*(1), 1–67.
Koenen, A., Kroemer, H. K., Grube, M., & Meyer zu Schwabedissen, H. E. (2011). Current understanding of hepatic and intestinal OATP-mediated drug-drug interactions. *Expert Review of Clinical Pharmacology, 4*(6), 729–742.
Kolhatkar, V., Diao, L., Acharya, C., Mackerell, A. D., Jr., & Polli, J. E. (2012). Identification of novel nonsteroidal compounds as substrates or inhibitors of hASBT. *Journal of Pharmaceutical Sciences, 101*(1), 116–126.
Kolhatkar, V., & Polli, J. E. (2012). Structural requirements of bile acid transporters: C-3 and C-7 modifications of steroidal hydroxyl groups. *European Journal of Pharmaceutical Sciences, 46*(1–2), 86–99.
König, J. (2011). Uptake transporters of the human OATP family: molecular characteristics, substrates, their role in drug-drug interactions, and functional consequences of polymorphisms. *Handbook of Experimental Pharmacology, 201*, 1–28.
König, J., Klatt, S., Dilger, K., & Fromm, M. F. (2012). Characterization of ursodeoxycholic and norursodeoxycholic acid as substrates of the hepatic uptake transporters OATP1B1, OATP1B3, OATP2B1 and NTCP. *Basic & Clinical Pharmacology & Toxicology, 111*(2), 81–86.
Kosters, A., & Karpen, S. J. (2008). Bile acid transporters in health and disease. *Xenobiotica, 38*(7–8), 1043–1071.
Kotani, N., Maeda, K., Watanabe, T., Hiramatsu, M., Gong, L. K., Bi, Y. A., et al. (2011). Culture period-dependent changes in the uptake of transporter substrates in sandwich-cultured rat and human hepatocytes. *Drug Metabolism and Disposition, 39*(9), 1503–1510.
Kramer, W. (2011). Transporters, Trojan horses and therapeutics: suitability of bile acid and peptide transporters for drug delivery. *Journal of Biological Chemistry, 392*(1–2), 77–94.
Kramer, W., Corsiero, D., Friedrich, M., Girbig, F., Stengelin, S., & Weyland, C. (1998). Intestinal absorption of bile acids: paradoxical behaviour of the 14 kDa ileal lipid-binding protein in differential photoaffinity labelling. *Biochemical Journal, 333*(Pt 2), 335–341.
Kramer, W., Girbig, F., Glombik, H., Corsiero, D., Stengelin, S., & Weyland, C. (2001). Identification of a ligand-binding site in the Na^+/bile acid cotransporting protein from rabbit ileum. *Journal of Biological Chemistry, 276*(38), 36020–36027.
Kramer, W., Girbig, F., Gutjahr, U., & Kowalewski, S. (1995). Radiation-inactivation analysis of the Na^+/bile acid co-transport system from rabbit ileum. *Biochemical Journal, 306*(Pt 1), 241–246.
Kramer, W., Girbig, F., Gutjahr, U., Kowalewski, S., Jouvenal, K., Müller, G., et al. (1993). Intestinal bile acid absorption. Na(+)-dependent bile acid transport activity in rabbit small intestine correlates with the coexpression of an integral 93-kDa and a peripheral 14-kDa bile acid-binding membrane protein along the duodenum-ileum axis. *Journal of Biological Chemistry, 268*(24), 18035–18046.
Kramer, W., & Glombik, H. (2006). Bile acid reabsorption inhibitors (BARI): novel hypolipidemic drugs. *Current Medicinal Chemistry, 13*(9), 997–1016.
Kramer, W., Nicol, S. B., Girbig, F., Gutjahr, U., Kowalewski, S., & Fasold, H. (1992). Characterization and chemical modification of the Na(+)-dependent bile-acid transport system in brush-border membrane vesicles from rabbit ileum. *Biochimica et Biophysica Acta, 1111*(1), 93–102.
Kramer, W., Stengelin, S., Baringhaus, K. H., Enhsen, A., Heuer, H., Becker, W., et al. (1999). Substrate specificity of the ileal and the hepatic Na(+)/bile acid cotransporters of the rabbit. I. Transport studies with membrane vesicles and cell lines expressing the cloned transporters. *Journal of Lipid Research, 40*(9), 1604–1617.
Kramer, W., & Wess, G. (1996). Bile acid transport systems as pharmaceutical targets. *European Journal of Clinical Investigation, 26*(9), 715–732.

Kramer, W., Wess, G., Bewersdorf, U., Corsiero, D., Girbig, F., Weyland, C., et al. (1997). Topological photoaffinity labeling of the rabbit ileal Na^+/bile-salt-cotransport system. *European Journal of Biochemistry*, *249*(2), 456–464.
Kramer, W., Wess, G., Enhsen, A., Bock, K., Falk, E., Hoffmann, A., et al. (1994). Bile acid derived HMG-CoA reductase inhibitors. *Biochimica et Biophysica Acta*, *1227*(3), 137–154.
Kramer, W., Wess, G., Enhsen, A., Falk, E., Hoffmann, A., Neckermann, G., et al. (1997). Modified bile acids as carriers for peptides and drugs. *Journal of Control Release*, *46*(1–2), 17–30.
Kramer, W., Wess, G., Neckermann, G., Schubert, G., Fink, J., Girbig, F., et al. (1994). Intestinal absorption of peptides by coupling to bile acids. *Journal of Biological Chemistry*, *269*(14), 10621–10627.
Kramer, W., Wess, G., Schubert, G., Bickel, M., Girbig, F., Gutjahr, U., et al. (1992). Liver-specific drug targeting by coupling to bile acids. *Journal of Biological Chemistry*, *267*(26), 18598–18604.
Krieger, E., Joo, K., Lee, J., Lee, J., Raman, S., Thompson, J., et al. (2009). Improving physical realism, stereochemistry, and side-chain accuracy in homology modeling: four approaches that performed well in CASP8. *Proteins*, 77(Suppl. 9), 114–122.
Kroker, R., Anwer, M. S., & Hegner, D. (1978). The lack of active bile acid transport in AS-30 D ascites hepatoma cells. *Naunyn-Schmiedeberg's Archives of Pharmacology*, *303*(3), 299–301.
Kröncke, K. D., Fricker, G., Meier, P. J., Gerok, W., Wieland, T., & Kurz, G. (1986). Alpha-Amanitin uptake into hepatocytes. Identification of hepatic membrane transport systems used by amatoxins. *Journal of Biological Chemistry*, *261*(27), 12562–12567.
Kullak-Ublick, G. A., Beuers, U., & Paumgartner, G. (1996). Molecular and functional characterization of bile acid transport in human hepatoblastoma HepG2 cells. *Hepatology*, *23*(5), 1053–1060.
Kullak-Ublick, G. A., Glasa, J., Böker, C., Oswald, M., Grützner, U., Hagenbuch, B., et al. (1997). Chlorambucil-taurocholate is transported by bile acid carriers expressed in human hepatocellular carcinomas. *Gastroenterology*, *113*(4), 1295–1305.
Kullak-Ublick, G. A., Ismair, M. G., Kubitz, R., Schmitt, M., Häussinger, D., Stieger, B., et al. (2000). Stable expression and functional characterization of a Na+-taurocholate cotransporting green fluorescent protein in human hepatoblastoma HepG2 cells. *Cytotechnology*, *34*(1–2), 1–9.
Kurata, H., Suzuki, S., Ohhata, Y., Ikeda, T., Hasegawa, T., Kitayama, K., et al. (2004). A novel class of apical sodium-dependent bile acid transporter inhibitors: the amphiphilic 4-oxo-1-phenyl-1,4-dihydroquinoline derivatives. *Bioorganic & Medicinal Chemistry Letters*, *14*(5), 1183–1186.
Larkin, M. A., Blackshields, G., Brown, N. P., Chenna, R., McGettigan, P. A., McWilliam, H., et al. (2007). Clustal W and Clustal X version 2.0. *Bioinformatics*, *23*(21), 2947–2948.
Lazaridis, K. N., Pham, L., Tietz, P., Marinelli, R. A., deGroen, P. C., Levine, S., et al. (1997). Rat cholangiocytes absorb bile acids at their apical domain via the ileal sodium-dependent bile acid transporter. *Journal of Clinical Investigation*, *100*(11), 2714–2721.
Lazaridis, K. N., Tietz, P., Wu, T., Kip, S., Dawson, P. A., & LaRusso, N. F. (2000). Alternative splicing of the rat sodium/bile acid transporter changes its cellular localization and transport properties. *Proceedings of the National Academy of Sciences of the United States of America*, *97*(20), 11092–11097.
Letschert, K., Faulstich, H., Keller, D., & Keppler, D. (2006). Molecular characterization and inhibition of amanitin uptake into human hepatocytes. *Toxicological Sciences.*, *91*(1), 140–149.
Leuschner, U., Leuschner, M., Sieratzki, J., Kurtz, W., & Hübner, K. (1985). Gallstone dissolution with ursodeoxycholic acid in patients with chronic active hepatitis and two years follow-up. A pilot study. *Digestive Diseases and Sciences*, *30*(7), 642–649.
Lewis, M. C., Brieaddy, L. E., & Root, C. (1995). Effects of 2164U90 on ileal bile acid absorption and serum cholesterol in rats and mice. *Journal of Lipid Research*, *36*(5), 1098–1105.

Liang, D., Hagenbuch, B., Stieger, B., & Meier, P. J. (1993). Parallel decrease of Na(+)-taurocholate cotransport and its encoding mRNA in primary cultures of rat hepatocytes. *Hepatology*, *18*(5), 1162–1166.

Li, T., & Chiang, J. Y. (2012). Bile Acid signaling in liver metabolism and diseases. *Journal of Lipids*, *2012*, 754067.

Lidofsky, S. D., Fitz, J. G., Weisiger, R. A., & Scharschmidt, B. F. (1993). Hepatic taurocholate uptake is electrogenic and influenced by transmembrane potential difference. *American Journal of Physiology*, *264*(3 Pt 1), G478–G485.

Li, T., Holmstrom, S. R., Kir, S., Umetani, M., Schmidt, D. R., Kliewer, S. A., et al. (2011). The G protein-coupled bile acid receptor, TGR5, stimulates gallbladder filling. *Molecular Endocrinology*, *25*(6), 1066–1071.

Lindstedt, S. (1957). The turnover of cholic acid in man: bile acids and steroids. *Acta Physiologica Scandinavica*, *40*(1), 1–9.

Lionarons, D. A., Boyer, J. L., & Cai, S. Y. (2012). Evolution of substrate specificity for the bile salt transporter ABST (SLC10A2). *Journal of Lipid Research*, *53*(8), 1535–1542.

Lischka, K., Starke, D., Failing, K., Herling, A., Kramer, W., & Petzinger, E. (2003). Hepatobiliary elimination of bile acid-modified oligodeoxynucleotides in Wistar and TR- rats: evidence for mrp2 as carrier for oligodeoxynucleotides. *Biochemical Pharmacology*, *66*(4), 565–577.

Li, H., Xu, G., Shang, Q., Pan, L., Shefer, S., Batta, A. K., et al. (2004). Inhibition of ileal bile acid transport lowers plasma cholesterol levels by inactivating hepatic farnesoid X receptor and stimulating cholesterol 7 alpha-hydroxylase. *Metabolism*, *53*(7), 927–932.

Love, M. W., Craddock, A. L., Angelin, B., Brunzell, J. D., Duane, W. C., & Dawson, P. A. (2001). Analysis of the ileal bile acid transporter gene, SLC10A2, in subjects with familial hypertriglyceridemia. *Arteriosclerosis, Thrombosis, and Vascular Biology*, *21*(12), 2039–2045.

Lücke, H., Stange, G., Kinne, R., & Murer, H. (1978). Taurocholate–sodium co-transport by brush-border membrane vesicles isolated from rat ileum. *Biochemical Journal*, *174*(3), 951–958.

Lundåsen, T., Andersson, E. M., Snaith, M., Lindmark, H., Lundberg, J., Östlund-Lindqvist, A. M., et al. (2012). Inhibition of intestinal bile acid transporter slc10a2 improves triglyceride metabolism and normalizes elevated plasma glucose levels in mice. *PLoS One*, 7(5). e37787.

Makishima, M., Okamoto, A. Y., Repa, J. J., Tu, H., Learned, R. M., Luk, A., et al. (1999). Identification of a nuclear receptor for bile acids. *Science*, *284*(5418), 1362–1365.

Maldonado-Valderrama, J., Wilde, P., Macierzanka, A., & Mackie, A. (2011). The role of bile salts in digestion. *Advances in Colloid and Interface Science*, *165*(1), 36–46.

Mareninova, O., Shin, J. M., Vagin, O., Turdikulova, S., Hallén, S., & Sachs, G. (2005). Topography of the membrane domain of the liver Na^{+}-dependent bile acid transporter. *Biochemistry*, *44*(42), 13702–13712.

Marin, J. J.G., Romero, M. R., Vallejo, M., Perez, M. J., & Briz, O. (2005). Emerging interest in bile acid transporters in pathophysiology and pharmacology. *Medical Hypothesis and Research*, *2*(3), 425–448.

Marin, J. J., Serrano, M. A., el-Mir, M.Y., Eleno, N., & Boyd, C. A. (1990). Bile acid transport by basal membrane vesicles of human term placental trophoblast. *Gastroenterology*, *99*(5), 1431–1438.

Marion, T. L., Perry, C. H., St Claire, R. L., & Brouwer, K. L. (2012). Endogenous bile acid disposition in rat and human sandwich-cultured hepatocytes. *Toxicology and Applied Pharmacology*, *261*(1), 1–9.

Meier, P. J. (1995). Molecular mechanisms of hepatic bile salt transport from sinusoidal blood into bile. *American Journal of Physiology*, *269*(6 Pt 1), G801–G812.

Meier-Abt, F., Faulstich, H., & Hagenbuch, B. (2004). Identification of phalloidin uptake systems of rat and human liver. *Biochimica et Biophysica Acta*, *1664*(1), 64–69.

Meier, P. J., Eckhardt, U., Schroeder, A., Hagenbuch, B., & Stieger, B. (1997). Substrate specificity of sinusoidal bile acid and organic anion uptake systems in rat and human liver. *Hepatology*, *26*(6), 1667–1677.

Meier, Y., Eloranta, J. J., Darimont, J., Ismair, M. G., Hiller, C., Fried, M., et al. (2007). Regional distribution of solute carrier mRNA expression along the human intestinal tract. *Drug Metabolism and Disposition*, *35*(4), 590–594.

Meier, P. J., & Stieger, B. (2002). Bile salt transporters. *Annual Review of Physiology*, *64*, 635–661.

Mita, S., Suzuki, H., Akita, H., Hayashi, H., Onuki, R., Hofmann, A. F., et al. (2006). Inhibition of bile acid transport across Na^+/taurocholate cotransporting polypeptide (SLC10A1) and bile salt export pump (ABCB 11)-coexpressing LLC-PK1 cells by cholestasis-inducing drugs. *Drug Metabolism and Disposition*, *34*(9), 1575–1581.

Modica, S., Gadaleta, R. M., & Moschetta, A. (2010). Deciphering the nuclear bile acid receptor FXR paradigm. *Nuclear Receptor Signaling*, *8*. e005.

Montagnani, M., Abrahamsson, A., Galman, C., Eggertsen, G., Marschall, H. U., Ravaioli, E., et al. (2006). Analysis of ileal sodium/bile acid cotransporter and related nuclear receptor genes in a family with multiple cases of idiopathic bile acid malabsorption. *World Journal of Gastroenterology*, *12*(47), 7710–7714.

Montagnani, M., Love, M. W., Rossel, P., Dawson, P. A., & Qvist, P. (2001). Absence of dysfunctional ileal sodium-bile acid cotransporter gene mutations in patients with adult-onset idiopathic bile acid malabsorption. *Scandinavian Journal of Gastroenterology*, *36*(10), 1077–1080.

Montagnani, M., Marangoni, A., Roda, A., Azzaroli, F., Mazzella, G., Roda, E., et al. (2009). Generation of a novel antibody probe to the apical sodium-dependent bile acid transporter that inhibits ileal bile acid absorption. *Molecular Pharmaceutics*, *6*(3), 1012–1018.

Münter, K., Mayer, D., & Faulstich, H. (1986). Characterization of a transporting system in rat hepatocytes. Studies with competitive and non-competitive inhibitors of phalloidin transport. *Biochimica et Biophysica Acta*, *860*(1), 91–98.

Nagengast, F. M., Grubben, M. J., & van Munster, I. P. (1995). Role of bile acids in colorectal carcinogenesis. *European Journal of Cancer*, *31A*(7–8), 1067–1070.

Niemi, M., Pasanen, M. K., & Neuvonen, P. J. (2011). Organic anion transporting polypeptide 1B1: a genetically polymorphic transporter of major importance for hepatic drug uptake. *Pharmacological Reviews*, *63*(1), 157–181.

Niessen, J., Jedlitschky, G., Grube, M., Bien, S., Schwertz, H., Ohtsuki, S., et al. (2009). Human platelets express organic anion-transporting peptide 2B1, an uptake transporter for atorvastatin. *Drug Metabolism and Disposition*, *37*(5), 1129–1137.

Noe, J., Stieger, B., & Meier, P. J. (2002). Functional expression of the canalicular bile salt export pump of human liver. *Gastroenterology*, *123*(5), 1659–1666.

Novak, D. A., Ryckman, F. C., & Suchy, F. J. (1989). Taurocholate transport by basolateral plasma membrane vesicles isolated from human liver. *Hepatology*, *10*(4), 447–453.

Oelkers, P., Kirby, L. C., Heubi, J. E., & Dawson, P. A. (1997). Primary bile acid malabsorption caused by mutations in the ileal sodium-dependent bile acid transporter gene (SLC10A2). *Journal of Clinical Investigation*, *99*(8), 1880.

Ogundare, M., Theofilopoulos, S., Lockhart, A., Hall, L. J., Arenas, E., Sjövall, J., et al. (2010). Cerebrospinal fluid steroidomics: are bioactive bile acids present in brain? *Journal of Biological Chemistry*, *285*(7), 4666–4679.

Padda, M. S., Sanchez, M., Akhtar, A. J., & Boyer, J. L. (2011). Drug-induced cholestasis. *Hepatology*, *53*(4), 1377–1387.

Pan, W., Song, I. S., Shin, H. J., Kim, M. H., Choi, Y. L., Lim, S. J., et al. (2011). Genetic polymorphisms in Na^+-taurocholate co-transporting polypeptide (NTCP) and ileal apical sodium-dependent bile acid transporter (ASBT) and ethnic comparisons of functional variants of NTCP among Asian populations. *Xenobiotica*, *41*(6), 501–510.

Pauli-Magnus, C., & Meier, P. J. (2006). Hepatobiliary transporters and drug-induced cholestasis. *Hepatology*, *44*(4), 778–787.

Paumgartner, G. (2010). Pharmacotherapy of cholestatic liver diseases. *Journal of Digestive Diseases*, *11*(3), 119–125.

Petzinger, E. (1994). Transport of organic anions in the liver. An update on bile acid, fatty acid, monocarboxylate, anionic amino acid, cholephilic organic anion, and anionic drug transport. *Reviews of Physiology, Biochemistry & Pharmacology*, *123*, 47–211.

Petzinger, E., Blumrich, M., Bruhl, B., Eckhardt, U., Föllmann, W., Honscha, W., et al. (1996). What we have learned about bumetanide and the concept of multispecific bile acid/drug transporters from the liver. *Journal of Hepatology*, *24*(Suppl. 1), 42–46.

Petzinger, E., & Frimmer, M. (1980). Comparative studies on the uptake of 14C-bile acids and 3H-demethylphalloin in isolated rat liver cells. *Archives of Toxicology*, *44*(1–3), 127–135.

Petzinger, E., & Frimmer, M. (1988). Comparative investigations on the uptake of phallotoxins, bile acids, bovine lactoperoxidase and horseradish peroxidase into rat hepatocytes in suspension and in cell cultures. *Biochimica et Biophysica Acta*, *937*(1), 135–144.

Petzinger, E., & Geyer, J. (2006). Drug transporters in pharmacokinetics. *Naunyn-Schmiedeberg's Archives of Pharmacology*, *372*(6), 465–475.

Petzinger, E., Nickau, L., Horz, J. A., Schulz, S., Wess, G., Enhsen, A., et al. (1995). Hepatobiliary transport of hepatic 3-hydroxy-3-methylglutaryl coenzyme A reductase inhibitors conjugated with bile acids. *Hepatology*, *22*(6), 1801–1811.

Petzinger, E., Wickboldt, A., Pagels, P., Starke, D., & Kramer, W. (1999). Hepatobiliary transport of bile acid amino acid, bile acid peptide, and bile acid oligonucleotide conjugates in rats. *Hepatology*, *30*(5), 1257–1268.

Petzinger, E., Ziegler, K., & Frimmer, M. (1979). Inhibition of 3H-demethylphalloin uptake in isolated rat hepatocytes under various experimental conditions. *Naunyn-Schmiedeberg's Archives of Pharmacology*, *307*(3), 275–281.

Pizzagalli, F., Hagenbuch, B., Stieger, B., Klenk, U., Folkers, G., & Meier, P. J. (2002). Identification of a novel human organic anion transporting polypeptide as a high affinity thyroxine transporter. *Molecular Endocrinology*, *16*(10), 2283–2296.

Portincasa, P., Di Ciaula, A., Wang, H. H., Moschetta, A., & Wang, D. Q. (2009). Medicinal treatments of cholesterol gallstones: old, current and new perspectives. *Current Medicinal Chemistry*, *16*(12), 1531–1542.

Poupon, R. E., Poupon, R., & Balkau, B. (1994). Ursodiol for the long-term treatment of primary biliary cirrhosis. The UDCA-PBC Study Group. *New England Journal of Medicine*, *330*(19), 1342–1347.

Rais, R., Acharya, C., Mackerell, A. D., & Polli, J. E. (2010). Structural determinants for transport across the intestinal bile acid transporter using C-24 bile acid conjugates. *Molecular Pharmacology*, *7*(6), 2240–2254.

Rais, R., Acharya, C., Tririya, G., Mackerell, A. D., & Polli, J. E. (2010). Molecular switch controlling the binding of anionic bile acid conjugates to human apical sodium-dependent bile acid transporter. *Journal of Medicinal Chemistry*, *53*(12), 4749–4760.

Rais, R., Fletcher, S., & Polli, J. E. (2011). Synthesis and in vitro evaluation of gabapentin prodrugs that target the human apical sodium-dependent bile acid transporter (hASBT). *Journal of Pharmaceutical Sciences*, *100*(3), 1184–1195.

Redinger, R. N. (2003). The role of the enterohepatic circulation of bile salts and nuclear hormone receptors in the regulation of cholesterol homeostasis: bile salts as ligands for nuclear hormone receptors. *Canadian Journal of Gastroenterology*, *17*(4), 265–271.

Reichen, J., & Paumgartner, G. (1976). Uptake of bile acids by perfused rat liver. *American Journal of Physiology*, *231*(3), 734–742.

Renner, O., Harsch, S., Schaeffeler, E., Schwab, M., Klass, D. M., Kratzer, W., et al. (2009). Mutation screening of apical sodium-dependent bile acid transporter (SLC10A2): novel haplotype block including six newly identified variants linked to reduced expression. *Human Genetics*, *125*(4), 381–391.

Renner, O., Harsch, S., Schaeffeler, E., Winter, S., Schwab, M., Krawczyk, M., et al. (2009). A variant of the SLC10A2 gene encoding the apical sodium-dependent bile acid transporter is a risk factor for gallstone disease. *PLoS One, 4*(10). e7321.

Root, C., Smith, C. D., Sundseth, S. S., Pink, H. M., Wilson, J. G., & Lewis, M. C. (2002). Ileal bile acid transporter inhibition, CYP7A1 induction, and antilipemic action of 264W94. *Journal of Lipid Research, 43*(8), 1320–1330.

Ruifrok, P. G., & Meijer, D. K. (1982). Sodium ion-coupled uptake of taurocholate by rat-liver plasma membrane vesicles. *Liver, 2*(1), 28–34.

Russell, D. W. (2003). The enzymes, regulation, and genetics of bile acid synthesis. *Annual Review of Biochemistry, 72*, 137–174.

Russell, D.W. (2009). Fifty years of advances in bile acid synthesis and metabolism. *Journal of Lipid Research, 50*(Suppl.), S120–S125.

Russell, D.W., & Setchell, K. D. (1992). Bile acid biosynthesis. *Biochemistry, 31*(20), 4737–4749.

Saeki, T., Matoba, K., Furukawa, H., Kirifuji, K., Kanamoto, R., & Iwami, K. (1999). Characterization, cDNA cloning, and functional expression of mouse ileal sodium-dependent bile acid transporter. *Journal of Biochemistry (Tokyo), 125*(4), 846–851.

Sandker, G. W., Weert, B., Olinga, P., Wolters, H., Slooff, M. J., Meijer, D. K., et al. (1994). Characterization of transport in isolated human hepatocytes. A study with the bile acid taurocholic acid, the uncharged ouabain and the organic cations vecuronium and rocuronium. *Biochemical Pharmacology, 47*(12), 2193–2200.

Schiff, M. (1870). Gallenbildung, abhängig von der Aufsaugung der Gallenstoffe. *Pflügers Archiv European Journal of Physiology, 3*(1), 598–613.

Schmidt, D. R., Holmstrom, S. R., Fon, T. K., Bookout, A. L., Kliewer, S. A., & Mangelsdorf, D. J. (2010). Regulation of bile acid synthesis by fat-soluble vitamins A and D. *Journal of Biological Chemistry, 285*(19), 14486–14494.

Schroeder, A., Eckhardt, U., Stieger, B., Tynes, R., Schteingart, C. D., Hofmann, A. F., et al. (1998). Substrate specificity of the rat liver Na(+)-bile salt cotransporter in Xenopus laevis oocytes and in CHO cells. *American Journal of Physiology, 274*(2 Pt 1), G370–G375.

Schwarz, L. R., Burr, R., Schwenk, M., Pfaff, E., & Greim, H. (1975). Uptake of taurocholic acid into isolated rat-liver cells. *European Journal of Biochemistry, 55*(3), 617–623.

Sharma, R., Long, A., & Gilmer, J. F. (2011). Advances in bile acid medicinal chemistry. *Current Medicinal Chemistry, 18*(26), 4029–4052.

Sharma, B. K., Singh, P., Pilania, P., Sarbhai, K., & Prabhakar, Y. S. (2011). CP-MLR/PLS directed QSAR study on apical sodium-codependent bile acid transporter inhibition activity of benzothiepines. *Molecular Diversity, 15*(1), 135–147.

Shitara, Y., Li, A. P., Kato, Y., Lu, C., Ito, K., Itoh, T., et al. (2003). Function of uptake transporters for taurocholate and estradiol 17beta-D-glucuronide in cryopreserved human hepatocytes. *Drug Metabolism and Pharmacokinetics, 18*(1), 33–41.

Shneider, B. L., Dawson, P. A., Christie, D. M., Hardikar, W., Wong, M. H., & Suchy, F. J. (1995). Cloning and molecular characterization of the ontogeny of a rat ileal sodium-dependent bile acid transporter. *Journal of Clinical Investigation, 95*(2), 745–754.

Simren, M., Bajor, A., Gillberg, P. G., Rudling, M., & Abrahamsson, H. (2011). Randomised clinical trial: the ileal bile acid transporter inhibitor A3309 vs. placebo in patients with chronic idiopathic constipation–a double-blind study. *Alimentary Pharmacology and Therapeutics, 34*(1), 41–50.

Splinter, P. L., Lazaridis, K. N., Dawson, P. A., & LaRusso, N. F. (2006). Cloning and expression of SLC10A4, a putative organic anion transport protein. *World Journal of Gastroenterology, 12*(42), 6797–6805.

St-Pierre, M. V., Stallmach, T., Freimoser, G. A., Dufour, J. F., Serrano, M. A., Marin, J. J., et al. (2004). Temporal expression profiles of organic anion transport proteins in placenta and fetal liver of the rat. *American Journal of Physiology – Regulatory, Integrative and Comparative Physiology, 287*(6), R1505–R1516.

Stamp, D. (2003). Bile acids and esophageal cancer. *Gastroenterology, 124*(1), 271–272.
Stamp, D. H. (2006). Bile acids aided by acid suppression therapy may be associated with the development of esophageal cancers in westernized societies. *Medical Hypotheses, 66*(1), 154–157.
Stapelbroek, J. M., van Erpecum, K. J., Klomp, L. W., & Houwen, R. H. (2010). Liver disease associated with canalicular transport defects: current and future therapies. *Journal of Hepatology, 52*(2), 258–271.
Starke, D., Lischka, K., Pagels, P., Uhlmann, E., Kramer, W., Wess, G., et al. (2001). Bile acid-oligodeoxynucleotide conjugates: synthesis and liver excretion in rats. *Bioorganic & Medicinal Chemistry Letters, 11*(7), 945–949.
Stayrook, K. R., Bramlett, K. S., Savkur, R. S., Ficorilli, J., Cook, T., Christe, M. E., et al. (2005). Regulation of carbohydrate metabolism by the farnesoid X receptor. *Endocrinology, 146*(3), 984–991.
Stieger, B. (2011). The role of the sodium-taurocholate cotransporting polypeptide (NTCP) and of the bile salt export pump (BSEP) in physiology and pathophysiology of bile formation. *Handbook of Experimental Pharmacology, 201*, 205–259.
Stieger, B., Hagenbuch, B., Landmann, L., Höchli, M., Schroeder, A., & Meier, P. J. (1994). In situ localization of the hepatocytic Na^+/Taurocholate cotransporting polypeptide in rat liver. *Gastroenterology, 107*(6), 1781–1787.
Strange, R. C. (1984). Hepatic bile flow. *Physiological Reviews, 64*(4), 1055–1102.
Stravitz, R. T., Sanyal, A. J., Pandak, W. M., Vlahcevic, Z. R., Beets, J. W., & Dawson, P. A. (1997). Induction of sodium-dependent bile acid transporter messenger RNA, protein, and activity in rat ileum by cholic acid. *Gastroenterology, 113*(5), 1599–1608.
Suchy, F. J., Bucuvalas, J. C., Goodrich, A. L., Moyer, M. S., & Blitzer, B. L. (1986). Taurocholate transport and Na^+-K^+-ATPase activity in fetal and neonatal rat liver plasma membrane vesicles. *American Journal of Physiology, 251*(5 Pt 1), G665–G673.
Sugiyama, D., Kusuhara, H., Taniguchi, H., Ishikawa, S., Nozaki, Y., Aburatani, H., et al. (2003). Functional characterization of rat brain-specific organic anion transporter (Oatp14) at the blood-brain barrier: high affinity transporter for thyroxine. *Journal of Biological Chemistry, 278*(44), 43489–43495.
Sun, A. Q., Ananthanarayanan, M., Soroka, C. J., Thevananther, S., Shneider, B. L., & Suchy, F. J. (1998). Sorting of rat liver and ileal sodium-dependent bile acid transporters in polarized epithelial cells. *American Journal of Physiology, 275*(5 Pt 1), G1045–G1055.
Sun, A. Q., Balasubramaniyan, N., Chen, H., Shahid, M., & Suchy, F. J. (2006). Identification of functionally relevant residues of the rat ileal apical sodium-dependent bile acid cotransporter. *Journal of Biological Chemistry, 281*(24), 16410–16418.
Sundaram, S. S., Bove, K. E., Lovell, M. A., & Sokol, R. J. (2008). Mechanisms of disease: inborn errors of bile acid synthesis. *Nature Clinical Practice Gastroenterology & Hepatology, 5*(8), 456–468.
Swaan, P. W., Hillgren, K. M., Szoka, F. C., Jr., & Oie, S. (1997). Enhanced transepithelial transport of peptides by conjugation to cholic acid. *Bioconjugate chemistry, 8*(4), 520–525.
Swaan, P. W., Szoka, F. C., Jr., & Oie, S. (1997). Molecular modeling of the intestinal bile acid carrier: a comparative molecular field analysis study. *Journal of Computer-Aided Molecular Design, 11*(6), 581–588.
Swift, B., Pfeifer, N. D., & Brouwer, K. L. (2010). Sandwich-cultured hepatocytes: an in vitro model to evaluate hepatobiliary transporter-based drug interactions and hepatotoxicity. *Drug Metabolism Reviews, 42*(3), 446–471.
Takashima, K., Kohno, T., Mori, T., Ohtani, A., Hirakoso, K., & Takeyama, S. (1994). The hypocholesterolemic action of TA-7552 and its effects on cholesterol metabolism in the rat. *Atherosclerosis, 107*(2), 247–257.

Tamai, I., Nezu, J., Uchino, H., Sai, Y., Oku, A., Shimane, M., et al. (2000). Molecular identification and characterization of novel members of the human organic anion transporter (OATP) family. *Biochemical and Biophysical Research Communications*, *273*(1), 251–260.

Telford, D. E., Edwards, J. Y., Lipson, S. M., Sutherland, B., Barrett, P. H., Burnett, J. R., et al. (2003). Inhibition of both the apical sodium-dependent bile acid transporter and HMG-CoA reductase markedly enhances the clearance of LDL apoB. *Journal of Lipid Research*, *44*(5), 943–952.

Thomas, C., Gioiello, A., Noriega, L., Strehle, A., Oury, J., Rizzo, G., et al. (2009). TGR5-mediated bile acid sensing controls glucose homeostasis. *Cell Metabolism*, *10*(3), 167–177.

Tolle-Sander, S., Lentz, K. A., Maeda, D. Y., Coop, A., & Polli, J. E. (2004). Increased acyclovir oral bioavailability via a bile acid conjugate. *Molecular Pharmaceutics*, *1*(1), 40–48.

Tönjes, A., Wittenburg, H., Halbritter, J., Renner, O., Harsch, S., Stange, E. F., et al. (2011). Effects of SLC10A2 variant rs9514089 on gallstone risk and serum cholesterol levels – meta-analysis of three independent cohorts. *BMC Medical Genetics*, *12*, 149.

Trauner, M., Claudel, T., Fickert, P., Moustafa, T., & Wagner, M. (2010). Bile acids as regulators of hepatic lipid and glucose metabolism. *Journal of Digestive Diseases*, *28*(1), 220–224.

Trauner, M., Fickert, P., Halilbasic, E., & Moustafa, T. (2008). Lessons from the toxic bile concept for the pathogenesis and treatment of cholestatic liver diseases. *Wiener Medizinische Wochenschrift*, *158*(19–20), 542–548.

Trauner, M., Meier, P. J., & Boyer, J. L. (1998). Molecular pathogenesis of cholestasis. *The New England journal of medicine*, *339*(17), 1217–1227.

Tremont, S. J., Lee, L. F., Huang, H. C., Keller, B. T., Banerjee, S. C., Both, S. R., et al. (2005). Discovery of potent, nonsystemic apical sodium-codependent bile acid transporter inhibitors (Part 1). *Journal of Medicinal Chemistry*, *48*(18), 5837–5852.

Twisk, J., Hoekman, M. F., Mager, W. H., Moorman, A. F., de Boer, P. A., Scheja, L., et al. (1995). Heterogeneous expression of cholesterol 7 alpha-hydroxylase and sterol 27-hydroxylase genes in the rat liver lobulus. *Journal of Clinical Investigation*, *95*(3), 1235–1243.

Ugele, B., St-Pierre, M. V., Pihusch, M., Bahn, A., & Hantschmann, P. (2003). Characterization and identification of steroid sulfate transporters of human placenta. *American Journal of Physiology – Endocrinology And Metabolism*, *284*(2), E390–E398.

Venneman, N. G., Besselink, M. G., Keulemans, Y. C., Vanberge-Henegouwen, G. P., Boermeester, M. A., Broeders, I. A., et al. (2006). Ursodeoxycholic acid exerts no beneficial effect in patients with symptomatic gallstones awaiting cholecystectomy. *Hepatology*, *43*(6), 1276–1283.

Vicens, M., Macias, R. I., Briz, O., Rodriguez, A., El-Mir, M. Y., et al. (2007). Inhibition of the intestinal absorption of bile acids using cationic derivatives: mechanism and repercussions. *Biochemical Pharmacology*, *73*(3), 394–404.

Vicens, M., Medarde, M., Macias, R. I., Larena, M. G., Villafaina, A., Serrano, M. A., et al. (2007). Novel cationic and neutral glycocholic acid and polyamine conjugates able to inhibit transporters involved in hepatic and intestinal bile acid uptake. *Bioorganic and Medicinal Chemistry*, *15*(6), 2359–2367.

Visser, W. E., Wong, W. S., van Mullem, A. A., Friesema, E. C., Geyer, J., & Visser, T. J., (2010). Study of the transport of thyroid hormone by transporters of the SLC10 family. *Molecular and Cellular Endocrinology*, *315*(1–2), 138–145.

Wang, Y. D., Chen, W. D., Moore, D. D., & Huang, W. (2008). FXR: a metabolic regulator and cell protector. *Cell Research*, *18*(11), 1087–1095.

Wang, W., Xue, S., Ingles, S. A., Chen, Q., Diep, A. T., Frankl, H. D., et al. (2001). An association between genetic polymorphisms in the ileal sodium-dependent bile acid transporter gene and the risk of colorectal adenomas. *Cancer Epidemiology, Biomarkers & Prevention*, *10*(9), 931–936.

Wan, M. H., Huang, W., Latawiec, D., Jiang, K., Booth, D. M., Elliott, V., et al. (2012). Review of experimental animal models of biliary acute pancreatitis and recent advances in basic research. *HPB (Oxford)*, *14*(2), 73–81.

Weber, F. (1983). Absorption of fat-soluble vitamins. *International Journal for Vitamin and Nutrition Research – Supplement*, *25*, 55–65.
Wehner, F. (1993). Taurocholate depolarizes rat hepatocytes in primary culture by increasing cell membrane Na^+ conductance. *Pflügers Archiv European Journal of Physiology*, *424*(2), 145–151.
Weinman, S. A. (1997). Electrogenicity of Na(+)-coupled bile acid transporters. *Yale Journal of Biology & Medicine*, *70*(4), 331–340.
Weinman, S. A., Carruth, M. W., & Dawson, P. A. (1998). Bile acid uptake via the human apical sodium-bile acid cotransporter is electrogenic. *Journal of Biological Chemistry*, *273*(52), 34691–34695.
West, K. L., Ramjiganesh, T., Roy, S., Keller, B. T., & Fernandez, M. L. (2002). 1-[4-[4[(4R,5R)-3,3-Dibutyl-7-(dimethylamino)-2,3,4,5-tetrahydro-4-hydroxy -1,1-dioxido-1-benzothiepin-5-yl]phenoxy]butyl]-4-aza-1-azoniabicyclo[2.2. 2]octane methanesulfonate (SC-435), an ileal apical sodium-codependent bile acid transporter inhibitor alters hepatic cholesterol metabolism and lowers plasma low-density lipoprotein-cholesterol concentrations in guinea pigs. *Journal of Pharmacology and Experimental Therapeutics*, *303*(1), 293–299.
West, K. L., Zern, T. L., Butteiger, D. N., Keller, B. T., & Fernandez, M. L. (2003). SC-435, an ileal apical sodium co-dependent bile acid transporter (ASBT) inhibitor lowers plasma cholesterol and reduces atherosclerosis in guinea pigs. *Atherosclerosis*, *171*(2), 201–210.
Wong, B. S., Camilleri, M., McKinzie, S., Burton, D., Graffner, H., & Zinsmeister, A. R. (2011). Effects of A3309, an ileal bile acid transporter inhibitor, on colonic transit and symptoms in females with functional constipation. *American Journal of Gastroenterology*, *106*(12), 2154–2164.
Wong, M. H., Oelkers, P., Craddock, A. L., & Dawson, P. A. (1994). Expression cloning and characterization of the hamster ileal sodium-dependent bile acid transporter. *Journal of Biological Chemistry*, *269*(2), 1340–1347.
Wong, M. H., Oelkers, P., & Dawson, P. A. (1995). Identification of a mutation in the ileal sodium-dependent bile acid transporter gene that abolishes transport activity. *Journal of Biological Chemistry*, *270*(45), 27228–27234.
Wright, E. M., Loo, D. D., & Hirayama, B. A. (2011). Biology of human sodium glucose transporters. *Physiological Reviews*, *91*(2), 733–794.
Xia, X., Francis, H., Glaser, S., Alpini, G., & Lesage, G. (2006). Bile acid interactions with cholangiocytes. *World Journal of Gastroenterology*, *12*(22), 3553–3563.
Zahner, D., Eckhardt, U., & Petzinger, E. (2003). Transport of taurocholate by mutants of negatively charged amino acids, cysteines, and threonines of the rat liver sodium-dependent taurocholate cotransporting polypeptide Ntcp. *European Journal of Biochemistry*, *270*(6), 1117–1127.
Zhang, E. Y., Phelps, M. A., Banerjee, A., Khantwal, C. M., Chang, C., Helsper, F., et al. (2004). Topology scanning and putative three-dimensional structure of the extracellular binding domains of the apical sodium-dependent bile acid transporter (SLC10A2). *Biochemistry*, *43*(36), 11380–11392.
Zheng, X., Ekins, S., Raufman, J. P., & Polli, J. E. (2009). Computational models for drug inhibition of the human apical sodium-dependent bile acid transporter. *Molecular Pharmacology*, *6*(5), 1591–1603.
Zheng, X., Pan, Y., Acharya, C., Swaan, P. W., & Polli, J. E. (2010). Structural requirements of the ASBT by 3D-QSAR analysis using aminopyridine conjugates of chenodeoxycholic acid. *Bioconjugate chemistry*, *21*(11), 2038–2048.
Zheng, X., & Polli, J. E. (2010). Synthesis and in vitro evaluation of potential sustained release prodrugs via targeting ASBT. *International Journal of Pharmaceutics*, *396*(1–2), 111–118.

Ziegler, K., Petzinger, E., & Frimmer, M. (1980). Decreased phalloidin response, phallotoxin uptake and bile acid transport in hepatocytes prepared from wistar rats treated chronically with diethylnitrosamine. *Naunyn-Schmiedeberg's Archives of Pharmacology*, *310*(3), 245–247.

Ziegler, K., Petzinger, E., Grundmann, E., & Frimmer, M. (1979). Decreased sensitivity of isolated hepatocytes from baby rats, from regenerating and from poisoned livers to phalloidin. *Naunyn-Schmiedeberg's Archives of Pharmacologyl.*, *306*(3), 295–300.

Zimmerli, B., Valantinas, J., & Meier, P. J. (1989). Multispecificity of Na^+-dependent taurocholate uptake in basolateral (sinusoidal) rat liver plasma membrane vesicles. *Journal of Pharmacology and Experimental Therapeutics*, *250*(1), 301–308.

CHAPTER FIVE

H^+-Coupled Divalent Metal-Ion Transporter-1: Functional Properties, Physiological Roles and Therapeutics

Ali Shawki, Patrick B. Knight, Bryan D. Maliken, Eric J. Niespodzany, Bryan Mackenzie*

Department of Molecular & Cellular Physiology, University of Cincinnati College of Medicine, Cincinnati, Ohio, USA

*Corresponding author: E-mail: bryan.mackenzie@uc.edu

Contents

Current Topics in Membranes, Volume 70
ISSN 1063-5823
http://dx.doi.org/10.1016/B978-0-12-394316-3.00005-3

Abstract

Divalent metal-ion transporter-1 (DMT1) is a widely expressed, iron-preferring membrane transport protein. Animal models establish that DMT1 plays indispensable roles in intestinal nonheme-iron absorption and iron acquisition by erythroid precursor cells. Rare mutations in human DMT1 result in severe microcytic–hypochromic anemia. When we express DMT1 in RNA-injected *Xenopus* oocytes, we observe rheogenic Fe^{2+} transport that is driven by the proton electrochemical potential gradient. In that same preparation, DMT1 also transports cadmium and manganese but not copper. Whether manganese metabolism relies upon DMT1 remains unclear but DMT1 contributes to the effects of overexposure to cadmium and manganese in some tissues. There exist at least four DMT1 isoforms that arise from variant transcription of the *SLC11A2* gene. Whereas these isoforms display identical functional properties, N- and C-terminal variations contain cues that direct the cell-specific targeting of DMT1 isoforms to discrete subcellular compartments (plasma membrane, endosomes, and lysosomes). An iron-responsive element (IRE) in the mRNA 3′-untranslated region permits the regulation of some isoforms by iron status, and additional mechanisms by which DMT1 is regulated are emerging. Natural-resistance-associated macrophage protein-1 (NRAMP1)—the only other member of the mammalian SLC11 gene family—contributes to antimicrobial function by extruding from the phagolysosome divalent metal ions (e.g. Mn^{2+}) that may be essential cofactors for bacteria-derived enzymes or required for bacterial growth. The principal or only intestinal nonheme-iron transporter, DMT1 is a validated therapeutic target in hereditary hemochromatosis (HHC) and other iron-overload disorders.

1. INTRODUCTION AND HISTORY

DMT1 (NRAMP2, DCT1) was cloned by Hediger's group in 1997 by the functional screening of a complementary DNA library prepared with duodenal mRNA isolated from rats fed a low-iron diet (Gunshin et al., 1997). We found that expression of the rat DMT1 in RNA-injected *Xenopus* oocytes stimulated $^{55}Fe^{2+}$ uptake and Fe^{2+}-evoked currents. We also found that this widely expressed transporter was reactive with several other divalent metal ions. Andrews' group around the same time had used a positional cloning strategy to identify the defective gene responsible for the microcytic-anemia phenotype of the *mk* inbred mouse strain (Fleming et al., 1997). These two findings established DMT1 as a divalent metal-ion transporter with a central role in iron metabolism.

During the intervening 15 years, DMT1 has been the focus of intense research effort—among the reasons for which, DMT1 plays nonredundant roles in iron homeostasis. Its widespread expression—present in at least some cell types in every organ tested (Gunshin et al., 1997; Hubert &

Hentze, 2002; Wang et al., 2012)—places DMT1 at the nexus of multiple pathways of epithelial iron transport and cellular iron acquisition in health and disease, and the transport of other metals (e.g. cadmium, manganese) particularly in chronic exposure. DMT1 is a mechanistically complex transporter with multiple isoforms that function in varied environments. This article is an attempt to review just some of this breadth and complexity. We will review lessons learned from animal models and human mutants, studies of the functional properties, structure–function, regulation, and tissue-specific roles of DMT1, and the development of new therapeutic strategies.

2. ANIMAL MODELS

The Belgrade (*b*) rat and microcytic (*mk*) mouse inbred strains have long been studied by investigators interested in iron metabolism (Bannerman, 1976). Both strains harbor a mutation that causes a Gly185→Arg substitution in the putative fourth transmembrane region (TM4) of DMT1 (Fig. 1) and exhibit a severe hypochromic–microcytic anemia (Fleming et al., 1997, 1998). Tissue or vesicle preparations from the *b* rat and *mk* mouse exhibited deficiencies in intestinal iron transport (Edwards & Hoke, 1972; Knöpfel, Zhao, & Garrick, 2005). Bypassing the intestine by giving parenteral iron results in some improvement but does not cure the anemia, revealing some other role in red blood cell production. A similar phenotype was observed in the "chardonnay" zebrafish mutant in which ethylnitrosourea mutagenesis introduced an early stop codon in place of Lys-264 of DMT1 (Donovan et al., 2002).

The chardonnay mutant, when expressed in HEK293T cells, exhibited around 10% the iron-transport activity observed for wild-type zebrafish DMT1 (Donovan et al., 2002). Meanwhile, expression of the mouse G185R in several transfected cell lines (HEK293T, LLC-PK_1, and CHO) has revealed that the mutant protein is correctly targeted to the plasma membrane and endosomes; however, G185R-DMT1 protein levels at the plasma membrane are lower than those of wild type (Su, Trenor, Fleming, Fleming, & Andrews, 1998; Touret et al., 2004). In one of the studies just cited, metal-ion transport activity for the G185R mutant was only modestly reduced (Touret et al., 2004) whereas, in the other study, it was substantially reduced (Su et al., 1998). Mn^{2+} transport activity in brush-border membrane vesicles (BBMV) isolated from homozygous Belgrade rats (*b*/*b*) was <10% that of the heterozygote (+/*b*) despite the much higher

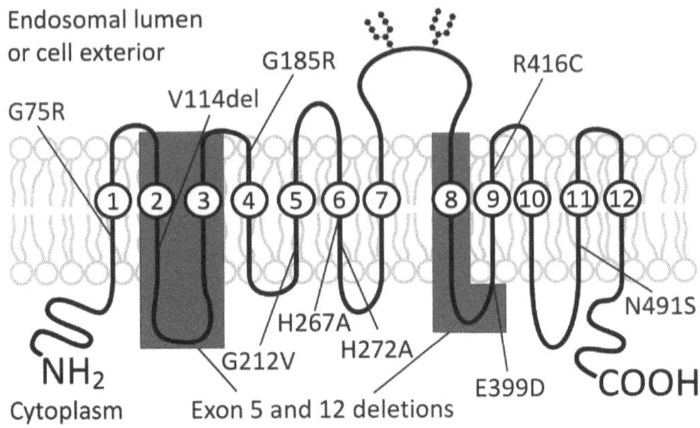

Figure 1 Predicted topology model of human DMT1, 1B/IRE(+) isoform (based on the hydrophobicity plot generated by using the Kyte–Doolittle algorithm at http://web.expasy.org/protscale/ and default values for predicting transmembrane regions) showing sites of the human mutations reported so far. Also labeled are amino acid residues corresponding to those mutated in rat and mouse DMT1 (including the G185R mutation found in the *b* and *mk* rodent models). See the color plate.

DMT1 protein levels detected in *b/b* intestine (Knöpfel, Zhao et al., 2005). Expression of the G185R mutant protein in *mk/mk* mouse duodenum is also dramatically increased compared with the heterozygote (+/*mk*) but is not correctly targeted to the BBM (Canonne-Hergaux et al., 2000).

Although the species and preparations used appear to differ with regard to the activity of the G185R-DMT1 protein, it is apparent that this mutant retains a fraction of wild-type metal-ion transport activity. Coupled with the expected upregulation of the protein due to iron deficiency, this residual activity may help to explain why the *b* and *mk* animals survive. Whereas homozygous *mk* mice generally survived to adulthood, mice in which the *SLC11A2* gene coding for DMT1 was globally inactivated ($SLC11A2^{-/-}$) did not survive more than 7 days (Gunshin, Fujiiwara et al., 2005). The $SLC11A2^{-/-}$ neonates exhibited a severe hypochromic–microcytic anemia. Total body iron was normal, indicating that maternofetal iron transfer was sufficient; however, liver iron stores were elevated, consistent with impaired erythroid iron utilization. Strengthening the conclusion of a critical defect in erythroid iron utilization, transfusion of red blood cells but not parenteral iron injections improved survival of these animals. In that same study, investigators also transplanted hematopoietic stem cells from $SLC11A2^{-/-}$ mice into lethal-dose-irradiated wild-type mice and found that the recipients exhibited defective erythropoiesis, consistent with iron deficiency (Gunshin, Fujiiwara et al., 2005).

To assess the role of DMT1 in the intestine, the same investigators generated a conditional intestine-specific knockout (*SLC11A2*$^{int/int}$) by crossing floxed DMT1 and villin-Cre transgenic lines (Gunshin, Fujiiwara et al., 2005). Although born healthy, the *SLC11A2*$^{int/int}$ mice soon developed progressive anemia and exhibited lower tissue-iron levels than did wild-type mice (Gunshin, Fujiiwara et al., 2005; Shawki et al., 2012). Parenteral iron injections rescued the *SLC11A2*$^{int/int}$ anemia phenotype, confirming a primary defect in intestinal iron absorption (Gunshin, Fujiiwara et al., 2005; Shawki et al., 2012).

These studies in animal models collectively demonstrate the importance of DMT1 in intestinal iron absorption and erythroid iron utilization. Whereas ablation of the *SLC11A2* gene may provide superior models in which to test the essential roles of DMT1, the *b* and *mk* rodent models are likely to be of continued utility in studies for which a less severe phenotype is desirable. Meanwhile, novel transgenic mouse models in which intestinal DMT1 is overexpressed may prove especially useful in expanding our understanding of how DMT1 is regulated (Barrientos De Renshaw, Bhutiani, Montross, & Andrews, 2011).

3. MUTATIONS IN HUMAN DMT1

Mutations in human DMT1 have been reported now in five unrelated cases of hypochromic–microcytic anemia, characterized by lowered values for hematocrit, blood hemoglobin concentration, mean corpuscular volume, and mean corpuscular hemoglobin content (Table 1). The first identified was a Czech proband homozygous for a G→C substitution in the last nucleotide of exon 12, with the following two consequences: (1) expression of a full-length DMT1 protein containing a conservative E399D substitution, and (2) increased frequency of skipping exon 12 from the normal 10–90% in the proband (Mims et al., 2005; Priwitzerova et al., 2004). Exon-12 deletion is predicted to result in a protein lacking putative TM8 and the cytosolic loop between TM8 and TM9 (Fig. 1). In experiments in which these constructs were expressed in mammalian cell lines or *Xenopus* oocytes, E399D-DMT1 functioned normally, whereas the exon-12 deletion product showed defects in expression, trafficking and transport activity (Gunshin, Jin et al., 2005; Lam-Yuk-Tseung, Mathieu, & Gros, 2005; Priwitzerova, Pospisilova, Sheftel, Divoky, & Ponka, 2005; Shawki, Illing, & Mackenzie, 2006).

Table 1 Mutations in human DMT1 and initial clinical features of the probands. Reference values are provided in the final row

Proband	Mutation: Gene	Mutation: Peptide	Mutation: Description	[Hb] (g/dL)	Mean corpuscular volume (fL)
Czech	G1285C	E399D / ex12 del	Homozygous; conservative; increased skipping of exon 12 from normal 10–90%	7.4	54
Italian	C1246T; CTT del	R416C /ex5 del	Compound heterozygous; 3-bp del in intron 4 leads to 30–35% skipping of exon 5	4.0	71
French [Paris]	G723T, $G^{428}T^{429}G^{430}$ del	G212V, V114 del	Compound heterozygous; in-frame deletion of Val-114	8.3	64
Ecuadorian	G311A	G75R	Homozygous	5.1	54
French [Rennes]	G723T, A1560G	G212V, N491S	Compound heterozygous	8.6	58
Reference values				9.0–22.5	70–126

*Papanikolaou et al., 2005.
†Children's Hospitals and Clinics of Minnesota accessed at http://www.childrensmn.org/manuals/lab/hematology/018981.asp.

Mean corpuscular hemoglobin (pg/red cell)	Hematocrit (%)	Urinary hepcidin (ng/mg creatinine)	Liver iron overload	Description of the impact of the mutation and appropriate references
15	25	4	+++	Exon-12 deletion disrupts the expression, stability, maturation and activity of the protein; however, E399D functions normally. Gunshin, Jin et al., 2005; Lam-Yuk-Tseung et al., 2005; Mims et al., 2005; Priwitzerova et al., 2004; Priwitzerova et al., 2005 Shawki et al., 2006
13	25	92–100	++	Deletion of exon 5 leads to loss of iron-transport activity; R416C exhibits defective processing, cell-surface expression and activity (slower kinetics). Iolascon et al., 2006; Lam-Yuk-Tseung, Camaschella, Iolascon, & Gros, 2006; Shawki et al., 2006.
15	26	19–43	+	Transport is activity disrupted in V114del but G212V is still capable of transporting iron. Beaumont et al., 2006; Shawki et al., 2008
22	27	N/A	Normal	G75R is postulated to be fully defective (Fig. 2). Blanco et al., 2009
18	N/A	N/A	+	N491S results in a truncated protein but G212V is still capable of transporting iron. Bardou-Jacquet et al., 2011; Shawki et al., 2008
23–40	28–67	10–200*		Reference values for hematological variables (age day 1 through >18 years in males and females)†

Three other probands are compound heterozygotes. Among them, there are five discrete mutations—one of these, a missense mutation leading to a Gly212→Val substitution in putative TM5 (Fig. 1), appears in two of the probands (Table 1) (Bardou-Jacquet et al., 2011; Beaumont et al., 2006; Iolascon et al., 2006). The parents (and siblings) of each of the compound heterozygous probands are healthy, indicating that mutations at both alleles contribute to the anemia phenotype. The mutant proteins are deficient in their transport activity or targeting to the endosomal or plasma membranes (summarized in Table 1).

Not only were these four probands anemic, all showed signs of hepatic iron overload—a finding which, at first, did not appear consistent with DMT1's perceived role in iron absorption. The first possible explanation offered was that intestinal heme-iron absorption could be upregulated, independently of DMT1 (Priwitzerova et al., 2004). In the study just cited, the authors suggested that iron overload was not apparent in the *b* and *mk* rodent models because rodents have poorly developed heme-absorptive pathways and heme iron is not typically a part of rodent chow. However, the *b* rat—in which a defect in intestinal metal-ion uptake has been established directly (Knöpfel, Zhao et al., 2005)—also displays hepatic iron loading (Thompson, Molina, Brain, & Wessling-Resnick, 2006).

It is plausible that the mutations affect the function of various DMT1 isoforms (Section 4.4.) differently, or that the severity of the mutation is more pronounced in the environment of the erythroid precursor cell endosome than it is at the intestinal brush border. A more straightforward explanation may be that the human mutations—like the rodent G185R mutation—possess some residual activity which, coupled with upregulation of the intestinal DMT1 isoform 1A/IRE(+) (Section 4.4) by an erythropoietic factor, permits some iron absorption to proceed in these probands. Because the absorbed iron is poorly utilized by erythroid precursors, much is diverted to the liver stores. Supporting this notion, urinary levels of the iron-regulatory hormone hepcidin, whose actions lead to a downregulation of iron absorption (Ganz & Nemeth, 2006), are low or markedly suppressed in these probands (Table 1).

In contrast, a fifth anemic proband, homozygous for a missense mutation leading to a G75R substitution in putative TM1 (Fig. 1), displayed no hepatic iron loading (Blanco, Kannengiesser, Grandchamp, Tasso, & Beaumont, 2009). Not discounting the possibility that iron loading may progress later in life, the authors in the study just cited postulated that the G75R mutant should be completely deficient in DMT1 functional activity and would more fully ablate iron absorption. To begin to test this hypothesis, we expressed wild-type DMT1 and the G75R mutant in RNA-injected *Xenopus* oocytes and measured $^{55}Fe^{2+}$ transport and currents.

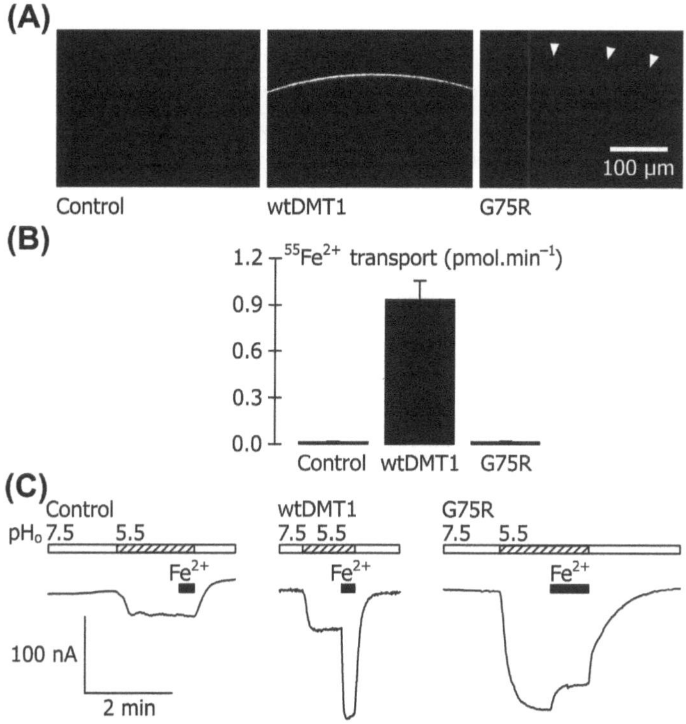

Figure 2 Impact of the G75R mutation on expression and activity of human DMT1 in RNA-injected *Xenopus* oocytes. A, Imaging of control oocytes, and wild-type DMT1 (wtDMT1)–enhanced green fluorescent protein (EGFP) or G75R–EGFP expression in RNA-injected *Xenopus* oocytes as described (Pinilla-Tenas et al., 2011) using the Zeiss LSM 7 DUO confocal laser-scanning microscope (excitation at 488 nm) fitted with the LD C-Apochromat 40×/1.1 W Korr objective to measure emission in the band 500–531 nm at a pinhole setting of 36 μm. Representative images are presented in which the optical slice (0.6 μm depth) approximately bisects the oocyte. Faint fluorescence at the cell perimeter of oocytes expressing G75R is marked by the white arrowheads. B, transport of 2 μM $^{55}Fe^{2+}$ was measured in control oocytes and oocytes expressing wild-type human DMT1 (wtDMT1) or G75R for 10 min at pH 5.5 in the presence of 1 mM L-ascorbic acid as described (Shawki & Mackenzie, 2010). Data are mean and S.D. for 10 to 11 oocytes per group. Analysis of variance ($P<0.001$) followed by Holm–Šidák pairwise comparisons revealed that G75R did not differ from control (adjusted $P=0.94$). C, Continuous currents recordings in a control oocyte and oocytes expressing wild-type human DMT1 or G75R and voltage clamped at −70 mV as described (Shawki & Mackenzie, 2010). Oocytes were continuously superfused for the periods shown by the bars with pH 7.5 medium (empty bars) or pH 5.5 medium (hatched bars) and 50 μM Fe^{2+} (black bars) in the presence of 1 mM L-ascorbic acid.

In oocytes expressing DMT–enhanced green fluorescent protein (EGFP), we observed strong EGFP fluorescence over the entire cell perimeter, whereas we detected only very faint fluorescence in oocytes expressing G75R–EGFP (Fig. 2A). We found no detectable iron-transport activity in the mutant (Fig. 2B). In oocytes expressing wild-type DMT1 superfused first with pH 7.5 medium, then pH 5.5, we observed a small inward current (Fig. 2C) that we have previously demonstrated to be a H^+ "leak" pathway (Gunshin et al., 1997; Mackenzie, Ujwal, Chang, Romero, & Hediger, 2006), and a large inward current upon the addition of 50 μM Fe^{2+}. In contrast, G75R mediated a very large inward H^+ leak current that was inhibited partly by the addition of Fe^{2+}. Therefore, although the G75R mutant was not processed to the oocyte plasma membrane to the same extent as was wild-type DMT1, G75R exhibited a partial reaction (H^+ transport) and was completely deficient in iron-transport activity.

For many years, investigators have known of human pedigrees containing multiple members displaying a severe microcytic-anemia phenotype, characterized by a deficiency in intestinal iron absorption that was only partially corrected with parenteral iron administration (Buchanan & Sheehan, 1981; Hartman & Barker, 1996). Whether or not such pedigrees harbor mutations in DMT1 is not known, but their phenotypes are strikingly similar to those described for the DMT1 mutants just described. In any event, mutations in human DMT1 are extremely rare—presumably a reflection of both its critical, nonredundant functional roles and its long phylogenetic ancestry (Chaloupka et al., 2005), such that mutations in DMT1 are not generally tolerated.

4. FUNCTIONAL PROPERTIES OF DMT1

4.1. Molecular Mechanisms of DMT1-Mediated Iron Transport

The proton electrochemical potential gradient provides the thermodynamic driving force for concentrative iron transport from the cell exterior or endosome into the cytoplasm, placing DMT1 among an important class of proteins we call cotransporters (symporters). When we expressed rat or human DMT1 in RNA-injected oocytes under voltage-clamp conditions, we found that Fe^{2+} evoked large inward currents (Gunshin et al., 1997; Illing, Shawki, Cunningham, & Mackenzie, 2012; Mackenzie et al., 2006; Mackenzie, Takanaga, Hubert, Rolfs, & Hediger, 2007; Shawki & Mackenzie, 2010). Such rheogenicity could conceivably arise from influx of Fe^{2+} alone; however, the observation of pre steady-state currents in the *absence* of metal-ion substrates pointed to the involvement of a second ligand. Pre steady-state currents, which decayed with time constants of 15–40 ms, were observed following step-changes in membrane potential (Cohen, Nevo,

& Nelson, 2003, Gunshin et al., 1997; Mackenzie et al., 2006, 2007). That these currents were pH-dependent was our first indication that DMT1 might be H^+-coupled. Pre steady-state currents have been observed for other ion-coupled transporters, e.g. the H^+/peptide cotransporter PEPT1 (Mackenzie, Loo, Fei et al., 1996), the Na^+/glucose cotransporters SGLT1 and SGLT3 (Loo, Hazama, Supplisson, Turk, & Wright, 1993; Mackenzie, Loo, Panayotova-Heiermann, & Wright 1996), and the Na^+/ascorbic acid cotransporter SVCT1 (Mackenzie, Illing, & Hediger, 2008). In the studies just cited, computer simulation of transport models have allowed us to attribute pre steady-state currents in cotransporters to two steps in the catalytic cycle, namely (1) reorientation of the empty, charged carrier within the membrane electric field, and (2) binding/dissociation of the driving ion partway within the membrane electric field (i.e. an "ion-well" effect). The magnitude of the pre steady-state currents correlates with plasma-membrane expression of the transporter and provides an estimate of the number of functional units (Mackenzie et al., 2007; Zampighi et al., 1995). When we expressed in oocytes a human DMT1 mutation (R416C) associated with disease, we observed altered kinetics of the pre steady-state currents (compared with wild type) consistent with the mutation having slowed the return of the empty carrier to the outward-facing state (Shawki et al., 2006).

$^{55}Fe^{2+}$ uptake and Fe^{2+}-evoked currents in oocytes expressing DMT1 were stimulated at low pH (at least to lower limits tested, pH 5.2–5.5) compared with pH 7.2–7.5 (Gunshin et al., 1997; Illing et al., 2012; Mackenzie et al., 2006, 2007). Such pH dependence of metal-ion transport was observed also in other preparations expressing DMT1 (Elisma & Jumarie, 2001; Forbes & Gros, 2003; Garrick et al., 2006; Lam-Yuk-Tseung, Govoni, & Gros, 2003; Linder, Moriya, Whon, Kassa, & Gilley, 2006; Picard, Govoni, Jabado, & Gros, 2000; Tandy et al., 2000; Xu, Jin, DeFelice, Andrews, & Clapham, 2004; Zhang, Canonne-Hergaux, Gruenheid, Gros, & Ponka, 2008) with one exception (Worthington, Browne, Battle, & Luo, 2000), which we have discussed elsewhere (Mackenzie et al., 2006). Kinetic analyses in DMT1-expressing oocytes indicated that low extracellular pH exerted the following two effects: (1) to increase the apparent affinity for Fe^{2+}, and (2) to provide the thermodynamic driving force for Fe^{2+} transport (Mackenzie et al., 2006). Conversely, we also found that Fe^{2+} transport was associated with a substantial H^+ influx (Gunshin et al., 1997; Mackenzie et al., 2006). Collectively, these experimental observations indicate that DMT1-mediated Fe^{2+} and H^+ fluxes are functionally coupled, i.e. DMT1 is a H^+/Fe^{2+} cotransporter (symporter). Whereas H^+ can substitute for Na^+

as the driving ion in the Na^+/glucose cotransporter SGLT1 (Hirayama, Loo, & Wright, 1994), Na^+ cannot substitute for H^+ as the driving ion in DMT1 (Mackenzie et al., 2007).

Conventional cotransporter models are deterministic—they comprise a series of ligand-induced conformational changes that result in alternating access to a substrate-binding region, and ion-coupling ratios are fixed. Experimental data for DMT1; however, are difficult to reconcile with such a model. We and others have expressed wild-type mammalian DMT1 or mutant proteins in *Xenopus* oocytes and observed significant uncoupled fluxes ("slippage") both of H^+ and Fe^{2+}, and variable H^+:Fe^{2+} coupling ratio (Mackenzie et al., 2006; Nevo & Nelson, 2004; Shawki, Illing, & Mackenzie, 2008). Development of models that can account for such slippage should direct experimental investigation into the molecular mechanisms of DMT1.

4.2. Metal-Ion Substrate Profile and Selectivity of DMT1

Which metal ions are transported by DMT1? That the question has been addressed in nearly as many reviews as there have been original research articles on the topic, we suspect, reflects both the high level of interest in the answer and the lack of complete consensus. A broad range of metal ions evoked currents in voltage-clamped *Xenopus* oocytes expressing rat or human DMT1 (Gunshin et al., 1997; Mackenzie et al., 2007). Many of these metal ions have been shown to inhibit radiotracer iron uptake in mammalian cell preparations in which DMT1 is constitutively or heterologously expressed (Bannon, Abounader, Lees, & Bressler, 2003; Conrad et al., 2000; Garrick & Dolan, 2002; Linder et al., 2006; Picard et al., 2000; Zhang et al., 2008). Nevertheless, currents and inhibition only demonstrate reactivity with the transporter, but do not demonstrate per se that the metal ion is transported by DMT1.

We performed a comprehensive substrate-profile analysis of DMT1 by using two methods that would directly test for metal-ion transport in the oocyte system—radiotracer assay and a fluorescence-based assay using the metal-sensitive fluorophore PhenGreen SK (Illing et al., 2012). We found that Cd^{2+}, Co^{2+}, Fe^{2+}, Mn^{2+}, Ni^{2+}, VO^{2+}, and Zn^{2+} are transported substrates of DMT1; however, we found no evidence of DMT1-mediated transport of Ca^{2+}, Cr^{2+}, Cr^{3+}, Cu^{1+}, Cu^{2+}, Fe^{3+}, Ga^{3+}, Hg^{2+}, or VO^{1+} (Shawki & Mackenzie, 2010; Illing et al., 2012). Our findings extended the catalog of DMT1 substrates and nonsubstrates and, to the extent that a few of these metals had been tested previously (see Illing et al. (2012) for references), were consistent with previous studies with the notable exception of copper. Other investigators have measured copper transport in the human intestinal cell model Caco-2 cells and concluded that DMT1 can transport cuprous ion (Cu^{1+}) (Arredondo, Muñoz, Mura, & Núñez, 2003; Espinoza et al., 2012), or have measured copper uptake in intestinal BBMV and concluded that DMT1 can transport cupric ion (Cu^{2+}) (Knöpfel, Smith, & Solioz, 2005). We found instead that DMT1 did not transport Cu^{1+} or Cu^{2+} in the oocyte preparation. One advantage of the oocyte system is its superior sensitivity—expression of DMT1 stimulated iron transport more than 700-fold over control (Illing et al., 2012)—whereas assigning transport activity to DMT1 is less clear in the mammalian systems. Still, it is possible that a subunit normally expressed in mammalian cells and required for copper transport is not expressed in the *Xenopus* oocyte, or that an isoform other than the 1A/IRE(+) expressed in the study just cited is capable of transporting copper. Clearly, further studies are needed if we are to conclude that copper is a transported substrate of DMT1 and, if so, to determine the valence of the transported species.

Assessing whether DMT1 plays important role in the absorption or cellular transport of each of its substrates requires that we first take into account the relative selectivity with which DMT1 reacts with and transports each of its substrates. We found that DMT1 exhibited a strong preference (selectivity) for Fe^{2+} ($K_{0.5}^{Fe} \approx 1$ μM) over any of its other physiological substrates (Illing et al., 2012). We ranked DMT1 substrates by using the "specificity constant" as follows: $Cd^{2+} > Fe^{2+} > Co^{2+}$, $Mn^{2+} >> Zn^{2+}$, Ni^{2+}, VO^{2+} (Fig. 3). With some knowledge of prevailing metal-ion concentrations in vivo, these data help us to predict DMT1's roles in transport of metals aside from iron, and such predictions can be tested in tissue preparations or animal models. Intestinal absorption of manganese and its uptake into reticulocytes is disrupted in the *b* rat indicating that DMT1 plays some role in manganese transport in those tissues (Chua & Morgan, 1997). Nevertheless, the involvement of DMT1 in the transport of manganese (moderate selectivity, $K_{0.5}^{Mn} \approx 4$ μM; Illing et al. (2012)) may be of greater pathophysiological significance than physiological, e.g. nasopulmonary absorption of manganese and its uptake into the brain in chronic manganese exposure (Section 6.4). We found zinc to be a very poor substrate ($K_{0.5}^{Zn} \approx 19$ μM)—and weak competitive inhibitor—of DMT1 (Illing et al., 2012). Because also the prevailing iron concentrations generally exceed those of zinc, we would expect iron to outcompete zinc under most conditions and DMT1 contribute little if at all to the absorption or cellular uptake of zinc. If DMT1 participates at all in zinc absorption, then it is not sufficient to compensate for the loss of the Zrt-/Irt-like transporter-4 (ZIP4), because hereditary defects in

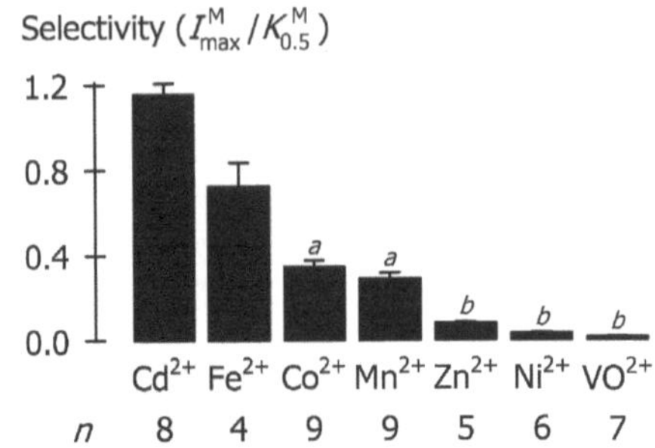

Figure 3 Metal-ion substrate selectivity of DMT1, expressed as the ratio $I^{M}_{max}/K^{M}_{0.5}$ (i.e. the specificity constant), derived from metal-ion evoked currents measured in RNA-injected oocytes expressing DMT1 at pH 5.5. Oocytes were voltage clamped at −70 mV. Data for each oocyte were obtained over a range of at least seven concentrations and were fit by a modified Michaelis–Menten function for which I^{M}_{max} is the derived maximal current and $K^{M}_{0.5}$ is the half-maximal metal-ion (M) concentration (or affinity constant). Fe^{2+} was superfused in the presence of 100 μM L-ascorbic acid. I^{M}_{max} was normalized by the current evoked by 50 μM Mn^{2+} in each individual oocyte. Data are mean ± SEM from *n* independent trials for each metal ion (M). Based on one-way ANOVA ($P<0.001$) followed by Holm–Šidák all-pairwise comparisons, all differed from one another ($P \leq 0.004$) except among those marked $^{a}P=0.64$ or $^{b}P \geq 0.61$. *This research was originally published in Illing et al., 2012; © the American Society for Biochemistry and Molecular Biology, used with permission*

ZIP4 result in acrodermatitis enteropathica (Wang, Zhou, Kuo, Zemansky, & Gitschier, 2002). We found that human DMT1 transports the toxic heavy metal Cd^{2+} more efficiently than it does Fe^{2+} (Illing et al., 2012), suggesting that DMT1 is a likely route of intestinal cadmium absorption and its cellular uptake (Section 6.2).

4.3. Structure–Function Analysis of DMT1 and Its Homolog NRAMP1

Human DMT1 shares 61% identity at the amino acid sequence level with NRAMP1, the only other member of the SLC11 gene family in mammals. DMT1 and NRAMP1 belong to an NRAMP superfamily in animals, plants and fungi, and extending to prokaryotes (Cellier, Bergevin, Boyer, & Richer, 2001; Chaloupka et al., 2005). Since there exists no crystal structure of mammalian DMT1, NRAMP1, or their orthologues, investigators have relied on the following approaches to study the structure–function relationships of SLC11 proteins: (1) predicting and testing membrane topology models, (2) site-directed mutagenesis and functional analysis, (3) structural analysis of short synthetic model peptides, and (4) modeling on known crystal structures of homologous proteins.

4.3.1. Predicting and Testing Membrane Topology Models

DMT1 is an integral membrane protein that is observed in vitro to undergo glycosylation (Gruenheid et al., 1999; Mackenzie et al., 2007). Kyte–Doolittle hydrophobicity analysis predicted 12 transmembrane (TM) regions for rat (Gunshin et al., 1997), mouse (Czachorowski, Lam-Yuk-Tseung, Cellier, & Gros, 2009; Lam-Yuk-Tseung et al., 2003) and human DMT1 (Fig. 1). Two consensus sites of *N*-linked glycosylation (at residues Asn-336, Asn-349 according to numbering for the human 1B isoforms) reside in the fourth extracellular loop between TM7 and TM8 (Mackenzie et al., 2007). To test the predicted membrane topology, hemagglutinin (HA) epitope tags were inserted into nonconserved regions of the protein, and the constructs expressed in Chinese hamster ovary (CHO) or pig kidney epithelial (LLC-PK_1) cells (Czachorowski et al., 2009; Picard et al., 2000). Tags inserted in the loops between TM5–6, TM7–8, and TM11–12 were accessible to an anti-HA antibody in nonpermeabilized cells, confirming that these loops are exofacial (cell exterior or endosomal lumen). Tags inserted into the N- or C-terminal regions, or between TM4–5, TM6–7, and TM10–11 were accessible only in permeabilized cells, confirming their intracytoplasmic location. Introduction of the HA epitope tag in other regions disrupted DMT1 transport activity in whole or in part, so the orientation of the loops between TM1–2, 2–3, 3–4, 8–9, and 9–10 could not be definitively assigned. The DMT1 homolog NRAMP1 is a phosphoglycoprotein (Vidal, Pinner, Lepage, Gauthier, & Gros, 1996) whose predicted membrane topology is very similar to that of DMT1. Consistent with that prediction, detection of HA epitope tags—used to drive the plasma-membrane expression of NRAMP1—in intact or permeabilized cells indicated that the loop between TM7–8 was extracellular and the N-terminus cytoplasmic (Forbes & Gros, 2003). Other investigators have proposed, however, that NRAMP1 be oriented oppositely (the C terminus directed into the phagosomal lumen and the TM7–8 loop cytoplasmic) (Kuhn, Lafuse, & Zwilling, 2001) and that NRAMP1 transport metal ions into the phagosome (or out of the cell if NRAMP1 was expressed on the plasma membrane) (Goswami et al., 2001; Kuhn et al., 2001; Techau et al., 2007).

4.3.2. Site-Directed Mutagenesis and Functional Analysis

We and other investigators have used site-directed mutagenesis in an effort to probe the structure–function of DMT1 and identify critical elements in the structure of DMT1. Strategies for targeting amino acid residues have included (1) identifying conserved residues, (2) predicting critical residues based on functional homology or structural modeling, and (3) identifying mutations associated with disease phenotypes in human patients or animal models. Around 30 amino acid residues are highly conserved among 145 NRAMP homologs and include a highly conserved DPGN motif (Chaloupka et al., 2005). Disruption of the DPGN motif in a distant NRAMP homolog, the *Escherichia coli* manganese transporter MntH, ablated metal-dependent H^+ transport in transfected *E. coli* cells (Chaloupka et al., 2005; Courville et al., 2008). In human DMT1, the corresponding D^{86}PGN motif is predicted to reside in putative TM1 (Fig. 1); however, in one study this motif was assigned to the first external loop between TM1 and 2 (Cohen et al., 2003). Mutations at Asp-86 or Gly-88 of mouse or rat DMT1 (numbered according to the 1B isoforms) abolished metal-ion uptake in RNA-injected *Xenopus* oocytes or transfected CHO cells (Cohen et al., 2003; Lam-Yuk-Tseung et al., 2003). Additional residues in that region have been proposed to participate in metal binding; however, metal-ion coordination by DMT1 may be a function of several discrete elements in the protein.

We targeted a pair of histidine residues in putative TM6 of rat DMT1 because histidine residues are known to participate in metal binding in other proteins, and contribute to regulating transport activity in other transporters. Whereas the His267→Ala substitution had little impact on DMT1 transport activity, a His272→Ala substitution uncoupled the H^+ and Fe^{2+} fluxes—a conclusion based on the observations that H272A supported a large H^+ leak that was inhibited by Fe^{2+} and conversely mediated $^{55}Fe^{2+}$ transport that was independent of pH (Mackenzie et al., 2006). Mutation of either His-267 or His-272 in mouse DMT1 affected the pH sensitivity of metal-ion transport in transfected CHO cells (Lam-Yuk-Tseung et al., 2003), leading these investigators to speculate that these two residues may form elements of the H^+-conductive pathway through DMT1. Other residues can also participate in H^+ transport or H^+-coupling; e.g. a Phe194→Ile substitution in rat DMT1 and the Gly212→Val human DMT1 mutation associated with a disease phenotype appear to reduce the proton slippage observed for wild-type DMT1 (Nevo & Nelson, 2004; Shawki et al., 2008). Most other human mutations (Section 3) and the G185R mutation found in the *b* rat and *mk* mouse appear to result in defects in protein processing or trafficking but, like G212V, some may also retain partial reactions. The mouse G185R mutant, however, resulted in an unexpected gain of function, involving a Ca^{2+} conductance not present in wild-type DMT1 (Xu et al., 2004).

4.3.3. Structural Analysis of Short Synthetic Model Peptides

To gain clues about DMT1 tertiary and quaternary structure, investigators have generated short synthetic model peptides, incorporated these into membrane-mimetic environments (solvents or micelles), and studied their structure by using circular dichroism spectropolarimetry and NMR spectroscopy. A peptide corresponding to TM4 of rat DMT1 assumed a central α-helical structure, randomly coiled N-terminal region, and (in SDS micelles at physiological pH 6.0–7.4) a disordered C terminus (Li, Li, Sun, & Qian, 2003, Li, Li, Qian, & Sun, 2004). Despite their being embedded in the SDS interior, the side chains of some hydrophilic residues were water-accessible and may therefore form part of a water-filled channel. The observation that the TM4 peptide self-assembled into trimers suggested to these investigators that TM4 of DMT1 may assemble with other TM regions or several DMT1 monomers to form part of a translocation pore that is lined by hydrophilic residues in TM4 (Thr-189, Asp-192, Thr-193, and Asp-200) (Li, Gu, & Sun, 2008). An identical peptide, except for the introduction of the G185R mutation, assumed the same structure as the wild-type TM4 peptide but self-assembled differently from the wild-type peptide, largely forming hexamers instead of trimers (Li et al., 2005). This observation led investigators to speculate that the introduction of the bulky, charged Arg residue in place of Gly at residue 185 should disrupt the quaternary structure of DMT1 or its multimeric assembly.

Synthetic peptides corresponding to TM1 and TM6 of rat DMT1 each formed discontinuous helical structures (i.e. α-helix–extended peptide–α-helix) (Wang, Song, Li, Wang, & Li, 2011; Xiao et al., 2010, Xiao, Wang, Li, & Li, 2011). Amino acid substitutions in TM1 disrupted metal binding in TM1. The DPGN motif in TM1 is thought to comprise a flexible linker region (extended peptide) between the two α-helices, and pH-dependent effects upon this region could regulate the relative orientation of the α-helices and thus affect metal binding. The flexible linker in TM6, like that in TM1, is predicted to undergo rapid, subtle H^+-induced conformational changes. Substitution of His-272 with Ala—which uncoupled the H^+ and Fe^{2+} fluxes through rat DMT1 expressed in *Xenopus* oocytes (Mackenzie et al., 2006)—resulted in an unfolding of the N-terminal α-helix in TM6 and decreased intermolecular interactions (Xiao et al., 2010, 2011). These observations confirm a critical role for His-272 in DMT1, and are consistent with the notion that this residue forms part of the mechanism by which Fe^{2+} transport is coupled to H^+ translocation.

4.3.4. Modeling on Known Crystal Structures of Homologous Proteins

While no crystal structure is available for DMT1, investigators have used a threading approach to predict structural features of DMT1 by using the primary sequences of DMT1 or SLC11 homologs to query known crystal structures of homologous proteins (Courville et al., 2008; Czachorowski et al., 2009). Despite very little similarity at the amino acid level, SLC11 transporters could be superimposed on the structures of the bacterial Na^+/Cl^--dependent leucine transporter (LeuT) of the SLC6 family (Yamashita, Singh, Kawate, Jin, & Gouaux, 2005) and the Na^+-coupled galactose transporter (vSGLT) of the SLC5 family (Faham et al., 2008). Modeling SLC11 homologs on the LeuT structure suggested an internal symmetry in which TM1–5 and TM6–10 fold similarly but are oriented oppositely (Courville et al., 2008). The LeuT fold also predicts (Courville et al., 2008) the discontinuous helical structures of TM1 and TM6 observed for the short synthetic model peptides (Wang, Song et al., 2011; Xiao et al., 2010, 2011) and places the D^{86}PG and (inverted) H^{267}PM motifs in the extended peptide regions of TM1 and TM6, respectively (Courville et al., 2008).

Thus, experimental results and model predictions provide evidence that TM1, TM4 and TM6 in particular contain structural elements that are critical for the catalytic activity of DMT1-mediated H^+-coupled metal-ion transport. That mutations or the introduction of HA epitope tags elsewhere in DMT1 disrupt DMT1 catalytic activity or the correct processing of the protein would suggest that additional critical elements await their discovery.

4.4. Multiple Isoforms of DMT1: Tissue Distribution, Subcellular Targeting, and Regulation

There exist at least four isoforms of the human DMT1 protein. The bulk of the protein (531 or more amino acid residues) is conserved among all isoforms; however, they differ in both their N and C termini as a result of variant transcription of the human *SLC11A2* gene (Table 2) at locus 12q13. Transcript variants arise from use of at least two, possibly three, discrete transcription initiation sites. Initiation from exon 1A, upstream of a second initiation site in exon 1B, adds 29 N-terminal amino acids not present in the 1B isoform (Hubert & Hentze, 2002). Both variants are spliced to exon 2 and, because exon 1B contains no start codon, translation of the 1B isoform begins with a start codon in exon 2. Orthologous transcript variants were also found in rat and mouse (Hubert & Hentze, 2002). Whereas the 1A isoform is found predominantly in intestine and kidney, 1B isoforms are widely expressed (Hubert & Hentze, 2002). In the NCBI Reference

Table 2 Multiple isoforms of human DMT1 arise from variant transcription of the *SLC11A2* gene. The *SLC11A2* gene coding for DMT1 is described at NCBI reference sequence NG_021139.1 (accessed on June 27, 2012). The 1A/IRE(−) isoform is presumptive since a full-length transcript variant that would encode this isoform has not been recorded so far

Transcript variant		Protein isoform					
Transcript number	**NCBI reference sequence**	**Isoform number**	**Name**	**NCBI reference sequence**	**Peptide length**	**N- and C-terminal sequences**	**Predominant localization**
1	NM_001174125.1	1	1A/IRE(+)	NP_001167596.1	590	N–MRKKQLK-TEAAPHCELKSYS-KNSATQVSTMV-LGPEQKMSDD…/…VSISKGLLTEE-ATRGYVK–C	Apical membrane, polarized epithelia
2 3	NM_001174126.1 NM_001174127.1	2	1B/IRE(−)	NP_001167597.1 NP_001167598.1	568	N–MVLGPEQK-MSDD…/…CHL-GLTAQPELYLLNT-MDADSLVSR–C	Early (recycling) endosomes, erythroid precursors
4 5 6	NM_000617.2 NM_001174128.1 NM_001174129.1	3	1B/IRE(+)	NP_000608.1 NP_001167599.1 NP_001167600.1	561	N–MVLGPEQK-MSDD…/…VSIS-KGLLTEEAT-RGYVK–C	Late endosomes and lysosomes, widespread cellular distribution
7	NM_001174130.1	4	Short-form IRE(+)	NP_001167601.1	557	N–MSTVDYL…/…VSISKGLLTEEAT-RGYVK–C	
8	NR_033421.1					Noncoding	
9	NR_033422.1					Noncoding	
			1A/IRE(−)		597	N–MRKKQLK-TEAAPHCELKSYS-KNSATQVSTM-VLGPEQKMSDD…/…CHLGL-TAQPELYLLNTM-DADSLVSR–C	Recycling endosomes*

*Based on expression of a GFP-tagged construct in cultured cells (Yanatori et al., 2010).

Sequence record for the human *SLC11A2* gene (NG_021139.1, version 27 June 2012), these exons—originally described as 1A, 1B, and 2—are annotated exons 1, 2A, and 4, respectively. Initiation of transcription from exon 5 codes a putative, shorter isoform with a distinct N terminus (MSTVDYL). All three 5′ variants are spliced to exon 6A, and transcription proceeds in common for most of the remainder of the coding region.

Alternative splicing at the 3′ end of the *SLC11A2* gene results in mRNA transcripts that differ in both their 3′-translated and 3′-untranslated regions (Canonne-Hergaux, Gruenheid, Ponka, & Gros, 1999; Lee, Gelbart, West, Halloran, & Beutler, 1998). One variant, IRE(+), contains in its 3′-untranslated region (3′-UTR) a stem-loop structure featuring a consensus IRE that is important for posttranscriptional mRNA stabilization according to cellular iron status (Section 5). Thus, the IRE(+) and IRE(−) isoforms are expected to differ mainly in their regulation by iron (Hubert & Hentze, 2002). The IRE(+) and IRE(−) transcripts also give rise to variability in the C-terminal amino-acid sequences (Table 2). Investigators have found a total of seven "coding" transcript variants for human DMT1, and these code for four isoforms (Table 2). A fifth isoform, 1A/IRE(−), is presumptive because a corresponding human mRNA that would code for it has not been reported in the database. No mouse tissue tested expressed exclusively 1A *and* IRE(−) transcripts (Hubert & Hentze, 2002), although a 1A-IRE(−) cDNA has been generated from rat kidney cortex RNA (Abouhamed et al., 2006). An engineered 1A/IRE(−) human construct has been expressed in RNA-injected *Xenopus* oocytes, and the expressed protein was functional (Mackenzie et al., 2007).

The DMT1 isoforms differ also in their tissue distribution and subcellular targeting (Table 2). Epithelial cell lines predominantly express the IRE(+) isoforms whereas blood cell lines express the IRE(−) isoforms (Tabuchi, Tanaka, Nishida-Kitayama, Ohno, & Kishi, 2002). The N and C termini of the DMT1 isoforms may contain signal sequences that direct the subcellular targeting of the protein. Mutational analyses in transfected epithelial cell lines (HEp-2 and MDCK) have revealed that disruption of the N-terminal region H^{13}CELKS (critically, Leu-16) prevents the insertion of the 1A isoform into the plasma membrane (Yanatori et al., 2010), and that the Y^{555}XLXX region in the C-terminal cytoplasmic tail of the IRE(−) isoform is required for its targeting to early endosomes (Tabuchi et al., 2002). In contrast, IRE(+) was localized to the apical plasma membrane, late endosomes, and lysosomes of polarized cells, and its apical-membrane insertion was dependent upon N-glycosylation of DMT1 (Tabuchi et al., 2002).

The N- and C-terminal sequence variations between the isoforms do not, however, alter the functional properties of the isoforms. We have expressed 1A/IRE(+), 1A/IRE(−), 1B/IRE(+), and 1B/IRE(−) isoforms of human DMT1 in *Xenopus* oocytes and found (1) that the cellular uptake of $^{55}Fe^{2+}$ correlated with the levels of expression at the plasma membrane, and (2) that the transport cycle turnover rate did not differ among the four isoforms, i.e. all four isoforms transport Fe^{2+} with equal efficiency (Mackenzie et al., 2007). We also found, after normalization for expression levels, that the 1A/IRE(+) and 1B/IRE(+) isoforms did not differ in their apparent affinity for Fe^{2+}, pH dependence, or relative ability to transport a range of metal ions (Mackenzie et al., 2007, 2010). Likewise, doxycycline-induced expression of the rodent 1A/IRE(+) and 1B/IRE(+) isoforms in stably transfected human embryonic kidney HEK293 cells revealed no differences in their ability to transport Fe^{2+} or Mn^{2+} (Garrick et al., 2006; Mackenzie et al., 2010).

5. REGULATION OF THE CELLULAR EXPRESSION OF DMT1

The dominant mechanism by which DMT1 is regulated at the cellular level is thought to be via post transcriptional mRNA stabilization of the IRE(+) isoform in response to cellular iron levels. Binding of the 3′-UTR mRNA IRE by iron-responsive binding protein-1 (IRP1) confers RNA stability under conditions of low iron (Gunshin et al., 2001). In contrast, the IRE(−) form is not subject to iron-dependent regulation (Rolfs et al., 2002; Zoller et al., 2001). Less is known about IRE/IRP interactions in DMT1 than for other IRE-containing transcripts, but control of expression of the DMT1 IRE(+) isoforms appears analogous to that of the transferrin receptor (TfR) (Pantopoulos, 2004; Wallander, Leibold, & Eisenstein, 2006). Other metals, such as Mn^{2+}, are also bound by IRPs (Kwik-Uribe, Reaney, Zhu, & Smith, 2003) and may exert some control over DMT1 expression.

The liver hormone hepcidin regulates iron metabolism predominantly via its interaction with the iron-export protein ferroportin (Ganz & Nemeth, 2006). Ferroportin binds hepcidin, triggering the internalization and degradation of this transporter. The resulting transient increase in intracellular iron is thought to act via the IRE/IRP system to downregulate expression of DMT1 in enterocytes. Nevertheless, novel mechanisms are emerging by which intestinal DMT1 may be regulated independently of ferroportin or independently of hepcidin. Two studies have proposed that hepcidin directly controls DMT1 expression. Hepcidin inhibited apical ^{55}Fe uptake in both Caco-2 cells and rat duodenal segments apparently by decreasing DMT1 transcription with no change in ferroportin mRNA or protein (Mena, Esparza, Tapia, Valdes, & Núñez, 2008). Other investigators using similar preparations also observed decreased DMT1 mRNA but attributed the loss of DMT1 transport activity mainly to its ubiquitin-dependent proteasomal degradation (Brasse-Lagnel et al., 2011). Such degradation may be facilitated by the adapter proteins Ndfip1 and Ndfip2 (Foot et al., 2008). Indeed, DMT1 levels in enterocytes were elevated in Ndfip1-null mice fed a low-iron diet more strongly than in wild-type mice (Foot et al., 2011). Ubiquitination and proteasomal degradation of DMT1 is reviewed elsewhere (Garrick et al., 2012), and data provided there indicate that ubiquitination by Parkin is specific to the 1B/IRE(−) isoform and not 1A/IRE(+).

Mutations in the hemochromatosis gene *HFE* are responsible for the majority of clinical cases of iron overload. Much emphasis has been placed on its controlling iron absorption by modulating the expression of hepcidin in the liver, but there is evidence also of a role for HFE in the enterocyte (reviewed, Fleming and Britton (2006)). Some colocalization of HFE with DMT1 is observed in rat enterocytes (West, Thomas, Sadlier, & Oates, 2006). Ectopic expression of HFE in Caco-2 cells increased the apical distribution of an HFE–β2 microglobulin complex that coprecipitated with DMT1, and direct interaction of DMT1 with HFE–β2 microglobulin was proposed to explain the decreased apical iron-uptake activity (Arredondo et al., 2006).

Transcript variants lacking the IRE are not subject to regulation by interactions of IRP at the translational level; however, a microRNA, miR-Let-7d, appears capable of inhibiting translation of the DMT1 IRE(−) isoform in erythroid cell lines CD34(+), K562, and HEL (Andolfo et al., 2010), suggesting that miR-Let-7d could regulate iron acquisition as erythroid cells differentiate. Endocytosis of DMT1 protein (corresponding to the IRE(+) isoform) from the brush-border membrane of enterocytes has been observed following iron feeding, a response that appears to be directed by peripheral benzodiazepine receptor-associated protein 7 (PAP7) based on the observation that DMT1 and PAP7 could be copurified from BBMV and were internalized together (Okazaki et al., 2012). In addition, siRNA inhibition of PAP7 in K562 cells caused a reduction in the plasma-membrane expression of DMT1 protein with no change in DMT1 mRNA (Okazaki et al., 2012). PAP7 expression is widespread (Okazaki et al., 2012) and appears to interact with DMT1 also in neurons (Cheah et al., 2006).

DMT1 is also regulated at the transcriptional level by hypoxia and stress-related signaling. Exposure of P19 embryonic carcinoma cells to nitric oxide decreased expression of the 1B but not 1A isoform (representing both IRE(+) and IRE(−) isoforms). This effect appears to have resulted from decreased binding of the nuclear factor NF-κB (p65 subunit) to the 1B promoter region of DMT1 (Paradkar & Roth, 2006). In the intestine, hypoxia regulates the expression of DMT1 via stabilization of the hypoxia-inducible transcription factor HIF2α, which can bind the 1A promoter region of DMT1, and intestine-specific ablation of HIF2α blunted the increased expression of DMT1 in response to iron deficiency (Mastrogiannaki et al., 2009; Shah, Matsubara, Ito, Yim, & Gonzalez, 2009).

6. TISSUE-SPECIFIC ROLES OF DMT1 IN PHYSIOLOGY AND PATHOPHYSIOLOGY

6.1. Erythroid Precursor Cells

Lethal-dose-irradiated wild-type mice into which hematopoietic stem cells were transplanted from $SLC11A2^{-/-}$ mice exhibited defective erythropoiesis, whereas those injected with hematopoietic stem cells of wild-type origin were normal (Gunshin, Fujiiwara et al., 2005), revealing a critical role for DMT1 in erythroid cells. The predominant DMT1 isoform in erythroid precursor cells, 1B/IRE(−), is localized to early (recycling) endosomes in which it participates in transferrin (Tf)-dependent iron acquisition (Canonne-Hergaux, Zhang, Ponka, & Gros, 2001; Gruenheid et al., 1999; Touret, Furuya, Forbes, Gros, & Grinstein, 2003). The transferrin cycle and role of DMT1 is illustrated in Fig. 4. DMT1 transports free Fe^{2+} from the endosome to the cytoplasm, a step that is thought to be energized by the H^+ electrochemical potential gradient established by the vacuolar H^+-ATPase (V-ATPase).

Some DMT1 is also detected on the plasma membrane of erythroid cells. Following endocytosis, the retromer cargo–recognition complex in maturing endosomes directs DMT1 to the recycling pathway (Tabuchi, Yanatori, Kawai, & Kishi, 2010) and, subsequently, the return of DMT1 to the plasmalemma is dependent on phosphatidylinositol 3-kinase activity, but not lipid rafts or caveolae (Touret et al., 2003). Although DMT1 retains some activity at pH 7.4 (Garrick et al., 2006; Mackenzie et al., 2006), plasmalemmal DMT1 is not expected to contribute significantly to iron acquisition in erythroid precursors because (1) most blood iron is associated with transferrin under normal conditions and (2) the endosomal pH ($\approx$6.0) is

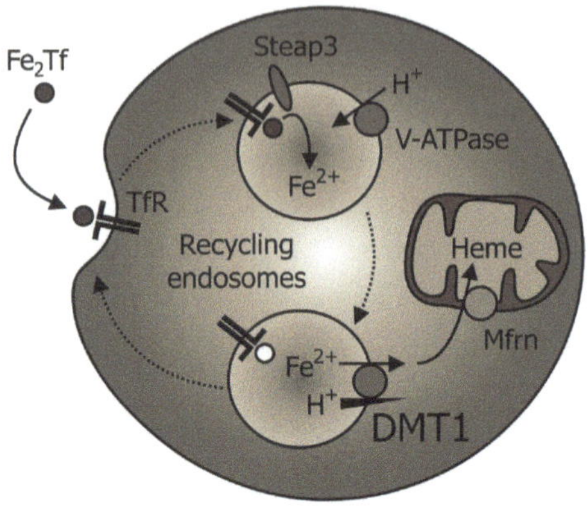

Figure 4 Model of iron uptake in erythroid precursor cells. Iron acquisition in erythroid precursor cells begins with the binding of holotransferrin (FeTf or Fe_2Tf) by transferrin receptor-1 (TfR) and their endocytosis. Acidification of the endosome by the vacuolar H^+-ATPase (V-ATPase) (Saroussi & Nelson, 2009) permits the dissociation of Fe^{3+} from the Tf–TfR complex and provides the proton-motive force to drive DMT1-mediated transport of iron, after its reduction by the ferrireductase six-transmembrane epithelial antigen of the prostate-3 (Steap3) (Knutson, 2007; Ohgami et al., 2005), from the endosome to the cytoplasm. Alternatively, Steap3 may interact directly with the Fe–Tf–TfR complex, and this may promote the dissociation of the reduced iron (Dhungana et al., 2004). The apotransferrin–TfR complex is returned to the plasma membrane, and cytoplasmic iron meanwhile is transported via mitoferrin (Shaw et al., 2006) into the mitochondrion for the production of heme. *(Redrawn and modified from Mackenzie & Hediger, 2004; © Springer Science + Business Media B.V., used with permission).* See the color plate.

more optimal for DMT1-mediated Fe^{2+} transport than is plasma pH (≈7.4) (Mackenzie et al., 2006).

Expression of DMT1 is detected in at least some cell types in every organ tested (Gunshin et al., 1997; Hubert & Hentze, 2002; Wang et al., 2012). We will see that, in most other cell types, DMT1 participates in Tf-dependent iron acquisition akin to that in the erythroid precursor except that, in many nonerythroid cells, this function may be served by the 1B/IRE(+) isoform predominantly observed in late endosomes and lysosomes (Table 2).

6.2. Epithelia

Intestinal DMT1 is critical for iron homeostasis in the mouse (Gunshin, Fujiiwara et al., 2005; Shawki et al., 2012), although defective intestinal iron transport has not yet been demonstrated in the $DMT1^{int/int}$ model as it has for the *mk* mouse (Edwards & Hoke, 1972) and *b* rat (Knöpfel, Zhao et al., 2005). Haploinsufficiency of intestinal absorption is suggested by the lower iron stores in the *SLC11A2*$^{+/-}$ heterozygote (Gunshin, Fujiiwara et al., 2005), but this possibility will be better tested in heterozygotes (DMT1$^{+/int}$) of the conditional intestinal knockout model. The 1A/IRE(+) isoform predominates in the intestine, and its expression is exquisitely upregulated by low-iron status (Canonne-Hergaux et al., 1999; Gunshin et al., 1997; Hubert & Hentze, 2002; Tabuchi et al., 2002) (Table 2). Intestinal DMT1 is also, inappropriately, upregulated in HHC and is expected to contribute to the hyperabsorption of iron observed in this and other iron-overload disorders (Section 7.2). DMT1 is expressed at the brush-border membrane of human and mouse duodenal enterocytes, and Caco-2 cells (Canonne-Hergaux et al., 1999; Griffiths, Kelly, Smith, & Cox, 2000; Tandy et al., 2000). DMT1 mediates the apical uptake of iron (reviewed, Mackenzie and Garrick (2005)) after its reduction by luminal ascorbic acid or surface ferrireductases, which may include DcytB (Cybrd1) and Steap2 (reviewed, Knutson (2007); McKie (2008)).

The acidic microclimate at the intestinal brush border, generated by the intestinal Na^+/H^+ exchangers, is expected to provide the H^+ to drive DMT1-mediated iron uptake (Mackenzie et al., 2011). Na^+/H^+ exchanger expression and activity is switched on toward weaning in the rat (Collins, Xu, Kiela, Zeng, & Ghishan, 1997) and iron absorption is not impaired in suckling Belgrade rat pups (Thompson, Molina, Donaghey, Brain, & Wessling-Resnick, 2007), suggesting that a transporter other than DMT1 serves iron absorption prior to weaning. DMT1 immunoreactivity (possibly the 1B/IRE(+) isoform) has also been identified on the basolateral membrane of enterocytes, and iron-transport activity in intestinal basolateral membrane vesicles or basolateral membranes of Caco-2 cells has been ascribed to DMT1 (Knöpfel, Zhao et al., 2005; Núñez et al., 2010). What roles DMT1 may serve on enterocyte basolateral membranes are not yet clear, but may include supplying iron to enterocytes during dietary iron deficiency or participating in the regulation of iron absorption.

DMT1 efficiently transports cadmium (Fig. 3) (Illing et al., 2012). Iron deficiency is a risk factor for cadmium intoxication (see the work just cited for references), suggesting that cadmium and iron share a common absorptive pathway. Indeed, investigators have ascribed to DMT1 the Cd^{2+} transport observed in Caco-2 cells (Bannon et al., 2003). Iron restriction increased both intestinal DMT1 mRNA and cadmium accumulation in rats and mice (Park, Cherrington, & Klaassen, 2002; Suzuki, Momoi, Hosoyamada, Kimura, & Shibasaki, 2008), implicating DMT1 in cadmium absorption.

Clouding that conclusion, in one of the studies just cited (Suzuki et al., 2008), *mk* mice accumulated cadmium to the same levels as did wild-type mice; however, direct measurement of rates of absorption may be needed to test the role of DMT1 in cadmium absorption and intoxication.

DMT1 is abundantly expressed in the kidney (Hubert & Hentze, 2002; Wang et al., 2012). One group has found that DMT1-IRE(+) immunoreactivity in rat kidney was strongest at the apical membranes of the collecting duct and the thick ascending limb of Henle's loop (Ferguson et al., 2001), roughly consistent with the major sites of absorption of a $^{55}FeCl_3$ dose delivered by intratubular microinjection (Wareing, Ferguson, Green, Riccardi, & Smith, 2000). Strong DMT1 immunoreactivity was also observed in the proximal tubule but confined to an intracellular location (Ferguson et al., 2001, 2003). Because no uptake of a $^{55}FeCl_3$ dose was observed in proximal tubule, we might surmise that an endosomal DMT1 isoform serves instead Tf-dependent iron uptake in the proximal tubule. Evidence for the presence of Tf in the glomerular filtrate is reviewed elsewhere (Smith & Thévenod, 2009), and internalization of Tf by megalin-dependent, cubilin-mediated endocytosis has been described for mammalian kidney proximal tubules (Kozyraki et al., 2001). A second group studying DMT1 localization—this time in mouse kidney—has observed strong DMT1-IRE(+) immunoreactivity in the proximal tubule and concluded that DMT1 is expressed at the apical membrane (Canonne-Hergaux & Gros, 2002). It is possible that the discrepancy in the subcellular localization of mouse and rat DMT1 in these studies is explained by a species difference, detection of different isoforms (the antibodies used were against the same region of the protein but not identical), or technical difficulties; however, Smith and Thévenod (2009) observed subapical DMT1 immunoreactivity also in the mouse and attributed this to expression of DMT1 in late endosomes and lysosomes. Immunogold electron microscopy of rat kidney cortex also places DMT1 in late endosomes and lysosomes, and DMT1 colocalized with the late endosomal/lysosomal markers LAMP1 in the WKPT-0293 Cl.2 cell line derived from rat proximal tubule (Abouhamed et al., 2006). Nevertheless, given the functional homology between the small intestine and renal proximal tubule, and the observation (Hubert & Hentze, 2002) that expression of the 1A/IRE(+) isoform is essentially limited to those tissues, the possibility should not be discounted that DMT1 appears on the apical membrane of proximal tubular cells at least in response to certain stimuli or regulators.

The relative importance of DMT1 in the renal reabsorption of iron (a topic that, historically, has not received much attention) is yet to be fully explored. Meanwhile, the literature supports an involvement of DMT1 in renal pathophysiology. Nephrotic syndromes may be associated with iron-induced damage resulting from increased glomerular permeability, increased transferrin uptake in the proximal tubule, and increased DMT1-mediated iron transport into the cytosol (reviewed, Smith and Thévenod (2009)). DMT1 is thought also to contribute to cadmium-induced nephrotoxicity (reviewed, Vesey (2010)) by transporting cadmium from the late endosome/lysosome to the cytosol following its liberation from metallothionein. Cadmium–metallothionein-1 internalization by WKPT-0293 Cl.2 proximal tubular cells, via the same megalin-dependent receptor-mediated endocytic pathway used by Tf, leads to cytotoxicity (Wolff, Abouhamed, Verroust, & Thévenod, 2006) that can be prevented by RNAi knockdown of late endosomal/lysosomal DMT1 (Abouhamed, Wolff, Lee, Smith, & Thévenod, 2007).

DMT1 is expressed in human term placenta (Georgieff, Wobken, Welle, Burdo, & Connor, 2000; Li et al., 2012) and mouse placenta at gestational day 18 (Balesaria et al., 2012). Those studies differ with regard to the precise localization of DMT1 in placenta, but endosomal DMT1 would appear to participate in transferrin-dependent uptake from the maternal side. Transcript levels of an IRE(+) isoform are responsive to iron (Gambling et al., 2001; Li et al., 2012). Total body iron was normal in *SLC11A2* global knockout mouse neonates, suggesting that fetal DMT1 is not critical for sufficient maternofetal iron transfer (Gunshin, Fujiiwara et al., 2005); however, a placenta-specific DMT1 knockout may be needed to assess precisely the contribution of DMT1 to placental iron transfer. Clearly, there is some redundancy among placental iron-transfer pathways, which may include Zrt-/Irt-like transporter-8 (ZIP8, SLC39A8) (Jenkitkasemwong, Wang, Mackenzie, & Knutson, 2012; Wang, He, Dong, Dalton, & Nebert, 2011).

Nasal and pulmonary epithelia express DMT1 that may provide additional routes of absorption of excess iron and of toxic heavy metals (e.g. cadmium, manganese) following occupational exposure (Ghio et al., 2005; Heilig, Molina, Donaghey, Brain, & Wessling-Resnick, 2005; Heiling et al., 2006; Ruvin Kumara & Wessling-Resnick, 2012; Thompson, Molina, Donaghey, Schwob et al., 2007), although DMT1-mediated clearance of metals from the lung may serve to minimize metal-related injury in that tissue (Ghio et al., 2005; Wang, Ghio et al., 2002).

6.3. Phagocytes—DMT1 and NRAMP1 in Iron Recycling and Host Defense

DMT1 has been observed to colocalize with Tf in early endosomes in RAW264.7 murine macrophages transfected with a c-Myc-tagged DMT1 construct (Gruenheid et al., 1999); however, following phagocytosis of opsonized red blood cells, the tag was also observed in erythrocyte-containing phagosomes (Jabado, Canonne-Hergaux, Gruenheid, Picard, & Gros, 2002). In the latter study, investigators also observed that endogenous DMT1 in TM4 and 15P-1 Sertoli cell lines were recruited to phagosomes containing latex beads or spermatozoids. It was therefore proposed that DMT1 participates in the recovery of iron from effete red blood cells in macrophages, and serve an analogous role in the recovery of iron (or other metals) from senescent sperm in Sertoli cells. If DMT1 plays a role in iron recycling, then it would appear to be shared with NRAMP1 because deficiencies in either protein reduced the efficiency of iron recycling in RAW264.7 macrophages (Soe-Lin et al., 2010). Supporting such a role for NRAMP1, investigators have observed (1) that iron recycling was accelerated in macrophages overexpressing NRAMP1 (Soe-Lin, Sheftel, Wasyluk, & Ponka, 2008), and (2) that, in mice deficient in NRAMP1, iron accumulated in the liver and spleen during erythrophagocytosis and hemolytic anemia, whereas this iron was recycled to the marrow and erythrocytes in wild-type mice (Soe-Lin et al., 2009).

A more recent study has relied on endogenous expression of SLC11 proteins in bone marrow-derived macrophages (BMDM) presented with artificially aged erythrocytes. This model may be superior to transfection of DMT1 and NRAMP1 in RAW264.7 macrophages, and more physiological than the use of opsonized red blood cells for phagocytosis. Whereas DMT1 protein was detected in both quiescent and activated BMDMs, it largely colocalized with TfR1 in recycling endosomes and was not recruited to the erythrophagosome (Delaby et al., 2012). NRAMP1 on the other hand was recruited to the erythrophagosome in BMDMs. Importantly, the observation that the putative heme transporter (heme-responsive gene-1, HRG1)—but neither heme oxygenase protein HO-1 nor HO-2—is recruited to the erythrophagosome (Delaby et al., 2012) prompts us to consider that heme may exit the erythrophagosome intact, and to question whether either NRAMP1 or DMT1 plays any role in iron exit from the phagosome. Whether they could serve that role in macrophages at other loci or participate in iron recycling at some other step, and whether NRAMP1 could transport metals other than Fe^{2+} out of the erythrophagosome, remain to be tested.

NRAMP1 was discovered in 1993 by positional cloning of a mouse gene affecting the capacity with which macrophages can fight bacterial pathogens (Vidal, Malo, Vogan, Skamene, & Gros, 1993). Polymorphisms in the *SLC11A1* gene coding for human NRAMP1 have been strongly associated with susceptibility to mycobacterial infections (tuberculosis, leprosy), and also associated with certain other infections, immune disorders and autoimmune disease (reviewed, Cellier, Courville, and Campion (2007); Mackenzie and Hediger (2004)). NRAMP1 expression is confined to phagocytes (neutrophils and macrophages) (Govoni, Gauthier, Billia, Iscove, & Gros, 1997; Canonne-Hergaux et al., 2002). A conserved N-terminal tyrosine-based motif (YGSI) directs NRAMP1 protein to the lysosomal membrane (Lam-Yuk-Tseung, Picard, & Gros, 2006). Upregulated in professional phagocytes, NRAMP1 contributes to antimicrobial function by extruding from the phagolysosome divalent metal ions (e.g. Mn^{2+}, Fe^{2+}) that may be essential cofactors for bacteria-derived enzymes or required for bacterial growth (Atkinson & Barton, 1999; Cellier et al., 2007; Cellier, 2012; Forbes & Gros, 2003; Jabado et al., 2000).

6.4. Brain

Iron transport and metabolism in the brain (reviewed, Burdo and Connor (2002); Moos and Morgan (2004a); Núñez et al. (2012); Rouault and Cooperman (2006); Snyder and Connor (2009)) warrant special attention because of the association of neurodegenerative disease and iron accumulation. The chief pathway by which iron enters the brain is via the brain capillary endothelial cells of the blood–brain barrier (BBB); the choroid plexus of the blood–cerebrospinal fluid barrier (BCB) presents a second pathway. At both sites, iron is taken up primarily via Tf-dependent endocytosis (Bradbury, 1997), and the identification of DMT1 in BBB and BCB (Burdo et al., 2001; Burdo, Simpson, Menzies, Beard, & Connor, 2004; Gunshin et al., 1997; Moos & Morgan, 2004b; Rouault, Zhang, & Jeong, 2009; Siddappa et al., 2003) implicates DMT1 in that pathway. Colocalization of DMT1 with early endosomal markers in neuronal and astrocyte cell lines (Lis, Barone, Paradkar, Plunkett, & Roth, 2004, Lis et al., 2005) suggests that DMT1 also participates in Tf-dependent uptake in these cells; although, in one study, DMT1 colocalized with late endosomal markers in neurons (Pelizzoni, Zacchetti, Smith, Grohovaz, & Codazzi, 2012). DMT1 is expressed in neurons throughout the brain, including cortex, cerebellum, and thalamus, anterior olfactory nucleus, substantia nigra, striatum, and hippocampal pyramidal and granule cells (Burdo et al., 2001; Gunshin et al., 1997; Knutson, Menzies, Connor, & Wessling-Resnick, 2004; Moos, Trinder, & Morgan, 2000; Williams, Wilson, & Bressler, 2000).

DMT1 protein is also observed on the plasma membrane of neurons (Lis et al., 2005; Roth et al., 2000) and astrocytes (Wang, Ong, & Connor, 2002), thereby prompting us to consider a role for DMT1 in uptake of non-transferrin-bound iron (NTBI). Although maximally stimulated at lower pH, DMT1 is still active at pH 7.4 (Garrick et al., 2006; Mackenzie et al., 2006). Brain interstitial fluid and cerebrospinal fluid contain substantial levels of NTBI (Moos & Morgan, 1998), much of it as Fe^{2+} because of the high levels of ascorbate in the brain; thus, DMT1 expressed at the plasma membrane could take up NTBI without the need for surface ferrireductases. Such a role has recently been ascribed to voltage-operated Ca^{2+} channels largely from the observation that iron uptake was inhibited by Ca^{2+} (Pelizzoni et al., 2011); however, because Ca^{2+} also inhibits DMT1 (Shawki & Mackenzie, 2010), this transporter should not be discounted. Hypoxia increased the expression of DMT1 on neuronal plasma membranes, and we should also consider a role for DMT1 in the transport of manganese at that location (Lis et al., 2005).

Neuronal iron accumulation is at least associated with prevalent neurodegenerative diseases including Parkinson disease, Alzheimer disease, amyotrophic lateral sclerosis, and Huntington disease. DMT1 is abundantly expressed in the substantia nigra pars compacta (SNpc) (Gunshin et al., 1997; Moos et al., 2000), a site of iron enrichment in healthy subjects. In Parkinson disease, substantial iron accumulation is observed in the SNpc, and this may be associated with the oxidative stress and neuronal damage characteristic of the disease (Snyder & Connor, 2009). Parkin, loss of which is characteristic of a familial form of Parkinson disease, directs the ubiquitination of DMT1 1B/IRE(−) and its proteasomal degradation (Garrick et al., 2012). There is no strong evidence with which to associate DMT1 mutations or single-nucleotide polymorphisms with the initiation of neurodegenerative diseases (Blasco et al., 2011; Jamieson et al., 2005); however, upregulation of DMT1 is implicated in their neuropathogenesis (Aguirre et al., 2012; Blasco et al., 2011; Huang, Ong, Go, & Connor, 2006; Núñez et al., 2012; Salazar et al., 2008; Song, Jiang, Wang, & Xie, 2007; Zheng et al., 2009) and, conversely, loss of DMT1 function in the *b* rat and *mk* mouse provides some protection against iron-mediated toxicity (Moos & Morgan, 2004b; Salazar et al., 2008).

Manganism—a syndrome resembling Parkinson disease—results from chronic exposure to manganese (e.g. via occupational exposure in miners and welders) (Roth, 2006; Rivera-Mancía, Ríos, & Montes, 2011). In addition to the typical routes by which iron may enter the brain, atmospheric manganese may gain access to the CNS via olfactory or trigeminal presynaptic nerve endings and retrograde axonal transport to the brain (Roth, 2006; Tjälve & Henriksson, 1999). Absorption of a nasally instilled manganese dose in olfactory epithelium required DMT1 and was increased by anemia (Thompson, Molina, Donaghey, Schwob et al., 2007). Within the brain, neuronal manganese uptake also appears to rely on DMT1, and this pathway may contribute to manganese-induced neurotoxicity (Au, Benedetto, & Aschner, 2008; Rivera-Mancía et al., 2011; Roth, 2006; Roth & Garrick, 2003).

6.5. Other Tissues

DMT1 is widely expressed in other tissues (Gunshin et al., 1997; Hubert & Hentze, 2002; Wang et al., 2012) in which DMT1 is thought to participate in Tf-dependent uptake in a manner similar to that in erythroid cells (Fig. 4). Such a role for DMT1 in liver (reviewed, Graham, Chua, Herbison, Olynyk, and Trinder (2007)) would appear redundant with that of Zrt-/Irt-like transporter-14 (ZIP14). ZIP14 is associated with TfR-containing endosomes in the hepatocellular carcinoma cell line HepG2, and knockdown of ZIP14 in those cells by siRNA reduced the assimilation of iron from ^{55}Fe–Tf by nearly 50% (Zhao, Gao, Enns, & Knutson, 2010); however, the pH-dependence of ZIP14-mediated iron-transport activity is much less suited to the acidic environment of the endosome than is DMT1 (Mackenzie et al., 2006; Pinilla-Tenas et al., 2011). DMT1 is considered nonessential in liver because the DMT1-null ($SLC11A2^{-/-}$) mouse and Belgrade rat exhibit hepatic iron loading (Gunshin, Fujiiwara et al., 2005; Thompson et al., 2006); however, we should bear in mind that this conclusion derives from animal models in which hepatic iron uptake may be primarily of NTBI (Thompson et al., 2006), and plasma NTBI is absent or very low under normal conditions. In contrast, Tf-dependent iron uptake in hepatocytes is tightly regulated by iron status (Trinder, Batey, Morgan, & Baker, 1990).

Nevertheless, DMT1 may also contribute to the cellular uptake of NTBI characteristic of iron-overload conditions (Section 8.2) because (1) DMT1 was observed to relocate to the plasma membrane of hepatocytes in iron overload (Trinder, Oates, Thomas, Sadleir, & Morgan, 2000), (2) hepatocytes and hepatoma (HLF) cells overexpressing DMT1 exhibit cell-membrane expression of DMT1 and increased NTBI uptake (Shindo et al., 2006), and (3) DMT1 retains iron-transport activity at plasma pH (Garrick et al., 2006; Mackenzie et al., 2006) so it is capable of functioning as an

NTBI transporter at the cell membrane. The extent to which DMT1 contributes to NTBI uptake in those organs most affected by iron overload (e.g. heart, liver, pancreas) requires further investigation, although clearly there is overlap with other pathways that may include ZIP8, ZIP14, and calcium channels (Kumfu et al., 2011; Liuzzi, Aydemir, Nam, Knutson, & Cousins, 2006; Oudit et al., 2003; Jenkitkasemwong et al., 2012; Pinilla-Tenas et al., 2011; Wang et al., 2012).

7. THERAPEUTICS AND NUTRITION

7.1. Iron Deficiency

Iron deficiency remains the most prevalent micronutrient deficiency worldwide and results in iron-deficiency anemia, as well as neurological and developmental defects in children. Iron fortification of infant milk formulas and cereals, and better diet, has decreased the incidence of iron deficiency over recent decades; nevertheless, a more precise understanding of the iron-absorptive machinery will lead to new strategies for improved iron nutrition.

Dietary calcium is known to reduce iron bioavailability, but the molecular basis of this interaction has been poorly understood. Although Ca^{2+} is not a transported substrate of DMT1, we have found that Ca^{2+} inhibited $^{55}Fe^{2+}$ transport activity and currents associated with DMT1 expression in *Xenopus* oocytes with inhibition constants (K_i) of 1–20 mM (Shawki & Mackenzie, 2010). Because intestinal luminal calcium concentrations periodically reach the millimolar range, dietary calcium (from sources such as milk) could substantially inhibit DMT1-mediated Fe^{2+} absorption. Calcium may also decrease intestinal brush-border iron uptake by stimulating the derecruitment of DMT1 from the apical membrane. Treatment of Caco-2 cells with 2.5 mM $CaCl_2$ for 4 h was sufficient to cause an 80% reduction in DMT1 immunoreactivity of the membrane fraction (Thompson, Sharp, Elliott, & Fairweather-Tait, 2010). Interactions of calcium with DMT1 therefore should be taken into consideration when developing strategies for improved iron nutrition, milk formulation, and iron fortification.

7.2. Iron-Overload Disorders

Since there exists no regulated mechanism for the excretion of iron, its homeostasis requires that absorption of the metal be tightly regulated. Excess iron (unless it is lost by cell desquamation or bleeding) is retained in the body. Accumulation in the liver, heart, and other vital organs can result in cirrhosis, hepatocellular carcinoma, cardiomyopathy, endocrine disorders, and arthritis. HHC is a disorder characterized by tissue-iron overload secondary to excessive dietary-iron absorption, and generally results from a disruption of the hepcidin–ferroportin axis regulating iron metabolism (Darshan et al., 2010; Fleming & Ponka, 2012). The most common form, HHC type 1, is an autosomal recessive disorder resulting from mutations in *HFE* (Harrison & Bacon, 2003). Intestinal expression of DMT1 is elevated in human patients and animal models of HHC type 1 in spite of increased serum iron or ferritin levels (Byrnes et al., 2002; Fleming et al., 1999). Less commonly, HHC can result also from mutations in the genes coding for hemojuvelin (type 2A), hepcidin (type 2B), transferrin receptor-2 (type 3) or ferroportin (type 4); the last of these being autosomal dominant. The mainstay of treatment for HHC is phlebotomy; however, the effectiveness of this approach is limited by poor patient compliance and the fact that phlebotomy itself can lead to upregulation of DMT1 mRNA in HHC patients (Dostalikova-Cimburova et al., 2012; Kelleher et al., 2004), presumably as a result of signaling from increased erythropoiesis post-phlebotomy.

Targeting DMT1 in the intestine may lead to superior therapies for the prevention of iron overload in HHC patients. Validation of intestinal DMT1 as a therapeutic target derives from at least the following three observations: (1) intestinal expression of DMT1 is upregulated in HHC patients; (2) the severity of the anemia phenotype of human probands carrying mutations in DMT1, demonstrating the critical role of DMT1 in human iron metabolism; and (3) the severe anemia in the intestine-specific DMT1-null mouse, demonstrating that DMT1 is the principal or only transporter serving intestinal uptake of nonheme iron, at least in the mouse post-weaning.

In thalassemia, an autosomal recessive disorder characterized by defective hemoglobin production, iron overload results from blood transfusions given to treat the primary lesion. Iron-chelation therapy is the main approach used to treat iron overload in these patients. Notably, increased intestinal iron absorption (presumably driven by the ineffective erythropoiesis) is also evident in thalassemia (Darshan et al., 2010; Hershko, 2010), so blockade of intestinal DMT1 may be of benefit in treating this disorder.

Screening approaches have identified a number of compounds that can inhibit DMT1-mediated iron uptake in mammalian cells. It is likely that some of these do not act directly upon DMT1. For example, two unrelated antioxidants, ebselen and pyrrolidine dithiobarbamate, inhibited DMT1-mediated Fe^{2+} uptake apparently by altering intracellular redox status because (1) both agents increased intracellular glutathione, and (2) ebselen did not inhibit transport of Mn^{2+} (the free metal is not redox active in physiological environments) (Wetli, Buckett, & Wessling-Resnick, 2006). A screen identified 10 compounds that could inhibit cellular uptake of NTBI in nontransfected HeLa cells (Brown, Buckett, & Wessling-Resnick, 2004). Of these, six were cytotoxic and of low apparent affinity (i.e. the concentration at which inhibition was half-maximal, $IC_{50} \geq 20$ μM), and may therefore be of limited utility. Two of the 10 also inhibited Tf-dependent iron uptake, so this group of compounds may serve as useful experimental tools with which to discriminate NTBI and Tf-dependent transport pathways. When one of these compounds, a polysulfonated dye (NSC306711), was tested in HEK293T cells stably transfected with DMT1, it reversibly and competitively inhibited DMT1-mediated iron uptake with $K_i \approx 7$ μM (Buckett & Wessling-Resnick, 2009); however, the action of NSC306711 is complex because this same compound also induced the internalization and degradation of TfR (Horonchik & Wessling-Resnick, 2008).

High-throughput screening has identified the triazinone pyrazolone as an inhibitor of iron transport in CHO cells transfected with DMT1 ($IC_{50} \approx 3$ μM) (Cadieux et al., 2012). Optimization efforts generated substituted pyrazoles with lower IC_{50} values (higher apparent affinity) but which did not completely inhibit iron transport. The same group has also identified a class of diaryl and tricyclic benzylisothiourea compounds that more potently inhibited iron transport with submicromolar IC_{50} values in CHO cells expressing DMT1 (Zhang et al., 2012). These compounds also boasted low cytotoxicity and low cell permeability—the latter property being significant in that it may be more desirable to block intestinal DMT1 using an agent that is not absorbed, thereby minimizing secondary effects. The mechanism of action of these blockers and the nature of inhibition has not been described; however, inhibition by the substituted pyrazoles did not appear to be mediated via redox effects or metal chelation (Cadieux et al., 2012).

The identification of DMT1 blockers offers both novel experimental tools that will help us more clearly define the roles of DMT1 in metal metabolism, and the prospect of improved therapies in iron-overload disorders.

ACKNOWLEDGMENTS

We thank François Canonne-Hergaux (INSERM, Toulouse, France) for helpful discussions and Colin J. Mitchell (University of Cincinnati) for assistance in confocal microscopy in *Xenopus* oocytes. Research in the authors' laboratory (including that reported in Fig. 2) is supported by PHS Grant R01 DK080047 (to B. Mackenzie) from the National Institute of Diabetes and Digestive and Kidney Diseases (NIDDK). E. J. Niespodzany was supported by the University of Cincinnati Medical Student Summer Research Program, funded in part by PHS Grant DK060444 (Short-Term Medical Student Training Grant, NIDDK). The content of this work is solely the responsibility of the authors and does not necessarily represent the official views of the NIDDK, or the National Institutes of Health.

REFERENCES

Abouhamed, M., Gburek, J., Liu, W., Torchalski, B., Wilhelm, A., Wolff, N. A., et al. (2006). Divalent metal transporter 1 in the kidney proximal tubule is expressed in late endosomes/lysosomal membranes: implications for renal handling of protein-metal complexes. *American Journal of Physiology Renal Physiology*, *290*, F1525–F1533.

Abouhamed, M., Wolff, N. A., Lee, W. -K., Smith, C. P., & Thévenod, F. (2007). Knockdown of endosomal/lysosomal divalent metal transporter 1 by RNA interference prevents cadmium-metallothionein-1 cytotoxicity in renal proximal tubule cells. *American Journal of Physiology Renal Physiology*, *293*, F705–F712.

Aguirre, P., Urrutia, P., Tapia, V., Villa, M., Paris, I., Segura-Aguilar, J. et al. (2012). The dopamine metabolite aminochrome inhibits mitochondrial complex I and modifies the expression of iron transporters DMT1 and FPN1. *Biometals*, 25, 795–803.

Andolfo, I., De, F. L., Asci, R., Russo, R., Colucci, S., Gorrese, M., et al. (2010). Regulation of divalent metal transporter 1 (DMT1) non-IRE isoform by the microRNA Let-7d in erythroid cells. *Haematologica*, *95*, 1244–1252.

Arredondo, M., Muñoz, P., Mura, C. V., & Núñez, M. T. (2003). DMT1, a physiologically relevant apical Cu^{1+} transporter of intestinal cells. *American Journal of Physiology Cell Physiology*, *284*, C1525–C1530.

Arredondo, M., Tapia, V., Rojas, A., Aguirre, P., Reyes, F., Marzolo, M. P., et al. (2006). Apical distribution of HFE-beta2-microglobulin is associated with inhibition of apical iron uptake in intestinal epithelia cells. *Biometals*, *19*, 379–388.

Atkinson, P. G. P., & Barton, C. H. (1999). High level expression of $Nramp1^{G169}$ in RAW264.7 cell transfectants: analysis of intracellular iron transport. *Immunology*, *96*, 656–662.

Au, C., Benedetto, A., & Aschner, M. (2008). Manganese transport in eukaryotes: the role of DMT1. *Neurotoxicology*, *29*, 569–576.

Balesaria, S., Hanif, R., Salama, M., Raja, K., Bayele, H. K., McArdle, H., et al. (2012). Fetal iron levels are regulated by maternal and fetal Hfe genotype and dietary iron. *Haematologica*, *97*, 661–669.

Bannerman, R. M. (1976). Genetic defects of iron transport. *Federation Proceedings*, *35*, 2281–2285.

Bannon, D. I., Abounader, R., Lees, P. S., & Bressler, J. P. (2003). Effect of DMT1 knockdown on iron, cadmium, and lead uptake in Caco-2 cells. *American Journal of Physiology Cell Physiology*, *284*, C44–C50.

Bardou-Jacquet, E., Island, M. L., Jouanolle, A. M., Détivaud, L., Fatih, N., Ropert, M., et al. (2011). A novel N491S mutation in the human *SLC11A2* gene impairs protein trafficking and in association with the G212V mutation leads to microcytic anemia and liver iron overload. *Blood Cells Molecules and Disease*, *47*, 243–248.

Barrientos De Renshaw, T., Bhutiani, M., Montross, L. K., & Andrews, N. C. (2011). Iron overload resulting from increased intestinal DMT1 expression. *American Journal of Hematology*, *86*, E43 (abstract).

Beaumont, C., Delaunay, J., Hetet, G., Grandchamp, B., de Montalembert, M., & Tchernia, G. (2006). Two new human DMT1 gene mutations in a patient with microcytic anemia, low ferritinemia, and liver iron overload. *Blood*, *107*, 4168–4170.

Blanco, E., Kannengiesser, C., Grandchamp, B., Tasso, M., & Beaumont, C. (2009). Not all DMT1 mutations lead to iron overload. *Blood Cells Molecules and Disease*, *43*, 199–201.

Blasco, H., Vourc'h, P., Nadjar, Y., Ribourtout, B., Gordon, P. H., Guettard, Y. O., et al. (2011). Association between divalent metal transport 1 encoding gene (SLC11A2) and disease duration in amyotrophic lateral sclerosis. *Journal of the Neurological Sciences, 303*, 124–127.
Bradbury, M. W. (1997). Transport of iron in the blood-brain-cerebrospinal fluid system. *Journal of Neurochemistry, 69*, 443–454.
Brasse-Lagnel, C., Karim, Z., Letteron, P., Bekri, S., Bado, A., & Beaumont, C. (2011). Intestinal DMT1 cotransporter is down-regulated by hepcidin via proteasome internalization and degradation. *Gastroenterology, 140*, 1261–1271.
Brown, J. X., Buckett, P. D., & Wessling-Resnick, M. (2004). Identification of small molecule inhibitors that distinguish between non-transferrin bound iron uptake and transferrin-mediated iron transport. *Chemistry and Biology, 11*, 407–416.
Buchanan, G. R., & Sheehan, R. G. (1981). Malabsorption and defective utilization of iron in three siblings. *Journal of Pediatrics, 98*, 723–728.
Buckett, P. D., & Wessling-Resnick, M. (2009). Small molecule inhibitors of divalent metal transporter-1. *American Journal of Physiology Gastrointestinal and Liver Physiology, 296*, G798–G804.
Burdo, J. R., & Connor, J. R. (2002). In D. M. Templeton (Ed.), *Iron transport in the central nervous system. Molecular and cellular iron transport* (pp. 487–508). New York: Marcel-Dekker.
Burdo, J. R., Menzies, S. L., Simpson, I. A., Garrick, L. M., Garrick, M. D., Dolan, K. G., et al. (2001). Distribution of divalent metal transporter 1 and metal transport protein 1 in the normal and Belgrade rat. *Journal of Neuroscience Research, 66*, 1198–1207.
Burdo, J. R., Simpson, I. A., Menzies, S., Beard, J., & Connor, J. R. (2004). Regulation of the profile of iron-management proteins in brain microvasculature. *Journal of Cerebral Blood Flow and Metabolism, 24*, 67–74.
Byrnes, V., Barrett, S., Ryan, E., Kelleher, T., O'Keane, C., Coughlan, B., et al. (2002). Increased duodenal DMT-1 expression and unchanged HFE mRNA levels in *HFE*-associated hereditary hemochromatosis and iron deficiency. *Blood Cells Molecules and Disease, 29*, 251–260.
Cadieux, J. A., Zhang, Z., Mattice, M., Brownlie-Cutts, A., Fu, J., Ratkay, L. G., et al. (2012). Synthesis and biological evaluation of substituted pyrazoles as blockers of divalent metal transporter 1 (DMT1). *Bioorganic and Medicinal Chemistry Letters, 22*, 90–95.
Canonne-Hergaux, F., Calafet, J., Richer, E., Cellier, M., Grinstein, S., Borregaard, N., et al. (2002). Expression and subcellular localization of NRAMP1 in human neutrophil granules. *Blood, 100*, 268–275.
Canonne-Hergaux, F., Fleming, M. D., Levy, J. E., Gauthier, S., Ralph, T., Picard, V., et al. (2000). The Nramp2/DMT1 iron transporter is induced in the duodenum of microcytic anemia *mk* mice but is not properly targeted to the intestinal brush border. *Blood, 96*, 3964–3970.
Canonne-Hergaux, F., & Gros, P. (2002). Expression of the iron transporter DMT1 in kidney from normal and anemic *mk* mice. *Kidney International, 62*, 147–156.
Canonne-Hergaux, F., Gruenheid, S., Ponka, P., & Gros, P. (1999). Cellular and subcellular localization of the Nramp2 iron transporter in the intestinal brush border and regulation by dietary iron. *Blood, 93*, 4406–4417.
Canonne-Hergaux, F., Zhang, A. -S., Ponka, P., & Gros, P. (2001). Characterization of the iron transporter DMT1 (NRAMP2/DCT1) in red blood cells of normal and anemic *mk/mk* mice. *Blood, 98*, 3823–3830.
Cellier, M. F. M. (2012). Nutritional immunity: homology modeling of Nramp metal import. *Advances in Experimental Medicine and Biology, 946*, 335–351.
Cellier, M. F. M., Bergevin, I., Boyer, E., & Richer, E. (2001). Polyphyletic origins of bacterial Nramp transporters. *Trends in Genetics, 17*, 365–370.
Cellier, M. F., Courville, P., & Campion, C. (2007). Nramp1 phagocyte intracellular metal withdrawal defense. *Microbes and Infection, 9*, 1662–1670.

Chaloupka, R., Courville, P., Veyrier, F., Knudsen, B., Tompkins, T. A., & Cellier, M. F. M. (2005). Identification of functional amino acids in the Nramp family by a combination of evolutionary analysis and biophysical studies of metal and proton cotransport *in vivo*. *Biochemistry, 44*, 726–733.

Cheah, J. H., Kim, S. F., Hester, L. D., Clancy, K. W., Patterson, S. E., III, Papadopoulos, V., et al. (2006). NMDA receptor-nitric oxide transmission mediates neuronal iron homeostasis via the GTPase Dexras1. *Neuron, 51*, 431–440.

Chua, A. C. G., & Morgan, E. H. (1997). Manganese metabolism is impaired in the Belgrade laboratory rat. *Journal of Comparative Physiology B, 167*, 361–369.

Cohen, A., Nevo, Y., & Nelson, N. (2003). The first external loop of the metal ion transporter DCT1 is involved in metal ion binding and specificity. *Proceedings of the National Academy of Sciences of the United States of America, 100*, 10694–10699.

Collins, J. F., Xu, H., Kiela, P. R., Zeng, J., & Ghishan, F. K. (1997). Functional and molecular characterization of NHE3 expression during ontogeny in rat jejunal epithelium. *American Journal of Physiology Cell Physiology, 273*, C1937–C1946.

Conrad, M. E., Umbreit, J. N., Moore, E. G., Hainsworth, L. N., Porubcin, M., Simovich, M. J., et al. (2000). Separate pathways for cellular uptake of ferric and ferrous iron. *American Journal of Physiology Gastrointestinal and Liver Physiology, 279*, G767–G774.

Courville, P., Urbankova, E., Rensing, C., Chaloupka, R., Quick, M., & Cellier, M. F. M. (2008). Solute carrier 11 cation symport requires distinct residues in transmembrane helices 1 and 6. *Journal of Biological Chemistry, 283*, 9651–9658.

Czachorowski, M., Lam-Yuk-Tseung, S., Cellier, M., & Gros, P. (2009). Transmembrane topology of the mammalian Slc11a2 iron transporter. *Biochemistry, 48*, 8422–8434.

Darshan, D., Frazer, D. M., & Anderson, G, J., (2010). Molecular basis of iron-loading disorders. *Expert Reviews in Molecular Medicine, 12*, e36.

Delaby, C., Rondeau, C., Pouzet, C., Willemetz, A., Pilard, N., Desjardins, M., et al. (2012). Subcellular localization of iron and heme metabolism related proteins at early stages of erythrophagocytosis. *PLoS One*, 7, e42199.

Dhungana, S., Taboy, C. H., Zak, O., Larvie, M., Crumbliss, A. L., & Aisen, P. (2004). Redox properties of human transferrin bound to its receptor. *Biochemistry, 43*, 205–209.

Donovan, A., Brownlie, A., Dorschner, M. O., Zhou, Y., Pratt, S. J., Paw, B. H., et al. (2002). The zebrafish mutant gene chardonnay (*cdy*) encodes divalent metal transporter 1 (DMT1). *Blood, 100*, 4655–4659.

Dostalikova-Cimburova, M., Kratka, K., Balusikova, K., Chmelikova, J., Hejda, V., Hnanicek, J. et al. (2012), Duodenal expression of iron transport molecules in patients with hereditary hemochromatosis or iron deficiency. *Journal of Cellular and Molecular Medicine*, 16, 1816–1826.

Edwards, J. A., & Hoke, J. E. (1972). Defect of intestinal mucosal iron uptake in mice with hereditary microcytic anemia. *Proceedings of the Society for Experimental Biology and Medicine, 141*, 81–84.

Elisma, F., & Jumarie, C. (2001). Evidence for cadmium uptake through Nramp2: metal speciation studies with Caco-2 cells. *Biochemical and Biophysical Research Communications, 285*, 662–668.

Espinoza, A., Le, B. S., Olivares, M., Pizarro, F., Ruz, M., & Arredondo, M. (2012). Iron, copper, and zinc transport: inhibition of divalent metal transporter 1 (DMT1) and human copper transporter 1 (hCTR1) by shRNA. *Biological Trace Element Research, 146*, 281–286.

Faham, S., Watanabe, A., Besserer, G. M., Cascio, D., Specht, A., Hirayama, B. A., et al. (2008). The crystal structure of a sodium galactose transporter reveals mechanistic insights into Na^+/sugar symport. *Science, 321*, 810–814.

Ferguson, C. J., Wareing, M., Delannoy, M., Fenton, R., McLarnon, S. J., Ashton, N., et al. (2003). Iron handling and gene expression of the divalent metal transporter, DMT1, in the kidney of the anemic Belgrade (*b*) rat. *Kidney International, 64*, 1755–1764.

Ferguson, C. J., Wareing, M., Ward, D. T., Green, R., Smith, C. P., & Riccardi, D. (2001). Cellular localization of divalent metal transporter DMT-1 in rat kidney. *American Journal of Physiology Renal Physiology, 280*, F803–F814.

Fleming, R. E., & Britton, R. S. (2006). Iron Imports. VI. HFE and regulation of intestinal iron absorption. *American Journal of Physiology Gastrointestinal and Liver Physiology, 290*, G590–G594.

Fleming, R. E., Migas, M. C., Zhou, X., Jiang, J., Britton, R. S., Brunt, E. M., et al. (1999). Mechanism of increased iron absorption in murine model of hereditary hemochromatosis: increased duodenal expression of the iron transporter DMT1. *Proceedings of the National Academy of Sciences of the United States of America, 96*, 3143–3148.

Fleming, R. E., & Ponka, P. (2012). Iron overload in human disease. *New England Journal of Medicine, 366*, 348–359.

Fleming, M. D., Romano, M. A., Su, M. A., Garrick, L. M., Garrick, M. D., & Andrews, N. C. (1998). *Nramp2* is mutated in the anemic Belgrade (*b*) rat: evidence of a role for Nramp2 in endosomal iron transport. *Proceedings of the National Academy of Sciences of the United States of America, 95*, 1148–1153.

Fleming, M. D., Trenor, C. C., Su, M. A., Foernzler, D., Beier, D. R., Dietrich, W. F., et al. (1997). Microcytic anaemia mice have a mutation in *Nramp2*, a candidate iron transporter gene. *Nature Genetics, 16*, 383–386.

Foot, N. J., Dalton, H. E., Shearwin-Whyatt, L. M., Dorstyn, L., Tan, S. S., Yang, B., et al. (2008). Regulation of the divalent metal ion transporter DMT1 and iron homeostasis by a ubiquitin-dependent mechanism involving Ndfips and WWP2. *Blood, 112*, 4268–4275.

Foot, N. J., Leong, Y. A., Dorstyn, L. E., Dalton, H. E., Ho, K., Zhao, L., et al. (2011). *Ndfip1*-deficient mice have impaired DMT1 regulation and iron homeostasis. *Blood, 117*, 638–646.

Forbes, J. R., & Gros, P. (2003). Iron, manganese, and cobalt transport by Nramp1 (Slc11a1) and Nramp2 (Slc11a2) expressed at the plasma membrane. *Blood, 102*, 1884–1892.

Gambling, L., Danzeisen, R., Gair, S., Lea, R. G., Charania, Z., Solanky, N., et al. (2001). Effect of iron deficiency on placental transfer of iron and expression of iron transport proteins in vivo and in vitro. *Biochemical Journal, 356*, 883–889.

Ganz, T., & Nemeth, E. (2006). Iron Imports. IV. Hepcidin and regulation of body iron metabolism. *American Journal of Physiology Gastrointestinal and Liver Physiology, 290*, G199–G203.

Garrick, M. D., & Dolan, K. G. (2002). An expression system for a transporter of iron and other metals. *Methods in Molecular Biology, 196*, 147–154.

Garrick, M. D., Kuo, H. C., Vargas, F., Singleton, S., Zhao, L., Smith, J. J., et al. (2006). Comparison of mammalian cell lines expressing distinct isoforms of divalent metal transporter 1 in a tetracycline-regulated fashion. *Biochemical Journal, 398*, 539–546.

Garrick, M. D., Zhao, L., Roth, J. A., Jiang, H., Feng, J., Foot, N. J., et al. (2012). Isoform specific regulation of divalent metal (ion) transporter (DMT1) by proteasomal degradation. *Biometals, 25*, 787–793.

Georgieff, M. K., Wobken, J. K., Welle, J., Burdo, J. R., & Connor, J. R. (2000). Identification and localization of divalent metal transporter-1 (DMT-1) in term human placenta. *Placenta, 21*, 799–804.

Ghio, A. J., Piantadosi, C. A., Wang, X., Dailey, L. A., Stonehuerner, J. D., Madden, M. C., et al. (2005). Divalent metal transporter-1 decreases metal-related injury in the lung. *American Journal of Physiology Lung Cellular and Molecular Physiology, 289*, L460–L467.

Goswami, T., Bhattacharjee, A., Babal, P., Searle, S., Moore, E., Li, M., et al. (2001). Natural-resistance-associated macrophage protein 1 is an H^+/bivalent cation antiporter. *Biochemical Journal, 354*, 511–519.

Govoni, G., Gauthier, S., Billia, F., Iscove, N. N., & Gros, P. (1997). Cell-specific and inducible *Nramp1* gene expression in mouse macrophages *in vitro* and *in vivo*. *Journal of Leukocyte Biology, 62*, 277–286.

Graham, R. M., Chua, A. C., Herbison, C. E., Olynyk, J. K., & Trinder, D. (2007). Liver iron transport. *World Journal of Gastroenterology, 13*, 4725–4736.

Griffiths, W. J. H., Kelly, A. L., Smith, S. J., & Cox, T. M. (2000). Localization of iron transport and regulatory proteins in human cells. *Quarterly Journal of Medicine, 93*, 575–587.

Gruenheid, S., Canonne-Hergaux, F., Gauthier, S., Hackam, D. J., Grinstein, S., & Gros, P. (1999). The iron transport protein NRAMP2 is an integral membrane glycoprotein that colocalizes with transferrin in recycling endosomes. *Journal of Experimental Medicine, 189*, 831–841.

Gunshin, H., Allerson, C. R., Polycarpou-Schwarz, M., Rofts, A., Rogers, J. T., Kishi, F., et al. (2001). Iron-dependent regulation of the divalent metal ion transporter. *FEBS Letters, 509*, 309–316.

Gunshin, H., Fujiwara, Y., Custodio, Á. O., Direnzo, C., Robine, S., & Andrews, N. C. (2005). Slc11a2 is required for intestinal iron absorption and erythropoiesis but dispensable in placenta and liver. *Journal of Clinical Investigation, 115*, 1258–1266.

Gunshin, H., Jin, J., Fujiwara, Y., Andrews, N. C., Mims, M., & Prchal, J. (2005). Analysis of the E399D mutation in *SLC11A2*. *Blood, 106*, 2221–2222.

Gunshin, H., Mackenzie, B., Berger, U. V., Gunshin, Y., Romero, M. F., Boron, W. F., et al. (1997). Cloning and characterization of a mammalian proton-coupled metal-ion transporter. *Nature, 388*, 482–488.

Harrison, S. A., & Bacon, B. R. (2003). Hereditary hemochromatosis: update for 2003. *Journal of Hepatology, 38*, S14–S23.

Hartman, K. R., & Barker, J. A. (1996). Microcytic anemia with iron malabsorption: an inherited disorder of iron metabolism. *American Journal of Hematology, 51*, 269–275.

Heilig, E., Molina, R., Donaghey, T., Brain, J. D., & Wessling-Resnick, M. (2005). Pharmacokinetics of pulmonary manganese absorption: evidence for increased susceptibility to manganese loading in iron deficient rats. *American Journal of Physiology Lung Cellular and Molecular Physiology, 288*, L887–L893.

Heilig, E. A., Thompson, K. J., Molina, R. M., Ivanov, A. R., Brain, J. D., & Wessling-Resnick, M. (2006). Manganese and iron transport across pulmonary epithelium. *American Journal of Physiology Lung Cellular and Molecular Physiology, 290*, L1247–L1259.

Hershko, C. (2010). Pathogenesis and management of iron toxicity in thalassemia. *Annals of the New York Academy of Science, 1202*, 1–9.

Hirayama, B. A., Loo, D. D. F., & Wright, E. M. (1994). Protons drive sugar transport through the Na^+/glucose cotransporter (SGLT1). *Journal of Biological Chemistry, 269*, 21407–21410.

Horonchik, L., & Wessling-Resnick, M. (2008). The small-molecule iron transport inhibitor ferristatin/NSC306711 promotes degradation of the transferrin receptor. *Chemistry and Biology, 15*, 647–653.

Huang, E., Ong, W. -Y., Go, M.-L., & Connor, J. R. (2006). Upregulation of iron regulatory proteins and divalent metal transporter-1 isoforms in the rat hippocampus after kainate induced neuronal injury. *Experimental Brain Research, 170*, 376–386.

Hubert, N., & Hentze, M. W. (2002). Previously uncharacterized isoforms of divalent metal transporter (DMT)-1: implications for regulation and cellular function. *Proceedings of the National Academy of Sciences of the United States of America, 99*, 12345–12350.

Illing, A. C., Shawki, A., Cunningham, C. L., & Mackenzie, B. (2012). Substrate profile and metal-ion selectivity of human divalent metal-ion transporter-1. *Journal of Biological Chemistry, 287*, 30485–30496.

Iolascon, A., d'Apolito, M., Servedio, V., Cimmino, F., Piga, A., & Camaschella, C. (2006). Microcytic anemia and hepatic iron overload in a child with compound heterozygous mutations in DMT1. *Blood, 107*, 349–354.

Jabado, N., Canonne-Hergaux, F., Gruenheid, S., Picard, V., & Gros, P. (2002). Iron transporter Nramp2/DMT-1 is associated with the membrane of phagosomes in macrophages and Sertoli cells. *Blood*, *100*, 2617–2622.

Jabado, N., Jankowski, A., Dougaparsad, S., Picard, V., Grinstein, S., & Gros, P. (2000). Natural resistance to intracellular infections: natural resistance-associated macrophage protein 1 (NRAMP1) functions as a pH-dependent manganese transporter at the phagosomal membrane. *Journal of Experimental Medicine*, *192*, 1237–1248.

Jamieson, S. E., White, J. K., Howson, J. M., Pask, R., Smith, A. N., Brayne, C., et al. (2005). Candidate gene association study of solute carrier family 11a members 1 (SLC11A1) and 2 (SLC11A2) genes in Alzheimer's disease. *Neuroscience Letters*, *374*, 124–128.

Jenkitkasemwong, S., Wang, C. -Y., Mackenzie, B., & Knutson, M. D. (2012). Physiologic implications of metal-ion transport by ZIP14 and ZIP8. *Biometals*, *25*, 643–655.

Kelleher, T., Ryan, E., Barrett, S., Sweeney, M., Byrnes, V., O'Keane, C., et al. (2004). Increased DMT1 but not IREG1 or HFE mRNA following iron depletion therapy in hereditary haemochromatosis. *Gut*, *53*, 1174–1179.

Knöpfel, M., Smith, C., & Solioz, M. (2005). ATP-driven copper transport across the intestinal brush border membrane. *Biochemical and Biophysical Research Communications*, *330*, 645–652.

Knöpfel, M., Zhao, L., & Garrick, M. D. (2005). Transport of divalent transition-metal ions is lost in small-intestinal tissue of b/b Belgrade rats. *Biochemistry*, *44*, 3454–3465.

Knutson, M. D. (2007). Steap proteins: implications for iron and copper metabolism. *Nutrition Reviews*, *65*, 335–340.

Knutson, M., Menzies, S., Connor, J., & Wessling-Resnick, M. (2004). Developmental, regional, and cellular expression of SFT/UbcH5A and DMT1 mRNA in brain. *Journal of Neuroscience Research*, *76*, 633–641.

Kozyraki, R., Fyfe, J., Verroust, P. J., Jacobsen, C., Dautry-Varsat, A., Gburek, J., et al. (2001). Megalin-dependent cubilin-mediated endocytosis is a major pathway for the apical uptake of transferrin in polarized epithelia. *Proceedings of the National Academy of Sciences of the United States of America*, *98*, 12491–12496.

Kuhn, D. E., Lafuse, W. P., & Zwilling, B. S. (2001). Iron transport into *Mycobacterium avium*-containing phagosomes from an *Nramp1*^{Gly169}-transfected RAW264.7 macrophage cell line. *Journal of Leukocyte Biology*, *69*, 43–49.

Kumfu, S., Chattipakorn, S., Srichairatanakool, S., Settakorn, J., Fucharoen, S., & Chattipakorn, N. (2011). T-type calcium channel as a portal of iron uptake into cardiomyocytes of β-thalassemic mice. *European Journal of Haematology*, 10–0609.

Kwik-Uribe, C. L., Reaney, S., Zhu, Z., & Smith, D. (2003). Alterations in cellular IRP-dependent iron regulation by in vitro manganese exposure in undifferentiated PC12 cells. *Brain Research*, *973*, 1–15.

Lam-Yuk-Tseung, S., Camaschella, C., Iolascon, A., & Gros, P. (2006). A novel R416C mutation in human DMT1 (SLC11A2) displays pleiotropic effects on function and causes microcytic anemia and hepatic iron overload. *Blood Cells Molecules and Disease*, *36*, 347–354.

Lam-Yuk-Tseung, S., Govoni, G., & Gros, P. (2003). Iron transport by NRAMP2/DMT1: pH regulation of transport by two histidines in transmembrane domain 6. *Blood*, *101*, 3699–3707.

Lam-Yuk-Tseung, S., Mathieu, M., & Gros, P. (2005). Functional characterization of the E399D DMT1/NRAMP2/SLC11A2 protein produced by an exon 12 mutation in a patient with microcytic anemia and iron overload. *Blood Cells Molecules and Disease*, *35*, 212–216.

Lam-Yuk-Tseung, S., Picard, V., & Gros, P. (2006). Identification of a tyrosine-based motif (YGSI) in the amino terminus of Nramp1 (Slc11A1) that is important for lysosomal targeting. *Journal of Biological Chemistry*, *281*, 31677–31688.

Lee, P. L., Gelbart, T., West, C., Halloran, C., & Beutler, E. (1998). The human *Nramp2* gene: characterization of the gene structure, alternative splicing, promoter region and polymorphisms. *Blood Cells Molecules and Disease, 24,* 199–215.

Li, Y. Q., Bai, B., Cao, X. X., Zhang, Y. H., Yan, H., Zheng, Q. Q., et al. (2012). Divalent metal transporter 1 expression and regulation in human placenta. *Biological Trace Element Research, 146,* 6–12.

Li, H., Gu, J. D., & Sun, H. (2008). Structure, topology and assembly of a 32-mer peptide corresponding to the loop 3 and transmembrane domain 4 of divalent metal transporter (DMT1) in membrane-mimetic environments. *Journal of Inorganic Chemistry, 283,* 33865–33873.

Li, F., Li, H., Hu, L., Kwan, M., Chen, G., He, Q. Y., et al. (2005). Structure, assembly, and topology of the G185R mutant of the fourth transmembrane domain of divalent metal transporter. *Journal of the American Chemical Society, 127,* 1414–1423.

Li, H., Li, F., Qian, Z. M., & Sun, H. (2004). Structure and topology of the transmembrane domain 4 of the divalent metal transporter in membrane-mimetic environments. *European Journal of Biochemistry, 271,* 1938–1951.

Li, H., Li, F., Sun, H., & Qian, Z. M. (2003). Membrane-inserted conformation of transmembrane domain 4 of divalent-metal transporter. *Biochemical Journal, 372,* 757–766.

Linder, M. C., Moriya, M., Whon, A., Kassa, A., & Gilley, C. (2006). Vesicular transport of Fe and interaction with other metal ions in polarized Caco2 cell monolayers. *Biological Research, 39,* 143–156.

Lis, A., Barone, T. A., Paradkar, P. N., Plunkett, R. J., & Roth, J. A. (2004). Expression and localization of different forms of DMT1 in normal and tumor astroglial cells. *Brain Research Molecular Brain Research, 122,* 62–70.

Lis, A., Paradkar, P. N., Singleton, S., Kuo, H. C., Garrick, M. D., & Roth, J. A. (2005). Hypoxia induces changes in expression of isoforms of the divalent metal transporter (DMT1) in rat pheochromocytoma (PC12) cells. *Biochemical Pharmacology, 69,* 1647–1655.

Liuzzi, J. P., Aydemir, F., Nam, H., Knutson, M. D., & Cousins, R. J. (2006). Zip14 (Slc39a14) mediates non-transferrin-bound iron uptake into cells. *Proceedings of the National Academy of Sciences of the United States of America, 103,* 13612–13617.

Loo, D. D. F., Hazama, A., Supplisson, S., Turk, E., & Wright, E. M. (1993). Relaxation kinetics of the Na^+/glucose cotransporter. *Proceedings of the National Academy of Sciences of the United States of America, 90,* 5767–5771.

Mackenzie, B., & Garrick, M. D. (2005). Iron Imports. II. Iron uptake at the apical membrane in the intestine. *American Journal of Physiology Gastrointestinal and Liver Physiology, 289,* G981–G986.

Mackenzie, B., & Hediger, M. A. (2004). SLC11 family of H^+-coupled metal-ion transporters NRAMP1 and DMT1. *Pflügers Archiv European Journal of Physiology, 447,* 571–579.

Mackenzie, B., Illing, A. C., & Hediger, M. A. (2008). Transport model of the human Na^+-coupled L-ascorbic acid (vitamin C) transporter SVCT1. *American Journal of Physiology Cell Physiology, 294,* C451–C459.

Mackenzie, B., Loo, D. D. F., Fei, Y. J., Liu, W., Ganapathy, V., Leibach, F. H., et al. (1996). Mechanisms of the human intestinal H^+-coupled oligopeptide transporter hPEPT1. *Journal of Biological Chemistry, 271,* 5430–5437.

Mackenzie, B., Loo, D. D.F., Panayotova-Heiermann, M., & Wright, E. M. (1996). Biophysical characteristics of the pig kidney Na^+/glucose cotransporter SGLT2 reveal a common mechanism for SGLT1 and SGLT2. *Journal of Biological Chemistry, 271,* 32678–32683.

Mackenzie, B., Shawki, A., Ghio, A. J., Stonehuerner, J. D., Zhao, L., Ghadersohi, S., et al. (2010). Calcium-channel blockers do not affect iron transport mediated by divalent metal-ion transporter-1. *Blood, 115,* 4148–4149.

Mackenzie, B., Shawki, A., Kim, R., Anthony, S. R., Knight, P. B., Bradford, E. M., et al. (2011). Intestinal brush-border Na^+/H^+ exchangers are required for iron homeostasis in the mouse. *FASEB Journal*, *25*, 238.1 (abstract).

Mackenzie, B., Takanaga, H., Hubert, N., Rolfs, A., & Hediger, M. A. (2007). Functional properties of multiple isoforms of human divalent metal-ion transporter 1 (DMT1). *Biochemical Journal*, *403*, 59–69.

Mackenzie, B., Ujwal, M. L., Chang, M.-H., Romero, M. F., & Hediger, M. A. (2006). Divalent metal-ion transporter DMT1 mediates both H^+-coupled Fe^{2+} transport and uncoupled fluxes. *Pflügers Archiv European Journal of Physiology*, *451*, 544–558.

Mastrogiannaki, M., Matak, P., Keith, B., Simon, M. C., Vaulont, S., & Peyssonnaux, C. (2009). HIF-2α, but not HIF-1α, promotes iron absorption in mice. *Journal of Clinical Investigation*, *38499*.

McKie, A. T. (2008). The role of Dcytb in iron metabolism: an update. *Biochemical Society Transactions*, *36*, 1239–1241.

Mena, N. P., Esparza, A., Tapia, V., Valdes, P., & Núñez, M. T. (2008). Hepcidin inhibits apical iron uptake in intestinal cells. *American Journal of Physiology Gastrointestinal and Liver Physiology*, *294*, G192–G198.

Mims, M. P., Guan, Y., Pospisilova, D., Priwitzerova, M., Indrak, K., Ponka, P., et al. (2005). Identification of a human mutation of DMT1 in a patient with microcytic anemia and iron overload. *Blood*, *105*, 1337–1342.

Moos, T., & Morgan, E. H. (1998). Evidence for low molecular weight, non-transferrin-bound iron in rat brain and cerebrospinal fluid. *Journal of Neuroscience Research*, *54*, 486–494.

Moos, T., & Morgan, E. H. (2004a). The metabolism of neuronal iron and its pathogenic role in neurological disease: review. *Annals of the New York Academy of Science*, *1012*, 14–26.

Moos, T., & Morgan, E. H. (2004b). The significance of the mutated divalent metal transporter (DMT1) on iron transport into the Belgrade rat brain. *Journal of Neurochemistry*, *88*, 233–245.

Moos, T., Trinder, D., & Morgan, E. H. (2000). Cellular distribution of ferric iron, ferritin, transferrin and divalent metal transporter 1 (DMT1) in substantia nigra and basal ganglia of normal and beta2-microglobulin deficient mouse brain. *Cellular and Molecular Biology*, *46*, 549–561.

Nevo, Y., & Nelson, N. (2004). The mutation F227I increases the coupling of metal-ion transport in DCT1. *Journal of Biological Chemistry*, *279*, 53056–53061.

Núñez, M. T., Tapia, V., Rojas, A., Aguirre, P., Gomez, F., & Nualart, F. (2010). Iron supply determines apical/basolateral membrane distribution of intestinal iron transporters DMT1 and ferroportin 1. *American Journal of Physiology Cell Physiology*, *298*, C477–C485.

Núñez, M. T., Urrutia, P., Mena, N., Aguirre, P., Tapia, V., & Salazar, J. (2012). Iron toxicity in neurodegeneration. *Biometals*, *25*, 761–776.

Ohgami, R. S., Campagna, D. R., Greer, E. L., Antiochos, B., McDonald, A., Chen, J., et al. (2005). Identification of a ferrireductase required for efficient transferrin-dependent iron uptake in erythroid cells. *Nature Genetics*, *37*, 1264–1269.

Okazaki, Y., Ma, Y., Yeh, M., Yin, H., Li, Z., Yeh, K.Y., et al. (2012). DMT1 (IRE) expression in intestinal and erythroid cells is regulated by peripheral benzodiazepine receptor-associated protein 7. *American Journal of Physiology Gastrointestinal and Liver Physiology*, *302*, G1180–G1190.

Oudit, G. Y., Sun, H., Trivieri, M. G., Koch, S. E., Dawood, F., Ackerley, C., et al. (2003). L-type Ca^{2+} channels provide a major pathway for iron entry into cardiomyocytes in iron-overload cardiomyopathy. *Nature Medicine*, *9*, 1187–1194.

Pantopoulos, K. (2004). Iron metabolism and the IRE/IRP regulatory system: an update. *Annals of the New York Academy of Science*, *1012*, 1–13.

Papanikolaou, G., Tzilianos, M., Christakis, J. I., Bogdanos, D., Tsimirika, K., Macfarlane, J., et al. (2005). Hepcidin in iron overload disorders. *Blood*, *105*, 4103–4105.

Paradkar, P. N., & Roth, J. A. (2006). Nitric oxide transcriptionally down-regulates specific isoforms of divalent metal transporter (DMT1) via NF-kappaB. *Journal of Neurochemistry*, *96*, 1768–1777.

Park, J. D., Cherrington, N. J., & Klaassen, C. D. (2002). Intestinal absorption of cadmium is associated with divalent metal transporter 1 in rats. *Toxicological Sciences*, *68*, 288–294.

Pelizzoni, I., Macco, R., Morini, M. F., Zacchetti, D., Grohovaz, F., & Codazzi, F. (2011). Iron handling in hippocampal neurons: activity-dependent iron entry and mitochondria-mediated neurotoxicity. *Aging Cell*, *10*, 172–183.

Pelizzoni, I., Zacchetti, D., Smith, C. P., Grohovaz, F., & Codazzi, F. (2012). Expression of divalent metal transporter 1 in primary hippocampal neurons: reconsidering its role in non-transferrin-bound iron influx. *Journal of Neurochemistry*, *120*, 269–278.

Picard, V., Govoni, G., Jabado, N., & Gros, P. (2000). Nramp 2 (DCT1/DMT1) expressed at the plasma membrane transports iron and other divalent cations into a calcein-accessible cytoplasmic pool. *Journal of Biological Chemistry*, *275*, 35738–35745.

Pinilla-Tenas, J. J., Sparkman, B. K., Shawki, A., Illing, A. C., Mitchell, C. J., Zhao, N., et al. (2011). Zip14 is a complex broad-scope metal-ion transporter whose functional properties support roles in the cellular uptake of zinc and nontransferrin-bound iron. *American Journal of Physiology Cell Physiology*, *301*, C862–C871.

Priwitzerova, M., Pospisilova, D., Prchal, J. T., Indrak, K., Hlobilkova, A., Mihal, V., et al. (2004). Severe hypochromic microcytic anemia caused by a congenital defect of the iron transport pathway in erythroid cells. *Blood*, *103*, 3991–3992.

Priwitzerova, M., Pospisilova, D., Sheftel, A. D., Divoky, V., & Ponka, P. (2005). Functional consequences of the human DMT1 mutation on protein expression and iron uptake. *Blood*, *106*, 3985–3987.

Rivera-Mancía, S., Ríos, C., & Montes, S. (2011). Manganese accumulation in the CNS and associated pathologies. *Biometals*, *24*, 811–825.

Rolfs, A., Bonkovsky, H. L., Kohlroser, J. G., McNeal, K., Sharma, A., Berger, U. V., et al. (2002). Intestinal expression of genes involved in iron absorption in humans. *American Journal of Physiology Gastrointestinal and Liver Physiology*, *282*, G598–G607.

Roth, J. A. (2006). Homeostatic and toxic mechanisms regulating manganese uptake, retention, and elimination. *Biological Research*, *39*, 45–57.

Roth, J. A., & Garrick, M. D. (2003). Iron interactions and other biological reactions mediating the physiological and toxic actions of manganese. *Biochemical Pharmacology*, *66*, 1–13.

Roth, J. A., Horbinski, C., Feng, L., Dolan, K. G., Higgins, D., & Garrick, M. D. (2000). Differential localization of divalent metal transporter 1 with and without iron response element in rat PC12 and sympathetic neuronal cells. *Journal of Neuroscience*, *20*, 7595–7601.

Rouault, T. A., & Cooperman, S. (2006). Brain iron metabolism. *Seminars in Pediatric Neurology*, *13*, 142–148.

Rouault, T. A., Zhang, D. L., & Jeong, S. Y. (2009). Brain iron homeostasis, the choroid plexus, and localization of iron transport proteins. *Metabolic Brain Disease*, *24*, 673–684.

Ruvin Kumara, V. M., & Wessling-Resnick, M. (2012). Olfactory ferric and ferrous iron absorption in iron-deficient rats. *American Journal of Physiology Lung Cellular and Molecular Physiology*, *302*, L1280–L1286.

Salazar, J., Mena, N., Hunot, S., Prigent, A., Alvarez-Fischer, D., Arredondo, M., et al. (2008). Divalent metal transporter 1 (DMT1) contributes to neurodegeneration in animal models of Parkinson's disease. *Proceedings of the National Academy of Sciences of the United States of America*, *105*, 18578–18583.

Saroussi, S., & Nelson, N. (2009). Vacuolar H^+-ATPase—an enzyme for all seasons. *Pflügers Archiv European Journal of Physiology*.

Shah, Y. M., Matsubara, T., Ito, S., Yim, S. H., & Gonzalez, F. J. (2009). Intestinal hypoxia-inducible transcription factors are essential for iron absorption following iron deficiency. *Cell Metabolism*.

Shaw, G. C., Cope, J. J., Li, L., Corson, K., Hersey, C., Ackermann, G. E., et al. (2006). Mitoferrin is essential for erythroid iron assimilation. *Nature*, *440*, 96–100.

Shawki, A., Anthony, S. R., Nose, Y., Barrientos De Renshaw, T., Thiele, D. J., & Mackenzie, B. (2012). Intestinal divalent metal-ion transporter-1 is critical for iron homeostasis but is not required for maintenance of Cu or Zn. *FASEB Journal*, *26*, 1112.2 (abstract).

Shawki, A., Illing, A. C., & Mackenzie, B. (2006). Molecular impact of human divalent metal-ion transporter DMT1 mutations associated with disease phenotypes. *FASEB Journal*, *20*, A1278 (abstract).

Shawki, A., Illing, A. C., & Mackenzie, B. (2008). Molecular impact of divalent metal-ion transporter (DMT1) mutations (V114del and G212V) found in a compound heterozygote with microcytic anemia and hepatic iron overload. *FASEB Journal*, *22*, 1192.3 (abstract).

Shawki, A., & Mackenzie, B. (2010). Interaction of calcium with the human divalent metal-ion transporter-1. *Biochemical and Biophysical Research Communications*, *393*, 471–475.

Shindo, M., Torimoto, Y., Saito, H., Motomura, W., Ikuta, K., Sato, K., et al. (2006). Functional role of DMT1 in transferrin-independent iron uptake by human hepatocyte and hepatocellular carcinoma cell, HLF. *Hepatology Research*, *35*, 152–162.

Siddappa, A. J., Rao, R. B., Wobken, J. D., Casperson, K., Leibold, E. A., Connor, J. R., et al. (2003). Iron deficiency alters iron regulatory protein and iron transport protein expression in the perinatal rat brain. *Pediatric Research*, *53*, 800–807.

Smith, C. P., & Thévenod, F. (2009). Iron transport and the kidney. *Biochimica et Biophysica Acta*, *1790*, 724–730.

Snyder, A. M., & Connor, J. R. (2009). Iron, the substantia nigra and related neurological disorders. *Biochimica et Biophysica Acta*, *1790*, 606–614.

Soe-Lin, S., Apte, S. S., Andriopoulos, B., Andrews, M. C., Schranzhofer, M., Kahawita, T., et al. (2009). Nramp1 promotes efficient macrophage recycling of iron following erythrophagocytosis in vivo. *Proceedings of the National Academy of Sciences of the United States of America*, *106*, 5960–5965.

Soe-Lin, S., Apte, S. S., Mikhael, M. R., Kayembe, L. K., Nie, G., & Ponka, P. (2010). Both Nramp1 and DMT1 are necessary for efficient macrophage iron recycling. *Experimental Hematology*, *38*, 609–617.

Soe-Lin, S., Sheftel, A. D., Wasyluk, B., & Ponka, P. (2008). Nramp1 equips macrophages for efficient iron recycling. *Experimental Hematology*, *36*, 929–937.

Song, N., Jiang, H., Wang, J., & Xie, J. X. (2007). Divalent metal transporter 1 up-regulation is involved in the 6-hydroxydopamine-induced ferrous iron influx. *Journal of Neuroscience Research*, *35*, 3118–3126.

Su, M. A., Trenor, C. C., Fleming, J. C., Fleming, M. D., & Andrews, N. C. (1998). The G185R mutation disrupts function of the iron transporter Nramp2. *Blood*, *92*, 2157–2163.

Suzuki, T., Momoi, K., Hosoyamada, M., Kimura, M., & Shibasaki, T. (2008). Normal cadmium uptake in microcytic anemia *mk/mk* mice suggests that DMT1 is not the only cadmium transporter in vivo. *Toxicology and Applied Pharmacology*, *227*, 462–467.

Tabuchi, M., Tanaka, N., Nishida-Kitayama, J., Ohno, H., & Kishi, F. (2002). Alternative splicing regulates the subcellular localization of divalent metal transporter 1 isoforms. *Molecular Biology of the Cell*, *13*, 4371–4387.

Tabuchi, M., Yanatori, I., Kawai, Y., & Kishi, F. (2010). Retromer-mediated direct sorting is required for proper endosomal recycling of the mammalian iron transporter DMT1. *Journal of Cell Science*, *123*, 756–766.

Tandy, S., Williams, M., Leggett, A., Lopez-Jimenez, M., Dedes, M., Ramesh, B., et al. (2000). Nramp2 expression is associated with pH-dependent iron uptake across the apical membrane of human intestinal Caco-2 cells. *Journal of Biological Chemistry*, *275*, 1023–1029.

Techau, M. E., Valdez-Taubas, J., Popoff, J. F., Francis, R., Seaman, M., & Blackwell, J. M. (2007). Evolution of differences in transport function in Slc11a family members. *Journal of Biological Chemistry, 282*, 35646–35656.

Thompson, K., Molina, R. M., Brain, J. D., & Wessling-Resnick, M. (2006). Belgrade rats display liver iron loading. *Journal of Nutrition, 136*, 3010–3014.

Thompson, K., Molina, R. M., Donaghey, T., Brain, J. D., & Wessling-Resnick, M. (2007). Iron absorption by Belgrade rat pups during lactation. *American Journal of Physiology Gastrointestinal and Liver Physiology, 293*, G640–G644.

Thompson, K., Molina, R. M., Donaghey, T., Schwob, J. E., Brain, J. D., & Wessling-Resnick, M. (2007). Olfactory uptake of manganese requires DMT1 and is enhanced by anemia. *FASEB Journal, 21*, 223–230.

Thompson, B. A.V., Sharp, P. A., Elliott, R., & Fairweather-Tait, S. J. (2010). Inhibitory effect of calcium on non-heme iron absorption may be related to translocation of DMT-1 at the apical membrane of enterocytes. *Journal of Agricultural and Food Chemistry, 58*, 8414–8417.

Tjälve, H., & Henriksson, J. (1999). Uptake of metals in the brain via olfactory pathways. *Neurotoxicology, 20*, 181–195.

Touret, N., Furuya, W., Forbes, J., Gros, P., & Grinstein, S. (2003). Dynamic traffic through the recycling compartment couples the metal transporter Nramp2 (DMT1) with the transferrin receptor. *Journal of Biological Chemistry, 278*, 25548–25557.

Touret, N., Martin-Orozco, N., Paroutis, P., Furuya, W., Lam-Yuk-Tseung, S., Forbes, J., et al. (2004). Molecular and cellular mechanisms underlying iron transport deficiency in microcytic anemia. *Blood, 104*, 1526–1533.

Trinder, D., Batey, R. G., Morgan, E. H., & Baker, E. (1990). Effect of cellular iron concentration on iron uptake by hepatocytes. *American Journal of Physiology, 259*, G611–G617.

Trinder, D., Oates, P. S., Thomas, C., Sadleir, J., & Morgan, E. H. (2000). Localisation of divalent metal transporter 1 (DMT1) to the microvillus membrane of rat duodenal enterocytes in iron deficiency, but to hepatocytes in iron overload. *Gut, 46*, 270–276.

Vesey, D. A. (2010). Transport pathways for cadmium in the intestine and kidney proximal tubule: focus on the interaction with essential metals. *Toxicology Letters, 198*, 13–19.

Vidal, S., Malo, D., Vogan, K., Skamene, E., & Gros, P. (1993). Natural resistance to infection with intracellular parasites: isolation of a candidate for *Bcg*. *Cell, 73*, 469–485.

Vidal, S. M., Pinner, E., Lepage, P., Gauthier, S., & Gros, P. (1996). Natural resistance to intracellular infections: Nramp1 encodes a membrane phosphoglycoprotein absent in macrophages from susceptible (Nramp1 D169) mouse strains. *Journal of Immunology, 157*, 3559–3568.

Wallander, M. L., Leibold, E. A., & Eisenstein, R. S. (2006). Molecular control of vertebrate iron homeostasis by iron regulatory proteins. *Biochimica et Biophysica Acta, 1763*, 668–689.

Wang, X., Ghio, A. J., Yang, F., Dolan, K. G., Garrick, M. D., & Piantadosi, C. A. (2002). Iron uptake and Nramp2/DMT1/DCT1 in human bronchial epithelial cells. *American Journal of Physiology Lung Cellular and Molecular Physiology, 282*, L987–L995.

Wang, B., He, L., Dong, H., Dalton, T. P., & Nebert, D. W. (2011). Generation of a *Slc39a8* hypomorph mouse: markedly decreased ZIP8 $Zn^{2+}/(HCO_3^-)_2$ transporter expression. *Biochemical and Biophysical Research Communications, 410*, 289–294.

Wang, C.-Y., Jenkitkasemwong, S., Duarte, S., Sparkman, B. K., Shawki, A., Mackenzie, B. et al. ZIP8 is an iron and zinc transporter whose cell-surface expression is upregulated by cellular iron loading. *Journal of Biological Chemistry,* 287, 34032–34043.

Wang, X. S., Ong, W. Y., & Connor, J. R. (2002). A light and electron microscopic study of divalent metal transporter-1 distribution in the rat hippocampus, after kainate-induced neuronal injury. *Experimental Neurology, 177*, 193–201.

Wang, D., Song, Y., Li, J., Wang, C., & Li, F. (2011). Structure and metal ion binding of the first transmembrane domain of DMT1. *Biochimica et Biophysica Acta, 1808*, 1639–1644.

Wang, K., Zhou, B., Kuo, Y. M., Zemansky, J., & Gitschier, J. (2002). A novel member of a zinc transporter family is defective in acrodermatitis enteropathica. *American Journal of Human Genetics, 71*, 66–73.
Wareing, M., Ferguson, C. J., Green, R., Riccardi, D., & Smith, C. P. (2000). *In vivo* characterization of renal iron transport in the anaesthetized rat. *Journal of Physiology, 524*, 581–586.
West, A. R., Thomas, C., Sadlier, J., & Oates, P. S. (2006). Haemochromatosis protein is expressed on the terminal web of enterocytes in proximal small intestine of the rat. *Histochemistry and Cell Biology, 125*, 283–292.
Wetli, H., Buckett, P. D., & Wessling-Resnick, M. (2006). Chemical genetic screening identifies the selanazal drug ebselen as a potent inhibitor of DMT1-mediated iron uptake. *Chemistry and Biology, 13*, 965–972.
Williams, K., Wilson, M. A., & Bressler, J. (2000). Regulation and developmental expression of the divalent metal-ion transporter in the rat brain. *Cellular and Molecular Biology, 46*, 563–571.
Wolff, N. A., Abouhamed, M., Verroust, P. J., & Thévenod, F. (2006). Megalin-dependent internalization of cadmium-metallothionein and cytotoxicity in cultured renal proximal tubule cells. *Journal of Pharmacology and Experimental Therapeutics, 318*, 782–791.
Worthington, M. T., Browne, L., Battle, E. H., & Luo, R. Q. (2000). Functional properties of transfected human DMT1 iron transporter. *American Journal of Physiology Gastrointestinal and Liver Physiology, 279*, G1265–G1273.
Xiao, S., Li, J., Wang, Y., Wang, C., Xue, R., Wang, S., et al. (2010). Identification of an "α-helix-extended segment-α-helix" conformation of the sixth transmembrane domain in DMT1. *Biochimica et Biophysica Acta, 1758*, 1556–1564.
Xiao, S., Wang, C., Li, J., & Li, F. (2011). Folding and assembly of TMD 6-related segments of DMT 1 in trifluoroethanol aqueous solution. *Journal of Peptide Science, 17*, 505–511.
Xu, H., Jin, J., DeFelice, L. J., Andrews, N. C., & Clapham, D. E. (2004). A spontaneous, recurrent mutation in divalent metal transporter-1 exposes a calcium entry pathway. *PLoS Biology, 2*, E50.
Yamashita, A., Singh, S. K., Kawate, T., Jin, Y., & Gouaux, E. (2005). Crystal structure of a bacterial homologue of Na^+/Cl^--dependent neurotransmitter transporters. *Nature, 437*, 215–223.
Yanatori, I., Tabuchi, M., Kawai, Y., Yasui, Y., Akagi, R., & Kishi, F. (2010). Heme and non-heme iron transporters in non-polarized and polarized cells. *BMC Cell Biology, 11*, 39.
Zampighi, G. A., Kreman, M., Boorer, K. J., Loo, D. D. F., Bezanilla, F., Chandy, G., et al. (1995). A method for determining the unitary functional capacity of cloned channels and transporters expressed in *Xenopus laevis* oocytes. *Journal of Membrane Biology, 148*, 65–78.
Zhang, A.-S., Canonne-Hergaux, F., Gruenheid, S., Gros, P., & Ponka, P. (2008). Use of *Nramp2*-transfected Chinese hamster ovary cells and reticulocytes from *mk/mk* mice to study iron transport mechanisms. *Experimental Hematology, 36*, 1227–1235.
Zhang, Z., Kodumuru, V., Sviridov, S., Liu, S., Chafeev, M., Chowdhury, S., et al. (2012). Discovery of benzylisothioureas as potent divalent metal transporter 1 (DMT1) inhibitors. *Bioorganic and Medicinal Chemistry Letters, 22*, 5108–5113.
Zhao, N., Gao, J., Enns, C. A., & Knutson, M. D. (2010). ZRT/IRT-like protein 14 (ZIP14) promotes the cellular assimilation of iron from transferrin. *Journal of Biological Chemistry, 285*, 32141–32150.
Zheng, W., Xin, N., Chi, Z. H., Zhao, B. L., Zhang, J., Li, J. Y., et al. (2009). Divalent metal transporter 1 is involved in amyloid precursor protein processing and Aβ generation. *FASEB Journal, 23*, 4207–4217.
Zoller, H., Koch, R. O., Theurl, I., Obrist, P., Pietrangelo, A., Montosi, G., et al. (2001). Expression of the duodenal iron transporters divalent-metal transporter 1 and ferroportin 1 in iron deficiency and iron overload. *Gastroenterology, 120*, 1412–1419.

CHAPTER SIX

Molecular Operation of the Cation Chloride Cotransporters: Ion Binding and Inhibitor Interaction

John A. Payne*
Department of Physiology and Membrane Biology, School of Medicine, University of California, Davis, California, USA
*Corresponding author: E-mail: japayne@ucdavis.edu

Contents

Abstract

The cation chloride cotransporters (CCCs) represent an important family of transporters that plays key roles in vectorial electrolyte movement across epithelia and in intracellular chloride homeostasis of neurons and muscle cells. The CCCs are composed of three broad groups, two of which include multiple isoforms: Na–Cl cotransporter (NCC; *SLC12A3*), Na–K–2Cl cotransporter (NKCC; *SLC12A1-2*), and K–Cl cotransporter (KCC; *SLC12A4-7*). The CCCs are inhibited by clinically relevant drugs, including loop diuretics that inhibit NKCC2 in the renal thick ascending limb and thiazide diuretics that inhibit NCC in the renal distal tubule. For many years, much research on this gene family has centered on understanding ion binding and inhibitor interaction which represent important features of the molecular operation of these transporters. Recently, high resolution structures of bacterial transport proteins related to the CCCs have become available, thus permitting structural context in which to evaluate previous ion and inhibitor studies of the CCCs. In this article, I review past molecular and structure–function studies that have provided key pieces of information about ion binding and inhibitor interaction primarily of NKCC for which we have the most information. I then place these findings into the structural context of recent homology models of NKCC

Current Topics in Membranes, Volume 70
ISSN 1063-5823
http://dx.doi.org/10.1016/B978-0-12-394316-3.00006-5

based on the outward-facing open and occluded conformations of the related bacterial transporters. These homology models provide our first glimpse into the fine details of the molecular operation of the CCCs.

1. OVERVIEW

Cation chloride cotransporters (CCCs) are ion carriers that mediate obligatorily coupled, cation (Na^+, K^+) and Cl^- movements across the plasma membranes of numerous cell types. Because each of these transporters moves an equal number of cations and anions, their operation alone does not result in any net charge movement across the membrane (i.e. they are electroneutral). This electroneutral nature of the CCCs is a key functional feature as these transporters can operate in the control of important physiological parameters (i.e. cell volume, ion concentrations) without directly affecting normal electrical activity of a cell. Clearly, this feature is of particular value to excitable cells such as neurons and muscle cells.

The CCC gene family (*SLC12*) is composed of three broad groups, two of which include multiple isoforms: the Na–Cl cotransporter (NCC; *SLC12A3*), the Na–K–2Cl cotransporter (NKCC; *SLC12A1-2*), and the K–Cl cotransporter (KCC; *SLC12A4-7*). Two additional homologs of the CCCs have been identified but their function is uncertain (*SLC12A8-9*). The CCCs were first characterized in single cells such as red blood cells and cultured cells (Dunham, Stewart, & Ellory, 1980; Ellory & Dunham, 1980; Geck, Pietrzyk, Burckhardt, Pfeiffer, & Heinz, 1980; Lauf & Theg, 1980; Schmidt & McManus, 1977a, 1977b, 1977c). These early studies established that both NKCC and KCC were inversely responsive to changes in cell volume and intracellular chloride concentration ($[Cl^-]_i$). NKCC is activated by cell shrinkage and/or low $[Cl^-]_i$, whereas KCC is switched on by cell swelling and/or high $[Cl^-]_i$. In most cells, these processes bring about corrective movements of salt and osmotically associated water that restore cell water content and $[Cl^-]_i$ at interdependent physiological set points. Subsequent studies have shown that the function of the CCCs goes far beyond cell volume regulation (e.g. Gamba, 2005). In epithelia, the CCCs play key roles in vectorial electrolyte transport and acid secretion. In neurons and cardiomyocytes, the CCCs are of paramount importance in intracellular Cl^- homeostasis (Prasad et al., 2008; Rivera et al., 1999; Shi, Jacob, Piwnica-Worms, & Lieberman, 1987). In fact, it can be argued that the primary function of the CCCs in all cells is control of intracellular $[Cl^-]$. By controlling $[Cl^-]_i$, the CCCs dictate the direction and magnitude of the Cl^- electrochemical gradient across the plasma membrane and therefore determine the electrical consequences of Cl^- channel operation (outward vs. inward current). Indeed, the CCCs are functionally coupled to Cl^- channels in all cells. For example, in epithelia, NKCC1 and NKCC2 operate in conjunction with the epithelial Cl^- channels, cystic fibrosis transmembrane conductance regulator (CFTR) and kidney-specific chloride channel (CLC-K2), respectively. By placing the NKCC and their coupled Cl^- channel in series on opposite surfaces of the epithelial cell (i.e. apical vs. basolateral membrane), these systems can help mediate ion secretion or absorption. In neurons, the level of expression of NKCC1 and KCC2 will dictate $[Cl^-]_i$ and hence the polarity of response (depolarizing vs. hyperpolarizing) to activation of the neuronal Cl^- channels, $GABA_A$ and glycine receptors (Payne, Rivera, Voipio, & Kaila, 2003).

The importance of the individual CCCs to whole body function can be easily illustrated by their specific pharmacologic or genetic inhibition. For example, inhibition of the renal absorptive isoform NKCC2 with "loop" diuretics (bumetanide, furosemide) or genetic loss of function mutation in NKCC2 (Bartter syndrome; Simon, Karet et al., 1996) leads to defective salt reabsorption in the thick ascending limb (TAL), resulting in significant fluid and electrolyte disorders (hypokalemia, alkalosis, salt wasting, and hypotension). Inhibition of NCC with thiazide diuretics or genetic loss of function mutation in NCC (Gitelman syndrome; Simon, Nelson-Williams et al., 1996) leads to defective salt reabsorption in the renal distal tubule, resulting in generally similar fluid and electrolyte disturbances as that observed with inhibition of NKCC2. Lastly, mice in which the neuronal isoform of KCC (KCC2) has been genetically inactivated die immediately after birth from consequences of aberrant spinal GABAergic and glycinergic neurotransmission due to improper Cl^- homeostasis of spinal neurons (Hubner et al., 2001).

Because the CCCs are electroneutral, the energy available for net transport is the sum of the transmembrane chemical potential differences of the transported ions (Cala, 1983). In the case of KCC under normal physiological conditions, transport is driven by the outward gradient of $[K^+]$ maintained by the Na,K-ATPase. Using the favorable energy in this transmembrane K^+ gradient ($\Delta\mu_K$), KCC can extrude Cl^- from the intracellular compartment against a Cl^- chemical potential difference ($\Delta\mu_{Cl}$). This is a clear example of secondary active transport as the driving force for net K + Cl cotransport ($\Delta\mu_{KCC} = \Delta\mu_K + \Delta\mu_{Cl}$) is secondarily dependent upon the K^+ chemical gradient maintained by the Na,K-ATPase.

While thermodynamics clearly dictate the free energy available for net transport, exactly how cotransporters utilize this free energy to effect ion translocation remains poorly understood. Our conceptual framework to address this question has been the familiar alternating access model (Fig. 1). In this model, the cotransporter oscillates between two major conformational states that allow an ion translocation pocket alternate access to the extracellular and intracellular compartments. Inherent in this model is the presence of two "gates" that allow access to the ion translocation pocket from only one side at a time. This latter feature prevents the transporter from behaving like a channel (with a continuous transmembrane pathway). An intermediate state that is often observed in transporters—and certainly seen for NKCC— is an occluded state where ions are bound within the translocation pocket and both gates are closed (Forbush & Haas, 1989). The transition between the major conformational states is highly dependent upon ion binding and debinding which ensures strict ion coupling.

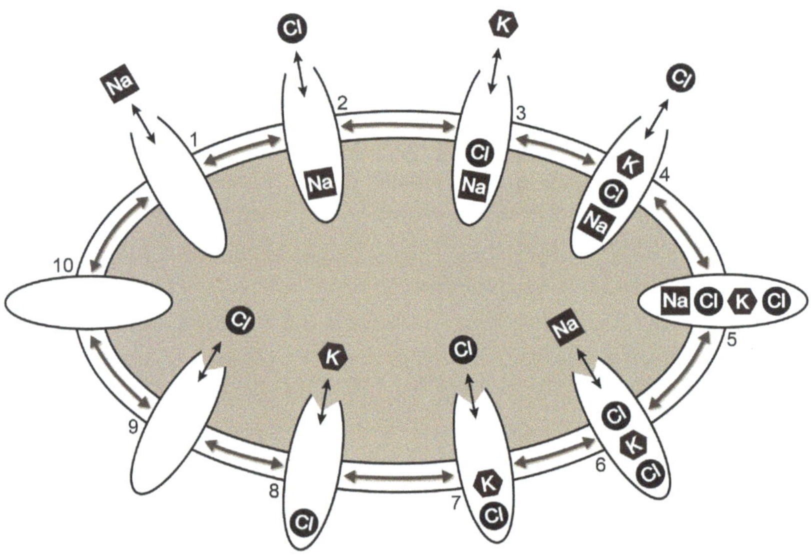

Figure 1 Alternating access model of avian red blood cell NKCC exhibiting ordered binding and glide symmetry (modified from Lytle, McManus, & Hass, 1998). Ten numbered conformational states of the transporter are shown. Ions bind in strict order from the extracellular compartment with the Na^+ ion first (1), a Cl^- ion second (2), the K^+ ion third (3), and the second Cl^- ion last (4). Once all four ions are bound, they become occluded with both extracellular and intracellular "gates" closed (5). In a major conformational step, the intracellular gate opens (6), permitting ions access to the cytosol. The ions then debind from the cotransporter in the same order in which they were bound with the Na^+ ion first off (6), a Cl^- ion second (7), the K^+ ion third (8), and the second Cl^- ion last (9). To permit net Na–K–2Cl cotransport, the empty transporter must transition back to the extracellular facing conformation ready for another cycle to start (10). See the color plate.

Net ion cotransport is made possible by the ability of both the empty and fully loaded transporter to move between the outward-facing and inward-facing conformations. Such an alternating access model has been applied to the NKCC in avian red blood cells, where Lytle, McManus, and Hass (1998) used cleverly designed ion transport experiments to determine that the reaction cycle for NKCC exhibited ordered ion binding and glide symmetry. That is, the ions are bound and released from the outward- and inward-facing conformations in strict order with the first ion bound from the outward-facing conformation being the first released from the inward-facing conformation (Fig. 1). Although such studies provided a foundation for understanding CCC operation, no molecular mechanism was implied. Even the molecular cloning of the CCCs, which provided us the primary structure of the proteins and identification of putative transmembrane segments, did little to inform us of how the polypeptides folded to produce the functional proteins. Structure–function experiments with cloned CCCs moved us closer to understanding specific regions involved in ion and inhibitor interactions as I will discuss in this review. Regardless, the most important developments in understanding the molecular mechanism of all transporters have come from recent developments in structural biology. High-resolution structures of transporter proteins have been obtained within the last decade, most of which have been of bacterial proteins (Krishnamurthy, Piscitelli, & Gouaux, 2009). Some of these high resolution structures are of bacterial proteins homologous to the CCCs (Fang et al., 2009; Gao et al., 2009; Kowalcykz et al., 2011; Shaffer, Goehring, Shankaranarayanan, & Gouaux, 2009). In this article, I review the early molecular and structure–function studies primarily of NKCC that have helped shape our understanding of CCC operation in terms of ion and inhibitor interactions. These studies are then placed in the context of recent high resolution structures of bacterial transport proteins related to the CCCs. With such an analysis, we can begin to understand in much greater detail how ions bind, how ions are translocated, and how inhibitors interact with the CCCs.

2. GENERAL STRUCTURE

The general structure and topology of the CCCs was initially inferred from hydropathy analysis following their molecular identification (Gamba et al., 1993, 1994; Gillen, Brill, Payne, & Forbush, 1996; Payne & Forbush, 1994; Payne, Stevenson, & Donaldson, 1996; Xu et al., 1994). Each of these studies predicted a central hydrophobic region accommodating 12 transmembrane segments flanked by large amino and carboxy termini (Fig. 2). Both amino and carboxy termini of shark rectal gland NKCC1 (saNKCC1) were predicted to be intracellular based on the fact that the amino terminus harbored a phosphorylation site (Thr^{189}; Lytle & Forbush, 1992; Xu et al., 1994) and the carboxy terminus harbored epitopes recognized by monoclonal antibodies generated against the partially purified saNKCC1 protein (Lytle, Xu, Biemesderfer, Haas, & Forbush, 1992). This general structure and topology has been supported by all subsequent studies and generally applies to all members of the gene family (e.g. Gerelsaikhan, Parvin, & Turner, 2006).

An additional feature related to the quaternary structure of the CCCs is the finding that they likely function as homodimers. Early evidence for oligomerization of the CCCs came primarily from biochemical studies using chemical cross-linking or purification of chimeras, deletion mutants, or epitope-tagged constructs expressed in HEK-293 cells or *Xenopus* oocytes (Blaesse et al., 2006; Casula et al., 2001; Moore-Hoon & Turner, 2000; Simard et al., 2004; Simard et al., 2007). One study measured fluorescence resonance energy transfer (FRET) between fluorescent proteins placed in the amino and carboxy termini of NKCC1 (Pedersen, Carmosino, & Forbush,

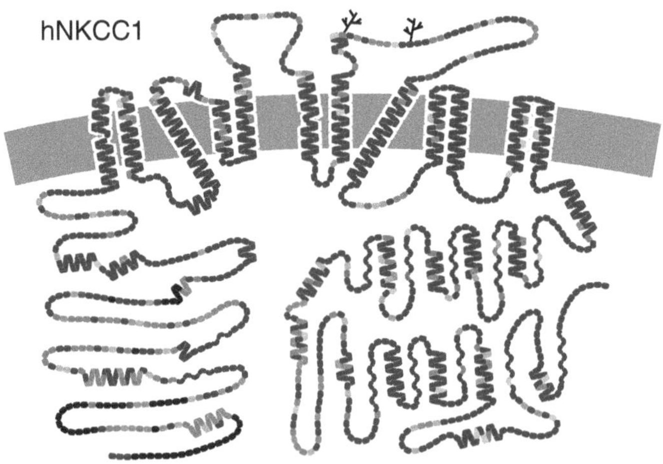

Figure 2 Two-dimensional representation of the structure of human NKCC1 (hNKCC1). The model is shown with a color overlay displaying amino acid identity to shark NKCC1 on a per residue basis. Branched lines are potential N-linked glycosylation sites between putative transmembrane segments 7 and 8. Predicted secondary structural elements are shown as helices for predicted α-helix and wavy lines for predicted ß-sheet. The figure was kindly provided by Bliss Forbush and created using DNAPLOT. See the color plate.

2008). The high ratio of intermolecular to intramolecular FRET of these fluorescent protein tagged-NKCC1 constructs indicated that oligomerization was the dominant form of the functioning transporter (Pedersen et al., 2008). Additional studies have pointed to the carboxy terminus as playing a critical role in dimer formation (Parvin, Gerelsaikhan, & Turner, 2007; Parvin & Turner, 2011). Recently, a high-resolution structure has been obtained for the carboxy terminus of a bacterial CCC protein that forms dimers in solution (Warmuth, Zimmermann, & Dutzler, 2009). As will be discussed further below, transmembrane segments (TMs) 11 and 12 have been shown in high-resolution structures of CCC-related proteins to be involved in coupling the two monomers (Fang et al., 2009; Gao et al., 2009).

3. ION BINDING AND INHIBITOR INTERACTION: STRUCTURE–FUNCTION STUDIES

Because ion translocation must be mediated by that portion of the protein embedded in the plane of the lipid bilayer, it was generally believed that ion binding and translocation by the CCCs occurs within the central hydrophobic region containing the 12 transmembrane segments as depicted in Fig. 2. This hypothesis was first tested by structure–function experiments using chimeras of saNKCC1 and human NKCC1 (hNKCC1; Isenring & Forbush, 1997). Although saNKCC1 and hNKCC1 share significant amino acid identity over their entire length (Fig. 2), they nonetheless exhibit marked differences in kinetics of ion transport and inhibition by bumetanide that mirrored functional studies of these proteins in their native tissue (McRoberts, Erlinger, Rindler, & Saier, 1982; Silva & Myers, 1986). That is, the Michaelis constant (K_m) for all ions and the inhibitory constant (K_i) for bumetanide are four- to nine-fold higher in saNKCC1 compared with hNKCC1 (Payne & Forbush, 1995). By interchanging the amino and carboxy termini between saNKCC1 and hNKCC1, Isenring and Forbush (1997) demonstrated that the large hydrophobic central domain encodes all of the information involved in determining the kinetic differences for ion transport as well as bumetanide inhibition between the two proteins. Based on these initial findings, Isenring and Forbush (1997) concluded that both ion and inhibitor binding involved primarily residues within the central transmembrane domain.

In subsequent studies, Isenring, Jacoby, Chang, and Forbush (1998); Isenring, Jacoby, Forbush (1998) combined the chimera approach with site-directed mutagenesis to further define the regions in NKCC1 that conferred differences between saNKCC1 and hNKCC1 in kinetics of ion transport.

These studies narrowed the search for residues specifying shark–human kinetic differences in ion transport to just three predicted transmembrane segments, i.e. TM2, TM4, and TM7 (Isenring, Jacoby, Chang et al. (1998); Isenring, Jacoby, Forbush (1998)). These studies showed that TM2 played an important role in shark–human species difference in cation affinities but not in Cl^- affinities (Isenring, Jacoby, Forbush (1998)). Only two pairs of residues in TM2 differ between saNKCC1 and hNKCC1, and site-directed mutagenesis experiments revealed that one set was involved in Na^+ affinity differences and the other set was responsible for Rb^+ affinity differences (Figs. 2 and 3; Isenring, Jacoby, Forbush (1998)). In contrast, both TM4 and TM7 were important in the shark–human affinity differences for all three ions. It was hypothesized that TM2, TM4, and TM7 participated in forming part of the ion translocation pathway (Isenring, Jacoby, Chang et al. (1998); Isenring, Jacoby, Forbush (1998)). Based on structural considerations, Isenring, Jacoby, Chang et al. (1998) predicted that at least one more TM would aid in forming the translocation "pore", and they suggested that TM8 and the highly conserved TM1 and TM6 were likely candidates.

The structure–function studies of Isenring, Jacoby, Chang et al. (1998); Isenring, Jacoby, Forbush (1998) also provided some insight into the interaction between NKCC and bumetanide. First, experiments with two shark–human chimeras with a junction at the beginning of TM3 revealed that the first two transmembrane domains of NKCC1 were important in determining bumetanide affinity differences but played a minor role in determining Cl^- affinity differences. Such findings contrast significantly with the long-held belief that bumetanide interacts with Cl^- binding sites (Forbush &

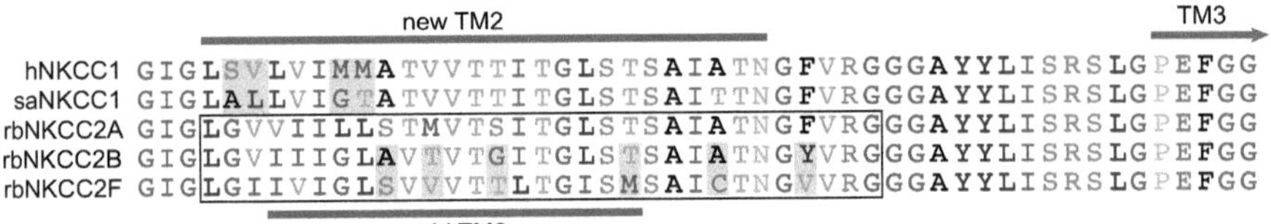

Figure 3 Multiple sequence alignment of human NKCC1 (hNKCC1), shark NKCC1 (saNKCC1), and rabbit NKCC2 splice variants (rbNKCC2A, rbNKCC2B, and rbNKCC2F). Early prediction for position of TM2 is shown (old TM2) along with current position of TM2 (new TM2) based on the hNKCC1 homology model of Somasekharan et al. (2012). The beginning of TM3 is also shown. The region encoded by alternatively splice 96-bp exon 4 of NKCC2 is shown boxed. Shaded amino acids are those mutated in studies by Isenring, Jacoby, Chang et al. (1998); Isenring, Jacoby, Forbush et al. (1998) and Gimenez and Forbush (2007). See the color plate.

Palfrey, 1983; Haas & Forbush, 1986). Second, bumetanide binding behavior of some chimeras did not fit well to the "principle of additivity" that applied so well to ion-transport behavior of the chimeras. That is, the ion affinities of the chimeras (and point mutants) always fell between the two extreme values of the parent transporters. With bumetanide binding, however, the K_i values of some of the chimeras were significantly lower than that of hNKCC1 or significantly greater than that of saNKCC1. These findings point to a much more complex model for inhibitor binding that might involve multiple domains of NKCC and/or conformational interactions (Isenring, Jacoby, Chang et al. (1998)).

The idea that TM2 was involved in ion binding of NKCC was first suggested by Payne and Forbush (1994) following the initial molecular characterization of NKCC2 from rabbit kidney. In the initial cloning of rabbit NKCC2, we identified three alternatively spliced exons that encoded (based on the then current models) most of TM2, as well as a number of amino acids that were predicted to be in the intracellular loop between TM2 and TM3 (Payne & Forbush, 1994). Remarkably, these three splice variants of NKCC2 (termed NKCC2A, NKCC2B, and NKCC2F; Fig. 3) exhibit a distinct axial distribution along the TAL with NKCC2F expressed primarily in the inner stripe of the outer medulla, NKCC2A expressed in the outer stripe of the outer medulla as well as the cortex, and NKCC2B expressed primarily at the end of the cortical TAL, i.e. the macula densa (Igarashi, Vaden Heuvel, Payne, & Forbush, 1995; Payne & Forbush, 1994; Yang, Huang, Singh, Schnermann, & Briggs, 1996). The fact that these three NKCC2 splice variants differed over a predicted TM segment and each displays a unique distribution along the TAL indicated the strong possibility that the transport characteristics (e.g. ion affinity differences) of each variant were specialized to provide optimal salt reabsorption along the TAL (Payne & Forbush, 1994, 1995). This hypothesis was tested by expressing the individual splice variants in *Xenopus* oocytes (Gimenez, Isenring, & Forbush, 2002). Gimenez et al. (2002) showed that each splice variant exhibited a distinct set of ion affinities that tended to matched the decrease in luminal concentration of Na^+, K^+, and Cl^- that occurs along the TAL where salt is reabsorbed "in excess" of water (i.e. the diluting segment). For example, variant NKCC2F found in the medulla where ion concentrations are highest exhibited the lowest affinities for Na^+, $K^+(Rb^+)$, and Cl^-. Gimenez et al. (2002) hypothesized that such fine tuning of ion affinity of the NKCC2 splice variants to luminal ion concentration permits uniform partitioning of salt reabsorption along the medullary and cortical regions of the TAL.

To identify the residues involved in generating the NKCC2 variant-specific ion affinity differences, Gimenez and Forbush (2007) used a mutagenic approach in which they progressively changed the rabbit NKCC2B variant into the NKCC2F variant. Variant B differs from variant F by six amino acids (three residues in TM2 and three residues in the intracellular loop, ICL1; Fig. 3). In converting variant B to variant F, Gimenez and Forbush (2007) found that none of the single amino acid substitutions were able to account for the full functional conversion of variant B to F as measured by reduction in Cl^- affinity. However, when they mutated together three residues that were in either TM2 or ICL1, each of the two triple mutants accounted for about half the difference in Cl^- affinity observed between variants B and F. The only double mutants that meaningfully reduced Cl^- affinity in the variant B–F conversion were those present in ICL1. Similar to the alteration in Cl^- affinity, each of the two triple mutants also accounted for about half the difference in Na^+ affinity observed between variants B and F (Gimenez & Forbush, 2007). It follows that these two sets of three residues (three in TM2 and three in ICL1) are involved in conferring kinetic differences between variants B and F. As mentioned above, Isenring, Jacoby, Forbush (1998) had shown earlier that two sets of residues in TM2 of NKCC1 (upstream of the residues that differed in the NKCC2 variants; Figs. 2 and 3) were involved in cation affinity differences between shark and human. Such studies along with others (Gagnon et al., 2004; Gagnon, Bergeron, Daigle, Lefoll, and Isenring, 2005) have provided support for the hypothesis that residues in TM2 are involved in forming part of the ion translocation pocket. Furthermore, the fact that residues within ICL1 altered ion affinity indicated the likelihood that ICL1 was membrane associated and also participated in forming the ion translocation pocket (Gagnon et al., 2004; Gimenez & Forbush, 2007).

Additional evidence for the involvement of TM2 in the ion translocation pocket has come from the identification of a patient with a mild phenotype of Bartter syndrome (Vargas-Poussou et al., 1998). This patient was heterozygous for a single-base transition at nucleotide 690 that happens to be within exon 4 of the NKCC2B splice variant. This nucleotide transition results in the substitution of a glycine by aspartic acid at position 224 (G224D) that corresponds to one of the three amino acids in TM2 identified by Gimenez and Forbush (2007) as forming part of the Cl^- binding pocket (rabbit NKCC2B; G243D). When the G243D mutation was introduced into rabbit NKCC2B and expressed in oocytes, Gimenez and Forbush (2007) found the protein to be nonfunctional even though it was produced at normal levels by the oocytes. Clearly, the introduction of a negative charge into an anion binding pocket would be expected to disrupt proper ion coordination and prevent transport. Because this particular Bartter syndrome patient had a defect only in the NKCC2B variant, and presumably both NKCC2A and NKCC2F remained functional, it is understandable that only a mild phenotype was observed. Loss of function of the NKCC2B variant that appears to be localized to macula densa cells would be expected to disrupt the important regulatory process of tubuloglomerular feedback; however, this loss would not completely prevent salt reabsorption in the cortical and medullary TAL where NKCC2A and NKCC2F variants are prevalent.

4. HOMOLOGY TO HIGH RESOLUTION STRUCTURES OF BACTERIAL TRANSPORT PROTEINS

The first high-resolution structures of sodium-coupled secondary active transporters were obtained for the bacterial aspartate transporter (GltPh; Yernool, Boudker, Jin, & Gouaux, 2004), the bacterial Na^+–H^+ exchanger (NhaA; Hunte et al., 2005), and the bacterial leucine transporter (LeuT; Yamashita, Singh, Kawate, & Gouaux, 2005). One of the remarkable discoveries that accompanied these early high-resolution bacterial structures was the fact that while each transporter displayed unique membrane protein folds, they all exhibited an internal twofold structural symmetry. For instance, the twofold structural symmetry in LeuT is provided by an inverted structural repeat of five transmembrane segments, TM1–5 and TM6–10 (5 + 5TM; Fig. 4). The presence of a repeated structural element

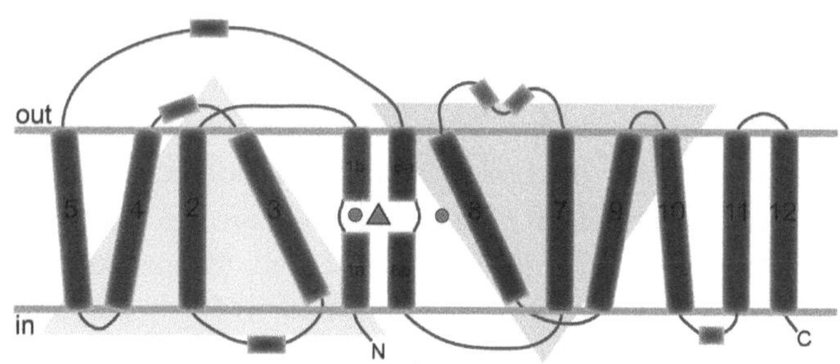

Figure 4 Schematic topology of the LeuT fold, showing 5 + 5 inverted repeat (TM1–5: green; TM6–10: red). TM1 and TM6 are discontinuous helices known to participate in substrate (yellow triangle) and ion (yellow circle) binding of LeuT. Intracellular and extracellular loops with short helical structures are shown as blue rectangles. *(Figure modified from Krishnamurthy et al. (2009))*. See the color plate.

in GltPh, NhaA, and LeuT was reminiscent of that found in a number of other transport proteins, including H^+–Cl^- antiporters of the ClC family (Dutzler, Campbell, Cadene, Chait, & MacKinnon, 2002) and water/glycerol channels of the aquaporin (AQP) family (Fu et al., 2000). Inverted structural repeats in a transporter impart a pseudo-symmetry that provides a rather elegant molecular mechanism to explain the alternating access model. By using two similar but inverted structural elements, two alternate but similar pathways can be created with opposite orientations in the membrane, i.e. outward- and inward-facing orientations. Conformational changes that are facilitated by ion/substrate binding at a central translocation pocket could be envisioned to provide alternating access to either the outward- or inward-facing orientations so as to obligatorily couple ion/substrate movement across the membrane.

The 5 + 5TM inverted repeat structural fold observed in LeuT was subsequently reported for a number of other bacterial transporters, including members of the solute-sodium symporters (vSGLT, Faham et al., 2008), nucleobase-cation symporters (Mhp1, Weyand et al., 2008), and amino acid polyamine–organocation (APC) transporters (AdiC, Fang et al., 2009; Kowalcykz et al., 2011; ApcT, Shaffer et al., 2009). It has become apparent that these functionally disparate transporter families that exhibit little to no sequence identity belong within a large single structural group of transporters (Krishnamurthy et al., 2009).

While we still do not have a high-resolution structure of the transmembrane region of a CCC protein, they are members of the APC family based on sequence homology to AdiC and ApcT for which high-resolution structures have been reported (Fang et al., 2009; Gao et al., 2009; Kowalcykz et al., 2011; Shaffer et al., 2009). Accordingly, the CCCs are expected to exhibit the 5 + 5TM inverted repeat structural fold (LeuT fold) and share general mechanistic properties. With such information, we can begin to place the findings of past studies into structural context of the LeutT fold and also begin to test new predictions about the molecular operation of the CCCs. A homology model of hNKCC1 based on the available structures of AdiC and ApcT has recently been presented (Somasekharan, Tanis, & Forbush, 2012) and is depicted as a homodimer in Fig. 5 and as monomers in panels of Fig. 6. In the outward-facing homology model of hNKCC1, a central cavity presumed to be the translocation pathway is clearly visible from the extracellular view (Fig.6B/D). This translocation pathway is lined by a set of inner TMs, including TMs 1 and 6 on one side and TMs 3, 8, and 10 lining the remainder of the pathway. The other seven TMs tend to

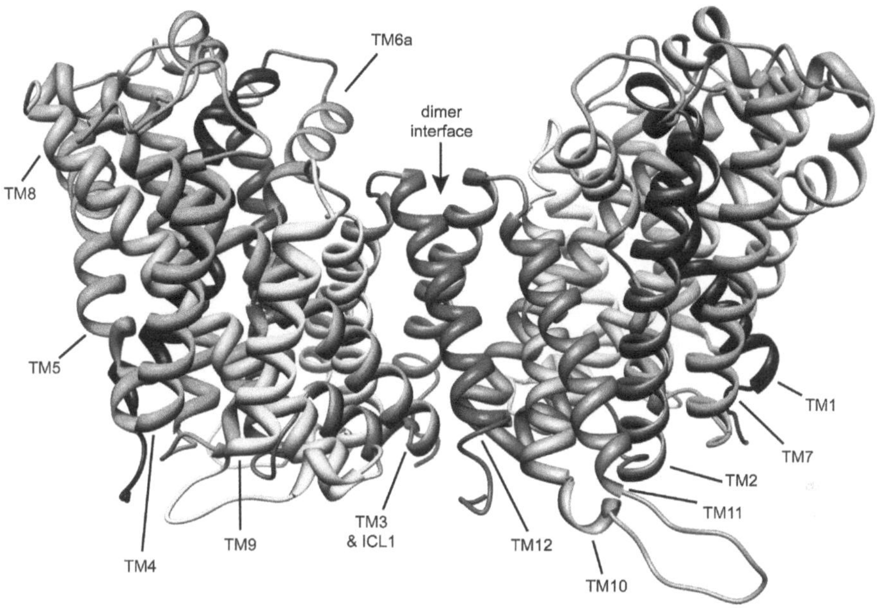

Figure 5 Membrane view of the outward-facing homology model of human NKCC1 from Somasekharan et al. (2012). Model is shown as a homodimer with TMs 11 and 12 of each monomer forming the transmembrane interface (arrow). TMs are shown in ribbon structure and labeled. Residues are rainbow colored from amino terminus (dark blue; TM1) to carboxy terminus (dark red; TM12) of model. Homology model was created using data file from Somasekharan et al. (2012) and the molecular modeling program *UCSF Chimera* (http://www.cgl.ucsf.edu/chimera). See the color plate.

wrap around the inner TMs. In comparing the outward-facing and occluded structures of hNKCC1, TM6 and TM10 appear to make the largest movements as they rotate and bend toward TMs 3, 8, and 1 (note difference in Fig. 6A/C). This concerted movement of helices appears to form the extracellular "gate" creating an occluded translocation pocket from the outside. The translocation pocket forms at the level of Met-382 (TM3) and the broken helix in TM6 down toward ICL1 which blocks access to the cytosol and appears to form the intracellular "gate" (Somasekharan et al., 2012). As one would expect, the position of the translocation pocket in hNKCC1 is similar to that found in AdiC (Gao et al., 2009), but the actual residues of hNKCC1 participating in ion binding are not known. Because the substrates transported by the APC family are quite diverse, it has been noted that NKCCs have likely evolved their own unique binding sites (Somasekharan et al., 2012). It is important to point out that such models are based

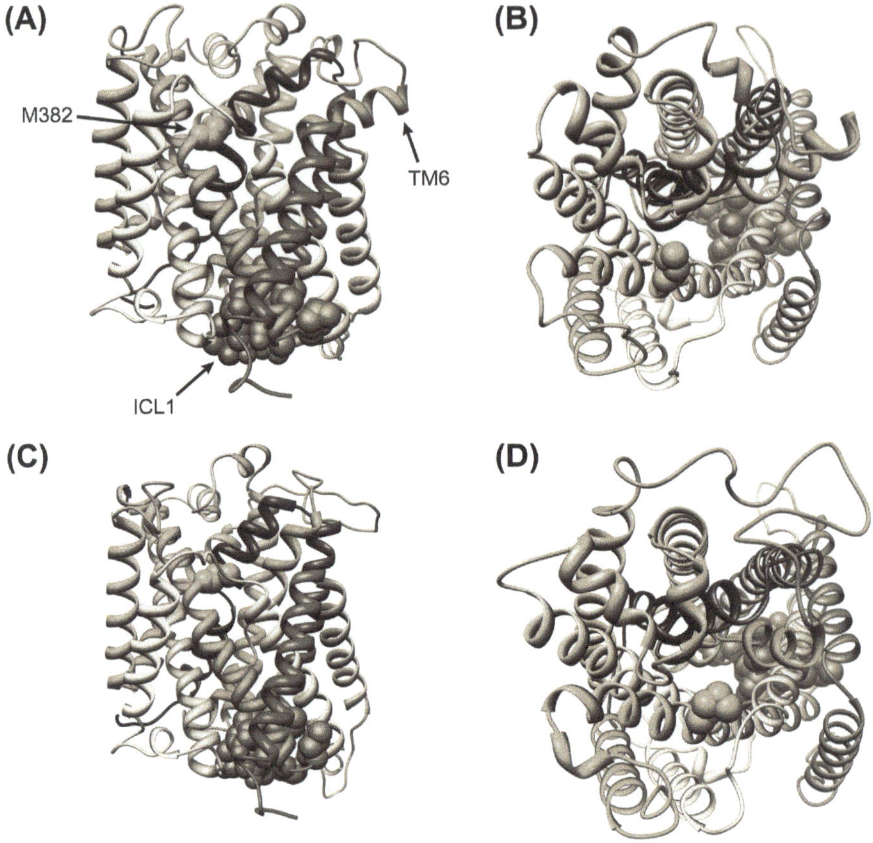

Figure 6 Homology models of the outward-facing open (A-B) and occluded states (C-D) of human NKCC1 from Somasekharan et al. (2012). (A) Membrane view of outward-facing open homology model. (B) Extracellular view of outward-facing open homology model with Met-382 and ICL1 shown as spheres. (C) Membrane view of occluded state homology model. (D) Extracellular view of occluded state homology model. (A-D) Met-382 and ICL1 shown as spheres. TMs are shown in ribbon structure and residues are rainbow colored from amino terminus (dark blue; TM1) to carboxy terminus (dark red; TM12) of model. Homology models were created using data files from Somasekharan et al. (2012) and the molecular modeling program *UCSF Chimera* (http://www.cgl.ucsf.edu/chimera). See the color plate.

upon alignment with distantly related proteins and therefore still require confirmation with structure–function studies or ultimately high resolution structure determination. In the following sections, I describe the findings from sequence data and structure–function studies that can now be placed into the structural context of the homology model of hNKCC1 presented by Somasekharan et al. (2012). I also discuss the data of Somasekharan et al. (2012) in which they tested the role of TM3 in the homology model.

4.1. Role of TM6 in the Translocation Pathway and the Extracellular Gate

In the LeuT fold, TM1 and TM6 are paired discontinuous helices in each of the two inverted repeats, and these two helices are generally believed to participate in forming the ion/substrate translocation pocket of LeuT fold proteins (Fig. 4; Krishnamurthy et al., 2009). Interrupted helices, like TM1 and TM6, are often used in transport proteins to create a polar environment capable of stabilizing ions/substrate in the lipid bilayer by using main chain hydrogen bonds and helical dipoles (Dutzler et al., 2002; Hunte et al., 2005). In the extracellular view of the homology model of hNKCC1, TM6 is seen clearly lining the translocation pathway (Fig. 6B). The critical involvement of TM6 in NKCC1 function is supported by a cysteine-scanning mutagenesis study of Dehaye, Nagy, Premkumar, and Turner (2003). They showed that mutation of two residues in TM6, P496C and A497C (equivalent to Pro-487 and Ala-488 in rat NKCC1 which Dehaye et al. (2003) used in their experiments), resulted in nonfunctional protein. Proline residues are well known as helix breakers because they lack hydrogen on the α-amino group, and thus cannot participate in the hydrogen bonding network of the helix. Hence, prolines are often found at the beginning or the end of an alpha helix or in turns. Fitting this role, Pro-496 is located near the end of the broken helix of TM6a along with Ala-497 next to it (TM6a in Fig. 4 and Fig. 6A/C). The mobility of TM6a appears to be critical in forming the occluded state of NKCC1 (Fig. 6A/C). Another cysteine mutation made by Dehaye et al. (2003) was notable, i.e. A492C (Ala-488 in rat NKCC1), which is present in the middle of the TM6a helix in the homology model. Transport activity of A492C, which exhibited normal function under control conditions, was inhibited by cysteine-modifying reagents. Furthermore, Dehaye et al. (2003) found that A492C was insensitive to inhibition by cysteine-modifying reagents in the absence of extracellular chloride, indicating that reactivity of this residue was dependent upon the conformational state of the transporter. One interpretation of this finding is that TM6a moves during ion binding. Indeed, such movement of TM6a is clearly evident in the conformational change from the outward-facing to the occluded state of hNKCC1 homology model (Fig. 6A/C). TM6a appears to straighten out as it bends near the critical P496-A497 residues (Fig. 6A/C). It is worth noting that the proline residue is highly conserved in the CCC family.

4.2. Understanding the Role of TM2

Site-directed mutagenesis studies of NKCC1/2 and alternative splicing in NKCC2 indicated that TM2 and the region immediately following it (ICL1) were intimately involved in the molecular operation of the transporter. Indeed, it was hypothesized that TM2 and possibly part of ICL1 were directly involved in forming the ion translocation pocket (Gimenez & Forbush, 2007). In examining the homology model for hNKCC1, TM2 is much longer than previously believed so that most of the region that is alternatively spliced in NKCC2 is now included in this extended transmembrane segment (new TM2 in Fig. 3). Remarkably, TM2 is not placed in a position where it is likely to have significant direct involvement in the translocation pathway. Somasekharan et al. (2012) have noted, however, that the intracellular end of TM2 might be capable of some direct interaction with the translocation pathway. Rather, it appears that most of TM2's ability to alter ion affinities is mediated through subtle indirect structural effects on TM6, which as discussed above is a helix that seems intimately involved in forming the ion translocation pocket. In the homology model, TM2 is seen to abut the back side of TM6, i.e. opposite side of the translocation pathway (Fig. 5).

4.3. ICL1 as the Intracellular Gate

The region from the end of TM2 to the beginning of TM3 (ICL1) is the most highly conserved region in the CCC family and contains the GGAYYLIS signature sequence (Fig. 3; Gimenez & Forbush, 2007). Glycine residues are small and help provide flexibility to peptides. Hence, the Gly–Gly motif of the signature sequence, which is strictly conserved in all CCC members, may provide a hinge or significant flexibility to this region of the protein (Gimenez & Forbush, 2007). An essential role of the Gly–Gly motif in the molecular mechanism of the CCCs is supported by the identification of a patient with antenatal Bartter syndrome in which a loss of function mutation in NKCC2 was localized to the second glycine of this motif (G243E; Vargas-Poussou et al., 1998). In the outward-facing and occluded state of the homology model of hNKCC1, the signature sequence is depicted as forming part of a small helix that appears to obstruct the intracellular pathway to the cytosol (Somasekharan et al., 2012; Fig. 6B and D). Importantly, there are well-conserved glycine residues at both ends of ICL1, i.e. immediately after TM2 ends and immediately before TM3 starts (Fig. 3). Given the potential flexibility of ICL1, it is likely that this region represents the intracellular gate that can swing open with a conformational change to permit intracellular access (Somasekharan et al., 2012).

4.4. Role of TM3 in the Translocation Pathway and the Extracellular Gate

Recently, Somasekharan et al. (2012) undertook a cysteine- and tryptophan-scanning mutagenesis study of TM3. In the homology model of hNKCC1, TM3 is shown lining the translocation pathway. Individual mutagenesis of three residues (Phe-372, Asn-376, and Met-382) that line one face of TM3 and are predicted to project into the translocation pathway resulted in significant reductions in ion affinities (Somasekharan et al., 2012). In contrast, Ala-379, which also projects into the translocation pathway, exhibited significant steric constraints as mutation to cysteine, tryptophan, or leucine caused loss of function whereas glycine or serine mutations were functional but with reduced ion affinities (Somasekharan et al., 2012). Somasekharan et al. (2012) also found that mutations of Tyr-383 (Y383F, Y383C, Y383W) resulted in a nonfunctional protein, indicating a critical role of this residue. They noted that the position of Tyr-383 in hNKCC1 corresponds to residues in both AdiC and in LeuT that help form their respective substrate binding sites (Somasekharan et al., 2012). All of these data are consistent with the homology model of hNKCC1 where TM3 helps line the translocation pathway.

Binding of bumetanide to NKCC requires the presence of all three transported ions and is competitive with one of the Cl^- binding sites (Forbush & Palfrey, 1983; Haas & Forbush, 1986). Accordingly, bumetanide is believed to bind within the translocation pocket with three of the ions (1Na^+, 1K^+, and 1Cl^-) in an occluded state. Mutagenesis studies by Somasekharan et al. (2012) support such an interaction of bumetanide with hNKCC1. Mutation of Met-382 with tryptophan resulted in a near complete block of bumetanide inhibition, and Somasekharan et al. (2012) interpreted the bumetanide-insensitive nature of the M382W mutation as resulting from the bulky tryptophan hindering entry into the translocation pathway. Met-382 is predicted to be located at the external opening of the central cavity in the outward-facing conformation. This places Met-382 at a key position controlling entry into the central cavity, and it may represent part of the extracellular gate. Consistent with the key role of Met-382 in the extracellular gate is the finding that reactivity of cysteine-modifying reagents is dramatically reduced in a M382C mutant when the cotransporter is prebound with bumetanide (i.e. in the bumetanide-occluded state) (Somasekharan et al., 2012). Interestingly, tryptophan mutations of Phe-372 and Ile-371, which are located near the intracellular end of TM3, dramatically reduced bumetanide affinity (Somasekharan et al., 2012). This is consistent with the interpretation that bumetanide binds deep within the translocation pathway.

4.5. Dimerization Interfaces and Role in Regulation

In the LeuT fold, TM11 and TM12 do not participate in either of the two repeats. The structure of AdiC, however, shows these two helices forming most of the transmembrane dimerization interface (Fang et al., 2009; Gao et al., 2009). In addition to this transmembrane interface, there also appears to be a cytosolic interface conferred by the carboxy terminus of the CCCs. Such a cytosolic interface is supported by high-resolution structure of the carboxy terminus of a bacterial CCC protein (*Methanosarcina acetivorans*; MaCCC) which forms dimers in solution (Warmuth et al., 2009). Cross-linking studies also support the involvement of the carboxy terminus of NKCC1 in dimer formation (Parvin et al., 2007; Parvin & Turner, 2011). Recent work examining movements of the carboxy terminus of NKCC1 following activation have provided significant insight into the molecular mechanism of CCC regulation (Monette & Forbush, 2012). Monette and Forbush (2012) placed the FRET pairs, cyan fluorescence protein (CFP) and yellow fluorescence protein (YFP), at different locations in the carboxy terminus of NKCC1 and monitored FRET changes upon activation. Singly tagged constructs (which measure intermolecular movement) exhibited larger fractional FRET changes upon activation than doubly tagged constructs (which measure intramolecular movement). The data indicate that the tagged positions within the carboxy termini of the NKCC1 homodimer make large movements away from each other upon activation (Monette & Forbush, 2012). Monette and Forbush (2012) proposed that in the inactive state of NKCC1 the carboxy termini of the two dimers are in a conformation where they interact through a similar cytosolic interface as that in MaCCC as well interact with their respective transmembrane domain intracellular loops to inhibit transport. In such a model, activation of NKCC1 would result from the carboxy termini of the dimers moving apart from one another and removing the transport inhibition from the transmembrane domain. There is still much that is unknown about the molecular mechanism of regulation of the CCCs. It seems possible that because TM12 is linked to the large carboxy terminus in all CCCs, movements of the carboxy terminus upon activation–deactivation might also be transmitted through TM12 to the transmembrane domain to control transport.

5. FUTURE PERSPECTIVES

Since their original description in single cells, the CCCs have proven to be a versatile and important transport group, mediating critical ion movements in epithelia, neurons, and muscle cells. After more than 30 years of research on these transporters, we are now finally able to begin visualizing their molecular structure and their dynamic movements as they progress through conformational states. It is pleasing to see how well earlier biophysical and functional data fit the current structural models of the CCCs. There is still much to be learned about the molecular mechanism of the CCCs. Importantly, we need to know much more about other members of the CCC family—KCC and NCC. Some key questions are 1) how do the different CCCs operate with different ion stoichiometries and ion specificities (i.e. where are the ion binding sites?)? 2) where/how do the different diuretics (bumetanide, furosemide, and thiazide) interact to inhibit the different CCCs?, 3) how are regulatory events at the amino and carboxy termini translated into changes in ion transport rate? These clearly are challenges for future research but having a structural model moves us much closer to answering these questions.

ACKNOWLEDGMENTS

The author is grateful to Chris Lytle, Peter Cala, and Vladimir Yarov-Yarovoy for thoughtful discussions and for reading and commenting on the manuscript. The author is indebted to Bliss Forbush for discussing his homology model of hNKCC1 and for kindly providing Fig. 2. Vladimir Yarov-Yarovoy also kindly helped with the *UCSF Chimera* program used to create the models shown in Figs. 5 and 6. The author's research has been funded by the NIH and by AHA National and Western States.

REFERENCES

Blaesse, P., Guillemin, I., Schindler, J., Schweizer, M., Delpire, E., Khiroug, L., et al. (2006). Oligomerization of KCC2 correlates with development of inhibitory neurotransmission. *Journal of Neuroscience*, *26*(41), 10407–10419.

Cala, P. M. (1983). Volume regulation by red blood cells: mechanisms of ion transport. *Molecular Physiology*, *4*, 33–52.

Casula, S., Shmukler, B. E., Wilhelm, S., Stuart-Tilley, A. K., Su, W., Chernova, M. N., et al. (2001). A dominant negative mutant of the KCC1 K-Cl cotransporter: both N- and C-terminal cytoplasmic domains are required for K-Cl cotransport activity. *Journal of Biological Chemistry*, *276*(45), 41870–41878.

Dehaye, J. P., Nagy, A., Premkumar, A., & Turner, R. J. (2003). Identification of a functionally important conformation-sensitive region of the secretory Na^+-K^+-$2Cl^-$ cotransporter (NKCC1). *Journal of Biological Chemistry*, *278*(14), 11811–11817.

Dunham, P. B., Stewart, G. W., & Ellory, J. C. (1980). Chloride-activated passive potassium transport in human erythrocytes. *Proceedings of the National Academy of Sciences of the United States of America*, *77*, 1711–1715.

Dutzler, R., Campbell, E. B., Cadene, M., Chait, B. T., & MacKinnon, R. (2002). X-ray structure of a ClC chloride channel at 3.0 reveals the molecular basis of anion selectivity. *Nature, 415*, 287–294.

Ellory, J. C. & Dunham, P. (1980). Volume-dependent passive potassium transport in LK sheep red cells. Membrane Transport in Erythrocytes, Alfred Benzon Symposium 14. U.V. Lassen, Ussing, H. H. & Wieth, J. O. Munksgaard: 409–427.

Faham, S., Watanabe, A., Besserer, G. M., Cascio, D., Specht, A., Hirayama, B. A., et al. (2008). The crystal structure of a sodium galactose transporter reveals mechanistic insight into Na^+/sugar symport. *Science, 321*, 810–814.

Fang, Y., Jayaram, H., Shane, T., Kolmakova-Partensky, L., Wu, F., Williams, C., et al. (2009). Structure of a prokaryotic virtual proton pump at 3.2Å resolution. *Nature, 460*, 1040–1043.

Forbush, B., & Haas, M. (1989). Evidence for occlusion of anions and cations on the Na, K, Cl-cotransport system isolated from shark rectal gland. *Biophysical Journal, 55*, 422a.

Forbush, B., & Palfrey, H. C. (1983). [^{3}H]Bumetanide binding to membranes isolated from dog kidney outer medulla. Relationship to Na, K, 2Cl co-transport system. *Journal of Biological Chemistry, 258*(19), 11787–11792.

Fu, D., Libson, A., Miercke, L. J. W., Weitzman, C., Nollert, P., Krucinski, J., et al. (2000). Structure of a glycerol-conducting channel and the basis for its selectivity. *Science, 290*, 481–486.

Gagnon, E., Bergeron, M. J., Brunet, G. M., Daigle, N. D., Simard, C. F., & Isenring, P. (2004). Molecular mechanism of Cl^- transport by the renal Na^+-K^+-Cl^- cotransporter: identification of an intracellular locus that may form part of a high affinity Cl^- binding site. *Journal of Biological Chemistry, 279*(7), 5648–5654.

Gagnon, E., Bergeron, M. J., Daigle, N. D., Lefoll, M. -H., & Isenring, P. (2005). Molecular mechanism of cation transport by the renal Na^+-K^+-Cl^- cotransporter: structural insight into the operatin characteristics of the ion transport sites. *Journal of Biological Chemistry, 280*(37), 32555–32563.

Gamba, G. (2005). Molecular physiology and pathophysiology of electroneutral cation-chloride cotransporters. *Physiological Reviews, 85*, 423–493.

Gamba, G., Miyanoshita, A., Lombardi, M., Lytton, J., Lee, W. -S., Hediger, M., et al. (1994). Molecular cloning, primary structure, and characterization of two members of the mammalian electroneutral sodium-(potassium)-chloride cotransporter family expressed in kidney. *Journal of Biological Chemistry, 269*(26), 17713–17722.

Gamba, G., Saltzberg, S. N., Lambardi, M., Miyanoshita, A., Lytton, J., Hediger, M. A., et al. (1993). Primary structure and functional expression of a cDNA encoding the thiazide-sensitive, electroneutral sodium-chloride cotransporter. *Proceedings of the National Academy of Sciences of the United States of America, 90*, 2749–2753.

Gao, X., Lu, R., Zhou, L., Dang, S., Sun, L., Li, X., et al. (2009). Structure and mechanism of an amino acid antiporter. *Science, 324*, 1565–1568.

Geck, P., Pietrzyk, C., Burckhardt, B. -C., Pfeiffer, B., & Heinz, E. (1980). Electrically silent cotransport of Na, K, and Cl in Ehrlich cells. *Biochimica et Biophysica Acta, 600*, 432–447.

Gerelsaikhan, T., Parvin, M. N., & Turner, R. J. (2006). Biogenesis and topology of the secretory Na^+-K^+-Cl^- cotransporter (NKCC1) studied in intact mammalian cells. *Biochemistry, 45*, 12060–12067.

Gillen, C. M., Brill, S., Payne, J. A., & Forbush, B. (1996). Molecular cloning and functional expression of the K-Cl cotransport from rabbit, rat, and human: a new member of the cation-chloride cotransporter family. *Journal of Biological Chemistry, 271*(27), 16237–16244.

Gimenez, I., & Forbush, B. (2007). The residues determining differences in ion affinities among the alternative splice variants F, A, and B of the mammalian renal Na-K-Cl cotranspoter (NKCC2). *Journal of Biological Chemistry, 282*(9), 6540–6547.

Gimenez, I., Isenring, P., & Forbush, B. (2002). Spatially distributed alternative splice variants of the renal Na-K-Cl cotransporter exhibit dramatically different affinities for the transported ions. *Journal of Biological Chemistry, 277*(11), 8767–8770.

Haas, M., & Forbush, B. (1986). [^{3}H]Bumetanide binding to duck red cells. Correlation with inhibition of (Na+K+2Cl) co-transport. *Journal of Biological Chemistry, 261*(18), 8434–8441.

Hubner, C.A., Stein, V., Hermans-Borgmeyer, I., Meyer, T., Ballanyi, K., & Jentsch, T. J. (2001). Disruption of KCC2 reveals an essential role of K-Cl cotransport already in early synaptic inhibition. *Neuron, 30*, 515–524.

Hunte, C., Screpanti, E., Venturi, M., Rimon, A., Padan, E., & Michel, H. (2005). Structure of a Na^+/H^+ antiporter and insights into mechanism of action and regulation by pH. *Nature, 435*, 1197–1202.

Igarashi, P., Vanden Heuvel, G. B., Payne, J. A., & Forbush, B. (1995). Cloning, embryonic expression, and alternative splicing of a murine kidney-specific Na-K-Cl cotransporter. *American Journal of Physiology Renal Physiology, 269*, F405–F418.

Isenring, P., & Forbush, B. (1997). Ion and bumetanide binding by the Na-K-Cl cotransporter: importance of transmembrane domains. *Journal of Biological Chemistry, 272*(39), 24556–24562.

Isenring, P., Jacoby, S. C., Chang, J., & Forbush, B. (1998). Mutagenic mapping of the Na-K-Cl cotransporter for domains involved in ion transport and bumetanide binding. *The Journal of General Physiology, 112*, 549–558.

Isenring, P., Jacoby, S. C., & Forbush, B. (1998). The role of transmembrane domain 2 in cation transport by the Na-K-Cl cotransporter. *Proceedings of the National Academy of Sciences of the United States of America, 95*(12), 7179–7184.

Kowalcykz, L., Ratera, M., Paladino, A., Bartoccioni, P., Errasti-Murugarren, E., Valencia, E., et al. (2011). Molecular basis of substrate-induced permeation by an amino acid antiporter. *Proceedings of the National Academy of Sciences of the United States of America, 108*(10), 3935–3940.

Krishnamurthy, H., Piscitelli, C. L., & Gouaux, E. (2009). Unlocking the molecular secrets of sodium-coupled transporters. *Nature, 459*, 347–355.

Lauf, P. K., & Theg, B. E. (1980). A chloride dependent K^+ flux induced by N-ethylmaleimide in genetically low K^+ sheep and goat erythrocytes. *Biochemical and Biophysical Research Communications, 92*(4), 1422–1428.

Lytle, C., & Forbush, B. (1992). The Na-K-Cl cotransport protein of shark rectal gland. II. Regulation by direct phophorylation. *Journal of Biological Chemistry, 267*(35), 25438–25443.

Lytle, C., McManus, T. J., & Hass, M. (1998). A model of Na-K-2Cl cotransport based on ordered ion binding and glide symmetry. *American Journal of Physiology Cell Physiology, 274*, C299–C309.

Lytle, C., Xu, J. -C., Biemesderfer, D., Haas, M., & Forbush, B. (1992). The Na-K-Cl cotransporter protein of shark rectal gland. I. Development of monoclonal antibodies, immunoaffinity purification, and partial biochemical characterization. *Journal of Biological Chemistry, 267*(35), 25428–25437.

McRoberts, J. A., Erlinger, S., Rindler, M. J., & Saier, M. H. J. (1982). Furosemide-sensitive salt transport in the Madin-Darby canine kidney cell line. *Journal of Biological Chemistry, 257*(5), 2260–2266.

Monette, M.Y., & Forbush, B. (2012). Regulatory activation is accompanied by movement in the C terminus of the Na-K-Cl cotransporter (NKCC1). *Journal of Biological Chemistry, 287*(3), 2210–2220.

Moore-Hoon, M. L., & Turner, R. J. (2000). The structural unit of the secretory Na^+-K^+-Cl^- cotransporter (NKCC1) is a homodimer. *Biochemistry, 39*, 3718–3724.

Parvin, M. N., Gerelsaikhan, T., & Turner, R. J. (2007). Regions in the cytosolic C-terminus of the secretory Na^+-K^+-$2Cl^-$ cotransporter NKCC1 are required for its homodimerization. *Biochemistry, 46*, 9630–9637.

Parvin, M. N., Turner, R. J. (2011). Identification of key residues involved in the dimerization of the secretory Na^+-K^+-Cl^- cotransporter NKCC1. *Biochemistry, 50*, 9857–9864.

Payne, J. A., & Forbush, B. (1994). Alternatively spliced isoforms of the putative renal Na-K-Cl cotransporter are differentially distributed within the rabbit kidney. *Proceedings of the National Academy of Sciences of the United States of America*, *91*, 4544–4548.

Payne, J. A., & Forbush, B. (1995). Molecular characterization of the epithelial Na-K-Cl cotransporter isoforms. *Current Opinion in Cell Biology*, 7(4), 493–503.

Payne, J. A., Rivera, C., Voipio, J., & Kaila, K. (2003). Cation-chloride cotransporters in neuronal communication, development, and trauma. *Trends Neurosciences*, *26*(4), 199–206.

Payne, J. A., Stevenson, T. J., & Donaldson, L. F. (1996). Molecular characterization of a putative K-Cl cotransporter in rat brain: a neuronal-specific isoform. *Journal of Biological Chemistry*, *271*(27), 16245–16252.

Pedersen, M., Carmosino, M., & Forbush, B. (2008). Intramolecular and intermolecular fluorescence resonance energy transfer in fluorescent protein-tagged Na-K-Cl cotransporter (NKCC1). *Journal of Biological Chemistry*, *283*(5), 2663–2674.

Prasad, V., Bodi, I., Meyer, J. W., Wang, Y., Ashraf, M., Engle, S. J., et al. (2008). Impaired cardiac contractility in mice lacking both the AE3 Cl^-/HCO_3^- exchanger and the NKCC1 Na^+-K^+-$2Cl^-$ cotransporter. *Journal of Biological Chemistry*, *283*, 31303–31314.

Rivera, C., Voipio, J., Payne, J. A., Ruusuvuori, E., Lahtinen, H., Lamsa, K., et al. (1999). A K-Cl cotransporter, KCC2, is the switch to hyperpolarizing GABA action during neuronal maturation. *Nature*, *397*, 251–255.

Schmidt, W. F., & McManus, T. J. (1977a). Ouabain-insensitive salt and water movements in duck red cells. I. Kinetics of cation transport under hypertonic conditions. *The Journal of General Physiology*, *70*, 59–79.

Schmidt, W. F., & McManus, T. J. (1977b). Ouabain-insensitive salt and water movements in duck red cells. II. Norepinephrine stimulation of sodium plus potassium cotransport. *The Journal of General Physiology*, *70*, 81–97.

Schmidt, W. F., & McManus, T. J. (1977c). Ouabain-insensitive salt and water movements in duck red cells. III. The role of chloride in the volume response. *The Journal of General Physiology*, *70*, 99–121.

Shaffer, P. L., Goehring, A., Shankaranarayanan, A., & Gouaux, E. (2009). Structure and mechanism of a Na^+-independent amino acid transporter. *Science*, *325*, 1010–1014.

Shi, L., Jacob, R., Piwnica-Worms, D., & Lieberman, M. (1987). (Na+K+2Cl) cotransport in cultured embryonic chick heart cells. *American Journal of Physiology Cell Physiology*, *253*, C721–C730.

Silva, P., & Myers, M. A. (1986). Stoichiometry of sodium chloride transport by rectal gland of *Squalus acanthias*. *American Journal of Physiology Renal Physiology*, *250*, F516–F519.

Simard, C. F., Bergeron, M. J., Frenette-Cotton, R., Carpentier, G. A., Pelchat, M. -E., Caron, L., et al. (2007). Homooligomeric and heterooligomeric associations between K^+-Cl^- cotransporter isoforms and between K^+-Cl^- and Na^+-K^+-Cl^- cotransporters. *Journal of Biological Chemistry*, *282*(25), 18083–18093.

Simard, C. F., Brunet, G. M., Daigle, N. D., Montminy, V., Caron, L., & Isenring, P. (2004). Self-interacting domains in the C terminus of a cation-Cl^- cotransporter described for the first time. *Journal of Biological Chemistry*, *279*(39), 40769–40777.

Simon, D. B., Karet, F. E., Hamdan, J. M., Di Pietro, A., Sanjad, S. A., & Lifton, R. P. (1996). Bartter's syndrome, hypokalaemic alkalosis with hypercalciuria, is caused by mutations in the Na-K-2Cl cotransporter NKCC2. *Nature Genetics*, *13*, 183–188.

Simon, D. B., Nelson-Williams, C., Bia, M. J., Ellison, D., Karet, F. E., Molina, A. M., et al. (1996). Gitelman's variant of Bartter's syndrome, inherited hypokalaemic alkalosis, is caused by mutations in the thiazide-sensitive Na-Cl cotransporter. *Nature Genetics*, *12*, 24–30.

Somasekharan, S., Tanis, J., & Forbush, B. (2012). Loop diuretic and ion-binding residues revealed by scanning mutagenesis of transmembrane helix 3 (TM3) of Na-K-Cl cotransporters (NKCC1). *Journal of Biological Chemistry*, *287*(21), 17308–17317.

Vargas-Poussou, R., Feldmann, D., Vollmer, M., Konrad, M., Kelly, L., van den Heuvel, L. P. W. J., et al. (1998). Novel molecular variants of the Na-K-2Cl cotransporter gene are responsible for antenatal Bartter syndrome. *American Journal of Human Genetics, 62*, 1332–1340.

Warmuth, S., Zimmermann, I., & Dutzler, R. (2009). X-ray structure of the C-terminal domain of a prokaryotic cation -chloride cotransporter. *Structure, 17*, 538–546.

Weyand, S., Shimamura, T., Yajim, S., Suzuki, S., Mirza, O., Krusong, K., et al. (2008). Structure and molecular mechanism of a nucleobase-cation-symport-1 family transporter. *Science, 322*, 709–713.

Xu, J. C., Lytle, C., Zhu, T., Payne, J. A., Benz, E., & Forbush, B. (1994). Molecular cloning and functional expression of the bumetanide-sensitive Na-K-Cl cotransporter. *Proceedings of the National Academy of Sciences of the United States of America, 91*, 2201–2205.

Yamashita, A., Singh, S. K., Kawate, T., & Gouaux, E. (2005). Crystal structure of a bacterial homologue of Na^+/Cl^--dependent neurotransmitter transporters. *Nature, 437*, 215–223.

Yang, T., Huang, Y. G., Singh, I., Schnermann, J., & Briggs, J. P. (1996). Localization of bumetanide- and thiazide-sensitive Na-K-Cl cotransporters along the rat nephron. *American Journal of Physiology Renal Physiology, 271*, F931–F939.

Yernool, D., Boudker, O., Jin, Y., & Gouaux, E. (2004). Structure of a glutamate transporter homologue from *Pyrrococcus horikoshii*. *Nature, 431*, 811–818.

CHAPTER SEVEN

Sodium-Sulfate/Carboxylate Cotransporters (SLC13)

Daniel Markovich*
Molecular Physiology Group, School of Biomedical Sciences, University of Queensland, Brisbane St Lucia, QLD, Australia
*Corresponding author: E-mail: d.markovich@uq.edu.au

Contents

Abstract

The SLC13 gene family is comprised of five sequence related proteins that are found in animals, plants, yeast and bacteria. Proteins encoded by the SLC13 genes are divided into the following two groups of transporters with distinct anion specificities: the Na^+-sulfate (NaS) cotransporters and the Na^+-carboxylate (NaC) cotransporters. Members of this gene family (in ascending order) are: SLC13A1 (NaS1), SLC13A2 (NaC1), SLC13A3 (NaC3), SLC13A4 (NaS2) and SLC13A5 (NaC2). SLC13 proteins encode plasma membrane polypeptides with 8–13 putative transmembrane

Current Topics in Membranes, Volume 70
ISSN 1063-5823
http://dx.doi.org/10.1016/B978-0-12-394316-3.00007-7

domains, and are expressed in a variety of tissues. They are all Na^+-coupled symporters with strong cation preference for Na^+, and insensitive to the stilbene 4, 4′-diisothiocyanatostilbene-2, 2′-disulphonic acid (DIDS). Their Na^+:anion coupling ratio is 3:1, indicative of electrogenic properties. They have a substrate preference for divalent anions, which include tetra-oxyanions for the NaS cotransporters or Krebs cycle intermediates (including mono-, di- and tricarboxylates) for the NaC cotransporters. This review will describe the molecular and cellular mechanisms underlying the biochemical, physiological and structural properties of the SLC13 gene family.

1. INTRODUCTION

The SLC13 gene family is comprised of structurally related membrane transport proteins with unique substrate specificities (Markovich, 2001; Markovich & Murer, 2004). They can be divided into the following two distinct groups: the Na^+-sulfate (NaS) cotransporters and the Na^+-carboxylate (NaC) cotransporters (Table 1). This review will focus on the functional mechanisms of the SLC13 transporters, including ion-coupling patterns, substrate specificities, regulatory domains, tissue distributions/expression patterns, physiological/pathological roles, as well as biochemical and structural information. The focus of this review will be on the vertebrate members of this family, with greatest emphasis on mammalian transporters, and will only briefly mention nonvertebrate, lower eukaryotic or prokaryotic SLC13 (gene-like) members. The review is divided into the following sections: the first part will discuss general features common to all vertebrate members of this gene family; the second part will focus on the NaS cotransporters (SLC13A1 and SLC13A4); the third part is devoted to the NaC cotransporters (SLC13A2, SLC13A3 and SLC13A5). Finally, future perspectives for these transporters will be discussed at the end of this review.

2. SLC13 GENE FAMILY

2.1. Discovery and Vertebrate Orthologs

Prior to the genome sequencing era when DNA sequence information was unavailable, the first several complementary Deoxyribo Nuclic

Table 1 Properties of the SLC13—Na^+-sulfate/carboxylate cotransporter family
SLC13—Na^+-sulfate/carboxylate cotransporter family

Human gene name	New protein name*	Former protein name & aliases	Predominant substrates	Transport type; coupling ions†	Tissue distribution and cellular/subcellular expression	Link to disease	Human gene locus	Sequence accession ID	Number of transcripts and splice variants, and their specific features
SLC13A1	NaS1	NaSi-1, Na-sulfate	Sulfate, selenate, thiosulfate	C; $Na^+/SO_4^{2-}/S_2O_3/SeO_4^{2-}$	Kidney, proximal tubular cells, brush border membrane	None	7q31–q32	AF260824	2.2 kb & 2.7 kb transcripts (both functional, with a 3′ UTR 0.5 kb longer variant)
SLC13A2	NaC1	NaDC-1, SDCT1, NaDC-2	Succinate, citrate, α-keto glutarate	C; Na^+/succinate/citrate/	Kidney, intestine, brush border membrane	None	17p11.1–q11.1	U26209	One 2.8 kb transcript only
SLC13A3	NaC3	NaDC-3, SDCT2	Succinate, citrate, α-ketoglutarate	C; Na^+/succinate/citrate/	Kidney proximal tubule basolateral membrane, liver, pancreas, brain, placenta	None	20q12–q13.1	AF154121	One 3.6 kb transcript only
SLC13A4	NaS2	SUT-1	Sulfate	C; Na^+/SO_4^{2-}	Placenta, tonsillar high-endothelial venules, testis, heart	None	7q33	AF169301	2.9 kb (functional) and 4.5 kb (function unknown) transcripts
SLC13A5	NaC2	NaCT	Citrate	C; Na^+/citrate/succinate/pyruvate	Liver, brain, testis	None	12p12–13	AY151833	One 3.2 kb transcript only

*Proteins with less than 55% amino acid identity belong to a separate family.
†C: Cotransporter (Function based on results obtained with heterologous expression systems).

Acids (cDNAs) encoding SLC13 gene members were isolated by expression cloning using *Xenopus laevis* oocytes (Markovich, 2008; Markovich, Werner, & Murer, 1999). The first cDNA to be cloned was Slc13a1, which encoded a rat NaS cotransporter [rNaS1 (Markovich, Forgo, Stange, Biber, & Murer, 1993)], followed by Slc13a2, a rabbit Na^+-dicarboxylate cotransporter [rbNaC1 (Pajor, 1995)], then the *Xenopus* Slc13a2 [xNaC1 (Bai & Pajor, 1997)] and the winter flounder Slc13a3 [fNaC3 (Steffgen et al., 1999)]. All other members of the SLC13 family, including orthologs of the above-mentioned transporters, were isolated by either homology hybridization screening or PCR techniques (or a combination of both), using available DNA sequences (for probes or primers) from other orthologs or known expressed sequence tag (EST) databases. These included the following mammalian SLC13 cDNAs (in no particular order): the mouse Slc13a1 [mNaS1 (Beck & Markovich, 2000)], human SLC13A1 [hNaS1 (Lee, Beck, & Markovich, 2000)], human SLC13A2 [hNaC1 (Pajor, 1996)], rat Slc13a2 [rNaC1 (Chen, Shayakul, Berger, Tian, & Hediger, 1998; Khatri, Kovacs, & Forstner, 1996; Sekine et al., 1998)], mouse Slc13a2 [mNaC1 (Pajor & Sun, 2000)], human SLC13A3 [hNaC3 (Wang et al., 2000)], rat Slc13a3 [rNaC3 (Chen, Tsukaguchi, Chen, Berger, & Hediger, 1999; Kekuda et al., 1999)], mouse Slc13a3 [mNaC3 (Pajor, Gangula, & Yao, 2001)], human SLC13A4 [hNaS2 (Girard, Baekkevold, Feliu, Brandtzaeg, & Amalric, 1999)], rat SLC13A4 [rNaS2 (Dawson, Pirlo et al., 2005)], mouse SLC13A4 [mNaS2 (Dawson, Pirlo et al., 2006)], human SLC13A5 [hNaC2 (Inoue, Zhuang, & Ganapathy, 2002)] and mouse Slc13a5 [mNaC2 (Inoue, Zhuang, Maddox, Smith, & Ganapathy, 2002)].

2.2. Phylogenetic Origins

SLC13 genes belong to a large superfamily of transporters found in prokaryotes and eukaryotes, in both plant and animal kingdoms (Saier et al., 1999). Genome sequencing projects of eukaryotic and prokaryotic species have identified SLC13-like genes in many species, including Gram-negative bacteria, cyanobacteria, archaea, plant chloroplasts, yeast and animals (Saier et al., 1999). However, the precise molecular transport mechanisms of the encoded proteins have yet to be characterized in these species. In the nematode worm *Caenorhabditis elegans*, the following orthologs of NaC1 have been isolated: ceNaDC1 encodes a low-affinity Na^+-succinate cotransporter, and ceNaDC2 encodes a high-affinity Na^+-succinate cotransporter (Fei, Inoue, & Ganapathy, 2003), which share 39% and 36% amino acid identity with hNaC1, respectively. In the vinegar fly *Drosophila melanogaster*, a NaC1 ortholog named *Indy* has also been isolated, and encodes a Na^+-independent succinate transporter (Inoue, Fei et al., 2002) that shares 36% amino acid identity with hNaC1. Plants have SLC13-like proteins that are H^+-coupled symporters, rather than Na^+-coupled transporters (Leustek & Saito, 1999); the cation preference appears to be a function of their external environment (H^+-rich soils). In bacteria and yeast species, SLC13 orthologs have been identified; however, their precise transport mechanisms remain to be elucidated at the molecular level (Saier et al., 1999).

2.3. Functional Features

For most animal SLC13 transporters, functional characterization was performed in *Xenopus* oocytes using either radiotracer flux measurements or two-electrode voltage-clamp techniques (Chen et al., 1998, 1999; Markovich, 2000, 2001; Pajor, 1999b, 2000, 2006). A wide range of kinetic values has been calculated for substrates transported by the SLC13 proteins, arising from varied assays performed by independent groups. However, the following general transport properties have been determined by the different laboratories. All mammalian SLC13 proteins encode DIDS-insensitive (in mM range) Na^+-coupled cotransporters that move various anions including tetra-oxyanions (for SLC13A1 and SLC13A4) and Krebs cycle intermediates (for SLC13A2, SLC13A3 and SLC13A5). The stoichiometry of the cation:anion coupling ratio has been determined experimentally to be 3:1 (i.e. a coupling of $3Na^+$ to one (generally divalent) anion, making the transport process electrogenic, with a net transfer of one positive charge across the cell membrane). The transport mechanisms appear to be ordered, with Na^+ binding to the transporter prior to the anion (Pajor, 2000, 2006).

2.4. Biochemical and Structural Properties

Based on hydropathy analyses of deduced primary amino acid sequences, proteins encoded by the mammalian SLC13 family members contain between 8 and 13 putative transmembrane domains (TMDs), with proteins ranging in length from 572 to 627 amino acids (Markovich & Murer, 2004). The mammalian NaS1 transporters are approximately 66.1 kDa in size, and comprise 13 TMDs predicted by TopPred2 (von Heijne, 1992), with an intracellular NH_2 terminus and an extracellular COOH terminus (Fig. 1). hNaS1 and rNaS1 proteins contain 595 amino acids, whereas mNaS1 is shorter by one amino acid (Glu at position 314). The functional significance of this single-residue loss is currently unknown. NaS1 proteins have one biochemically confirmed extracellular N-glycosylation site near the C-terminus (Fig. 1) that is critical for transport activity in *Xenopus* oocytes (Li & Pajor, 2003a) but not essential for plasma membrane sorting in mammalian cells (Regeer, Nicke, & Markovich, 2007). Alanine substitution of hNaS1 Ser-260 resulted in an increased affinity constant (K_m) for both sulfate and Na^+, whereas alanine replacement of hNaS1 Ser-288 resulted in a broadened cation selectivity (Li & Pajor, 2003b). These data indicate that the two serines (Ser-260 and Ser-288) on putative TMD5 of hNaS1 might play roles in cation and/or substrate binding, and that these TMDs may participate in the conformational movement during its transport cycle (Li & Pajor, 2003b). The native form of the NaS1 protein appears to be a dimer, and its apical membrane sorting in renal epithelial cells is brefeldin A-sensitive and independent of lipid rafts or glycosylation (Regeer et al.,

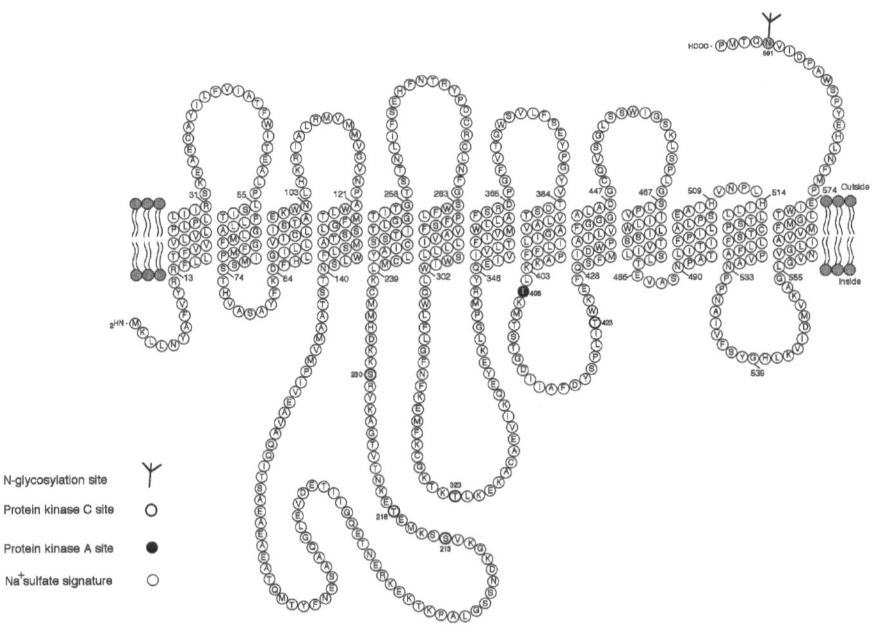

Figure 1 ***Secondary structure model of the NaS1 protein.*** Topology is based on the prediction by the TopPred2 program. The N-terminus is intracellular and the C-terminus is extracellular. Putative consensus sites are indicated.

2007). NaS2 was predicted to have 12 TMDs with both termini being extracellular (Girard et al., 1999); however, no biochemical evidence exists to approve or disprove this model.

rbNaC1 has been proposed to contain 11 TMDs, with an intracellular NH_2 terminus and an extracellular COOH terminus that contains one N-glycosylation site at Asn-578. This site appears to be conserved among all SLC13A2 orthologs, suggestive of an extracellular COOH terminus (Pajor, 2000). mNaC1 contains two consensus N-glycosylation sites (Asn-570 and Asn-580) on its extracellular COOH terminus (Pajor & Sun, 2000). NaC3 proteins conform to the 11 TMD model with two consensus N-glycosylation sites (Asn-584 and Asn-594) on the extracellular COOH terminus (Chen et al., 1999). Based on hydropathy analysis, NaC2 proteins (SLC13A5) have been proposed to contain 11 putative TMDs (Inoue, Zhuang, Maddox et al., 2002).

All SLC13 family proteins contain numerous predicted consensus sites for phosphorylation (protein kinase C, cAMP and cGMP, casein kinase II, tyrosine kinase sites) and N-myristoylation sites, for which the functional significances have yet to be determined. In addition, there is a 17 amino acid long consensus sequence motif (TSFAFLLPVANPPNAIV) called the "sodium:sulfate symporter family signature" (PROSITE PS01271) situated at amino acids 523–539 of rNaS1. This motif, for which the significance is yet unknown (Markovich, 2001), is highly conserved among the proteins SLC13A1 (100% identity), SLC13A2 (71–82% identity), SLC13A3 (59–77%), SLC13A4 (77% identity) and SLC13A5 (77–82% identity). This motif is also well conserved in SLC13-like genes in lower eukaryotes and prokaryotes.

Since NaS1 and NaC1 proteins share significant (>40%) amino acid identities [for review see Markovich and Murer (2004)] but have differing substrate specificities, NaC1/NaS1 chimeric proteins were generated in order to determine the location of substrate binding domains. Chimeric proteins containing TMDs 1–4 of rNaS1 (amino acids 1–139) and TMDs 5–11 of rbNaC1 (amino acids 141–593) led to Na^+-succinate cotransport, but not NaS cotransport, suggesting that the substrate recognition site of NaC1 is found in the C-terminal portion (TMDs 5–11) of the protein, past amino acid 141 (Pajor, Sun, Bai, Markovich, & Sule, 1998). The substrate-binding site and a cation-binding site are proposed to be close together in the C-terminal portion of NaC1, with TMDs 8 and 9 being suggested as candidate structures for the formation of the translocation pathway (Griffith & Pajor, 1999).

2.5. Pathological Implications

Several human diseases/disorders arise from disturbances in SO_4^{2-} metabolism or transport, including metachromatic leukodystrophy, Hunter syndrome, Morquio syndrome, Maroteaux–Lamy syndrome, multiple-sulfohydrolase deficiency and osteochondrodysplasias (Markovich, 2001). However, no direct or indirect links have been made with disorders that may arise from NaS1 (SLC13A1) and/or NaS2 (SLC13A4) transporters (Markovich, 2001). For the NaC transporters (SLC13A2, SLC13A3 and SLC13A5), their substrate preferences and renal localization may suggest possible involvement in kidney stone formation (Bergeron, Bürzle, Kovacs, Simonin, & Hediger, 2011). The NaC1 (SLC13A2) may be beneficial in clarifying the pathophysiology underlying nephrolithiasis because it transports citrate, which could inhibit kidney stone formation (Bergeron et al., 2011).

Recent efforts using animal models have been utilized to investigate the precise roles of SLC13 proteins in the body. Mouse models have provided clues to the physiological roles of these transporters. The mNaS1 (Slc13a1) gene knockout (KO) mouse has extensive phenotypes, including hyposulfatemia, growth retardation, reduced fertility, spontaneous seizures, altered metabolism and behavior, reduced detoxification capacity and altered gastrointestinal function (Dawson, Beck, & Markovich, 2003; Dawson, Stesne, Markovich, 2005; Dawson, Gardiner, Grimmond, Markovich, 2006; Lee, Dawson, Hewavitharana, Shaw, & Markovich, 2006; Lee et al., 2007). On the other hand, the mNaC1 (Slc13a2) KO mouse has a mild phenotype with increased urinary excretion of citrate, succinate, α-ketoglutarate, fumarate and malate (Ho et al., 2007), indicating a defect in kidney re-absorption of these Krebs cycle intermediates. The KO mouse has no changes in growth or serum citrate levels, and no report of kidney stones. The mechanisms underlying these pathophysiological states are presently being investigated. Further studies will be needed to determine the precise physiological and pathophysiological roles of these transporters *in vivo*.

3. NaS COTRANSPORTERS

3.1. Members and Tissue Specificities

The three most studied mammalian NaS1 (SLC13A1) orthologs are human NaS1 [hNaS1 (Lee et al., 2000)], rat NaS1 [rNaS1 (Markovich et al., 1993)] and mouse NaS1 [mNaS1 (Beck & Markovich, 2000)]. hNaS1 was shown to be expressed in the kidney, rNaS1 mRNA was detected in kidney and small intestine (Markovich et al., 1993) and mNaS1 was strongly detected in kidney, ileum, duodenum/jejunum and colon, with weaker signals in cecum, testis, adrenal and adipose tissue.

The three most studied mammalian NaS2 (SLC13A4) members are human NaS2 [hNaS2 (Girard et al., 1999; Markovich, Regeer, Kunzelmann, & Dawson, 2005), rat NaS2 [rNaS2 (Dawson, Pirlo et al., 2005)] and mouse NaS2 [mNaS2 (Dawson, Pirlo et al., 2006)]. hNaS2 was detected in human placenta, heart, testis, tonsillar epithelium (Girard et al., 1999), as well as in brain, thymus and liver (Markovich et al., 2005). rNaS2 was detected in placenta, brain and liver (Dawson, Pirlo et al., 2005), whereas mNaS2 was detected in placenta, brain, lung, eye, heart, testis, thymus and liver (Dawson, Pirlo et al., 2006). The functional roles of these proteins in these tissues are related to cellular sulfate transport.

3.2. Functional Characteristics

NaS1 members encode electrogenic, pH-insensitive, Na^+-dependent SO_4^{2-} transporters (Na^+–SO_4^{2-} cotransporters/symporters), with substrate preferences for the anions sulfate (K_m = 93 μM), thiosulfate (K_m = 84 μM), selenate (K_m = 580 μM) and the cation Na^+ (K_m = 16–24 mM) (Busch et al., 1994; Markovich et al., 1993). Significant *cis*-inhibition for NaS1-induced sulfate transport was observed with thiosulfate, selenate, tungstate and molybdate for all NaS1orthologs, and with succinate and citrate for hNaS1 only (Lee et al., 2000).

hNaS2 was shown to mediate Na^+-dependent SO_4^{2-} uptake (Girard et al., 1999) that was inhibited by thiosulfate, phosphate, molybdate, selenate and tungstate, but not by oxalate, citrate, succinate, phenol red or DIDS (Markovich et al., 2005). hNaS2 has a K_m for sulfate of 0.38 mM and a Hill coefficient of $n = 1.6 \pm 0.6$ based on Na^+ kinetics, suggesting a $Na{:}SO_4^{2-}$ stoichiometry of 2:1 (Markovich et al., 2005). Expression of rNaS2 protein in *Xenopus* oocytes led to Na^+-dependent transport of sulfate that was inhibited by phosphate, thiosulfate, tungstate, selenate, oxalate, chromate, arsenate and molybdate, but not by citrate, succinate or DIDS (Dawson, Pirlo et al., 2005; Miyauchi et al., 2006). Transport kinetics of rNaS2 determined a K_m for sulfate between 150 μM and ≈1 mM (Dawson, Pirlo et al., 2005; Miyauchi et al., 2006). Na^+ kinetics determined a Hill coefficient of $n = 3.0 \pm 0.7$, suggesting a $Na{:}SO_4^{2-}$ stoichiometry of 3:1 (Dawson, Pirlo et al., 2005). rNaS2-induced transport appears electrogenic because K^+-induced depolarization inhibits uptake (Miyauchi et al., 2006), although electrophysiological studies are needed to confirm this.

3.3. Regulation of NaS Expression

NaS1 mRNA and protein levels in vivo are downregulated in the renal cortex by high-sulfate diet (Markovich et al., 1998), hypothyroidism (Sagawa, Murer, & Morris, 1999), vitamin D depletion (Fernandes et al., 1997), glucocorticoids (Sagawa, Darling, Murer, & Morris, 2000), hypokalemia (Markovich, Wnag et al., 1999), metabolic acidosis (Puttaparthi et al., 1999) and nonsteroidal anti-inflammatory drugs (NSAIDs) (Sagawa, Benincosa, Murer, & Morris, 1998), and upregulated by low-sulfate diet (Sagawa, DuBois, Almon, Murer, & Morris, 1998), thyroid hormone (Dawson & Markovich, 2002), vitamin D supplementation (Fernandes et al., 1997; Dawson & Markovich, 2002), growth hormone (Sagawa, Han et al., 1999), chronic renal failure (Fernandes et al., 2001) and during post-natal growth (Markovich & Fogelis, 1999). The mNaS1 gene (Slc13a1) is transcriptionally upregulated by vitamin D and thyroid hormone binding to VDRE (at −525 bp) and T_3RE (at −436 bp) on the Slc13a1 promoter, respectively (Dawson & Markovich, 2002), and downregulated by glucocorticoids (Markovich, 2000). The hNaS1 gene (SLC13A1) is significantly upregulated by xenobiotic 3-methycholanthrene binding to an XRE (at −2052 bp) on the SLC13A1 promoter (Lee & Markovich, 2004). rNaS1-induced sulfate transport is acutely inhibited in *Xenopus* oocytes by pharmacological activators of PKA (8br-cAMP) and PKC (dioctanoyl glycerol; DOG) (Markovich, 2000). rNaS1-induced sulfate transport in *Xenopus* oocytes is inhibited by the heavy metals mercury, lead, chromium and cadmium (Markovich & Knight, 1998). Thus, heavy metals may inhibit renal brush border NaS cotransport via the NaS1 protein, and this blockade may be responsible for sulfaturia following heavy metal intoxication. No published studies to date have examined NaS2 (SLC13A4) regulation.

3.4. Physiological Implications

The NaS1 proteins encode Na^+–SO_4^{2-} cotransporters located on the apical or brush border membranes (BBM) of epithelial cells (Fig. 2) lining the renal proximal tubule and small/large intestine (Markovich, 2001, 2011a, 2011b). Inorganic sulfate is absorbed in the lower small intestine (ileum), and up to 80% of filtered sulfate is reabsorbed in the proximal tubule. Due to its cellular localization and regulation under a variety of dietary and hormonal conditions (listed above), NaS1 has been proposed to be a major regulator of serum sulfate concentrations at 0.3 mM in humans (Besseghir

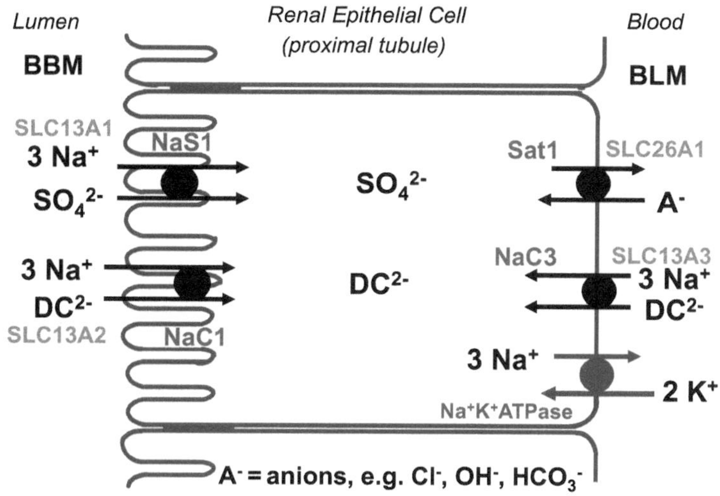

Figure 2 ***Model of a renal proximal tubular cell with sulfate and dicarboxylate transport systems.*** The NaC1 and NaS1 proteins are localized to the luminal brush border membrane (BBM), whereas NaC3 and sat-1 proteins are localized to the basolateral membrane (BLM). Protein names are shown in red, gene names are indicated in green, and substrates are shown in black. See the color plate.

& Roch-Ramel, 1987; Markovich, 2001). The NaS2 (SLC13A4) protein encodes a Na^+–SO_4^{2-} cotransporter that is expressed in high endothelial venules where it is suggested to play an important role in sulfation, as well as in the placenta where it possibly facilitates sulfate transfer across the trophoblast from mother to fetus (Girard et al., 1999). The role of NaS2 in other tissues has yet to be determined.

4. NaC COTRANSPORTERS

4.1. Members and Tissue Specificities

The five major vertebrate NaC1 (SLC13A2) orthologs mostly studied are human NaC1 (Pajor, 1996), rabbit NaC1 (Pajor, 1995), mouse NaC1 (Pajor & Sun, 2000), rat NaC1 (Chen et al., 1998; Khatri et al., 1996; Sekine et al., 1998), and *Xenopus* NaC1 (Bai & Pajor, 1997). Northern blotting detected hNaC1 mRNA in kidney and intestine (Pajor, 1996); rbNaC1 mRNA was strongly expressed in kidney and jejunum, with weaker signals in liver (Pajor, 1995). Northern blotting detected mNaC1 mRNA in kidney and intestine (Pajor & Sun, 2000), and rNaC1 mRNA in small intestine, large intestine and kidney (Sekine et al., 1998).

Immunocytochemistry detected rNaC1 protein in the luminal membranes of the renal superficial cortex and the outer stripe of the outer medulla (Sekine et al., 1998). In situ hybridization detected rNaC1 in the S3 segments of the kidney proximal tubules (outer stripe of the outer medulla), enterocytes lining the intestinal villi, hepatocytes in the liver, alveolar and bronchiole epithelial cells in the lung, and tubular epithelium of the epididymus (Chen et al., 1998). Northern blotting detected xNaC1 only in the *Xenopus* intestine (Bai & Pajor, 1997).

The two mammalian NaC2 (SLC13A5) orthologs primarily studied are human NaC2 (Inoue, Zhuang et al., 2002) and mouse NaC2 (Inoue, Zhuang, Maddox et al., 2002). Northern blotting detected hNaC2 and mNaC2 mRNAs predominantly in liver, with weaker signals in brain and testis (Inoue, Zhuang et al., 2002; Inoue, Zhuang, Maddox et al., 2002).

The three major mammalian NaC3 (SLC13A3) orthologs studied are human hNaC3 (Wang et al., 2000), mouse NaC3 (Pajor et al., 2001), and rat NaC3 (Chen et al., 1999; Kekuda et al., 1999). Northern blotting detected hNaC3 in human brain, kidney, liver, pancreas and placenta (Wang et al., 2000), and mNaC3 mRNA in mouse kidney and brain (Pajor et al., 2001). Northern blotting and RT-PCR detected rNaC3 mRNA in rat kidney, liver, brain and placenta (Chen et al., 1999; Kekuda et al., 1999). In situ hybridization detected rNaC3 in the S3 segments of the renal proximal tubule, in hepatocytes surrounding the central vein and in the meningeal layers of the brain (Chen et al., 1999). These proteins in these tissues contribute to cellular (mono-, di- and/or tri-) carboxylate transport.

4.2. Functional Characteristics

NaC1 (SLC13A2) members encode an electrogenic, pH-independent (pH 6–8), low-affinity Na^+-dicarboxylate cotransporter with substrate preferences for the anions succinate (K_m = 180–1100 μM) and citrate (K_m = 320–6800 μM) (Pajor, 1999b, 2000). Significant *cis*-inhibition for rNaDC-1-induced succinate transport was observed (in decreasing order of potency) with succinate, fumarate, malate, oxaloacetate, and citrate (Sekine et al., 1998).

NaC2 (SLC13A5) members encode an electrogenic, low-affinity Na^+-tricarboxylate cotransporter. hNaC2 has substrate preference for citrate (K_m = 650 μM), and is unable to transport succinate or malate (Inoue, Zhuang et al., 2002). However, mNaC2 transport specificity was broader, including the tricarboxylate citrate (K_m = 18 μM), monocarboxylate pyruvate, and dicarboxylate succinate (Inoue, Zhuang, Maddox et al., 2002). Based on competition studies, hNaC2-induced citrate transport was significantly inhibited (in decreasing order of potency) by unlabeled citrate, succinate, malate and fumarate (Inoue, Zhuang et al., 2002), and mNaC2-induced citrate transport was significantly inhibited (in decreasing order of potency) by unlabeled citrate, succinate, fumarate, *cis*-aconitate, and α-ketoglutarate (Inoue, Zhuang, Maddox et al., 2002).

NaC3 (SLC13A3) members encode an electrogenic, pH-dependent, high-affinity Na^+-dicarboxylate cotransporter (Burckhardt & Burckhardt, 2003) with substrate preferences for the anions succinate (K_m = 15–102 μM) (Chen et al., 1999; Wang et al., 2000), dimethyl succinate (K_m = 7–112 μM) (Burckhardt et al., 2002; Chen et al., 1999; Wang et al., 2000), citrate (K_m = 220 μM) (Chen et al., 1999), N-acetyl-aspartate (K_m = 300 μM) (Huang et al., 2000). For rbNaC1, rNaC1, rNaC3 and fNaC3, lithium can act as a competitive inhibitor of Na^+ (Pajor, 2000). Significant *cis*-inhibition for hNaC3-induced succinate transport was observed (in decreasing order of potency) with dimethylsuccinate, succinate, glutarate, α-ketoglutarate, fumarate, aspartate, citrate, malonate, glutamate, maleate and pyruvate (Wang et al., 2000).

4.3. Regulation of NaC Expression

Unlike the NaS transporters, limited data exist for the regulation of NaC transporters. NaC1 mRNA and protein levels were shown to be upregulated in vivo by chronic metabolic acidosis (Aruga et al., 2000) and potassium depletion (Levi, McDonald, Preisig, & Alpern, 1991). rbNaC1-mediated succinate transport was shown to be acutely inhibited in *Xenopus* oocytes by the two activators of protein kinase C, phorbol 12-myristate, 13-acetate (PMA) and sn-1, 2-DOG; inhibition was independent of the two PKC consensus sites on rbNaC1 (Pajor & Sun, 1999). Serum- and glucocorticoid-inducible kinases SGK1, SGK3 and protein kinase B were shown to increase the transport activity and trafficking of NaC1 in *Xenopus* oocytes (Boehmer et al., 2004).

4.4. Physiological Implications

NaC1 proteins encode low-affinity Na^+-dicarboxylate cotransporters located on the brush border membrane (BBM) of epithelial cells (Fig. 2) in the renal proximal tubule and intestinal enterocytes, in hepatocytes, alveolar and bronchiolar epithelium, and in the tubular epithelium of the epididymus (Chen et al., 1998; Pajor, 1999a, 1999b, 2000, 2006; Sekine et al., 1998). Plasma citric acid cycle intermediates are freely filtered at the glomerulus and up to 65% are reabsorbed by NaC1 in the proximal tubule. These intermediated are then metabolized within the proximal tubular cells or get exchanged with organic anions during organic-anion secretion (Pajor, 1999b). In the intestine, NaC1 absorbs dietary dicarboxylates, including citrate and other Krebs cycle intermediates, from the intestinal lumen. Absorbed citrate is mainly utilized in the liver and kidney, although less is known about its metabolism in other organs. NaC1 in the head of the epididymus is thought to contribute to a low-pH environment and the nutritional needs of sperm (Chen et al., 1998). The precise role of NaC1 in the liver and respiratory tissue is yet unknown. The mNaC1 (Slc13a2) KO mouse has provided some insights into the physiological role of NaC1. The KO mouse displays increased urinary excretion of citrate, succinate, α-ketoglutarate, fumarate, and malate (Ho et al., 2007), indicating a defect in kidney re-absorption of those Krebs cycle intermediates.

NaC2 proteins are Na^+-citrate cotransporters expressed in the testis, brain and liver (Inoue, Zhuang et al., 2002; Inoue, Zhuang, Maddox et al., 2002). Their function is to facilitate the utilization of circulating citrate for the generation of metabolic energy, and for the synthesis of fatty acids and cholesterol (Inoue, Zhuang et al., 2002; Inoue, Zhuang, Maddox et al., 2002). As with NaC1, the function of NaC2 in the testis has been suggested to contribute to the nutritional requirements of sperm (Inoue, Zhuang et al., 2002; Inoue, Zhuang, Maddox et al., 2002).

NaC3 proteins are high-affinity Na^+-dicarboxylate cotransporters located on the basolateral membrane (BLM) of epithelial cells (Fig. 2) lining the renal proximal tubule, on the apical membranes of the placenta, in hepatocytes, and in meningeal layers of the brain and brain synaptosomes (Chen et al., 1999; Kekuda et al., 1999; Pajor et al., 2001). The high-affinity NaC3 is ideally suited to facilitate peritubular uptake of Krebs cycle intermediates through the BLM of renal proximal tubular cells because the plasma dicarboxylate concentration is 50 μM or less. Possible roles of NaC3 in liver are to capture Krebs cycle intermediates from the systemic circulation, and to promote dicarboxylate-mediated glutamine synthesis in perivenous hepatocytes (Chen et al., 1999). Furthermore, NaC3 has been suggested to correspond to the liver sinusoidal carboxylate transporter, which may play a role in hepatic glutamine and ammonia metabolism and facilitate secretion of organic anions from hepatocytes (Chen et al., 1999). The expression of NaC3 mRNA in the meningeal layers of the brain supports the possibility that the cells in this region use dicarboxylates as nourishment (Chen et al., 1999).

5. FUTURE PERSPECTIVES

Since the initial cloning of the SLC13 gene family members from the early 1990s, a significant amount of research has been undertaken on this transporter gene superfamily. NaC transporters most likely play an important role in the excretion of organic anions and small therapeutic drugs, including gluconate, sulfate and cysteine conjugates, thus making them excellent tools for therapeutic targets (Pajor, 1999b, 2006). Organic anions and small therapeutic drugs are excreted from hepatocytes into the blood and secreted into kidney proximal tubules—findings that agree with the cellular locations of the NaC transporters. On the other hand, NaS transporters play important roles in cell growth, conjugation and detoxification of various endogenous and exogenous compounds because their preferred substrate, inorganic sulfate, is the fourth most abundant anion in mammalian plasma that is crucial for many physiological functions (Markovich, 2001; Markovich & Aronson, 2007; Markovich, 2012). Sulfate conjugation (sulfonation or sulfation) is an important step in the biotransformation of xenobiotics such as steroids, anti-inflammatory agents, adrenergic stimulants and blockers, analgesics and, in most cases, leads to an increase in their urinary excretion (Markovich, 2001). Sulfate is also required for the activation of many endogenous compounds such as heparin, heparan sulfate, dermatan sulfate and bile acids (Markovich, 2001). In addition, sulfation of structural components such as glycosaminoglycans and cerebroside sulfate is essential for the maintenance of normal structure and function of tissues (Markovich, 2001). It is obvious that NaS transporters are targets for therapeutic interventions to regulate the processes of sulfate conjugation of various molecules, as well as the detoxification of xenobiotics and various endogenous compounds that get excreted via the kidneys. Future research will be aimed at refining the molecular mechanisms and physiological functions of the NaC and NaS transporters *in vivo*. Also, the abilities of NaC and NaS transporters to act as possible drug targets for pharmaceutical intervention will be investigated in order to elucidate their roles in (patho)physiological states.

ACKNOWLEDGMENTS

Research in the authors' laboratory has been supported by the National Health and Medical Research Council, Cancer Council Queensland and the Australian Research Council.

REFERENCES

Aruga, S., Wehrli, S., Kaissling, B., Moe, O., Preisig, P., Pajor, A., et al. (2000). *Kidney International, 58*, 206–215.

Bai, L., & Pajor, A. M. (1997). *American Journal of Physiology, 273*, G267–G274.

Beck, L., & Markovich, D. (2000). *Journal of Biological Chemistry, 275*, 11880–11890.

Bergeron, M., Bürzle, M., Kovacs, G., Simonin, A., & Hediger, M. (2011). *Journal of Biological Chemistry, 286*, 11242–11253.

Besseghir, K., & Roch-Ramel, F. (1987). *Renal Physiology, 10*, 221–241.

Boehmer, C., Embark, H., Bauer, A., Palmada, M., Yun, C., Weinman, E., et al. (2004). *Biochemical and Biophysical Research Communications, 313*, 998–1003.

Burckhardt, B. C., & Burckhardt, G. (2003). *Reviews Physiology, Biochemistry and Pharmacology, 146*, 95–158.

Burckhardt, B. C., Drinkuth, B., Menzel, C., Konig, A., Steffgen, J., Wright, S. H., et al. (2002). *Journal of the American Society of Nephrology, 13*, 2628–2638.

Busch, A. E., Waldegger, S., Herzer, T., Biber, J., Markovich, D., Murer, H., et al. (1994). *Journal of Biological Chemistry, 269*, 12407–12409.

Chen, X. Z., Shayakul, C., Berger, U.V., Tian, W., & Hediger, M. A. (1998). *Journal of Biological Chemistry, 273*, 20972–20981.

Chen, X., Tsukaguchi, H., Chen, X. Z., Berger, U.V., & Hediger, M. A. (1999). *The Journal of Clinical Investigation, 103*, 1159–1168.

Dawson, P. A., Beick, L., & Markovich, D. (2003). *Proceedings of the National Academy of Sciences of the United States of America, 100*, 13704–13709.

Dawson, P. A., Gardiner, B., Grimmond, S., & Markovich, D. (2006). *Physiological Genomics, 26*, 116–124.

Dawson, P. A., & Markovich, D. (2002). *Pflugers Archiv, 444*, 353–359.

Dawson, P. A., Pirlo, K. J., Steane, S. E., Kunzelmann, K., Chien, Y. J., & Markovich, D. (2006). *Genes and Genetic Systems, 81*, 265–272.

Dawson, P. A., Pirlo, K. J., Steane, S. E., Nguyen, K. A., Kunzelmann, K., Chien, Y. J., et al. (2005). *Pflugers Archiv, 450*, 262–268.

Dawson, P. A., Steane, S. E., & Markovich, D. (2005). *Behavioural Brain Research, 159*, 15–20.

Fei, Y. J., Inoue, K., & Ganapathy, V. (2003). *Journal of Biological Chemistry, 278*, 6136–6144.

Fernandes, I., Hampson, G., Cahours, X., Morin, P., Coureau, C., Couette, S., et al. (1997). *The Journal of Clinical Investigation, 100*, 2196–2203.

Fernandes, I., Laouairi, D., Tutt, P., Hampson, G., Friedlander, G., & Silve, C. (2001). *Kidney International, 59*, 210–221.

Girard, J. P., Baekkevold, E. S., Feliu, J., Brandtzaeg, P., & Amalric, F. (1999). *Proceedings of the National Academy of Sciences of the United States of America, 96*, 12772–12777.

Griffith, D. A., & Pajor, A. M. (1999). *Biochemistry, 38*, 7524–7531.

Ho, H., Ko, B., Cheung, A., Lam, A., Tam, S., Chung, S., et al. (2007). *Kidney International, 72*, 63–71.

Huang, W., Wang, H., Kekuda, R., Fei, Y., Friedrich, A., Wang, J., et al. (2000). *The Journal of Pharmacology and Experimental Therapeutics, 295*, 392–403.

Inoue, K., Fei, Y. J., Huang, W., Zhuang, L., Chen, Z., & Ganapathy, V. (2002). *Biochemistry Journal, 367*, 313–319.

Inoue, K., Zhuang, L., & Ganapathy, V. (2002). *Biochemical and Biophysical Research Communications, 299*, 465–471.

Inoue, K., Zhuang, L., Maddox, D. M., Smith, S. B., & Ganapathy, V. (2002). *Journal of Biological Chemistry, 277*, 39469–39476.

Kekuda, R., Wang, H., Huang, W., Pajor, A. M., Leibach, F. H., Devoe, L. D., et al. (1999). *Journal of Biological Chemistry, 274*, 3422–3429.

Khatri, I. A., Kovacs, S.V., & Forstner, J. F. (1996). *Biochimica et Biophysica Acta, 1309*, 58–62.

Lee, A., Beck, L., & Markovich, D. (2000). *Genomics, 70*, 354–363.
Lee, S., Dawson, P. A., Hewavitharana, A. K., Shaw, P. N., & Markovich, D. (2006). *Hepatology, 43*, 1241–1247.
Lee, S., Kesby, J. P., Muslim, M. D., Steane, S. E., Eyles, D. W., Dawson, P. A., et al. (2007). *Neuroreport, 18*, 1981–1985.
Lee, A., & Markovich, D. (2004). *Pflugers Archiv, 448*, 490–499.
Leustek, T., & Saito, K. (1999). *Plant Physiology, 120*, 637–644.
Levi, M., McDonald, L. A., Preisig, P. A., & Alpern, R. J. (1991). *American Journal of Physiology, 261*, F767–F773.
Li, H., & Pajor, A. M. (2003a). *American Journal of Physiology Cell Physiology, 285*, C1188–C1196.
Li, H., & Pajor, A. M. (2003b). *Journal of Biological Chemistry, 278*, 37204–37212.
Markovich, D. (2000). *European Journal of Cell Biology, 79*, 531–538.
Markovich, D. (2001). *Physiological Reviews, 81*, 1499–1533.
Markovich, D. (2008). *Nature Protocols, 3*, 1975–1980.
Markovich, D. (2011a). *Archivum Immunologiae et Therapiae Experimentalis (Warsz), 59*, 113–116.
Markovich, D. (2011b). *American Journal of Physiology Renal Physiology, 300*, F1267–F1270.
Markovich, D. (2012). *Physiology (Bethesda), 27*, 7–14.
Markovich, D., & Aronson, P. S. (2007). *Annual Review of Physiology, 69*, 361–375.
Markovich, D., & Fogelis, T. S. (1999). *Pediatric Nephrology, 13*, 806–811.
Markovich, D., Forgo, J., Stange, G., Biber, J., & Murer, H. (1993). *Proceedings of the National Academy of Sciences of the United States of America, 90*, 8073–8077.
Markovich, D., & Knight, D. (1998). *American Journal of Physiology, 274*, F283–F289.
Markovich, D., & Murer, H. (2004). *Pflugers Archiv, 447*, 594–602.
Markovich, D., Murer, H., Biber, J., Sakhaee, K., Pak, C., & Levi, M. (1998). *Journal of the American Society of Nephrology, 9*, 1568–1573.
Markovich, D., Regeer, R. R., Kunzelmann, K., & Dawson, P. A. (2005). *Biochemical and Biophysical Reserach Communications, 326*, 729–734.
Markovich, D., Wang, H., Puttaparthi, K., Zajicek, H., Rogers, T., Murer, H., et al. (1999). *Kidney International, 55*, 244–251.
Markovich, D., Werner, A., & Murer, H. (1999). Expression cloning with *Xenopus* oocytes. In F. Hildebrandt, & P. Igarashi (Eds.), *Techniques in molecular medicine* (pp. 310–318). Heidelberg: Springer Verlag.
Miyauchi, S., Srinivas, S. R., Fei, Y. J., Gopal, E., Umapathy, N. S., Wang, H., et al. (2006). *Placenta, 27*, 550–559.
Pajor, A. M. (1995). *Journal of Biological Chemistry, 270*, 5779–5785.
Pajor, A. M. (1996). *American Journal of Physiology, 270*, F642–F648.
Pajor, A. M. (1999a). *Seminars in Nephrology, 19*, 195–200.
Pajor, A. M. (1999b). *Annual Review of Physiology, 61*, 663–682.
Pajor, A. M. (2000). *The Journal of Membrane Biology, 175*, 1–8.
Pajor, A. M. (2006). *Pflugers Archiv, 451*, 597–605.
Pajor, A., Gangula, R., & Yao, X. (2001). *American Journal of Physiology Cell Physiology, 280*, C1215–C1223.
Pajor, A. M., & Sun, N. (1999). *Biochimica et Biophysica Acta, 1420*, 223–230.
Pajor, A. M., & Sun, N. N. (2000). *American Journal of Physiology Renal Physiology, 279*, F482–F490.
Pajor, A. M., Sun, N., Bai, L., Markovich, D., & Sule, P. (1998). *Biochimica et Biophysica Acta, 1370*, 98–106.
Puttaparthi, K., Markovich, D., Halaihel, N., Wilson, P., Zajicek, H. K., Wang, H., et al. (1999). *American Journal of Physiology, 276*, C1398–C1404.
Regeer, R. R., Nicke, A., & Markovich, D. (2007). *The International Journal of Biochemistry and Cell Biology, 39*, 2240–2251.

Sagawa, K., Benincosa, L. J., Murer, H., & Morris, M. E. (1998). *The Journal of Pharmacology and Experimental Therapeutics*, *287*, 1092–1097.
Sagawa, K., Darling, I. M., Murer, H., & Morris, M. E. (2000). *The Journal of Pharmacology and Experimental Therapeutics*, *294*, 658–663.
Sagawa, K., DuBois, D. C., Almon, R. R., Murer, H., & Morris, M. E. (1998). *The Journal of Pharmacology and Experimental Therapeutics*, *287*, 1056–1062.
Sagawa, K., Han, B., DuBois, D. C., Murer, H., Almon, R. R., & Morris, M. E. (1999). *The Journal of Pharmacology and Experimental Therapeutics*, *290*, 1182–1187.
Sagawa, K., Murer, H., & Morris, M. E. (1999). *American Journal of Physiology*, *276*, F164–F171.
Saier, M. H.J., Eng, B. H., Fard, S., Garg, J., Haggerty, D. A., Hutchinson, W. J., et al. (1999). *Biochimica et Biophysica Acta.*, *1422*, 1–56.
Sekine, T., Cha, S. H., Hosoyamada, M., Kanai, Y., Watanabe, N., Furuta, Y., et al. (1998). *American Journal of Physiology*, *275*, F298–F305.
Steffgen, J., Burckhardt, B. C., Langenberg, C., Kuhne, L., Muller, G. A., Burckhardt, G., et al. (1999). *Journal of Biological Chemistry*, *274*, 20191–20196.
von Heijne, G. (1992). *Journal of Molecular Biology*, *225*, 487–494.
Wang, H., Fei, Y., Kekuda, R., Yang-Feng, T., Devoe, L., Leibach, F., et al. (2000). *American Journal of Physiology Cell Physiology*, *278*, C1019–C1030.

CHAPTER EIGHT

Recent Advances in Structural Biology of Peptide Transporters

Tomohiro Terada[*,1], **Ken-ichi Inui**[†,1]
[*]Department of Pharmacy, Shiga University of Medical Science Hospital, Shiga, Japan
[†]Kyoto Pharmaceutical University, Kyoto, Japan
[1]Corresponding authors: E-mail: teradat@belle.shiga-med.ac.jp; inui@mb.kyoto-phu.ac.jp.

Contents

Abstract

Peptide transporters localized at brush-border membranes of intestinal and renal epithelial cells mediate the membrane transport of di- and tripeptides, and play important roles in protein absorption and the conservation of peptide-bound amino nitrogen. Peptide-like drugs that show structural similarities to di- and tripeptides are also recognized by peptide transporters. The energy for transport of small peptides and peptide-like drugs is provided by the proton gradient across the cell membrane. Since the cloning of H^+/peptide cotransporter (PEPT1, SLC15A1), there have been advances in the molecular biology, biochemistry, biophysics and structural determination of PEPT1. By integrating these advances, much effort has been made to understand the relationship between structure and function. *In silico* experimental strategies are classified as (1) construction of kinetic models, (2) computer modeling of PEPT1 structure and (3) homology modeling of PEPT1 with crystal structures of bacterial transporters. The hypotheses regarding the structure–function relationship produced by these strategies have been confirmed by in vitro mutagenesis including cysteine-scanning mutagenesis. Recently, the crystal structure of $PepT_{So}$, a functionally similar prokaryotic

Current Topics in Membranes, Volume 70
ISSN 1063-5823
http://dx.doi.org/10.1016/B978-0-12-394316-3.00008-9

homolog of the mammalian peptide transporters from *Shewanella oneidensis*, was classified, and the previous hypotheses regarding the structure–function relationship of PEPT1 have been re-evaluated. This review highlights the recent advances in our knowledge of the structural biology of PEPT1.

1. INTRODUCTION

Dietary protein undergoes a series of degradative steps, resulting in a mixture of free amino acids and small peptides. Then, these products are taken up by the intestinal epithelial cells and delivered into circulation. A large number of studies have provided evidence that the absorption of protein digestion products in the small intestine occurs primarily in the form of small peptides via peptide transporters. Transport studies using brush-border membrane vesicles or the human intestinal cell line Caco-2 demonstrated that the peptide transporters utilize an inwardly directed proton (H^+) electrochemical gradient as a driving force (Daniel, 2004), and that this transporter can accept not only di- and tripeptides but also foreign compounds structurally resembling small peptides, such as oral β-lactam antibiotics (Terada & Inui, 2004).

In 1994, a cDNA encoding the H^+/peptide cotransporter (PEPT1, SLC15A1) was identified from a rabbit small intestinal cDNA library by expression cloning (Fei et al., 1994). Subsequently, a cDNA encoding the renal peptide transporter PEPT2 (SLC15A2) was also isolated (Liu et al., 1995). PEPT1 is localized to brush-border membranes of intestinal and renal epithelial cells (Ogihara et al., 1996), whereas PEPT2 is preferentially expressed in the kidney and located at brush-border membranes of renal epithelial cells (Shen et al., 1999). Fundamental functions related to substrate specificity and driving force have been molecularly confirmed using in vitro gene expression systems. Furthermore, novel implications of peptide transporters have been clarified after cDNA cloning. For example, PEPT1 can transport nonpeptidic compounds (Han et al., 1998) and bacterium-derived chemotactic peptides (Merlin et al., 1998). These discoveries led to the development of new research areas of intestinal PEPT1 in drug design/delivery and intestinal immunology. Construction of Pept1 or Pept2 knockout mice has provided direct in vivo evidence of physiological roles, such as protein homeostasis (Kamal, Keep, & Smith, 2008; Nässl et al., 2011). In addition to these findings, biochemical and biophysical aspects have been unveiled by protein engineering, computer modeling and crystallization. In particular, the crystal structure of $PepT_{So}$ and $PepT_{St}$, functionally similar prokaryotic homologs of the mammalian peptide transporters from bacteria, were recently reported (Newstead, 2011; Newstead et al., 2011; Solcan et al., 2012).

In this review, we will introduce current topics regarding the functional characteristics of PEPT1 and PEPT2 with a focus on biophysics/mechanisms along with the associated structural aspects.

2. FUNCTIONAL CHARACTERISTICS OF PEPT1 AND PEPT2

2.1. Substrate Specificity

Both PEPT1 and PEPT2 can transport di- and tripeptides with different molecular sizes and charges, but not free amino acids and tetrapeptides. Not only physiological small peptides but also bacterium-derived small peptides such as *N*-formylmethionyl-leucyl-phenylalanine and muramyl dipeptide are transported by PEPT1 (Ingersoll et al., 2012). Numerous studies have clarified that the transport of bacterial small peptides by PEPT1 in the colon, where PEPT1 is abnormally expressed in some pathophysiological conditions, leads to activation of downstream proinflammatory responses (Ingersoll et al., 2012). β-Lactam antibiotics and bestatin, which are structurally similar to small peptides, are found to be transported by PEPT1 and PEPT2 (Terada & Inui, 2004). Midodrine, an antihypotension prodrug for combining glycine via a peptide bond with an active drug, is also demonstrated to be a substrate for PEPT1 (Tsuda et al., 2006).

It had been believed that the presence of peptide bond(s) is the most important factor in the recognition of substrates by peptide transporters. However, structural requirements of PEPT1 were re-evaluated, and it was demonstrated that even compounds without peptide bond(s) can be accepted as substrates. For example, 4-aminophenylacetic acid (Temple et al., 1998), δ-amino levulinic acid (Döring, Walter et al., 1998), ω-amino fatty acids such as 8-amino octanoic acid (Döring, Will et al., 1998), amino acid aryl amide (Börner et al., 1998) and valacyclovir (Han et al., 1998) can be accepted as substrates. Figure 1 shows the typical PEPT1 substrates, classified into peptidic and nonpeptidic structures.

2.2. Substrate Affinity

Functional expression studies have clearly established that PEPT1 is a low-affinity transporter. PEPT1 is mainly expressed in the small intestine, where

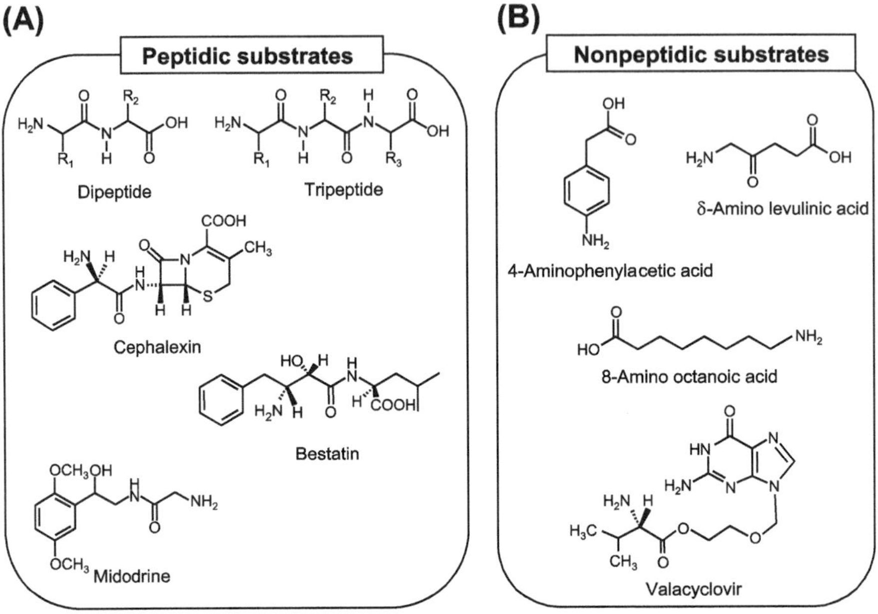

Figure 1 Chemical structures of various substrates for peptide transporters. (A) Typical substrates with peptide bond(s). Di- and tripeptides are physiological substrates. Peptide-like drugs such as cephalexin, bestatin and midodrine are transported by peptide transporters (Terada & Inui, 2004). (B) Typical substrates without a peptide bond. It has been demonstrated that 4-aminophenylacetic acid (Temple et al., 1998), δ-amino levulinic acid (Döring, Walter et al., 1998), 8-amino octanoic acid (Döring, Will et al., 1998) and valacyclovir (Han et al., 1998) can be transported by PEPT1.

large amounts of small peptides exist. Therefore, the functional properties of low affinity and high capacity of PEPT1 are physiologically reasonable for efficient absorption of di- and tripeptides from the lumen. In contrast to PEPT1, PEPT2 has been demonstrated to be the high-affinity transporter for di- and tripeptides (Ramamoorthy et al., 1995) and for amino β-lactam antibiotics (Terada, Saito, Mukai, & Inui, 1997). These differences in substrate affinity are not necessarily limited to substrates with peptide bond(s). Nonpeptidic compounds such as valacyclovir and δ-amino levulinic acid are preferentially recognized by PEPT2 rather than PEPT1 (Terada et al., 2000). There are some exceptions to this relationship. Namely, PEPT1 shows higher affinity for ceftibuten, anionic β-lactam antibiotic (Terada, Saito, Mukai, & Inui, 1997) and 8-amino octanoic acid (Terada et al., 2000). A feature of these compounds is the lack of α- or β-amino carbonyl function. Thus, it has been suggested that the α- or β-amino carbonyl structure of substrates is the key determinant of substrate affinity (Terada et al., 2000; Theis, Hartrodt, Kottra, Neubert, & Daniel, 2002).

2.3. Driving Force and Stoichiometry

PEPT1 and PEPT2 can transport a wide variety of substrates in an electrogenic mode as a consequence of H^+ and substrate cotransport. These electrogenic characteristics of PEPT1 and PEPT2 are commonly observed, but it has been unclear how carrier proteins can transport differently charged substrates in the same general transport mode.

Several studies have shown that the transport of zwitterionic dipeptides by PEPT1 is electrogenic, and this occurred at a proton to substrate flux coupling ratio of 1:1 (Amasheh et al., 1997; Fei et al., 1994; Mackenzie, Fei, Ganapathy, & Leibach, 1996; Steel et al., 1997). Cationic dipeptides can be transported in neutral and positively charged forms, resulting in excess transport current compared with that of neutral substrates (Mackenzie, Fei, Ganapathy, & Leibach, 1996). On the other hand, in the case of anionic dipeptides, the situation is more complicated, and several explanatory hypotheses have been suggested as follows: (1) electrogenic transport of anionic dipeptides may be due to the cotransport of one substrate molecule together with two protons (Amasheh et al., 1997; Kottra, Stamfort, & Daniel, 2002; Steel et al., 1997), or (2) it could result from the cotransport of one proton per substrate molecule and simultaneous countertransport of one negatively charged counterion such as OH^- or HCO_3^- (Mackenzie, Fei, Ganapathy, & Leibach, 1996). (3) An alternative explanation could be the preferential transport of only the zwitterionic form of the substrate with one proton (Fei et al., 1999; Wenzel et al., 1996). Therefore, the stoichiometry of anionic dipeptides to H^+ is still controversial. As described below, we hypothesized from a kinetic model that the stoichiometry of anionic dipeptides to H^+ depends on the extracellular pH (Irie, Terada, Katsura, Matsuoka, & Inui, 2005). In the case of PEPT2, a proton to substrate stoichiometry of 3:1 was proposed for the influx of dipeptides (Chen, Zhu, Smith, & Hediger, 1999).

2.4. Kinetic Model of the Transport Process via PEPT1

Although various transport characteristics have been delineated, fundamental mechanisms of the transport process via PEPT1 have been poorly understood. For example, with the uptake of most neutral substrates, a bell-shaped pH profile with maximal activity at pH 5.5–6.0 is commonly observed (Terada, Saito, Mukai, & Inui, 1997). However, it has not been elucidated why the uptake of these substrates is reduced under more acidic conditions, such as at pH 5.0, despite an increase in the H^+ gradient. In the case of other charged substrates, different pH profiles were shown compared with those of neutral substrates. The transport mechanisms of neutral, anionic and cationic substrates via PEPT1 have been investigated (Chen, Steel, & Hediger, 2000; Kottra, Stamfort, & Daniel, 2002). Chen et al. (2000) proposed that two histidine residues are intimately involved in the binding of the coupling ion H^+ and the binding of transportable substrates.

On the basis of these reports, we previously constructed a kinetic model of PEPT1 with a focus on the relationship between the transport of H^+ and substrates (Irie et al., 2005). Figure 2A shows the scheme for the recognition patterns of PEPT1 in terms of charges of substrates. In this model, it was defined that all substrates share the same substrate-binding site (Sawada et al., 1999; Wenzel et al., 1996), and that empty PEPT1 has one negative charge (Mackenzie, Loo et al., 1996). It was also hypothesized that PEPT1 has one H^+-binding site to which H^+ is bound before substrate, except for the case of cationic substrates (described below). At high pH, as most PEPT1 proteins do not have H^+ at the H^+-binding site (Fig. 2A, left), neutral and anionic substrates could not interact. However, the possibility that cationic substrates can interact with empty PEPT1 and be transported without H^+ was not excluded, because the intracellular acidification, as a consequence of the transport of cationic substrates, was observed (Kottra, Stamfort, & Daniel, 2002; Steel et al., 1997). At moderate pH values, as most PEPT1 proteins have H^+ at the H^+-binding site, both cationic and neutral substrates can be recognized and transported (Fig. 2A, middle). As described above, it was assumed that anionic substrates cannot access the substrate-binding site lacking H^+, and that the protonated substrate-binding site can accept only negatively charged substrates (Fig. 2A, right). On the basis of the above hypotheses, kinetic and computational models were constructed (Fig. 2B). When transport characteristics of PEPT1 were simulated by this model, the voltage dependence, hyperbolic saturation and pH profile of various charged substrates can be reproduced (Irie et al., 2005), suggesting the validity of the constructing kinetic model.

3. STRUCTURES OF PEPT1 AND PEPT2

3.1. Amino Acid Sequences and Secondary Structure

PEPT1 and PEPT2 consist of 707–710 and 729 amino acid residues, respectively, with several putative glycosylation and phosphorylation sites.

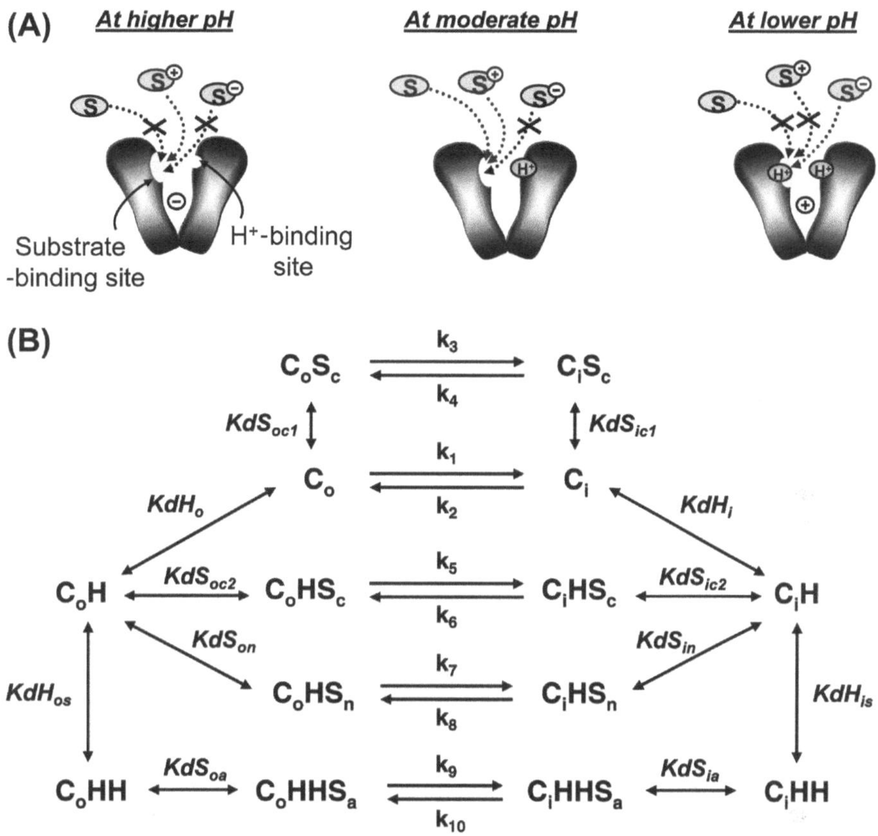

Figure 2 (A) Scheme for the recognition patterns of PEPT1 for neutral and charged substrates. Left, neither neutral substrates nor anionic ones can be recognized by empty PEPT1, which is the principle state at higher pH. Empty PEPT1 can accept only cationic substrates. Middle, at moderate pH, H^+ binds to an H^+-binding site, and PEPT1 with H^+ on the H^+-binding site can recognize neutral and cationic substrates, but not anionic ones. Right, anionic substrates can be accepted by only a protonated substrate-binding site at lower pH. (B) 14-State model for PEPT1. The transport mechanism assumed was represented by a 14-state model. Co and Ci represent empty PEPT1 facing the exterior and interior sides, respectively, and CoX and CiX stand for PEPT1 carrying X. S and H are substrate and H^+, respectively. KdX represents the dissociation constant of X, and KdHo and KdHos stand for the dissociation constants of H^+ to the H^+- and substrate-binding sites, respectively. The subscripts a, n and c stand for anionic, neutral and cationic charges of substrates, respectively. For color version of this figure, the reader is referred to the online version of this book.

Hydropathy analysis suggested that peptide transporter proteins contain 12 transmembrane domains (TMDs), with both the C and the N termini localized inside the cell (Fig. 3). A unique structural feature of PEPT1 is the large extracellular loop between TMD 9 and 10. Furthermore, epitope tagging analyses showed that the C terminus of PEPT1 is intracellular, and that the loops between TMDs 3–4 and 9–10 are extracellular (Covitz, Amidon, & Sadée, 1998). These findings are in agreement with the hydropathy plot (Fei et al., 1994). The overall amino acid identity between PEPT1 and PEPT2 is approximately 50%, and the amino acid sequence in the intra- or extracellular loops is more divergent than that in TMDs.

3.2. Computer Modeling of PEPT1 Structure

At the end of the 1990s, Dr Lee's laboratory group proposed the putative structure model of human PEPT1 based on the pairwise calculations and amphipathicity (Bolger et al., 1998; Lee et al., 1999; Yeung et al., 1998;). They evaluated all pairwise interactions for each helix–helix combination, under the assumption that only neighboring helices, for example, 1–2, 2–3 and so on, would contact each other. The model indicated that seven amphipathic TMDs (1, 3, 5, 7, 8, 9 and 10) form an aqueous channel through which the substrate is transported, and that Arg282 (TMD7), Asp341 (TMD8) and Glu595 (TMD10) oriented the center of the cavity (Fig. 4A). The functional importance of these residues was confirmed by site-directed mutagenesis experiments (Bolger et al., 1998; Yeung et al., 1998). However, there are limitations to this approach, which assumes that only TMDs adjacent in sequence are also physically adjacent, and that the TMDs are predicted to cross the lipid bilayer perpendicular to the surface. These assumptions were challenged by other researchers as described below.

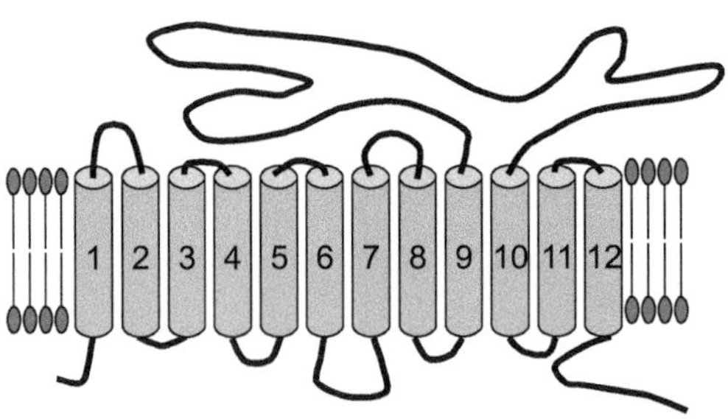

Figure 3 (A) Putative secondary structure of PEPT1. Hydropathy analyses of rabbit (Fei et al., 1994) and human (Liang et al., 1995) isoforms of PEPT1 indicated that the protein consists of 12 TMDs.

In 2003, crystal structures for the *Escherichia coli* proton-coupled lactose permease (LacY, Abramson et al., 2003) and the glycerol-3-phosphate/inorganic phosphate antiporter (GlpT, Huang, Lemieux, Song, Auer, & Wang, 2003) with 12 TMDs were clarified. LacY, GlpT and PEPT1 belong to the same major facilitator superfamily (Saier, Tran, & Barabote, 2006)

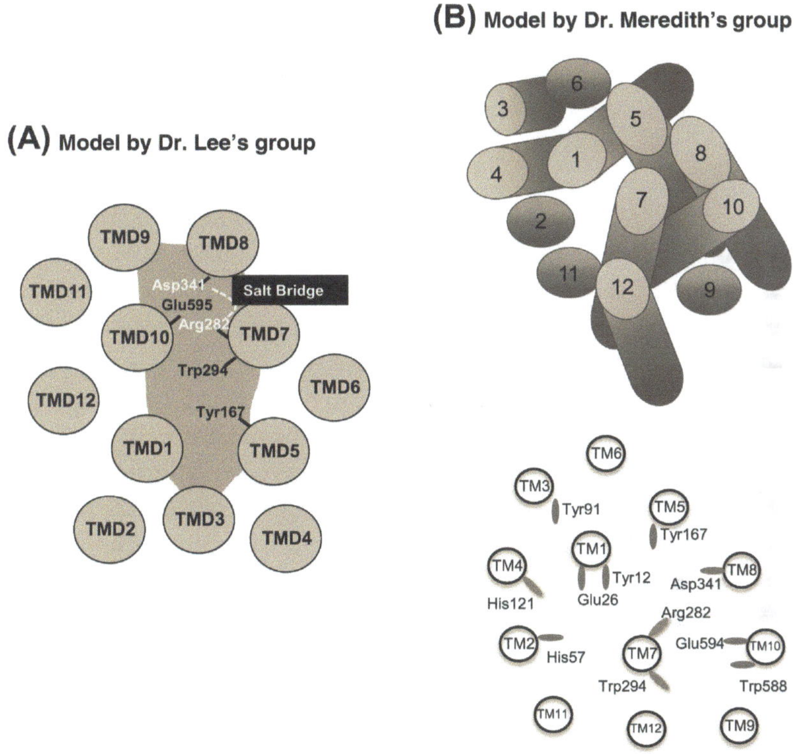

Figure 4 (A) Putative arrangement of 12 TMDs of human PEPT1 proposed by Dr Lee's laboratory group. The scheme shows the perspective from above the cell. The putative hydrophilic transport pathway is shown in gray. Essential amino acid residues confirmed by in vitro mutagenesis are labeled. Among them, Arg282 (TMD7) and Asp341 (TMD8) may form the salt bridge to drive the conformational changes needed for substrate transport and transporter resetting (Kulkarni et al., 2007; Pieri et al., 2008). *(Redrawn from previous reports (Bolger et al., 1998; Lee et al., 1999; Yeung et al., 1998))* (B) Putative arrangement of 12 TMDs of rabbit PEPT1 proposed by Dr Meredith's laboratory group. Upper, scheme obtained by homology modeling of rabbit PEPT1 without the large extracellular loop with the crystal structures of LacY and GlpT, bacterial transporters. Many of the TMDs do not lie perpendicular to the plane of the membrane, but are diagonally oriented. Bottom, scheme of arrangement of TMD and essential amino acid residues obtained by helical wheel analysis. *(Redrawn from a previous report (Meredith & Price, 2006)).* See the color plate.

and appear to be structurally similar. In 2006, utilizing homology modeling to LacY and GlpT, Meredith and Price (2006) developed a structural model of a modified rabbit PEPT1 lacking the large extracellular loop between TMD 9 and 10. The structural model by this approach is quite different from that by Lee's group. In particular, TMDs that are adjacent in the sequence show little or no apparent interaction, and many of the TMDs do not lie perpendicular to the plane of the membranes, but are diagonally oriented (Fig. 8.4B). As a result of being diagonally oriented, most of the TMDs are predicted to be longer than the minimum number of amino acid residues needed to cross a lipid bilayer as an α-helix. The limitation of this approach is modeling with a modified PEPT1 lacking a large extracellular loop.

In 2008, Pedretti et al. (2008) built a model for full human PEPT1 using fragment-based homology to LacY. They also assessed the constructed model by investigating the interaction capacities with a set of 50 known PEPT1 substrates. Because the structural features originated from LacY, this model is similar to that of Meredith & Price, especially in TMDs, but quite different in the extra- and intracellular loops. In fact, the limitation of this approach is imprecise modeling of the extra- and intracellular loops because there are no equivalent structures in LacY.

The last two structural models were also challenged after crystallization of $PepT_{So}$, a functionally similar prokaryotic homolog of PEPT1 and PEPT2 from *Shewanella oneidensis*. The history and merits/demerits of computer modeling of the PEPT1 structure were excellently presented in a recent review (Foley, Rajamanickam, Bailey, & Meredith, 2010).

3.3. Crystal Structure

However, in 2011, Newstead et al. (2011) reported the first crystal structure of $PepT_{So}$. Figure 5A shows a topology model of the crystal structure. Figure 5B shows a side view in the plane of the membrane, and the view from the extracellular side of the membrane is shown in Fig. 5C.

$PepT_{So}$ contains 14 TM helices with two six-helical bundles formed by helices H1–H6 (N-terminal domain) and H7–H12 (C-terminal domain), and forms a "V"-shaped transporter. $PepT_{So}$ has two additional TM helices (HA and HB) that are inserted into the cytoplasmic loop connecting the N- and C-terminal bundles, but their role is currently unclear. A large extracellular loop between TMD 9 and 10 in PEPT proteins does not exist in $PepT_{So}$.

In the structure of $PepT_{So}$, hydrophilic central and extracellular cavities are located in the membranes (Fig. 5B). These cavities are believed to be binding sites or transport routes of both H^+ and substrates. The extracellular cavity is located at the interface between the N- and C-terminal domains, and is roughly cone-shaped with the apex at the bottom near the central cavity. The dimensions of the extracellular cavity is approximately ~16 × 8 × 8 Å. The central cavity is located at the center of the membrane, and close to the extracellular space by a gate made of helices H1, H2, H7 and H8, which pack closely together (Fig. 5C). The dimensions of the

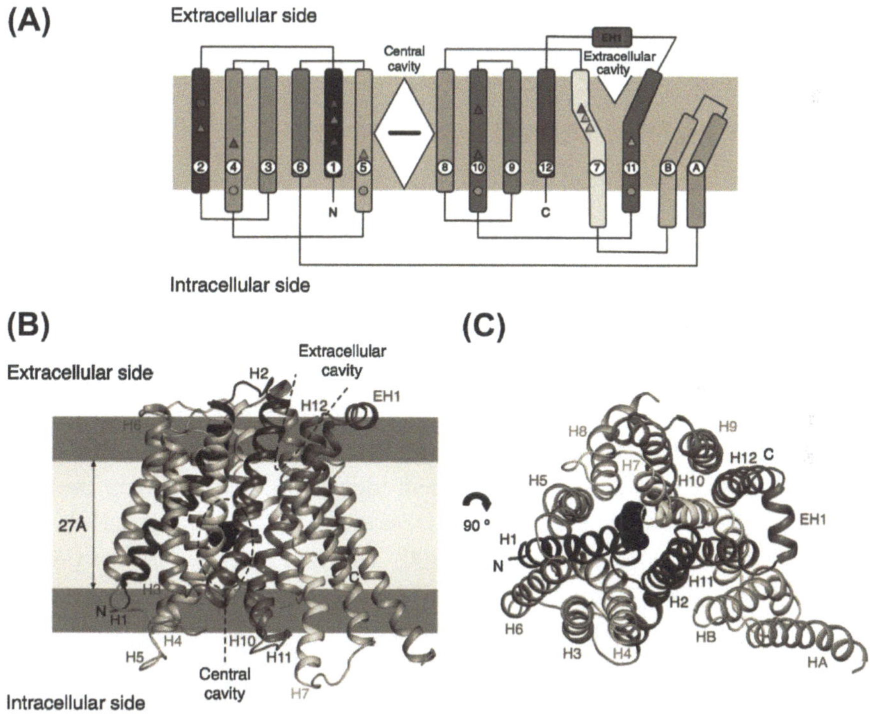

Figure 5 Structure of $PepT_{So}$. (A) The central and extracellular cavities are shown as a closed diamond and open triangle, respectively. A bound substrate in the central cavity is represented as a black horizontal bar. Residues lining the central cavity are highlighted by triangles. Residues forming the intracellular gate are denoted by circles. Residues His61 is highlighted by a square. (B) $PepT_{So}$ structure viewed in the plane of the membrane. Two hydrophilic cavities present in the structure are outlined by dashed lines. The hydrophobic core of the membrane (pale yellow) is distinguished from the interfacial region (light gray). N and C represent the N and C termini, respectively. Bound substrate is shown in black. Helices are labeled. (C) View from the extracellular side of the membrane. *(Reproduced from a previous report (Newstead et al., 2011) with permission).* See the color plate.

central cavity are approximately 13 × 12 × 11 Å, which may be sufficient to accommodate both di- and tripeptides, but not single amino acids or larger tetrapeptides. Exit of bound peptide to the intracellular side is currently believed to be restricted by side-chain interactions between two helix hairpins formed by helices H4–H5 and H10–H11. On the basis of structures of two bacterial transporters, it is supposed that opening of the intracellular gate of $PepT_{So}$ is controlled by the movement of helix hairpin H10–H11.

Very recently, Dr Newstead's laboratory group has determined the crystal structure of $PepT_{St}$ from the bacterium *Streptococcus thermophilus* (Solcan et al., 2012). In contrast to the occluded structure of $PepT_{So}$, the structure of $PepT_{St}$ reveals an inward-facing conformation. Nevertheless, the overall structure of $PepT_{St}$ was similar to that of $PepT_{So}$.

4. ESSENTIAL AMINO ACID RESIDUES

4.1. Site-Directed Mutagenesis

Since the brush-border membrane vesicle studies, it has been recognized that histidine residues of peptide transporters are involved in transport. In particular, it was assumed that histidine residues are essential for the substrate recognition (Kato, Maegawa, Okano, Inui, & Hori, 1989). After molecular identification of PEPT1, the critical importance of His57 (TMD2) and His121 (TMD4) in PEPT1 has been clarified by various investigators (Chen, Steel, & Hediger, 2000; Fei et al., 1997; Terada, Saito, Mukai, & Inui, 1996; Uchiyama, Kulkarni, Davies, & Lee, 2003), and His57 has been identified as the primary protonation site (Chen, Steel, & Hediger, 2000; Fei et al., 1997; Uchiyama, Kulkarni, Davies, & Lee, 2003). Chen et al. (2000) also demonstrated that aromatic residues Tyr56 (TMD2) and Tyr64 (TMD2) neighboring His57 are required for normal PEPT1 transport activities. Because tyrosine residues were considered to contribute to stabilizing positive charges within the membrane electric field, it was proposed that His57, which is flanked by tyrosine residues, participates in H^+ binding. These findings are based on the construction of a kinetic model of the transport process by PEPT1 as described in Section 2.4.

As described in Section 3.2, Dr Lee's laboratory group proposed the functional importance of TMDs 1, 3, 5, 7, 8, 9, and 10, all of which are predicted to form a transport pathway based on computational modeling. Initially, using site-directed mutagenesis, they demonstrated that Tyr167 (TMD5), Trp294 (TMD7) and Glu595 (TMD10) were essential for the transport process (Bolger et al., 1998; Lee et al., 1999; Yeung et al., 1998). Subsequently, with cysteine-scanning mutagenesis techniques, they proposed that Tyr91 (TMD3) (Links, Kulkarni, Davies, Lee, & Haworth, 2007), Tyr167, Gln171 and Ser174 in TMD5 (Kulkarni, Haworth, & Lee, 2003), and Arg282 (TMD7) (Kulkarni, Haworth, Uchiyama, & Lee, 2003) are involved in substrate interaction and translocation. Subsequent studies have demonstrated that Arg282 (TMD7) and Asp341 (TMD8) formed the salt-bridge (Bossi et al., 2011; Kulkarni et al., 2007; Meredith, 2004; Pieri, Hall, Price, Bailey, & Meredith, 2008), and it was proposed that the breaking and reforming of this salt bridge may drive the conformational changes needed for substrate transport and transporter resetting (Kulkarni et al., 2007; Pieri et al., 2008).

Several studies have addressed the functional consequences of single-nucleotide polymorphisms (SNPs) in PEPT1 and PEPT2. Among them, it was demonstrated that the Arg57His mutant of PEPT2 showed no transport activity despite proper membrane targeting (Terada, Irie, Okuda, & Inui, 2004).

4.2. Inside the Hydrophilic Cavities of the Crystal Structure

As described in Section 3.3, a central cavity and a smaller extracellular cavity in the crystal structure of $PepT_{So}$ are assumed to be route of the substrates. Similar cavities are found in $PepT_{St}$. On the basis of the locations and properties of amino acid residues in these cavities and previous in vitro mutagenesis data, several interesting mechanisms were proposed for the transport function of peptide transporters (Newstead, 2011; Newstead et al., 2011; Solcan et al., 2012).

A conserved His61 (His57 in TMD2 of PEPT1) is located at the bottom of the extracellular cavity. This location is within the interface of helices H2 and H7 that forms the extracellular gate. This region is supposed to contribute to the transporter's conformational change from the occluded to inward-facing state. As described in Section 4.1, His57 in TMD2 of PEPT1 was identified as the primary H^+-binding site (Fei et al., 1997; Uchiyama, Kulkarni, Davies, & Lee, 2003). In the vicinity of His61 (His57 in TMD2 of PEPT1), Asp316 (Asp298 in TMD7 of PEPT1) is located at the top of the central cavity. This residue is believed to be involved in the salt bridge interaction with Arg32 (Arg34 in TDM1 of PEPT1) to seal off the central cavity tightly from the external environment. Although the roles of these amino acid residues remain unclear, proton binding to the histidine residue may trigger the opening of the extracellular gate. In contrast to this hypothesis, recent functional studies based on the crystal structure of $PepT_{St}$ suggest that a conserved sequence motif (ExxERFxYY) on helix H1 is involved in the proton binding (Solcan et al., 2012).

In the central cavity, charged residues face inwardly to create a dipole across the cavity. One possible mechanism for peptide recognition is the following. The N terminus of a peptide interacts with Glu419 (Glu595 in TMD10 of PEPT1) within the C-terminal bundle, and the C terminus interacts with positively charged residues, Arg25 (Arg27 in TMD1 of PEPT1), Arg32 (Arg34 in TMD1 of PEPT1) and Lys127 (Lys140 in TMD1 of PEPT1) on the opposing N-terminal bundle. Arg57 in TMD2 of PEPT2 (identical to Arg27 in TMD1 of PEPT1) was clarified to be essential for transport activity (Terad, Irie, Okuda, & Inui, 2004). The dimensions for these amino acid residues are suggested to be only sufficient for di- and tripeptides. Larger tetrapeptides would be sterically restricted, whereas single amino acids would be unable to interact with both halves of the transporter, and presumably fail to trigger the required conformational changes to reorient the binding site across the membrane.

In the central cavity, one conserved Tyr154 (Tyr167 in TMD5 of PEPT1) is located close to Glu419 (Glu595 in TMD10 of PEPT1), and two conserved Tyr29 and Tyr68 (Tyr64 in TMD2 of PEPT1) are positioned close to the positively charged cluster in the N-terminal bundle. A recent study suggested that Tyr29 and Tyr68 contribute to determining peptide affinity (Solcan et al., 2012). In addition, most of the other residues in the binding site are conserved hydrophobic residues, including Ile157 (Ile170 in TMD5 of PEPT1), Trp312 (Trp294 in TMD7 of PEPT1), Phe315 (Phe297 in TMD7 of PEPT1) and Trp446 (Trp622 in TMD11 of PEPT1). The functional importance of Trp294 in TMD7 of PEPT1 was confirmed by in vitro mutagenesis (Bolger et al., 1998). These residues are likely to provide a suitable environment for peptide side chains, which are generally more hydrophobic than the peptide backbone. The presence of several possible hydrogen-bond donors and acceptors could be advantageous in adapting to peptides of various lengths, sequences and charges.

The crystal structures of $PepT_{So}$ and $PepT_{St}$ provided many direct and indirect ideas regarding the recognition and translocation mechanisms of peptide transporters. However, it should be noted that the present structures merely provide hints as to one possible arrangement.

5. FUTURE PERSPECTIVES

Recent advances in biochemical and structural aspects of peptide transporters enable us to integrate structure and function, and to understand the substrate recognition or translocation on a molecular level. However, fundamental features remain unclear, such as how peptide transporters can recognize thousands of substrates including nonpeptide compounds. Although the central cavity of $PepT_{So}$ is believed to fit chemical structures of di- and tripeptides, it would be quite interesting to determine how atypical substrates such as valacyclovir and ω-amino fatty acid fit. In addition, the large extracellular loop between TMD 9 and 10 is a unique structural feature of PEPT1, but this loop is not present in $PepT_{So}$ and $PepT_{St}$. We should keep this fact in mind when analyzing the structure–function relationships based on the crystal structures of $PepT_{So}$ and $PepT_{St}$.

ACKNOWLEDGMENT

This work was supported in part by a Grant-in-Aid for Scientific Research from the Ministry of Education, Culture, Sports, Science and Technology of Japan, and The Mochida Memorial Foundation for Medical and Pharmaceutical Research.

REFERENCES

Abramson, J., Smirnova, I., Kasho, V., Verner, G., Kaback, H. R., & Iwata, S. (2003). Structure and mechanism of the lactose permease of *Escherichia coli*. *Science*, *301*(5633), 610–615.

Amasheh, S., Wenzel, U., Boll, M., Dorn, D., Weber, W., Clauss, W., et al. (1997). Transport of charged dipeptides by the intestinal H^+/peptide symporter PepT1 expressed in xenopus laevis oocytes. *Journal of Membrane Biology*, *155*(3), 247–256.

Bolger, M. B., Haworth, I. S., Yeung, A. K., Ann, D., von Grafenstein, H., Hamm-Alvarez, S., et al. (1998). Structure, function, and molecular modeling approaches to the study of the intestinal dipeptide transporter PepT1. *Journal of Pharmaceutical Sciences*, *87*(11), 1286–1291.

Börner, V., Fei, Y. J., Hartrodt, B., Ganapathy, V., Leibach, F. H., Neubert, K., et al. (1998). Transport of amino acid aryl amides by the intestinal H^+/peptide cotransport system, PEPT1. *European Journal of Biochemistry*, *255*(3), 698–702.

Bossi, E., Renna, M. D., Sangaletti, R., D'Antoni, F., Cherubino, F., Kottra, G., et al. (2011). Residues R282 and D341 act as electrostatic gates in the proton-dependent oligopeptide transporter PepT1. *Journal of Physiology*, *589*(Pt 3), 495–510.

Chen, X. Z., Steel, A., & Hediger, M. A. (2000). Functional roles of histidine and tyrosine residues in the H^+-peptide transporter PepT1. *Biochemical and Biophysical Research Communications*, *272*(3), 726–730.

Chen, X. Z., Zhu, T., Smith, D. E., & Hediger, M. A. (1999). Stoichiometry and kinetics of the high-affinity H^+-coupled peptide transporter PepT2. *Journal of Biological Chemistry*, *274*(5), 2773–2779.

Covitz, K. M., Amidon, G. L., & Sadée, W. (1998). Membrane topology of the human dipeptide transporter, hPEPT1, determined by epitope insertions. *Biochemistry*, *37*(43), 15214–15221.

Daniel, H. (2004). Molecular and integrative physiology of intestinal peptide transport. *Annual Review of Physiology*, *66*, 361–384.

Döring, F., Walter, J., Will, J., Focking, M., Boll, M., Amasheh, S., et al. (1998). Delta-aminolevulinic acid transport by intestinal and renal peptide transporters and its physiological and clinical implications. *Journal of Clinical Investigation*, *101*(12), 2761–2767.

Döring, F., Will, J., Amasheh, S., Clauss, W., Ahlbrecht, H., & Daniel, H. (1998). Minimal molecular determinants of substrates for recognition by the intestinal peptide transporter. *Journal of Biological Chemistry*, *273*(36), 23211–23218.

Fei, Y. J., Kanai, Y., Nussberger, S., Ganapathy, V., Leibach, F. H., Romero, M. F., et al. (1994). Expression cloning of a mammalian proton-coupled oligopeptide transporter. *Nature*, *368*(6471), 563–566.

Fei, Y. J., Liu, W., Prasad, P. D., Kekuda, R., Oblak, T. G., Ganapathy, V., et al. (1997). Identification of the histidyl residue obligatory for the catalytic activity of the human H^+/peptide cotransporters PEPT1 and PEPT2. *Biochemistry*, *36*(2), 452–460.

Fei, Y. J., Nara, E., Liu, J. C., Boyd, C. A., Ganapathy, V., & Leibach, F. H. (1999). Preferential recognition of zwitterionic dipeptides as transportable substrates by the high-affinity peptide transporter PEPT2. *Biochimica et Biophysica Acta*, *1418*(2), 344–351.

Foley, D. W., Rajamanickam, J., Bailey, P. D., & Meredith, D. (2010). Bioavailability through PepT1: the role of computer modelling in intelligent drug design. *Current Computer-Aided Drug Design*, *6*(1), 68–78.

Han, H., de Vrueh, R. L., Rhie, J. K., Covitz, K. M., Smith, P. L., et al. (1998). 5'-amino acid esters of antiviral nucleosides, acyclovir, and AZT are absorbed by the intestinal PEPT1 peptide transporter. *Pharmaceutical Research*, *15*(8), 1154–1159.

Huang, Y., Lemieux, M. J., Song, J., Auer, M., & Wang, D. N. (2003). Structure and mechanism of the glycerol-3-phosphate transporter from *Escherichia coli*. *Science*, *301*(5633), 616–620.

Ingersoll, S. A., Ayyadurai, S., Charania, M. A., Laroui, H., Yan, Y., & Merlin, D. (2012). The role and pathophysiological relevance of membrane transporter PepT1 in intestinal inflammation and inflammatory bowel disease. *American Journal of Physiology Gastrointestinal and Liver Physiology*, *302*(5), G484–G492.

Irie, M., Terada, T., Katsura, T., Matsuoka, S., & Inui, K. (2005). Computational modelling of H^+-coupled peptide transport via human PEPT1. *Journal of Physiology*, *565*(Pt 2), 429–439.

Kamal, M. A., Keep, R. F., & Smith, D. E. (2008). Role and relevance of PepT2 in drug disposition, dynamics, and toxicity. *Drug Metabolism and Pharmacokinetics*, *23*(4), 236–242.

Kato, M., Maegawa, H., Okano, T., Inui, K., & Hori, R. (1989). Effect of various chemical modifiers on H^+ coupled transport of cephradine via dipeptide carriers in rabbit intestinal brush-border membranes: role of histidine residues. *Journal of Pharmacology and Experimental Therapeutics*, *251*(2), 745–749.

Kottra, G., Stamfort, A., & Daniel, H. (2002). PEPT1 as a paradigm for membrane carriers that mediate electrogenic bidirectional transport of anionic, cationic, and neutral substrates. *Journal of Biological Chemistry*, *277*(36), 32683–32691.

Kulkarni, A. A., Davies, D. L., Links, J. S., Patel, L. N., Lee, V. H., & Haworth, I. S. (2007). A charge pair interaction between Arg282 in transmembrane segment 7 and Asp341 in transmembrane segment 8 of hPepT1. *Pharmaceutical Research*, *24*(1), 66–72.

Kulkarni, A. A., Haworth, I. S., & Lee, V. H. (2003). Transmembrane segment 5 of the dipeptide transporter hPepT1 forms a part of the substrate translocation pathway. *Biochemical and Biophysical Research Communications*, *306*(1), 177–185.

Kulkarni, A. A., Haworth, I. S., Uchiyama, T., & Lee, V. H. (2003). Analysis of transmembrane segment 7 of the dipeptide transporter hPepT1 by cysteine-scanning mutagenesis. *Journal of Biological Chemistry*, *278*(51), 51833–51840.

Lee, V. H., Chu, C., Mahlin, E. D., Basu, S. K., Ann, D. K., Bolger, M. B., et al. (1999). Biopharmaceutics of transmucosal peptide and protein drug administration: role of transport mechanisms with a focus on the involvement of PepT1. *Journal of Controlled Release*, *62*(1–2), 129–140.

Liang, R., Fei, Y. J., Prasad, P. D., Ramamoorthy, S., Han, H., Yang-Feng, T. L., et al. (1995). Human intestinal H^+/peptide cotransporter. Cloning, functional expression, and chromosomal localization. *Journal of Biological Chemistry*, *270*(12), 6456–6463.

Links, J. L., Kulkarni, A. A., Davies, D. L., Lee, V. H., & Haworth, I. S. (2007). Cysteine scanning of transmembrane domain three of the human dipeptide transporter: implications for substrate transport. *Journal of Drug Targeting, 15*(3), 218–225.

Liu, W., Liang, R., Ramamoorthy, S., Fei, Y. J., Ganapathy, M. E., Hediger, M. A., et al. (1995). Molecular cloning of PEPT2, a new member of the H^+/peptide cotransporter family, from human kidney. *Biochimica et Biophysica Acta, 1235*(2), 461–466.

Mackenzie, B., Fei, Y. J., Ganapathy, V., & Leibach, F. H. (1996). The human intestinal H^+/ oligopeptide cotransporter hPEPT1 transports differently-charged dipeptides with identical electrogenic properties. *Biochimica et Biophysica Acta, 1284*(2), 125–128.

Mackenzie, B., Loo, D. D., Fei, Y., Liu, W. J., Ganapathy, V., Leibach, F. H., et al. (1996). Mechanisms of the human intestinal H^+-coupled oligopeptide transporter hPEPT1. *Journal of Biological Chemistry, 271*(10), 5430–5437.

Meredith, D. (2004). Site-directed mutation of arginine 282 to glutamate uncouples the movement of peptides and protons by the rabbit proton-peptide cotransporter PepT1. *Journal of Biological Chemistry, 279*(16), 15795–15798.

Meredith, D., & Price, R. A. (2006). Molecular modeling of PepT1–towards a structure. *Journal of Membrane Biology, 213*(2), 79–88.

Merlin, D., Steel, A., Gewirtz, A. T., Si-Tahar, M., Hediger, M. A., & Madara, J. L. (1998). hPepT1-mediated epithelial transport of bacteria-derived chemotactic peptides enhances neutrophil-epithelial interactions. *Journal of Clinical Investigation, 102*(11), 2011–2018.

Nässl, A. M., Rubio-Aliaga, I., Fenselau, H., Marth, M. K., Kottra, G., & Daniel, H. (2011). Amino acid absorption and homeostasis in mice lacking the intestinal peptide transporter PEPT1. *American Journal of Physiology Gastrointestinal and Liver Physiology, 301*(1), G128–G137.

Newstead, S. (2011). Towards a structural understanding of drug and peptide transport within the proton-dependent oligopeptide transporter (POT) family. *Biochemical Society Transactions, 39*(5), 1353–1358.

Newstead, S., Drew, D., Cameron, A. D., Postis, V. L., Xia, X., Fowler, P. W., et al. (2011). Crystal structure of a prokaryotic homologue of the mammalian oligopeptide-proton symporters, PepT1 and PepT2. *EMBO Journal, 30*(2), 417–426.

Ogihara, H., Saito, H., Shin, B. C., Terada, T., Takenoshita, S., Nagamachi, Y., et al. (1996). Immuno-localization of H+/peptide cotransporter in rat digestive tract. *Biochemical and Biophysical Research Communications, 220*(3), 848–852.

Pedretti, A., De Luca, L., Marconi, C., Negrisoli, G., Aldini, G., & Vistoli, G. (2008). Modeling of the intestinal peptide transporter hPepT1 and analysis of its transport capacities by docking and pharmacophore mapping. *ChemMedChem, 3*(12), 1913–1921.

Pieri, M., Hall, D., Price, R., Bailey, P., & Meredith, D. (2008). Site-directed mutagenesis of arginine282 suggests how protons and peptides are co-transported by rabbit PepT1. *International Journal of Biochemistry & Cell Biology, 40*(4), 721–730.

Ramamoorthy, S., Liu, W., Ma, Y. Y., Yang-Feng, T. L., Ganapathy, V., & Leibach, F. H. (1995). Proton/peptide cotransporter (PEPT2) from human kidney: functional characterization and chromosomal localization. *Biochimica et Biophysica Acta, 1240*(1), 1–4.

Saier, M. H., Jr., Tran, C. V., & Barabote, R. D. (2006). TCDB: the transporter classification database for membrane transport protein analyses and information. *Nucleic Acids Research, 34*(Database issue), D181–D186.

Sawada, K., Terada, T., Saito, H., Hashimoto, Y., & Inui, K. (1999). Recognition of l-amino acid ester compounds by rat peptide transporters PEPT1 and PEPT2. *Journal of Pharmacology and Experimental Therapeutics, 291*(2), 705–709.

Shen, H., Smith, D. E., Yang, T., Huang, Y. G., Schnermann, J. B., Brosius, F., et al. (1999). Localization of PEPT1 and PEPT2 proton-coupled oligopeptide transporter mRNA and protein in rat kidney. *The American Journal of Physiology, 276*(5 Pt 2), F658–F665.

Solcan, N., Kwok, J., Fowler, P. W., Cameron, A. D., Drew, D., Iwata, S., et al. (2012). Alternating access mechanism in the POT family of oligopeptide transporters. *EMBO Journal, 31*(16), 3411–3421.

Steel, A., Nussberger, S., Romero, M. F., Boron, W. F., Boyd, C. A., & Hediger, M. A. (1997). Stoichiometry and pH dependence of the rabbit proton-dependent oligopeptide transporter PepT1. *Journal of Physiology, 498*(Pt 3), 563–569.

Temple, C. S., Stewart, A. K., Meredith, D., Lister, N. A., Morgan, K. M., Collier, I. D., et al. (1998). Peptide mimics as substrates for the intestinal peptide transporter. *Journal of Biological Chemistry, 273*(1), 20–22.

Terada, T., & Inui, K. (2004). Peptide transporters: structure, function, regulation and application for drug delivery. *Current Drug Metabolism, 5*(1), 85–94.

Terada, T., Irie, M., Okuda, M., & Inui, K. (2004). Genetic variant Arg57His in human H^+/peptide cotransporter 2 causes a complete loss of transport function. *Biochemical and Biophysical Research Communications, 316*(2), 416–420.

Terada, T., Saito, H., Mukai, M., & Inui, K. (1996). Identification of the histidine residues involved in substrate recognition by a rat H^+/peptide cotransporter, PEPT1. *FEBS Letters, 394*(2), 196–200.

Terada, T., Saito, H., Mukai, M., & Inui, K. (1997). Recognition of β-lactam antibiotics by rat peptide transporters, PEPT1 and PEPT2, in LLC-PK1 cells. *American Journal of Physiology, 273*(5 Pt 2), F706–F711.

Terada, T., Sawada, K., Irie, M., Saito, H., Hashimoto, Y., & Inui, K. (2000). Structural requirements for determining the substrate affinity of peptide transporters PEPT1 and PEPT2. *Pflügers Archiv: European Journal of Physiology, 440*(5), 679–684.

Theis, S., Hartrodt, B., Kottra, G., Neubert, K., & Daniel, H. (2002). Defining minimal structural features in substrates of the H^+/peptide cotransporter PEPT2 using novel amino acid and dipeptide derivatives. *Molecular Pharmacology, 61*(1), 214–221.

Tsuda, M., Terada, T., Irie, M., Katsura, T., Niida, A., Tomita, K., et al. (2006). Transport characteristics of a novel peptide transporter 1 substrate, antihypotensive drug midodrine, and its amino acid derivatives. *Journal of Pharmacology and Experimental Therapeutics, 318*(1), 455–460.

Uchiyama, T., Kulkarni, A. A., Davies, D. L., & Lee, V. H. (2003). Biophysical evidence for His57 as a proton-binding site in the mammalian intestinal transporter hPepT1. *Pharmaceutical Research, 20*(12), 1911–1916.

Wenzel, U., Gebert, I., Weintraut, H., Weber, W. M., Clauss, W., & Daniel, H. (1996). Transport characteristics of differently charged cephalosporin antibiotics in oocytes expressing the cloned intestinal peptide transporter PepT1 and in human intestinal Caco-2 cells. *Journal of Pharmacology and Experimental Therapeutics, 277*(2), 831–839.

Yeung, A. K., Basu, S. K., Wu, S. K., Chu, C., Okamoto, C. T., Hamm-Alvarez, S. F., et al. (1998). Molecular identification of a role for tyrosine 167 in the function of the human intestinal proton-coupled dipeptide transporter (hPepT1). *Biochemical and Biophysical Research Communications, 250*(1), 103–107.

CHAPTER NINE

The SLC16A Family of Monocarboxylate Transporters (MCTs)—Physiology and Function in Cellular Metabolism, pH Homeostasis, and Fluid Transport

Jeffrey Adijanto*, Nancy J. Philp*
Department of Pathology, Anatomy and Cell Biology, Thomas Jefferson University, Philadelphia, PA, USA
*Corresponding authors: E-mail: jadijan1@gmail.com; nancy.philp@jefferson.edu

Contents

Abstract

The SLC16A family of monocarboxylate transporters (MCTs) is composed of 14 members. MCT1 through MCT4 (MCTs 1–4) are H^+-coupled monocarboxylate transporters, MCT8 and MCT10 transport thyroid hormone and aromatic amino acids, while the substrate specificity and function of other MCTs have yet to be determined. The focus of this review is on MCTs 1–4 because their role in lactate transport is intrinsically linked to cellular metabolism in various biological systems, including skeletal muscle, brain, retina, and testis. Although MCTs 1–4 all transport lactate, they differ in their transport kinetics and vary in tissue and subcellular distribution, where they facilitate "lactate-shuttling" between glycolytic and oxidative cells within tissues and across blood–tissue barriers.

Current Topics in Membranes, Volume 70
ISSN 1063-5823
http://dx.doi.org/10.1016/B978-0-12-394316-3.00009-0

However, the role of MCTs 1–4 is not confined to cellular metabolism. By interacting with bicarbonate transport proteins and carbonic anhydrases, MCTs participate in the regulation of pH homeostasis and fluid transport in renal proximal tubule and corneal endothelium, respectively. Here, we provide a comprehensive review of MCTs 1–4, linking their cellular distribution to their functions in various parts of the human body, so that we can better understand the physiological roles of MCTs at the systemic level.

1. INTRODUCTION

The SLC16 family of monocarboxylate transporters, MCTs, is composed of 14 members. They share similarities in sequence and protein structure, but each member differs from others in the substrates that they transport. Lactate, among other monocarboxylates, is the primary substrate for MCTs 1–4. MCT6 was shown to transport synthetic drugs such as probenecid and bumetanide. MCT8 transports thyroid hormones (T_3 and T_4) and mutations in this transporter (Allan–Herndon–Dudley Syndrome) cause crippling defects in brain and motor functions throughout development because of the lack of T3 uptake by neurons (Friesema et al., 2004; Schwartz et al., 2005). MCT10 (or T-type amino acid transporter-1; TAT1) is also capable of transporting thyroid hormones (Friesema et al., 2008), but unlike MCT8, its transport substrate includes other aromatic amino acids and that its gene expression is found primarily in the kidney, skeletal muscle, and small intestine (Kim et al., 2002). MCT9 was recently identified as a carnitine transporter through a meta-analysis of 14 GWAS studies of patients with elevated serum uric acid (Kolz et al., 2009). This finding was also independently verified in a separate study that evaluated the genetic links to human metabolism by coupling GWAS analysis with a high-throughput screening of blood metabolites (Suhre et al., 2011). Beyond the abovementioned MCTs, however, the transport substrates of the other MCTs (MCTs 5, 7, 11–14) are unknown. Of all MCTs, MCTs 1–4 are the most widely studied and understood. This is not surprising, as lactate has well-established roles in acid–base balance, with the earliest studies dating back to almost half a century ago. The advent of a noninvasive fluorescence-based intracellular pH indicator, BCECF (Rink, Tsien, & Pozzan, 1982), allowed physiologists to easily demonstrate that transmembrane lactate transport is accompanied by intracellular acidification. This observation gave rise to an explosion of studies that demonstrated the existence and function of H^+-coupled lactate transporters in various biological systems (Giasson & Bonanno, 1994; Kenyon, Yu, La Cour, & Miller, 1994; Poole & Halestrap, 1991; Roth & Brooks, 1990; Wang, Levi, & Halestrap, 1994).

For many decades, lactate was regarded as nothing more than a metabolic waste product of glycolysis. The accumulation of lactate within cells and their microenvironment, such as in muscle fibers during exercise, was thought to cause lactic acidosis with detrimental consequences to normal physiological function. However, with accumulating evidence demonstrating that lactate can be utilized as a metabolic substrate for oxidative respiration, there came a realization that lactate is more than simply a waste product (Gladden, 2004)—a radical idea at the time that raised much interest in the roles of MCTs 1–4 in metabolism in many parts of the human body, including the muscle, brain, kidney, liver, retina, and testis, all of which will be discussed in detail below (Section 2). Although MCTs are important for normal metabolic function, they are also implicated in disease. In cancer cells, for example, MCT1 and MCT4 are commonly upregulated to adapt to a change in metabolism and to maintain pH homeostasis. In addition to regulating cellular metabolism, MCTs also participate in fluid transport in corneal endothelium of the eye (Section 4.2) and acid–base balance in the proximal convoluted tubule (PCT) (Section 4.3) of the kidney. These processes are achieved by functional interaction between MCTs and HCO_3^- transport mechanisms (Section 3). With most of the functions of MCTs 1–4 revolving around lactate, it is easy to overlook the fact that MCTs 1–4 are just as capable of transporting other substrates, such as acetate, propionate, and butyrate—all of which are absorbed from the colon lumen via Na^+-coupled MCTs (SMCTs) and MCT1 (Section 4.4).

Although the crystal structure of MCTs has not been determined, hydrophobicity plots of MCTs 1–14 predict that these proteins consist of 12 transmembrane domains (TMs) with cytoplasmic N and C termini (Halestrap, 2012). The coupling of H^+ to monocarboxylate transport is unique to MCTs 1–4, and the structural property that is responsible for this feature, a lysine residue in TM1 (Lys38; Fig. 1A), has only been recently identified in rat MCT1 (Wilson, Meredith, Bunnun, Sessions, & Halestrap, 2009). In their model, Lys38 is normally uncharged in the hydrophobic environment of the ion pore, but it can bind to an extracellular H^+ to trigger the open conformation, in which the same lysine residue could subsequently bind a lactate ion. Upon binding, both the H^+ and lactate ions are transferred to an aspartate–arginine ion pair (Asp302–Arg306; TM8) located deeper in the channel pore (Manoharan, Wilson, Sessions, & Halestrap, 2006). These residues were identified in rat MCT1, but they are also found in MCTs 1–4 of both rat and human (Fig. 1B). Although MCTs

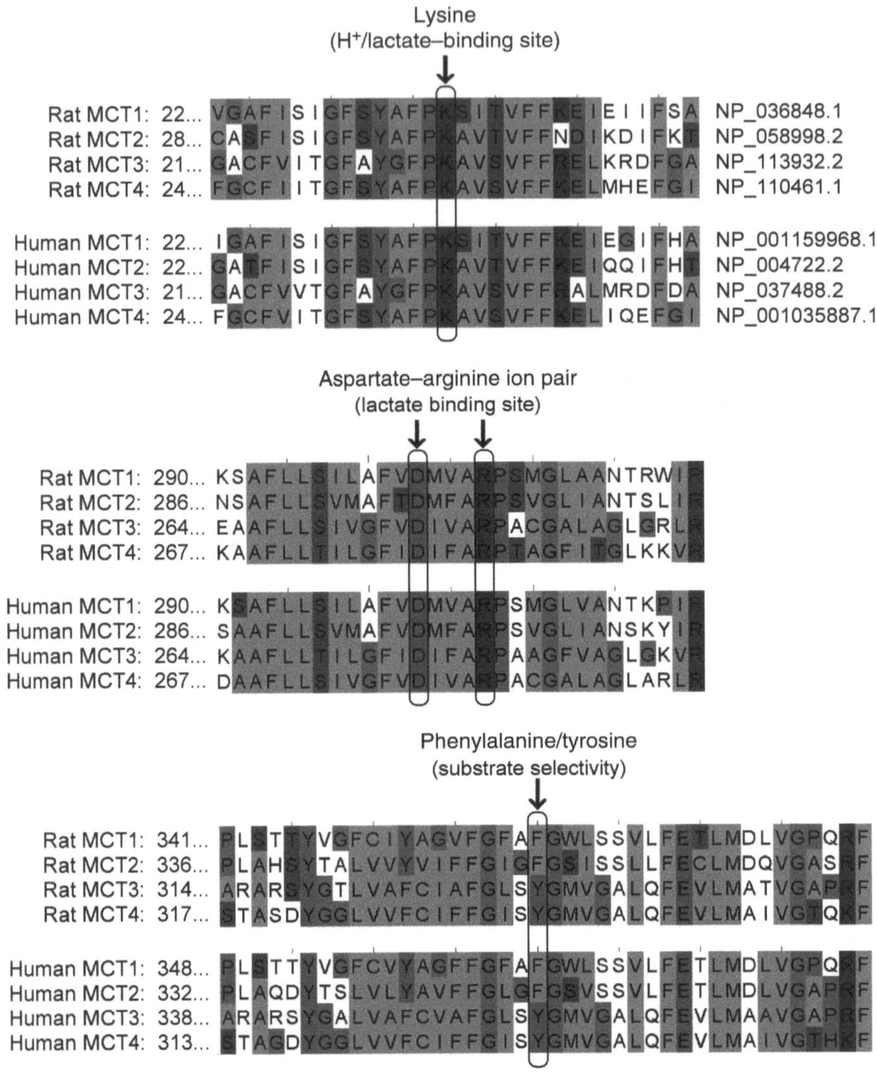

Figure 1 ***Amino acid sequence alignment of MCTs 1–4 for rat and human.*** (A) Lysine 38 (Rat MCT1) in TM1 binds H^+ and lactate and this residue is conserved between MCTs 1–4 of Rat and human. (B) The H^+ and lactate are subsequently transferred to an aspartate–arginine ion pair (D302–R306 in Rat MCT1) located deeper in the channel pore (TM8). (C) The substrate selectivity of the MCT is conferred by a phenylalanine (MCT1 and MCT2; F360 in rat MCT1) or tyrosine (MCT3 and MCT4; Y333 in rat MCT3) residue in TM10. The ascension numbers of the MCTs 1–4 protein sequences are provided. The sequence alignments are in ClustalX color scheme and generated using MUSCLE (Multiple Sequence Alignment) (http://www.ebi.ac.uk/Tools/msa/muscle/). See the color plate.

1–4 transport lactate and other similar substrates, they differ in substrate binding affinity and selectivity. An early study by Garcia et al. identified an amino acid residue that regulates the substrate selectivity of MCT1—a phenylalanine residue in TM10, Phe360 (Garcia, Brown, Pathak, & Goldstein, 1995). It is interesting to note that in MCTs 3 and 4, however, a tyrosine, rather than phenylalanine, is found in this position (Fig. 1C). This switch may account for the differences in lactate-binding affinities among MCTs 1–4, which seem to correlate with their function when expressed in specific cell types within a biological system. For example, MCT1 and MCT2, with their high affinity for lactate ($K_m \approx$ 1–3 mM), possess the ability to mediate rapid lactate uptake into cells (Halestrap & Price, 1999). As such, MCT1 and MCT2 are commonly expressed in tissues that utilize lactate as a substrate for oxidative metabolism, such as slow-twitch type I muscle fibers (MCT1) and neurons (MCT2). On the other hand, MCT3 and MCT4 have lower affinities for lactate (MCT3, $K_m \approx 6$ mM; MCT4, $K_m \approx 30$ mM), but they possess a wide range of transport rates and are, therefore, useful for maintaining lactate equilibrium in cells. MCT4 is commonly expressed in glycolytic cell types, such as type II fast-twitch skeletal muscle cells and astrocytes, where it mediates lactic acid extrusion. MCT3 is preferentially expressed in the retinal pigment epithelium, where it also mediates lactic acid extrusion.

With specific MCTs being differentially expressed in different cell types, or even in different membranes of a polarized epithelium, it is obvious that MCTs, similar to many other transporters, are under strict regulatory control. SLC16A1 (MCT1) and SLC16A3 (MCT4) carry multiple splice variants and differ only at the 5′-UTR region, suggesting that MCT1 and MCT4 transcripts can be subject to both transcriptional and post-transcriptional regulation. This level of control may be necessary because MCT1 and MCT4 are widely expressed in many different cell types, thus requiring additional mechanisms to regulate their relative abundance in a cell-specific manner. Furthermore, MCT1 transcripts can also be regulated by microRNAs, which target the mRNA at its 3′-UTR for translational repression or mRNA degradation. This mechanism is particularly important in beta-cells, where MCT1 is not normally expressed (Pullen, da Silva Xavier, Kelsey, & Rutter, 2011). On the other hand, no other transcript variants for SLC16A7 (MCT2) and SLC16A8 (MCT3) have been found, consistent with the more limited tissue distributions of MCT2 and MCT3.

Although the protein abundance of MCTs 1–4 can be regulated at the genetic level, the proper folding and trafficking of the MCTs 1–4 proteins to the plasma membrane requires each MCT to be assembled with either CD147/basigin (encoded by Bsg) or embigin (encoded by Emb), both of which are highly glycosylated single-pass membrane proteins that serve as chaperones for proper targeting of MCTs to the plasma membrane (Kirk et al., 2000; Philp, Ochrietor, Rudoy, Muramatsu, & Linser, 2003; Wilson et al., 2005). CD147 is widely expressed in tissues and forms a heterodimer with MCT1, 3, or 4. Embigin, on the other hand, associates primarily with MCT2, but it can also interact with MCT1 in some tissues. The dependence of MCTs 1–4 on their accessory proteins has been demonstrated in in vitro studies, but the physiological importance of this interaction in MCT maturation and trafficking is perhaps best demonstrated in the Bsg-null mouse (Nakai, Chen, & Nowak, 2006; Philp et al., 2003), which present with blindness, sterility, immunodeficiency, and problems with learning and memory (Hori et al., 2000; Igakura et al., 1998; Naruhashi et al., 1997). These phenotypes, as is discussed in the following sections (Sections 2 and 4), are direct pathological manifestations caused by the absence of MCTs. The underlying mechanisms behind these conditions will become readily apparent when these MCTs are placed in the context of their physiological functions in various cell types within their native biological systems.

2. PHYSIOLOGY AND FUNCTION OF MCTS 1–4 IN CELLULAR METABOLISM

2.1. Skeletal Muscle

The concept of lactate-shuttling is perhaps best established in skeletal muscle, which is composed of two major cell types with vastly different metabolic properties: type I oxidative fibers and type II glycolytic fibers (reviewed in (Schiaffino & Reggiani, 2011)). Anatomically, type I and II muscle fibers are closely apposed onto each other and separated by a small volume of extracellular space that is occupied by neurovascular structures. Type I oxidative fibers can be distinguished from type II fibers by the higher mitochondrial density, as well as the higher expression and activity of enzymes involved in oxidative metabolism (Essen, Jansson, Henriksson, Taylor, & Saltin, 1975), such as H^+-ATPase, succinic dehydrogenase, and cytochrome oxidase (COX). In contrast, type II fibers maintain higher expression and activity of glycolytic enzymes, such as phosphofructokinase (PFK) and lactate dehydrogenase (LDH) (Lowry et al., 1978; Spamer & Pette, 1977). More recent studies using PCR and microarray analysis further confirmed the differential expression of glycolytic genes in type I vs type II muscle fibers (Chemello et al., 2011; Plomgaard et al., 2006). Together, these studies establish type II muscles as a major site for lactate production; thus, it is perhaps not surprising that they also express high levels of MCT4 to facilitate lactate extrusion (Fishbein, Merezhinskaya, & Foellmer, 2002; Manning Fox, Meredith, & Halestrap, 2000). On the other hand, type I fibers specifically express MCT1, and this is consistent with the role of MCT1 in mediating lactate uptake for oxidative metabolism (Hashimoto, Masuda, Taguchi, & Brooks, 2005). The close anatomical relationship between type I and type II fibers and their vastly different metabolic activities support the notion of lactate-shuttling and metabolic coupling between the two muscle fibers (reviewed in Bergersen, 2007). In this scheme, as illustrated in Fig. 2, lactic acid generated by type II muscles is released into the extracellular space, where it can be taken up by type I fibers for oxidative metabolism. Excess lactic acid is removed via the vasculature and can be delivered to cardiac muscles for oxidative metabolism, or to the kidneys and liver for gluconeogenesis. In the blood circulation, efficient lactate uptake by erythrocytes is achieved via MCT1 expression at the plasma membrane. At the liver, erythrocytes release lactate, which is taken up by hepatocytes via MCT2 and converted to glucose via gluconeogenesis (Garcia et al., 1995).

2.2. Neurons and Astrocytes

In the field of neuroenergetics, a similar lactate-shuttling mechanism between astrocytes and neurons has been established. In this model, the highly glycolytic astrocytes generate and release lactic acid that is taken up by neurons, which are capable of using lactate as an energy substrate to fuel neuronal processes such as excitatory synaptic activity (Rouach, Koulakoff, Abudara, Willecke, & Giaume, 2008) and long-term memory formation (Suzuki et al., 2011). Although the astrocyte-neuron lactate shuttle (ANLS) model (Tsacopoulos & Magistretti, 1996) has not been without controversy (Mangia, Simpson, Vannucci, & Carruthers, 2009; Simpson, Carruthers, & Vannucci, 2007), a comparison between ANLS and the more established lactate shuttle in skeletal muscle reveals many similarities. The highly glycolytic nature of astrocytes (Bittner et al., 2010; Itoh et al., 2003) and their need for lactate extrusion is consistent with their expression of MCT4 (Bergersen, 2007; Rafiki, Boulland, Halestrap, Ottersen, & Bergersen, 2003). Neurons, on the other hand, specifically express MCT2 to facilitate lactate uptake for oxidative metabolism. However, it is important to note that in this system, expression of MCTs within astrocytes and neurons is polarized to specific structural processes. Astrocytes that are in close proximity to the vasculature

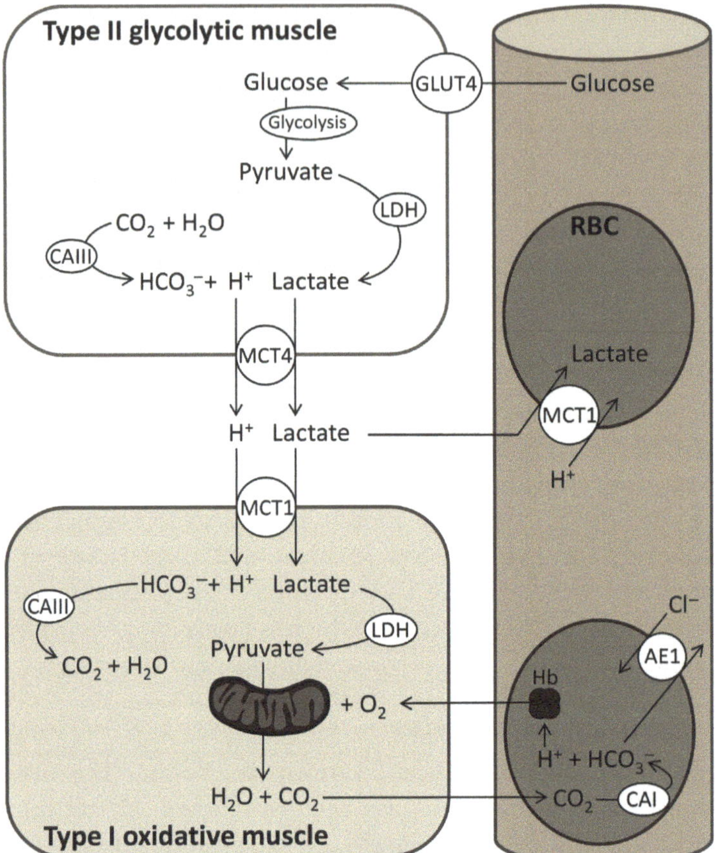

Figure 2 ***Metabolic coupling in skeletal muscle.*** Type II muscle cells are highly adapted for glycolysis. They express high levels of GLUT4 for efficient glucose uptake from the surrounding vasculature and subsequent conversion into lactic acid, which is transported out of the cell via MCT4. Muscle cells also express cytosolic CA III, which helps facilitate lactate transport. Type I muscle fibers, on the other hand, are highly oxidative and express MCT1 to mediate lactate uptake for oxidative metabolism. Excess lactate is cleared by the vasculature.

extend vascular processes (endfeet) that closely appose to intraparenchymal blood vessels. Within the confines of these structures, lactate transport out of the astrocytic vascular processes is mediated by MCT4 (Rafiki et al., 2003), and the transfer of lactate into the blood circulation is mediated by MCT1 on the endothelial surface of the blood vessels. Adjacent astrocytes interact with each other via gap junctions, which allows for the transfer of ions and glucose within the astrocytic network (astrocyte syncytium) (Rouach et al., 2008). Neuronal synapses are embedded within the astrocytic network and are ensheathed by fine perisynaptic astrocyte processes that are densely populated with MCT4. Within the neurons, MCT2 is specifically expressed in the postsynaptic terminals, where it is also host to a high density of mitochondria (Bergersen et al., 2001; Rafiki et al., 2003). This strategic placement of MCT2 allows lactate released from the adjacent astrocytes to be directly utilized for oxidative metabolism, which is needed to support the high metabolic demands of postsynaptic action potentials (MacAskill & Kittler, 2009)—a process that is responsible for a large fraction of total energy utilization in the brain (Attwell & Gibb, 2005).

2.3. Kidney and Liver

As demonstrated in muscle and in the brain, the differential expression of MCT isoforms correlates with their roles within different cell types where they facilitate lactate transport in the direction that can be predicted by the metabolic profile of the cells. While type I muscles and neurons express MCT1 and MCT2 to facilitate lactate uptake to support metabolism, liver hepatocytes express MCT2 to mediate uptake of lactate as a substrate for gluconeogenesis (Garcia et al., 1995). This function is particularly important at times of physical activity, during which blood lactic acid levels can increase from 1 mM (at rest) up to 14–18 mM (Greenhaff et al., 1994). In addition to the liver, the kidney also participates in lactate gluconeogenesis and is responsible for the clearance of approximately 30% of blood lactate (reviewed in (Bellomo, 2002)). This process, however, occurs exclusively in the polarized epithelium of the PCT (Schmid et al., 1977). Its apical membrane is adorned with numerous microvilli to enhance the reabsorption of ions, glucose, and amino acids from the glomerular filtrate. Its basolateral membrane is directly adjacent to the peritubular capillaries, which carry away reabsorbed ions and amino acids. The PCT reabsorbs lactate from the glomerular filtrate via sodium-linked monocarboxylate transporters (SMCT1 and SMCT2) at its apical membrane, and extrudes lactate into the peritubular capillaries via MCT1 at the basolateral membrane (Becker et al., 2010; Gopal et al., 2007). Although lactate extrusion is more commonly mediated by MCT4, the need of the PCT to reabsorb all lactate from the glomerular filtrate makes MCT1, with its higher lactate affinity, a better candidate for this role. However, at times of intense physical activity, when blood lactate can reach up to 18 mM, the increased lactate load in the PCT can easily saturate the low maximal transport capacity of MCT1 ($V_{max} \approx 5$ mM), resulting in lactic acid accumulation in PCT cells. In the event of intracellular acidosis, the PCT can adapt by increasing gluconeogenic activity to convert more lactate into glucose (reviewed in (Bellomo, 2002)), which is subsequently released into the peritubular blood supply via GLUT2.

2.4. Photoreceptors and Retinal Pigment Epithelium

The retina is among the most glycolytically active tissues in the body, converting more than 80% of all glucose that it consumes into lactic acid (Wang, Kondo, & Bill, 1997; Wang, Tornquist, & Bill, 1997; Winkler, Starnes, Twardy, Brault, & Taylor, 2008). However, not all cells in the retina are glycolytic. Photoreceptors are dependent on oxidative metabolism to satisfy the high energy demands of visual phototransduction. This is facilitated by the high levels of GLUT1 and MCT1 at photoreceptor inner segments (Gospe, Baker, & Arshavsky, 2010; Philp et al., 2003), which also contain the highest density of mitochondria (Stone, van Driel, Valter, Rees, & Provis, 2008). Müller glia cells, on the other hand, are the most glycolytically active cells in the retina, consistent with their high expression of MCT4 (Chidlow, Wood, Graham, & Osborne, 2005). As shown in Fig. 3, the respective localization of MCT1 and MCT4 at the photoreceptors and the neighboring Müller cells suggests the existence of a lactate-shuttle mechanism similar to that in muscle and brain. In support of this model, studies have demonstrated that Müller cells metabolize glucose rapidly to generate lactate (Poitry, Poitry-Yamate, Ueberfeld, MacLeish, & Tsacopoulos, 2000; Winkler, Arnold, Brassell, & Puro, 2000), which upon its extrusion into the subretinal space (SRS) via MCT4, can be taken up by photoreceptor cells for oxidative metabolism (Poitry-Yamate, Poitry, & Tsacopoulos, 1995).

The retinal pigment epithelium (RPE) is a monolayer of highly pigmented cells that form the outer blood–retina barrier and performs many critical functions that support photoreceptor health and integrity (reviewed in (Strauss, 2005)). Anatomically, the RPE lies between the photoreceptor cells and their main blood supply (choroidal vasculature), where it plays a fundamental role in SRS pH homeostasis by transporting excess lactic acid and other metabolic wastes (i.e. CO_2 and H_2O) from the SRS to the choroidal blood supply. Each photoreceptor outer segment is ensheathed by the numerous microvillis that extend from the RPE apical membrane. This anatomical configuration significantly enhances the total surface area for lactate uptake by the RPE, which express high levels of MCT1 at its apical membrane to facilitate this process (Philp, Yoon, & Grollman, 1998). Upon entry into the cell, lactate may exit the RPE basolateral membrane via MCT3 into the choroidal blood vessels (Philp et al., 1998; Philp, Yoon,

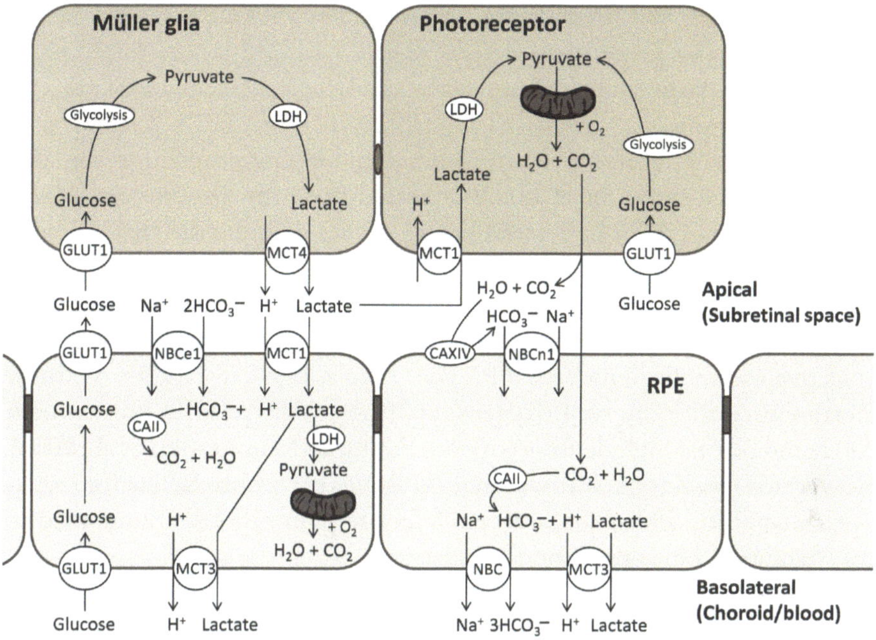

Figure 3 ***Metabolic coupling in the retina and the role of the retinal pigment epithelium (RPE) in lactate and pH homeostasis.*** Müller cells are highly glycolytic. They express high levels of GLUT1 to mediate rapid uptake of glucose, which is converted into lactic acid that is subsequently extruded into the subretinal space via MCT4. Lactate in the subretinal space can be taken up by either photoreceptor or RPE cells. Unlike Müller cells, photoreceptor cells possess a high capacity for both glycolysis and oxidative metabolism. Photoreceptor cells take up glucose and lactic acid at their inner segment via GLUT1 and MCT1, respectively, and convert them to pyruvate for entry into the TCA cycle. Excess lactate in the subretinal space is cleared by the RPE, which mediates transepithelial lactate transport to the choroidal blood vessels. Lactate entry across the apical membrane is mediated by MCT1, and intracellular lactate may either be converted to pyruvate for oxidative metabolism or be transported out of the basolateral membrane via MCT3. Experimental evidence supports an acid–base-coupled mechanism between MCT1 and NBCs (NBCe1 and NBCn1) at the apical membrane, and this process is facilitated by cytosolic CA II. At the basolateral membrane, MCT3 may be acid–base coupled to an electrogenic NBC, the activity of which is driven by HCO_3^- that is generated from intracellular CO_2, as catalyzed by CA II.

& Lombardi, 2001). This transepithelial lactate transport helps prevent SRS acidosis, which could compromise photoreceptor health and function. The importance of MCT3 in regulating transepithelial lactate transport can be inferred from studies of the MCT3 knockout mouse, which presented with impaired retinal function and abnormally high levels of lactate (≈fivefold) in the outer retina (Daniele, Sauer, Gallagher, Pugh, & Philp, 2008). These observations are consistent with earlier studies of transepithelial lactate transport, which showed that the flux of [^{14}C]-labeled lactate tracer across native bovine RPE can be influenced by the lactate gradient across the epithelium (Kenyon et al., 1994). Furthermore, in vivo studies demonstrated that changes in photoreceptor metabolism and lactate production can alter the lactate concentration of blood collected from the vortex vein (Wang, Tornquist et al., 1997). Collectively, these studies demonstrate that a significant fraction of lactate that enters the RPE from the apical membrane is transported out of its basolateral membrane. It is also of interest to consider the possibility that lactate may be utilized in the RPE as a substrate for oxidative metabolism. Indeed, RPE cells have a high capacity for oxidative metabolism—they are well oxygenated (Wangsa-Wirawan & Linsenmeier, 2003) and possess high densities of mitochondria (Maminishkis et al., 2006). However, it remains unclear whether RPE cells can utilize lactate for oxidative metabolism, and if so, how much lactate is metabolized compared to that transported into the blood circulation.

2.5. Sertoli and Germ Cells

In the male reproductive system, sperm development occurs in the seminiferous tubules, which are the main functional units of the testis and are lined with a monolayer of Sertoli cells. Sertoli cells form a polarized epithelium with tight junctions that serve as the blood–testis barrier (BTB), but they also play a particularly important role in spermatogenesis (Rato et al., 2012). At any given time, a single Sertoli cell is a host to 30–50 germ cells at various developmental stages, from the newly differentiated spermatogonia, multiple intermediate stages of spermatocytes (meiosis), to the immature spermatozoa (mid-piece and tail development) (Cheng, Wong, Yan, & Mruk, 2010). Sertoli cells extend lateral cytoplasmic protrusions that ensheathe each germ cell, establishing a large surface area that allows for efficient transfer of nutrients to nourish the developing germ cells. As germ cells progress through the different stages of development, they migrate across the BTB and ascend the lateral junctions of Sertoli cells toward the tubule lumen (Cheng & Mruk, 2009). During this process, they undergo a shift in cellular metabolism to become increasingly dependent on lactate, instead of glucose, as the primary substrate. Spermatogonia localized outside the BTB utilize glucose, but meiotic and postmeiotic germ cells within the BTB rely almost exclusively on lactate for cellular metabolism (Boussouar & Benahmed, 2004; Rato et al., 2012). It is well established that germ cells derive their source of lactate from Sertoli cells, and that the transfer of lactate is facilitated by their complementary expression of MCTs. Sertoli cells express MCT4 to release lactate, which is subsequently taken up by differentiating germ cells via MCT1 and MCT2 (Galardo, Riera, Pellizzari, Cigorraga, & Meroni, 2007; Mannowetz, Wandernoth, & Wennemuth, 2012). With the established importance of lactate in germ cell metabolism, one may predict that defects in this lactate-shuttle mechanism would block sperm maturation and cause male sterility. Indeed, these phenotypes are observed in the Bsg-null mice (Igakura et al., 1998), which lack MCT1 and MCT4 expression because of the loss of CD147.

2.6. Cancer and Stroma

Although lactate transport via MCT1 and MCT4 is important in metabolism in a variety of biological systems under normal physiological conditions (Section 2.1–2.5), MCT1 and MCT4 are commonly overexpressed by cancer cells to maintain intracellular lactate and pH homeostasis. Unlike muscle or neuronal cells, however, cancer cells are derived from random genetic mutations that confer proliferative, metabolic, and survival advantages to the developing cancer cell (reviewed in (Hanahan & Weinberg, 2011)). Although cancers may arise from different genomic mutations, a near universal phenotype observed among cancer cells is their ability to metabolize glucose rapidly via glycolysis, even in the presence of oxygen. This phenomenon, recognized as the Warburg effect, can be induced by various oncogenic pathways that are commonly activated in cancers (reviewed in (Ferreira, Hebrant, & Dumont, 2011; Gatenby & Gillies, 2004)). For example, loss of P53 function, a tumor suppressor that is frequently mutated in cancer (Olivier, Hussain, Caron de Fromentel, Hainaut, & Harris, 2004), results in increased glycolysis due to upregulation of GLUT1 (Zawacka-Pankau et al., 2011) as well as downregulation of the TP53-inducible glycolysis regulator (TIGAR) (Bensaad et al., 2006). In many cancers, glycolytic activity is enhanced by ectopic expression of c-MYC, which also inhibits oxidative respiration by inducing the expression of LDHA and pyruvate kinase (reviewed in (Ferreira et al., 2011; Gordan, Thompson, & Simon, 2007)). Other cancers have found various ways to hijack the hypoxia-inducible factor 1 (HIF-1) machinery, which also controls the expression of proteins involved in glycolysis, such as GLUT1 and hexokinases (HK1 and HK2) (reviewed in (Semenza, 2010)). These examples demonstrate that various genetic mutations can alter the metabolism of cancer cells via different mechanisms, but the phenotypic endpoint is usually the transformation of cells into a highly proliferative cell type that relies extensively on glycolysis for cellular metabolism. With upregulation of glycolysis comes increased production of lactic acid, thus necessitating an effective lactate extrusion mechanism to avoid lactic acidosis, which could act in a feedback mechanism to inhibit glycolysis (Chen et al., 2008). In this context, many cancer cells overexpress MCT1 (reviewed in Pinheiro et al., 2012), which may be initiated by the loss of P53 tumor suppressor (Boidot et al., 2012) or overexpression of MYC (Kang, Im, Go, & Han, 2009) or RAS oncogenes (Le Floch et al., 2011). MCT4 is also commonly upregulated in cancer, and its expression is regulated by HIF-1α (Gallagher, Castorino, Wang, & Philp, 2007; Le Floch et al., 2011; Ullah, Davies, & Halestrap, 2006). With established roles of MCTs in lactate extrusion and pH homeostasis, it is perhaps not surprising that blocking MCT1 and MCT4 in cancer cells, either by RNAi or small molecule inhibitors was able to inhibit glycolysis and slow tumor growth (Le Floch et al., 2011).

Over the course of cancer progression, cancer cells lose the ability to check for spontaneous genetic mutations, and the increased rate of mutation allows for rapid generation of genetically different cancer cell subpopulations that are constantly engaged in a fight for survival—a process analogous to Darwinian evolution, but only in the microscopic scale (Bernards & Weinberg, 2002; Johnston, 2009). However, recent studies have demonstrated that these cancer subpopulations can also function in a symbiotic relationship. In this model, hypoxic (HIF-1α positive) cells at the poorly vascularized tumor core metabolize glucose into lactic acid, which is subsequently taken up by well-oxygenated cells at the tumor periphery (Fig. 4) (Sonveaux et al., 2008). In support of this model, an in vivo study of tumor metabolism utilizing metabolic imaging techniques (DCE-MRI & FDG-PET) demonstrated that glucose uptake in well-vascularized tumor periphery regions is lower than in the poorly vascularized and hypoxic tumor core (Galie et al., 2007). In addition, cells at the tumor periphery expresses high levels of MCT1 to facilitate lactate entry for oxidative metabolism (Sonveaux et al., 2008). This process was verified in the same study by NMR evaluation of [^{13}C]-lactate conversion into [^{13}C]-glutamate by tumor cells. Other studies demonstrated that MCT4, along with GLUT1 and LDHA, was expressed at high levels in hypoxic tumor regions further away from the vasculature (Meijer et al., 2011; Rademakers, Lok, van der Kogel, Bussink, & Kaanders, 2011), consistent with HIF-1α-induced expression of MCT4 and other glycolytic genes (reviewed in (Semenza, 2010)). In light of the Warburg effect, the notion of cancer cells metabolizing lactate via oxidative respiration may seem almost paradoxical, but this

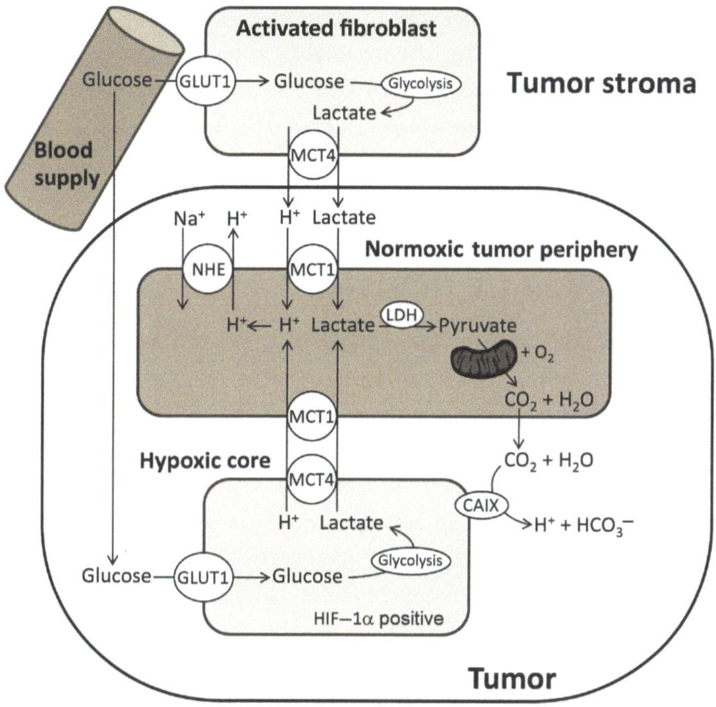

Figure 4 ***Inter-tumor lactate shuttle and tumor-stromal interactions.*** The tumor core, being furthest away from the blood supply, is hypoxic—a physiological condition that promotes the stabilization of HIF-1α, a transcription factor that regulates GLUT1 and MCT4 expression, as well as other proteins involved in glycolysis. As such, cells in the hypoxic core generate and extrude lactate that is subsequently taken up by the cells at the normoxic tumor periphery that express high levels of MCT1. These tumor periphery cells are closer to the blood supply, and are therefore well supplied with oxygen needed to metabolize lactate via the TCA cycle. Tumor cells also receive their lactate supply from fibroblast cells in the tumor stroma. In this scheme, tumor cells secrete growth factors to activate adjacent stromal fibroblast cells, turning them into a glycolytic powerhouse that generates lactate, which is subsequently released, via MCT4, into the tumor microenvironment.

phenomenon is observed in less-invasive cancer cell lines, such as SiHa (cervical cancer) (Sonveaux et al., 2008) and MCF-7 (breast cancer) (Hussien & Brooks, 2010; Whitaker-Menezes et al., 2011). The ability of tumor cells to utilize lactate for oxidative metabolism is also highlighted in recent studies of metabolic coupling in tumor–stromal interactions (Pavlides et al., 2009; Whitaker-Menezes et al., 2011; Witkiewicz et al., 2012). In this model, also illustrated in Fig. 4, activated fibroblasts in the tumor stroma maintain high rates of aerobic glycolysis and generate lactate and ketone bodies to fuel the oxidative metabolism of tumor cells (Pavlides et al., 2009). Consistent with their glycolytic phenotype, tumor stromal fibroblasts express high levels of MCT4 as observed in cancer-fibroblast cocultures and in human breast cancer samples (Whitaker-Menezes et al., 2011). The oxidative tumor cells, on the other hand, express high levels of MCT1.

3. INTERACTIONS BETWEEN MCTS AND BICARBONATE TRANSPORT MECHANISMS

MCT-mediated transfer of lactate between cells and their microenvironment often involves large acid loads that can inadvertently disrupt cellular function. To avoid such an event, most biological systems employ a set of pH-regulatory systems to help buffer transient or chronic changes in intracellular pH. As in other mammals, CO_2/HCO_3^- buffering is the major pH-regulatory system in humans, and it is mediated by the activities of carbonic anhydrases (CAs) and HCO_3^- transporters (reviewed in (Casey, 2006)). CAs are zinc metalloenzymes that catalyze the interconversion of CO_2 and HCO_3^- to equilibrium (reviewed in (Supuran, 2008)). To date, 16 isoforms of CAs have been identified. Each member can be distinguished from others by its enzymatic reaction rate and its cellular localization in the cytosol (CA I, CA II, CA III, CA VII, CA XIII), plasma membrane (CA IV, CA IX, CA XII, CA XV), or mitochondria (CA VA, CA VB). Alternatively, CAs may also be secreted (CA VI). Intracellular CAs respond to an acidic challenge by neutralizing the H^+ load with HCO_3^-, generating CO_2 that can easily diffuse across the lipid bilayer and out of most cells. The resultant decrease in intracellular HCO_3^- creates a concentration gradient for HCO_3^- transport into the cell via Na^+-linked HCO_3^- transporters (e.g. Na^+/HCO_3^- cotransporters or NBCs, and Na^+-driven Cl^- HCO_3^- exchangers or NCBE) or Cl^-/HCO_3^- anion exchangers (AEs) (reviewed in (Cordat & Casey, 2009)). In humans, two of the four NBCs (NBCe1 (SLC4A4) and NBCe2 (SLC4A5)) are electrogenic transporters that deliver one Na^+ with two or three HCO_3^- (reviewed in (Boron, Chen, & Parker, 2009)). This property endows these transporters with high sensitivity to changes in intracellular HCO_3^-, and also allows them to be driven by the membrane potential against a large inward Na^+ gradient (Becker & Deitmer, 2004). The other NBCs (NBCn1 (SLC4A7) and NBCn2 (SLC4A10)) and AEs (AE1, AE2, and AE3), on the other hand, being electroneutral transporters, are controlled by the Na^+ and Cl^- gradient, respectively; thus, they normally mediate HCO_3^- transport into the cell.

Although CAs and HCO_3^- transporters are well-established components of the CO_2/HCO_3^- buffering mechanism, work from several laboratories also demonstrated protein–protein interactions between CAs and NBCs (Alvarez, Loiselle, Supuran, Schwartz, & Casey, 2003; Loiselle, Morgan, Alvarez, & Casey, 2004; Orlowski, de Giusti, Morgan, Aiello, & Alvarez, 2012), as well as between CAs and AEs (Casey, Sly, Shah, & Alvarez, 2009; Morgan, Pastorekova, Stuart-Tilley, Alper, & Casey, 2007; Sterling, Alvarez, & Casey, 2002; Sterling, Reithmeier, & Casey, 2001; Svichar et al., 2009; Vince & Reithmeier, 2000). The same authors also demonstrate that these interactions could enhance HCO_3^- transport (Schueler, Becker, McKenna, & Deitmer, 2011; Sterling et al., 2001), giving rise to the notion that physical interaction between CAs and HCO_3^- transporters would place them in close proximity to form a "bicarbonate transport metabolon," a configuration that allows for optimal HCO_3^- transport activity. However, this "metabolon" theory has been challenged by a study designed to test the functional interaction between CA II and NBCe1 in the *Xenopus* oocyte system, from which the authors concluded that CA II could not enhance NBCe1 activity, even when the two proteins are experimentally fused (Lu et al., 2006). The controversy may derive from comparing the effect of CA II on NBCe1 activity under different experimental conditions—where Alvarez et al. (2003) subject cells to large pH loads to stimulate NBCe1, Lu and colleagues evaluated steady-state electrical conductance of NBCe1 at normal physiological pH. These observations suggest that the role of CAs in HCO_3^- transport may be more important under conditions of increased metabolic activity.

Since MCT-mediated lactate transport is H^+ linked and may therefore be enhanced by CA activity, one may ask if a physical or functional interaction could exist between CAs and MCTs. In this regard, studies by Becker and colleagues demonstrated that CA II or CA IV can directly bind to MCT1, MCT2, or MCT4 and enhance lactate transport although this effect was achieved via a mechanism independent of CA activity (Becker & Deitmer, 2008; Becker, Hirnet, Fecher-Trost, Sultemeyer, & Deitmer, 2005; Becker, Klier, Schuler, McKenna, & Deitmer, 2011; Klier et al., 2011). In addition, MCTs can also functionally interact with NBCs in an acid–base coupling mechanism that enhances the activities of both MCT and NBC (Becker, Broer, & Deitmer, 2004). In this system, the influx of H^+/lactate via MCT1 induces intracellular acidification and activates NBC-mediated HCO_3^- transport, which buffers the accumulating H^+ at the inner surface of the plasma membrane. Although a direct protein–protein interaction between MCTs and NBCs would place them in close proximity that could further enhance lactate transport, such an interaction has yet to be found. Nonetheless, recent studies on the corneal endothelium demonstrate that lactate transport via MCT1 is facilitated by the activities of NBCe1 and CA II, suggesting that a functional interaction can occur between MCTs, NBCs, and CAs (Nguyen & Bonanno, 2011, 2012). Because other biological systems such as the PCT in the kidney, RPE in the eye, and muscles in the heart also coexpress MCTs with various isoforms of HCO_3^- transporters (NBCs and AEs) and carbonic anhydrases, a tripartite interaction could exist between these proteins that allows them to function in a synergistic manner. On the basis of the localization and expression of different isoforms of MCTs and HCO_3^- transporters in various biological systems (reviewed in (Cordat & Casey, 2009)), we can predict how MCTs and HCO_3^- transporters could interact with each other to perform a specific function in each individual system.

4. PHYSIOLOGY AND FUNCTION OF MCTS 1–4 IN EPITHELIAL AND ENDOTHELIAL CELLS

4.1. Retinal Pigment Epithelium: pH Homeostasis and Metabolic Acid Removal

As previously discussed (Section 2.4), the neural retina is a metabolically robust tissue that generates large quantities of both CO_2/HCO_3^- and lactic acid, both of which are released into the photoreceptor extracellular milieu (SRS) where they are subsequently removed by the RPE and transported into the choroidal blood supply. Earlier studies demonstrated that CO_2 and lactate levels in the SRS could reach up to 10% and 13 mM, respectively, both of which are significantly higher than that in the choroidal blood vessels (5% and 1 mM, respectively) (Adler & Southwick, 1992; Adijanto, Banzon, Jalickee, Wang, & Miller, 2009). As the RPE is exposed to large loads of CO_2 and lactate on a normal basis, it requires an efficient transport system that may utilize HCO_3^- and lactate transporters in a synergistic manner. Earlier studies on the RPE have established the presence and membrane-specific localization of HCO_3^- transporters (NBCs and AEs), which have important roles in pH buffering (reviewed in (Hughes, Gallemore, & Miller, 1998)). The RPE expresses the electrogenic NBCe1 (1:2, Na^+/HCO_3^-) and electroneutral NBCn1 at its apical membrane (Adijanto et al., 2009). Under normal physiological conditions, these transporters mediate HCO_3^- transport into the cell, driven by the strong Na^+-gradient across the

RPE apical membrane (125–130 mM). The activities of these transporters are facilitated by the catalytic activities of CA II in the cytosol (Adijanto et al., 2009) and CA XIV at the apical membrane (Nagelhus et al., 2005). Thus, the apical localization of MCT1, NBCe1 & NBCn1, and CAII & CA XIV suggests that these transporters and enzymes could work synergistically as an acid–base coupled "lactate/bicarbonate transport metabolon." At the RPE basolateral membrane, HCO_3^- efflux is mediated by a Cl^-/HCO_3^- exchanger, AE2, and an electrogenic NBC that has a 1:3 (Na^+/HCO_3^-) stoichiometry, which allows the basolateral membrane potential (~−55 mV) to drive Na^+/HCO_3^- efflux against the strong inward Na^+ gradient. As illustrated in Fig. 3, the strategic placement of these HCO_3^- transporters and MCTs at the apical and basolateral membranes allows the RPE to couple transepithelial lactate transport with HCO_3^- transport. It is important to realize that transepithelial lactate and HCO_3^- transport carries a broader physiological significance than simply a mechanism for metabolic acid clearance. Earlier studies of the RPE demonstrated that transepithelial Na^+/HCO_3^- transport is a significant driving force for transepithelial fluid transport, a critical process that maintains photoreceptor-RPE adhesion and preserves photoreceptor health (Adijanto et al., 2009). In light of the close association between HCO_3^- and fluid transport, one may ask if lactate could drive fluid transport by interacting with HCO_3^- transporters. Consistent with this notion, earlier studies by Hamann and colleagues demonstrated that lactate entry across the RPE apical membrane could induce cell swelling (Hamann, la Cour, Lui, Bundgaard, & Zeuthen, 2000; Hamann, Kiilgaard, la Cour, Prause, & Zeuthen, 2003). It follows that fluid entry across the RPE apical membrane would be balanced by an equivalent fluid efflux from the basolateral membrane. However, such an effect of lactate on transepithelial fluid transport has not been demonstrated.

4.2. Corneal Endothelium: Fluid Transport

The role of MCTs in transepithelial fluid transport may perhaps be better appreciated in the cornea as studies demonstrate that the accumulation of lactate and fluid within the corneal stroma can easily disrupt corneal transparency (Klyce, 1981; Riley & Winkler, 1990). The stroma is the thickest component of the cornea, representing ~90% of cornea thickness, and is sandwiched between the surface corneal epithelium and the corneal endothelium, separated only by their respective basement membranes. The stroma contains glycosaminoglycans (GAGs) that are embedded within multiple layers of structurally uniform and highly organized collagen fibers, the integrity of which is critical for corneal transparency. GAGs are highly hydrophilic in nature, and it is this property that bestows the stroma with the ability to maintain hydration by passively drawing fluid from its surrounding. With this ability, however, comes a natural tendency to swell excessively, which is followed by a concomitant loss of corneal transparency. Under normal physiological conditions, this potentially catastrophic event is averted by the action of the corneal endothelium, which maintains normal corneal thickness by actively transporting fluid from the stroma to the anterior chamber. As illustrated in Fig. 5, this process is driven by transepithelial HCO_3^- transport (reviewed in (Bonanno, 2003, 2012)), which is mediated by the pancreatic variant of NBCe1 (or NBCe1-B; 1:2 Na^+:HCO_3^- stoichiometry) and AE2 at the basolateral membrane. The major HCO_3^- transport mechanism at the apical membrane is less clear although several lines

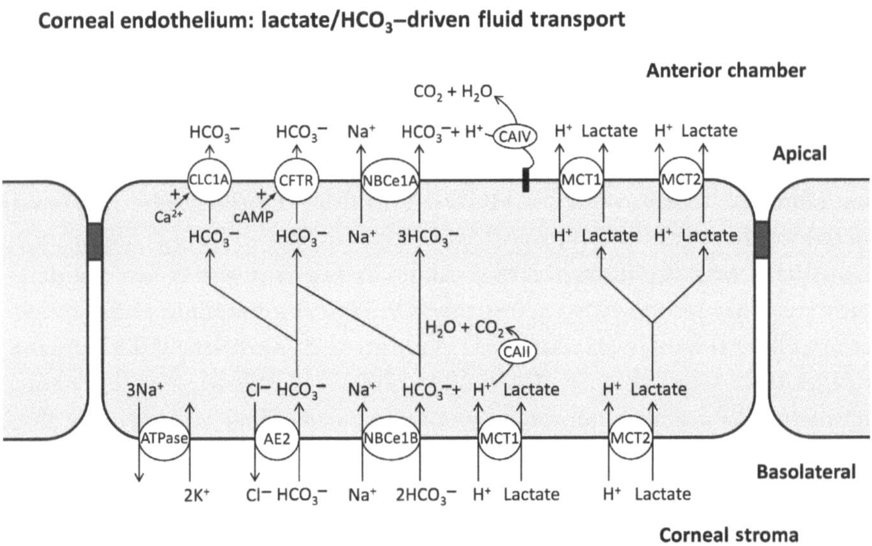

Figure 5 ***Lactate/HCO$_3^-$-driven fluid transport in corneal endothelium.*** The primary responsibility of the corneal endothelium is to dehydrate the corneal stroma and maintain corneal transparency. This dehydration process is mediated by transendothelial HCO_3^- transport, a mechanism that is composed of various HCO_3^- transporters and exchangers, functioning in a synergistic manner with MCT-mediated lactate transport. At the basolateral membrane, HCO_3^- entry from the corneal stroma is mediated by a Cl^-/HCO_3^- exchanger (AE2) and the pancreas isoform of NBCe1 (or NBCe1-B; 1:2 Na^+/HCO_3^- stoichiometry). At the apical membrane, HCO_3^- efflux is mediated by the kidney isoform of NBCe1 (or NBCe1-A; 1:3 Na^+/HCO_3^- stoichiometry) and anion channels (Ca^{2+}-activated ClC1A and cAMP-activated CFTR), and their activities may be facilitated by cytosolic CA II and the extracellular CA IV. Lactate is transported across the corneal endothelium via MCT1 and MCT2 at both its apical and basolateral membranes.

of evidence implicate the kidney variant of NBCe1 (or NBCe1-A; 1:3 stoichiometry) (Diecke, Wen, Sanchez, Kuang, & Fischbarg, 2004) and anion channels (CLC1A and CFTR) in this role (Li, Allen, Sun, Cui, & Bonanno, 2008; Zhang, Li, Xie, & Bonanno, 2006; Zhang, Xie, Sun, & Bonanno, 2002). The large volumes of fluid exchanged across cells in the cornea are matched by a high glycolytic activity, with more than 80% of all glucose consumed in the cornea being converted into lactic acid. This metabolic profile is a direct consequence of the low mitochondrial density in the cornea—a sacrifice of energy efficiency for transparency. As a result, there exists a high lactate gradient from the cornea (~13 mM) to the anterior chamber (~7 mM) (Klyce, 1981), and the removal of lactate is the responsibility of the corneal endothelium, which expresses MCTs to perform this important task. However, the localization and identity of these MCTs remain controversial, which may be due to comparisons across different species. In both mouse (unpublished data) and rat corneal endothelium (Chidlow et al., 2005), MCT1 and MCT2 exhibit unpolarized localization to both membranes, whereas MCT4 is absent. In bovine corneal endothelium, MCT2 is also found at both membranes, but MCT1 and MCT4 are reportedly found at the basolateral and apical membranes, respectively (Nguyen & Bonanno, 2011). Perhaps more interesting is the demonstration that MCT1 at the basolateral membrane is functionally dependent on the activities of NBCe1-A and cytosolic CAs, thus suggesting that these proteins may function synergistically to mediate acid–base coupled lactate/HCO_3^- transport (Nguyen & Bonanno, 2011). Although one may postulate the existence of a similar lactate/HCO_3^- transport system at the apical membrane that involves MCT1 (or MCT2), NBCe1-B, and CAs (cytosolic CA II and extracellular CA IV), such a mechanism has yet to be demonstrated. Synergistic interactions between the two transport mechanisms would enhance lactate-driven fluid transport across the corneal endothelium, and this process may therefore have an important role in maintaining proper corneal thickness and transparency.

4.3. Proximal Convoluted Tubule: Acid–Base Balance

Because AEs, NBCs, and CAs are almost ubiquitously expressed, the concept of "bicarbonate transport metabolon" may find applications in lactate shuttle, HCO_3^- and fluid transport, and acid–base balance in many biological systems. The role of HCO_3^- transport in acid–base balance is perhaps most important in the kidney, which is responsible for the regulation of CO_2/HCO_3^- balance and pH in the blood (reviewed in (Boron, 2006)).

The PCT is the main site for HCO_3^- reabsorption, transferring approximately 80% of all HCO_3^- in the glomerular filtrate back into the blood circulation. As illustrated in Fig. 6, CA IV at the apical surface of the PCT converts HCO_3^- into CO_2, which subsequently diffuses into the cell. The PCT expresses high levels of cytosolic CA II, which converts the entering CO_2 back to H^+ and HCO_3^-. Intracellular H^+ is transported out of the apical membrane via V-type H^+ ATPases and the Na^+/H^+ exchanger (NHE3), whereas the HCO_3^- is transported across the basolateral membrane via the electrogenic Na^+/HCO_3^- cotransporter NBCe1-A (using a 1:3 Na^+/HCO_3^- stoichiometry). Together, these enzymes and transporters

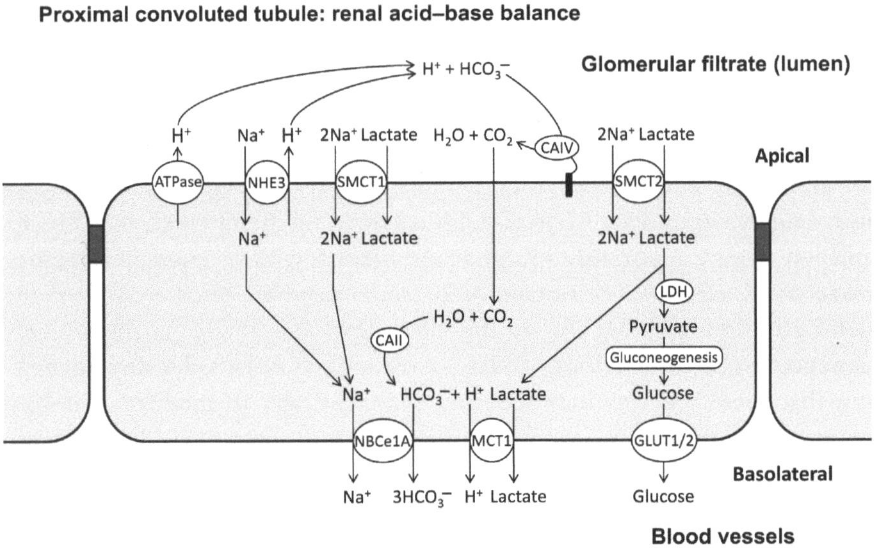

Figure 6 ***Acid–base balance in the proximal convoluted tubule (PCT).*** Of all segments in the renal tubule, the PCT is where most HCO_3^- and lactate from the glomerular filtrate are reabsorbed. In addition, the PCT is one of the only two sites (the other being the liver) in the human body where gluconeogenesis can occur. The process of lactate reabsorption from the glomerular filtrate (facing the apical membrane) is mediated by Na^+-linked lactate cotransporters (SMCT1 and SMCT2). The high Na^+-gradient needed to drive lactate uptake is sustained by the Na^+/K^+ ATPase at the basolateral membrane. Intracellular lactate can either be converted to glucose via gluconeogenesis or be transported across the basolateral membrane via MCT1 into the blood vessels. Unlike lactate, however, there are no mechanisms for carrier-mediated HCO_3^- transport at the apical membrane. Instead, HCO_3^- is converted by CA IV to its nonionic form, CO_2, which can passively diffuse into the cell. Intracellular CO_2 is rapidly reconverted by cytosolic CA II to H^+ and HCO_3^-, the latter of which is transported out of the basolateral membrane into the blood vessels via NBCe1-A (1:3 Na^+/HCO_3^- stoichiometry). Excess H^+ is transported out of the apical membrane via a V-type H^+-ATPase and a Na^+/H^+ exchanger (NHE3).

orchestrate net transepithelial Na^+ and HCO_3^- transport into the blood stream. Mutations in any one of these proteins cause major defects in renal function. Patients with mutations in NBCe1 present with permanent isolated proximal renal tubular acidosis (pRTA), and have significantly lower blood pH (≈7.1–7.2) and HCO_3^- levels (≈5–13 mM) compared to normal (pH ≈ 7.4; HCO_3^- ≈ 23 mM) (Igarashi et al., 1999, 2001; Inatomi et al., 2004). Similarly, patients with CA II deficiency also present with pRTA (Roth, Venta, Tashian, & Sly, 1992; Sly, Hewett-Emmett, Whyte, Yu, & Tashian, 1983). These severe renal function defects are also replicated in mice carrying genetic mutations in CA II or NBCe1-A (Gawenis et al., 2007; Lien & Lai, 1998; Lo et al., 2011). These studies demonstrate that a functional interaction between CA II and NBCe1-A in the PCT is essential to HCO_3^- reabsorption, which may be further enhanced by a direct protein–protein interaction between CA II and NBCe1-A (Gross, Pushkin, Abuladze, Fedotoff, & Kurtz, 2002; Pushkin et al., 2004). As previously discussed (Section 2.3), the PCT is also a major site for lactate reabsorption, driven by Na^+-linked lactate transporters (SMCT1 and SMCT2) at the apical membrane and MCT1 at the basolateral membrane. With the coexpression of MCT1 and NBCe1-A at the basolateral membrane, together with CA II recruited by NBCe1-A, it seems likely that these three proteins can form a functional complex. Furthermore, lactate entry at the apical membrane via SMCT1 and SMCT2 provides an additional supply of Na^+ to drive basolateral NBCe1-A activity. Thus, interactions between lactate and HCO_3^- transport mechanisms would allow lactate reabsorption in the PCT to stimulate HCO_3^- transport. From a teleological point of view, such a mechanism carries significant physiological importance during exercise, as it allows the increase in blood lactate to signal for an increase in HCO_3^- reabsorption in the kidney to help buffer the elevated levels of metabolic acids (i.e. CO_2 and lactic acid) in the blood, thus avoiding systemic acidosis.

4.4. Colon Epithelium: Short Chain Fatty Acid (SCFA) Transport

MCTs 1–4 carry out their physiological function by mediating lactate transport, but they are just as capable of transporting other nonlactate substrates. The colon epithelium is a good example of an epithelium that expresses MCTs primarily to mediate absorption of SCFAs (i.e. acetate, propionate, and butyrate) from the intestinal lumen. SCFAs are the product of carbohydrate fermentation by the large and diverse population of colonic bacteria that resides within a mucin- and glycoprotein-rich mucus layer, which lines and protects the surface of the colon epithelium. Of all SCFAs in the colon lumen, acetate is present at the highest concentration (~75 mM), followed by propionate and butyrate (~20 mM each). Among these three major SCFAs, butyrate is particularly important for colon health as it serves a dual function in colon epithelial cells, both as a primary metabolic substrate (Cummings, 1984) and as a molecular regulator of cellular differentiation (Scheppach, 1994). As illustrated in Fig. 7, SCFAs can enter the apical membrane of colon epithelial cells through the following ways: passive nonionic diffusion (Luciano, Reale, Rechkemmer, & von Engelhardt, 1984), Na^+-linked monocarboxylate cotransport (via SMCTs) (Iwanaga, Takebe, Kato, Karaki, & Kuwahara, 2006; Paroder et al., 2006), and SCFA/HCO_3^- exchange (Charney, Micic, & Egnor, 1998; von Engelhardt et al., 1994). Entry of SCFAs by

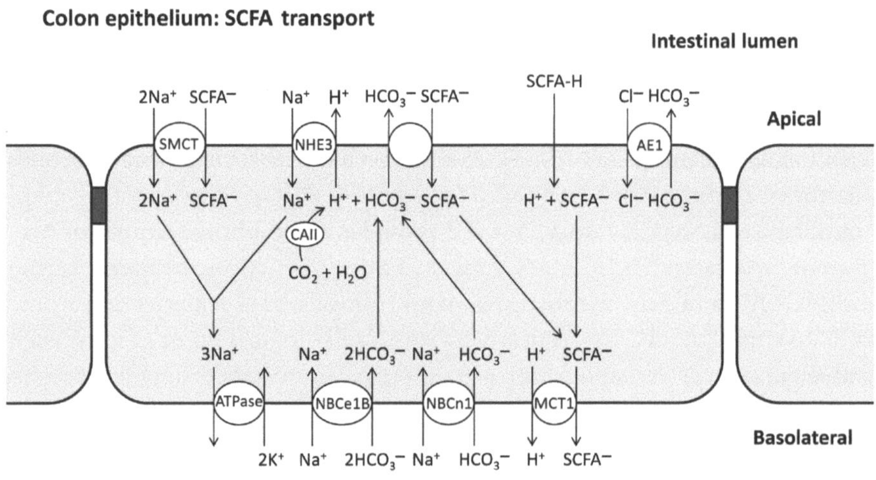

Figure 7 ***Short chain fatty acid transport by the colon epithelium.*** The colon epithelium is coated with a thick mucosal layer, which is host to a large and diverse population of bacteria that metabolize carbohydrates from the ileal effluent, generating short-chain fatty acids (SCFAs) such as acetate, propionate, and butyrate that are taken up by the colon epithelium for absorption into the blood circulation. SCFA can enter the apical membrane via three pathways: (1) nonionic diffusion of SCFA-H, (2) Na^+-linked SCFA cotransport (SMCT), or (3) SCFA/HCO_3^- exchange. The SCFA/HCO_3^- exchanger is functionally coupled to the Na^+/H^+ exchanger (NHE3), which maintains a favorable local HCO_3^- gradient by promoting the conversion of CO_2 to H^+ and HCO_3^-, the process of which is catalyzed by CA II. In addition, NBCe1 and NBCn1 at the basolateral membrane also provide HCO_3^- substrate to drive apical SCFA entry via the SCFA/HCO_3^- exchanger. Transport of SCFA across the basolateral membrane is mediated by MCT1.

either passive diffusion or SCFA/HCO_3^- exchanger causes intracellular acidification (Chu & Montrose, 1996), which in turn stimulates electroneutral Na^+ absorption via NHE2 and NHE3 at the apical membrane (Hoogerwerf et al., 1996; Sellin & De Soignie, 1998). At the basolateral membrane, MCT1 is the primary mechanism for SCFA transport into the colonic submucosal vasculature (Kirat et al., 2006; Ritzhaupt, Wood, Ellis, Hosie, & Shirazi-Beechey, 1998; Tamai et al., 1999), through which SCFAs are delivered directly to the liver via the portal system for gluconeogenesis.

In light of the interactions between MCTs and HCO_3^- transporters in other systems, it may be interesting to examine the role of HCO_3^- transport in SCFA absorption by the colon epithelium. The colon epithelium expresses a variety of HCO_3^- transporters, with AE1 and DRA (downregulated in adenoma) mediating Cl^-/HCO_3^- exchange at the apical membrane (Melvin, Park, Richardson, Schultheis, & Shull, 1999; Rajendran et al., 2000; Schweinfest et al., 2006), and Na^+-linked HCO_3^- cotransporters, NBCe1 and NBCn1, at the basolateral membrane (Feldman, 1994; Gawenis et al., 2007). The activities of these transporters, as is demonstrated in other epithelia, are facilitated by CAs (Lonnerholm, Selking, & Wistrand, 1985; Purkerson & Schwartz, 2005). From the localization of these HCO_3^- transporters, it may be readily apparent that the colon epithelium, unlike the RPE and PCT, is a HCO_3^--secreting epithelium. Because the colon epithelium transports HCO_3^- in the opposite direction as SCFAs, the acid–base coupling mechanism between MCTs and NBCs as previously described for the RPE, corneal endothelium, and PCT (see Sections 4.1–4.3) may not exist in this system. Nonetheless, HCO_3^- secretion is directly linked to SCFA absorption via the apical SCFA/HCO_3^- exchanger, the activity of which induces intracellular acidification (Sellin & De Soignie, 1998) that can promote basolateral SCFA efflux via MCT1 (Kirat & Kato, 2006). Furthermore, SCFA absorption across colon epithelium can be blocked by CA inhibitors (Kawamata, Hayashi, & Suzuki, 2007; von Engelhardt et al., 1994). Thus, CAs play an important role in transepithelial SCFA transport by maintaining favorable transmembrane HCO_3^- and pH gradients that facilitate the activities of the SCFA/HCO_3^- exchanger at the apical membrane and H^+/SCFA transport (via MCT1) at the basolateral membrane. In addition, CAs (cytosolic CA II in particular) may also directly bind and interact with MCT1 to enhance H^+/SCFA transport in a mechanism previously demonstrated by Becker et al. (2005).

5. FUTURE DIRECTIONS

Studies in the muscle and brain have established MCTs 1–4 as integral components of the lactate-shuttle network that are directly linked to cellular metabolism. With the demonstration that MCTs can form a functional complex with HCO_3^- transporters and carbonic anhydrases, it becomes clear that MCTs also participate in pH regulation and homeostasis, as well as transepithelial fluid transport in a variety of tissues, such as the kidney, cornea, and retina/RPE. Hence, it is not surprising that the lack of proper expression of MCT1, 3, and 4, as occurs in all tissues of the Bsg-null mice, is associated with major defects in visual function and spermatogenesis, among others. Although much more can be learned from studying the Bsg-null mice, one must be mindful of the fact that in this model, MCTs are absent throughout all stages of development. Furthermore, the absence of MCTs in all parts of the body could also cause global alterations in systemic acid–base balance and metabolism. All these factors promote the adaptation of various compensatory mechanisms, which may confound the analysis of MCT function. With these potential issues in mind, one may choose to develop a conditional MCT1 (or MCT4) knockout mouse model using cell-specific inducible-CRE promoters. A faster and cheaper alternative may be to take advantage of RNAi technology, which over the years has become increasingly potent and specific. Furthermore, the transfection of shRNA constructs in vivo via viral delivery (AAV or lentivirus), cationic lipid-based delivery (e.g. invivofectamine), and electroporation is increasingly becoming mainstream in biomedical research.

MCT1 and MCT4 have also garnered much attention from the cancer research community and private companies such as AstraZeneca, which saw the value in developing specific MCT inhibitors as anticancer drugs. A promising candidate (AR-C155858) with high sensitivity and specificity for MCT1 and MCT2 has recently been developed and is currently undergoing phase I/II clinical trial. This MCT1/2 inhibitor can also be a valuable tool to evaluate further the many roles of MCT1/2 in physiology and function, the knowledge of which is highly complementary to the design and development of drug-delivery strategies to target cancer cells specifically without adversely affecting the function of nearby tissues.

Another avenue of research focuses on evaluating the physiological substrates of other members of MCTs, which, unlike MCTs 1–4 and MCT8, have received little or no attention. Partly due to the lack of tools and good experimental readouts that allow for cost- and labor-efficient screening of candidate substrates. Furthermore, the crystal structure of not even one MCT has been solved, thus rendering *in silico* substrate prediction algorithms, such as those based upon thermodynamics or solute size, practically inapplicable. Alternatively, one may attempt to determine the function of MCTs based upon their expression in different biological systems and their subcellular localization, as has been done for MCTs 1–4. However, such analysis can be difficult to interpret because of the lack of specific antibodies that are suitable for immunocytochemistry. Nonetheless, recent years have seen the development of new technologies and services that can be used for high-throughput screening of potential MCT substrates. For example, one can use a water-soluble tetrazolium (WST)-based technology to compare the metabolism of an array of substrates between cells with differential expression of a specific MCT. However, the major limitation of this technology is the requirement that the MCT transport substrate can be used in metabolic pathways to generate NADH or NADPH. Alternatively, the analysis of cellular and extracellular metabolite levels can be achieved using gas or liquid chromatography coupled with mass spectrometry, which can be accessed as a service from a research institution or a private company. When used in properly designed experiments, this chromatography/spectrometry approach can yield a wealth of useful information and clues about the identity of the MCT substrate.

6. CONCLUSION

Since Poole and Halestrap published their first paper on the isolation of the erythrocyte lactate transporter (Poole & Halestrap, 1988), which was subsequently found to be identical to MCT1 cloned by Garcia, Goldstein, Pathak, Anderson, and Brown (1994), there has been renewed interest in these transporters and their function in metabolism in both health and disease. The cloning of MCT1 (SCL16A1) paved the way for identification of other members of the MCT family and the development of antibodies, siRNA, and inhibitors for studying their tissue-specific expression and activity. A second key finding in understanding the regulation of expression and trafficking of MCTs came when Halestrap and colleagues demonstrated that MCT1 and MCT4 require an accessory protein for its maturation and trafficking to the plasma membrane (Kirk et al., 2000). Their results explained why CD147 was expressed at high levels in blood–tissue barriers, where it plays important roles in maintaining the health and function of the retina, brain, testis, and immune system, as verified by studies of the Bsg-null mice. These findings, in conjunction with early physiological studies, have provided us with greater insight into the functional roles of MCTs 1–4 in the context of their cell-specific expressions and their contributions to metabolic coupling and homeostasis in various biological systems. However, the functions of other MCT isoforms have yet to be established. Nonetheless, in the past decade, we have witnessed the development of increasingly sophisticated tools for effective gene knockdown of MCTs and high-throughput assays to efficiently evaluate the roles of MCTs. These technologies can be used to evaluate the roles of MCTs 1–4 in animal models at the systemic level, as well as to identify the substrates and functions of other less-characterized MCTs.

ACKNOWLEDGMENTS

The authors thank Melissa Liu and Lea Hecht for reading the manuscript and their insightful comments.

This work was funded by NIH grant EY-012042.

REFERENCES

Adijanto, J., Banzon, T., Jalickee, S., Wang, N. S., & Miller, S. S. (2009). CO_2-induced ion and fluid transport in human retinal pigment epithelium. *Journal of General Physiology*, *133*(6), 603–622.

Adler, A. J., & Southwick, R. E. (1992). Distribution of glucose and lactate in the interphotoreceptor matrix. *Ophthalmic Research*, *24*(4), 243–252.

Alvarez, B. V., Loiselle, F. B., Supuran, C. T., Schwartz, G. J., & Casey, J. R. (2003). Direct extracellular interaction between carbonic anhydrase IV and the human NBC1 sodium/bicarbonate co-transporter. *Biochemistry*, *42*(42), 12321–12329.

Attwell, D., & Gibb, A. (2005). Neuroenergetics and the kinetic design of excitatory synapses. *Nature Reviews Neuroscience*, *6*(11), 841–849.

Becker, H. M., Broer, S., & Deitmer, J. W. (2004). Facilitated lactate transport by MCT1 when coexpressed with the sodium bicarbonate cotransporter (NBC) in Xenopus oocytes. *Biophysical Journal*, *86*(1 Pt 1), 235–247.

Becker, H. M., & Deitmer, J. W. (2004). Voltage dependence of H^+ buffering mediated by sodium bicarbonate cotransport expressed in Xenopus oocytes. *Journal of Biological Chemistry*, *279*(27), 28057–28062.

Becker, H. M., & Deitmer, J. W. (2008). Nonenzymatic proton handling by carbonic anhydrase II during H^+-lactate cotransport via monocarboxylate transporter 1. *Journal of Biological Chemistry*, *283*(31), 21655–21667.

Becker, H. M., Hirnet, D., Fecher-Trost, C., Sultemeyer, D., & Deitmer, J. W. (2005). Transport activity of MCT1 expressed in Xenopus oocytes is increased by interaction with carbonic anhydrase. *Journal of Biological Chemistry*, *280*(48), 39882–39889.

Becker, H. M., Klier, M., Schuler, C., McKenna, R., & Deitmer, J. W. (2011). Intramolecular proton shuttle supports not only catalytic but also noncatalytic function of carbonic anhydrase II. *Proceedings of the National Academy of Sciences of the United States of America*, *108*(7), 3071–3076.

Becker, H. M., Mohebbi, N., Perna, A., Ganapathy, V., Capasso, G., & Wagner, C. A. (2010). Localization of members of MCT monocarboxylate transporter family Slc16 in the kidney and regulation during metabolic acidosis. *American Journal of Physiology. Renal Physiology*, *299*(1), F141–F154.

Bellomo, R. (2002). Bench-to-bedside review: lactate and the kidney. *Critical Care, 6*(4), 322–326.

Bensaad, K., Tsuruta, A., Selak, M. A., Vidal, M. N., Nakano, K., Bartrons, R., et al. (2006). TIGAR, a p.53-inducible regulator of glycolysis and apoptosis. *Cell, 126*(1), 107–120.

Bergersen, L. H. (2007). Is lactate food for neurons? Comparison of monocarboxylate transporter subtypes in brain and muscle. *Neuroscience, 145*(1), 11–19.

Bergersen, L., Waerhaug, O., Helm, J., Thomas, M., Laake, P., Davies, A. J., et al. (2001). A novel postsynaptic density protein: the monocarboxylate transporter MCT2 is co-localized with delta-glutamate receptors in postsynaptic densities of parallel fiber-Purkinje cell synapses. *Experimental Brain Research, 136*(4), 523–534.

Bernards, R., & Weinberg, R. A. (2002). A progression puzzle. *Nature, 418*(6900), 823.

Bittner, C. X., Loaiza, A., Ruminot, I., Larenas, V., Sotelo-Hitschfeld, T., Gutierrez, R., et al. (2010). High resolution measurement of the glycolytic rate. *Front Neuroenergetics, 2*.

Boidot, R., Vegran, F., Meulle, A., Le Breton, A., Dessy, C., Sonveaux, P., et al. (2012). Regulation of monocarboxylate transporter MCT1 expression by p.53 mediates inward and outward lactate fluxes in tumors. *Cancer Research, 72*(4), 939–948.

Bonanno, J. A. (2003). Identity and regulation of ion transport mechanisms in the corneal endothelium. *Progress in Retinal and Eye Research, 22*(1), 69–94.

Bonanno, J. A. (2012). Molecular mechanisms underlying the corneal endothelial pump. *Experimental Eye Research, 95*(1), 2–7.

Boron, W. F. (2006). Acid-base transport by the renal proximal tubule. *Journal of the American Society of Nephrology, 17*(9), 2368–2382.

Boron, W. F., Chen, L., & Parker, M. D. (2009). Modular structure of sodium-coupled bicarbonate transporters. *Journal of Experimental Biology, 212*(Pt 11), 1697–1706.

Boussouar, F., & Benahmed, M. (2004). Lactate and energy metabolism in male germ cells. *Trends in Endocrinology and Metabolism, 15*(7), 345–350.

Casey, J. R. (2006). Why bicarbonate? *Biochemistry and Cell Biology, 84*(6), 930–939.

Casey, J. R., Sly, W. S., Shah, G. N., & Alvarez, B.V. (2009). Bicarbonate homeostasis in excitable tissues: role of AE3 Cl^-/HCO_3- exchanger and carbonic anhydrase XIV interaction. *American Journal of Physiology. Cell Physiology, 297*(5), C1091–C1102.

Charney, A. N., Micic, L., & Egnor, R. W. (1998). Nonionic diffusion of short-chain fatty acids across rat colon. *American Journal of Physiology, 274*(3 Pt 1), G518–G524.

Chemello, F., Bean, C., Cancellara, P., Laveder, P., Reggiani, C., & Lanfranchi, G. (2011). Microgenomic analysis in skeletal muscle: expression signatures of individual fast and slow myofibers. *PLoS One, 6*(2), e16807.

Cheng, C.Y., & Mruk, D. D. (2009). An intracellular trafficking pathway in the seminiferous epithelium regulating spermatogenesis: a biochemical and molecular perspective. *Critical Reviews In Biochemistry and Molecular Biology, 44*(5), 245–263.

Cheng, C.Y., Wong, E. W., Yan, H. H., & Mruk, D. D. (2010). Regulation of spermatogenesis in the microenvironment of the seminiferous epithelium: new insights and advances. *Molecular and Cellular Endocrinology, 315*(1–2), 49–56.

Chen, J. L., Lucas, J. E., Schroeder, T., Mori, S., Wu, J., Nevins, J., et al. (2008). The genomic analysis of lactic acidosis and acidosis response in human cancers. *PLoS Genetics, 4*(12), e1000293.

Chidlow, G., Wood, J. P., Graham, M., & Osborne, N. N. (2005). Expression of monocarboxylate transporters in rat ocular tissues. *American Journal of Physiology. Cell Physiology, 288*(2), C416–C428.

Chu, S., & Montrose, M. H. (1996). Non-ionic diffusion and carrier-mediated transport drive extracellullar pH regulation of mouse colonic crypts. *Journal of Physiology, 494*(Pt 3), 783–793.

Cordat, E., & Casey, J. R. (2009). Bicarbonate transport in cell physiology and disease. *Biochemical Journal, 417*(2), 423–439.

Cummings, J. H. (1984). Colonic absorption: the importance of short chain fatty acids in man. *Scandinavian Journal of Gastroenterology Supplement, 93*, 89–99.
Daniele, L. L., Sauer, B., Gallagher, S. M., Pugh, E. N., Jr., & Philp, N. J. (2008). Altered visual function in monocarboxylate transporter 3 (Slc16a8) knockout mice. *American Journal of Physiology. Cell Physiology, 295*(2), C451–C457.
Diecke, F. P., Wen, Q., Sanchez, J. M., Kuang, K., & Fischbarg, J. (2004). Immunocytochemical localization of Na^+-HCO_3- cotransporters and carbonic anhydrase dependence of fluid transport in corneal endothelial cells. *American Journal of Physiology. Cell Physiology, 286*(6), C1434–C1442.
Essen, B., Jansson, E., Henriksson, J., Taylor, A. W., & Saltin, B. (1975). Metabolic characteristics of fibre types in human skeletal muscle. *Acta Physiologica Scandinavica, 95*(2), 153–165.
Feldman, G. M. (1994). HCO_3^- secretion by rat distal colon: effects of inhibitors and extracellular Na^+. *Gastroenterology, 107*(2), 329–338.
Ferreira, L. M., Hebrant, A., & Dumont, J. E. (2011). Metabolic reprogramming of the tumor. *Oncogene.*
Fishbein, W. N., Merezhinskaya, N., & Foellmer, J. W. (2002). Relative distribution of three major lactate transporters in frozen human tissues and their localization in unfixed skeletal muscle. *Muscle & Nerve, 26*(1), 101–112.
Friesema, E. C., Grueters, A., Biebermann, H., Krude, H., von Moers, A., Reeser, M., et al. (2004). Association between mutations in a thyroid hormone transporter and severe X-linked psychomotor retardation. *Lancet, 364*(9443), 1435–1437.
Friesema, E. C., Jansen, J., Jachtenberg, J. W., Visser, W. E., Kester, M. H., & Visser, T. J. (2008). Effective cellular uptake and efflux of thyroid hormone by human monocarboxylate transporter 10. *Molecular Endocrinology, 22*(6), 1357–1369.
Galardo, M. N., Riera, M. F., Pellizzari, E. H., Cigorraga, S. B., & Meroni, S. B. (2007). The AMP-activated protein kinase activator, 5-aminoimidazole-4-carboxamide-1-b-D-ribonucleoside, regulates lactate production in rat Sertoli cells. *Journal of Molecular Endocrinology, 39*(4), 279–288.
Galie, M., Farace, P., Nanni, C., Spinelli, A., Nicolato, E., Boschi, F., et al. (2007). Epithelial and mesenchymal tumor compartments exhibit in vivo complementary patterns of vascular perfusion and glucose metabolism. *Neoplasia, 9*(11), 900–908.
Gallagher, S. M., Castorino, J. J., Wang, D., & Philp, N. J. (2007). Monocarboxylate transporter 4 regulates maturation and trafficking of CD147 to the plasma membrane in the metastatic breast cancer cell line MDA-MB-231. *Cancer Research, 67*(9), 4182–4189.
Garcia, C. K., Brown, M. S., Pathak, R. K., & Goldstein, J. L. (1995). cDNA cloning of MCT2, a second monocarboxylate transporter expressed in different cells than MCT1. *Journal of Biological Chemistry, 270*(4), 1843–1849.
Garcia, C. K., Goldstein, J. L., Pathak, R. K., Anderson, R. G., & Brown, M. S. (1994). Molecular characterization of a membrane transporter for lactate, pyruvate, and other monocarboxylates: implications for the Cori cycle. *Cell, 76*(5), 865–873.
Gatenby, R. A., & Gillies, R. J. (2004). Why do cancers have high aerobic glycolysis? *Nature Reviews Cancer, 4*(11), 891–899.
Gawenis, L. R., Bradford, E. M., Prasad, V., Lorenz, J. N., Simpson, J. E., Clarke, L. L., et al. (2007). Colonic anion secretory defects and metabolic acidosis in mice lacking the NBC1 Na^+/HCO_3^- cotransporter. *Journal of Biological Chemistry, 282*(12), 9042–9052.
Giasson, C., & Bonanno, J. A. (1994). Facilitated transport of lactate by rabbit corneal endothelium. *Experimental Eye Research, 59*(1), 73–81.
Gladden, L. B. (2004). Lactate metabolism: a new paradigm for the third millennium. *Journal of Physiology, 558*(Pt 1), 5–30.
Gopal, E., Umapathy, N. S., Martin, P. M., Ananth, S., Gnana-Prakasam, J. P., Becker, H., et al. (2007). Cloning and functional characterization of human SMCT2 (SLC5A12) and expression pattern of the transporter in kidney. *Biochimica et Biophysica Acta, 1768*(11), 2690–2697.

Gordan, J. D., Thompson, C. B., & Simon, M. C. (2007). HIF and c-Myc: sibling rivals for control of cancer cell metabolism and proliferation. *Cancer Cell, 12*(2), 108–113.

Gospe, S. M., 3rd, Baker, S. A., & Arshavsky, V. Y. (2010). Facilitative glucose transporter Glut1 is actively excluded from rod outer segments. *Journal of Cell Science, 123*(Pt 21), 3639–3644.

Greenhaff, P. L., Nevill, M. E., Soderlund, K., Bodin, K., Boobis, L. H., Williams, C., et al. (1994). The metabolic responses of human type I and II muscle fibres during maximal treadmill sprinting. *Journal of Physiology, 478*(Pt 1), 149–155.

Gross, E., Pushkin, A., Abuladze, N., Fedotoff, O., & Kurtz, I. (2002). Regulation of the sodium bicarbonate cotransporter kNBC1 function: role of Asp(986), Asp(988) and kNBC1-carbonic anhydrase II binding. *Journal of Physiology, 544*(Pt 3), 679–685.

Halestrap, A. P. (2012). The monocarboxylate transporter family–Structure and functional characterization. *IUBMB Life, 64*(1), 1–9.

Halestrap, A. P., & Price, N. T. (1999). The proton-linked monocarboxylate transporter (MCT) family: structure, function and regulation. *Biochemical Journal, 343*(Pt 2), 281–299.

Hamann, S., Kiilgaard, J. F., la Cour, M., Prause, J. U., & Zeuthen, T. (2003). Cotransport of H^+, lactate, and H_2O in porcine retinal pigment epithelial cells. *Experimental Eye Research, 76*(4), 493–504.

Hamann, S., la Cour, M., Lui, G. M., Bundgaard, M., & Zeuthen, T. (2000). Transport of protons and lactate in cultured human fetal retinal pigment epithelial cells. *Pflügers Archiv European Journal of Physiology, 440*(1), 84–92.

Hanahan, D., & Weinberg, R. A. (2011). Hallmarks of cancer: the next generation. *Cell, 144*(5), 646–674.

Hashimoto, T., Masuda, S., Taguchi, S., & Brooks, G. A. (2005). Immunohistochemical analysis of MCT1, MCT2 and MCT4 expression in rat plantaris muscle. *Journal of Physiology, 567*(Pt 1), 121–129.

Hoogerwerf, W. A., Tsao, S. C., Devuyst, O., Levine, S. A., Yun, C. H., Yip, J. W., et al. (1996). NHE2 and NHE3 are human and rabbit intestinal brush-border proteins. *American Journal of Physiology, 270*(1 Pt 1), G29–G41.

Hori, K., Katayama, N., Kachi, S., Kondo, M., Kadomatsu, K., Usukura, J., et al. (2000). Retinal dysfunction in basigin deficiency. *Investigative Ophthalmology & Visual Science, 41*(10), 3128–3133.

Hughes, B. A., Gallemore, R. P., & Miller, S. S. (1998). Transport mechanisms in the retinal pigment epithelium. In M. F. Marmor, & T. J. Wolfensberger (Eds.), *The retinal pigment epithelium* (pp. 103–134). New York: Oxford University Press.

Hussien, R., & Brooks, G. A. (2010). Mitochondrial and plasma membrane lactate transporter and lactate dehydrogenase isoform expression in breast cancer cell lines. *Physiological Genomics, 43*(5), 255–264.

Igakura, T., Kadomatsu, K., Kaname, T., Muramatsu, H., Fan, Q. W., Miyauchi, T., et al. (1998). A null mutation in basigin, an immunoglobulin superfamily member, indicates its important roles in peri-implantation development and spermatogenesis. *Developmental Biology, 194*(2), 152–165.

Igarashi, T., Inatomi, J., Sekine, T., Cha, S. H., Kanai, Y., Kunimi, M., et al. (1999). Mutations in SLC4A4 cause permanent isolated proximal renal tubular acidosis with ocular abnormalities. *Nature Genetics, 23*(3), 264–266.

Igarashi, T., Inatomi, J., Sekine, T., Seki, G., Shimadzu, M., Tozawa, F., et al. (2001). Novel nonsense mutation in the Na^+/HCO_3- cotransporter gene (SLC4A4) in a patient with permanent isolated proximal renal tubular acidosis and bilateral glaucoma. *Journal of the American Society of Nephrology, 12*(4), 713–718.

Inatomi, J., Horita, S., Braverman, N., Sekine, T., Yamada, H., Suzuki, Y., et al. (2004). Mutational and functional analysis of SLC4A4 in a patient with proximal renal tubular acidosis. *Pflügers Archiv European Journal of Physiology, 448*(4), 438–444.

Itoh, Y., Esaki, T., Shimoji, K., Cook, M., Law, M. J., Kaufman, E., et al. (2003). Dichloroacetate effects on glucose and lactate oxidation by neurons and astroglia in vitro and on glucose utilization by brain in vivo. *Proceedings of the National Academy of Sciences of the United States of America*, *100*(8), 4879–4884.

Iwanaga, T., Takebe, K., Kato, I., Karaki, S., & Kuwahara, A. (2006). Cellular expression of monocarboxylate transporters (MCT) in the digestive tract of the mouse, rat, and humans, with special reference to slc5a8. *Biomedical Research*, *27*(5), 243–254.

Johnston, L. A. (2009). Competitive interactions between cells: death, growth, and geography. *Science*, *324*(5935), 1679–1682.

Kang, K. W., Im, Y. B., Go, W. J., & Han, H. K. (2009). C-myc amplification altered the gene expression of ABC- and SLC-transporters in human breast epithelial cells. *Molecular Pharmacology*, *6*(2), 627–633.

Kawamata, K., Hayashi, H., & Suzuki, Y. (2007). Propionate absorption associated with bicarbonate secretion in vitro in the mouse cecum. *Pflügers Archiv European Journal of Physiology*, *454*(2), 253–262.

Kenyon, E., Yu, K., La Cour, M., & Miller, S. S. (1994). Lactate transport mechanisms at apical and basolateral membranes of bovine retinal pigment epithelium. *American Journal of Physiology*, *267*(6 Pt 1), C1561–C1573.

Kim, D. K., Kanai, Y., Matsuo, H., Kim, J. Y., Chairoungdua, A., Kobayashi, Y., et al. (2002). The human T-type amino acid transporter-1: characterization, gene organization, and chromosomal location. *Genomics*, *79*(1), 95–103.

Kirat, D., & Kato, S. (2006). Monocarboxylate transporter 1 (MCT1) mediates transport of short-chain fatty acids in bovine caecum. *Experimental Physiology*, *91*(5), 835–844.

Kirat, D., Masuoka, J., Hayashi, H., Iwano, H., Yokota, H., Taniyama, H., et al. (2006). Monocarboxylate transporter 1 (MCT1) plays a direct role in short-chain fatty acids absorption in caprine rumen. *Journal of Physiology*, *576*(Pt 2), 635–647.

Kirk, P., Wilson, M. C., Heddle, C., Brown, M. H., Barclay, A. N., & Halestrap, A. P. (2000). CD147 is tightly associated with lactate transporters MCT1 and MCT4 and facilitates their cell surface expression. *EMBO Journal*, *19*(15), 3896–3904.

Klier, M., Schuler, C., Halestrap, A. P., Sly, W. S., Deitmer, J. W., & Becker, H. M. (2011). Transport activity of the high-affinity monocarboxylate transporter MCT2 is enhanced by extracellular carbonic anhydrase IV but not by intracellular carbonic anhydrase II. *Journal of Biological Chemistry*, *286*(31), 27781–27791.

Klyce, S. D. (1981). Stromal lactate accumulation can account for corneal oedema osmotically following epithelial hypoxia in the rabbit. *Journal of Physiology*, *321*, 49–64.

Kolz, M., Johnson, T., Sanna, S., Teumer, A., Vitart, V., Perola, M., et al. (2009). Meta-analysis of 28,141 individuals identifies common variants within five new loci that influence uric acid concentrations. *PLoS Genetics*, *5*(6), e1000504.

Le Floch, R., Chiche, J., Marchiq, I., Naiken, T., Ilk, K., Murray, C. M., et al. (2011). CD147 subunit of lactate/H^+ symporters MCT1 and hypoxia-inducible MCT4 is critical for energetics and growth of glycolytic tumors. *Proceedings of the National Academy of Sciences of the United States of America*, *108*(40), 16663–16668.

Li, J., Allen, K. T., Sun, X. C., Cui, M., & Bonanno, J. A. (2008). Dependence of cAMP meditated increases in Cl^- and HCO(3)− permeability on CFTR in bovine corneal endothelial cells. *Experimental Eye Research*, *86*(4), 684–690.

Lien, Y. H., & Lai, L. W. (1998). Respiratory acidosis in carbonic anhydrase II-deficient mice. *American Journal of Physiology*, *274*(2 Pt 1), L301–L304.

Loiselle, F. B., Morgan, P. E., Alvarez, B. V., & Casey, J. R. (2004). Regulation of the human NBC3 Na^+/HCO_3^- cotransporter by carbonic anhydrase II and PKA. *American Journal of Physiology. Cell Physiology*, *286*(6), C1423–C1433.

Lonnerholm, G., Selking, O., & Wistrand, P. J. (1985). Amount and distribution of carbonic anhydrases CA I and CA II in the gastrointestinal tract. *Gastroenterology*, *88* (5 Pt 1), 1151–1161.

Lowry, C. V., Kimmey, J. S., Felder, S., Chi, M. M., Kaiser, K. K., Passonneau, P. N., et al. (1978). Enzyme patterns in single human muscle fibers. *Journal of Biological Chemistry, 253*(22), 8269–8277.

Lo, Y. F., Yang, S. S., Seki, G., Yamada, H., Horita, S., Yamazaki, O., et al. (2011). Severe metabolic acidosis causes early lethality in NBC1 W516X knock-in mice as a model of human isolated proximal renal tubular acidosis. *Kidney International, 79*(7), 730–741.

Luciano, L., Reale, E., Rechkemmer, G., & von Engelhardt, W. (1984). Structure of zonulae occludentes and the permeability of the epithelium to short-chain fatty acids in the proximal and the distal colon of guinea pig. *Journal of Membrane Biology, 82*(2), 145–156.

Lu, J., Daly, C. M., Parker, M. D., Gill, H. S., Piermarini, P. M., Pelletier, M. F., et al. (2006). Effect of human carbonic anhydrase II on the activity of the human electrogenic Na/HCO_3 cotransporter NBCe1-A in Xenopus oocytes. *Journal of Biological Chemistry, 281*(28), 19241–19250.

MacAskill, A. F., & Kittler, J. T. (2009). Control of mitochondrial transport and localization in neurons. *Trends in Cell Biology, 20*(2), 102–112.

Maminishkis, A., Chen, S., Jalickee, S., Banzon, T., Shi, G., Wang, F. E., et al. (2006). Confluent monolayers of cultured human fetal retinal pigment epithelium exhibit morphology and physiology of native tissue. *Investigative Ophthalmology & Visual Science, 47*(8), 3612–3624.

Mangia, S., Simpson, I. A., Vannucci, S. J., & Carruthers, A. (2009). The in vivo neuron-to-astrocyte lactate shuttle in human brain: evidence from modeling of measured lactate levels during visual stimulation. *Journal of Neurochemistry, 109*(Suppl. 1), 55–62.

Manning Fox, J. E., Meredith, D., & Halestrap, A. P. (2000). Characterisation of human monocarboxylate transporter 4 substantiates its role in lactic acid efflux from skeletal muscle. *Journal of Physiology, 529*(Pt 2), 285–293.

Mannowetz, N., Wandernoth, P., & Wennemuth, G. (2012). Basigin interacts with both MCT1 and MCT2 in murine spermatozoa. *Journal of Cellular Physiology, 227*(5), 2154–2162.

Manoharan, C., Wilson, M. C., Sessions, R. B., & Halestrap, A. P. (2006). The role of charged residues in the transmembrane helices of monocarboxylate transporter 1 and its ancillary protein basigin in determining plasma membrane expression and catalytic activity. *Molecular Membrane Biology, 23*(6), 486–498.

Meijer, T. W., Schuurbiers, O. C., Kaanders, J. H., Looijen-Salamon, M. G., de Geus-Oei, L. F., Verhagen, A. F., et al. (2011). Differences in metabolism between adeno- and squamous cell non-small cell lung carcinomas: spatial distribution and prognostic value of GLUT1 and MCT4. *Lung Cancer, 76*(3), 316–323.

Melvin, J. E., Park, K., Richardson, L., Schultheis, P. J., & Shull, G. E. (1999). Mouse down-regulated in adenoma (DRA) is an intestinal Cl(−)/HCO(3)(−) exchanger and is up-regulated in colon of mice lacking the NHE3 Na(+)/H(+) exchanger. *Journal of Biological Chemistry, 274*(32), 22855–22861.

Morgan, P. E., Pastorekova, S., Stuart-Tilley, A. K., Alper, S. L., & Casey, J. R. (2007). Interactions of transmembrane carbonic anhydrase, CAIX, with bicarbonate transporters. *American Journal of Physiology. Cell Physiology, 293*(2), C738–C748.

Nagelhus, E. A., Mathiisen, T. M., Bateman, A. C., Haug, F. M., Ottersen, O. P., Grubb, J. H., et al. (2005). Carbonic anhydrase XIV is enriched in specific membrane domains of retinal pigment epithelium, Muller cells, and astrocytes. *Proceedings of the National Academy of Sciences of the United States of America, 102*(22), 8030–8035.

Nakai, M., Chen, L., & Nowak, R. A. (2006). Tissue distribution of basigin and monocarboxylate transporter 1 in the adult male mouse: a study using the wild-type and basigin gene knockout mice. *The Anatomical Record. Part A, Discoveries in Molecular, Cellular, and Evolutionary Biology, 288*(5), 527–535.

Naruhashi, K., Kadomatsu, K., Igakura, T., Fan, Q.W., Kuno, N., Muramatsu, H., et al. (1997). Abnormalities of sensory and memory functions in mice lacking Bsg gene. *Biochemical and Biophysical Research, 236*(3), 733–737.

Nguyen, T. T., & Bonanno, J. A. (2011). Bicarbonate, NBCe1, NHE, and carbonic anhydrase activity enhance lactate-H+ transport in bovine corneal endothelium. *Investigative Ophthalmology & Visual Science, 52*(11), 8086–8093.

Nguyen, T. T., & Bonanno, J. A. (2012). Lactate-H^+ transport is a significant component of the in vivo corneal endothelial pump. *Investigative Ophthalmology & Visual Science, 53*(4), 2020–2029.

Olivier, M., Hussain, S. P., Caron de Fromentel, C., Hainaut, P., & Harris, C. C. (2004). TP53 mutation spectra and load: a tool for generating hypotheses on the etiology of cancer. *IARC Scientific Publications, 157*, 247–270.

Orlowski, A., de Giusti, V. C., Morgan, P. E., Aiello, E. A., & Alvarez, B.V. (2012). Binding of carbonic anhydrase IX to extracellular loop 4 of the NBCe1 Na^+/HCO_3^- cotransporter enhances NBCe1-mediated HCO_3^- influx in the rat heart. *American Journal of Physiology. Cell Physiology.*

Paroder, V., Spencer, S. R., Paroder, M., Arango, D., Schwartz, S., Jr., Mariadason, J. M., et al. (2006). Na(+)/monocarboxylate transport (SMCT) protein expression correlates with survival in colon cancer: molecular characterization of SMCT. *Proceedings of the National Academy of Sciences of the United States of America, 103*(19), 7270–7275.

Pavlides, S., Whitaker-Menezes, D., Castello-Cros, R., Flomenberg, N., Witkiewicz, A. K., Frank, P. G., et al. (2009). The reverse Warburg effect: aerobic glycolysis in cancer associated fibroblasts and the tumor stroma. *Cell Cycle, 8*(23), 3984–4001.

Philp, N. J., Ochrietor, J. D., Rudoy, C., Muramatsu, T., & Linser, P. J. (2003). Loss of MCT1, MCT3, and MCT4 expression in the retinal pigment epithelium and neural retina of the 5A11/basigin-null mouse. *Investigative Ophthalmology & Visual Science, 44*(3), 1305–1311.

Philp, N. J., Yoon, H., & Grollman, E. F. (1998). Monocarboxylate transporter MCT1 is located in the apical membrane and MCT3 in the basal membrane of rat RPE. *American Journal of Physiology, 274*(6 Pt 2), R1824–R1828.

Philp, N. J., Yoon, H., & Lombardi, L. (2001). Mouse MCT3 gene is expressed preferentially in retinal pigment and choroid plexus epithelia. *American Journal of Physiology. Cell Physiology, 280*(5), C1319–C1326.

Pinheiro, C., Longatto-Filho, A., Azevedo-Silva, J., Casal, M., Schmitt, F. C., & Baltazar, F. (2012). Role of monocarboxylate transporters in human cancers: state of the art. *Journal of Bioenergetics and Biomembranes, 44*(1), 127–139.

Plomgaard, P., Penkowa, M., Leick, L., Pedersen, B. K., Saltin, B., & Pilegaard, H. (2006). The mRNA expression profile of metabolic genes relative to MHC isoform pattern in human skeletal muscles. *Journal of Applied Physiology, 101*(3), 817–825.

Poitry-Yamate, C. L., Poitry, S., & Tsacopoulos, M. (1995). Lactate released by Muller glial cells is metabolized by photoreceptors from mammalian retina. *Journal of Neuroscience, 15*(7 Pt 2), 5179–5191.

Poitry, S., Poitry-Yamate, C., Ueberfeld, J., MacLeish, P. R., & Tsacopoulos, M. (2000). Mechanisms of glutamate metabolic signaling in retinal glial (Muller) cells. *Journal of Neuroscience, 20*(5), 1809–1821.

Poole, R. C., & Halestrap, A. P. (1988). Reconstitution of the L-lactate carrier from rat and rabbit erythrocyte plasma membranes. *Biochemical Journal, 254*(2), 385–390.

Poole, R. C., & Halestrap, A. P. (1991). Reversible and irreversible inhibition, by stilbenedisulphonates, of lactate transport into rat erythrocytes. Identification of some new high-affinity inhibitors. *Biochemical Journal, 275*(Pt 2), 307–312.

Pullen, T. J., da Silva Xavier, G., Kelsey, G., & Rutter, G. A. (2011). MiR-29a and miR-29b contribute to pancreatic beta-cell-specific silencing of monocarboxylate transporter 1 (Mct1). *Molecular and Cellular Biology, 31*(15), 3182–3194.

Purkerson, J. M., & Schwartz, G. J. (2005). Expression of membrane-associated carbonic anhydrase isoforms IV, IX, XII, and XIV in the rabbit: induction of CA IV and IX during maturation. *American Journal of Physiology – Regulatory, Integrative and Comparative Physiology*, *288*(5), R1256–R1263.

Pushkin, A., Abuladze, N., Gross, E., Newman, D., Tatishchev, S., Lee, I., et al. (2004). Molecular mechanism of kNBC1-carbonic anhydrase II interaction in proximal tubule cells. *Journal of Physiology*, *559*(Pt 1), 55–65.

Rademakers, S. E., Lok, J., van der Kogel, A. J., Bussink, J., & Kaanders, J. H. (2011). Metabolic markers in relation to hypoxia; staining patterns and colocalization of pimonidazole, HIF-1alpha, CAIX, LDH-5, GLUT-1, MCT1 and MCT4. *BMC Cancer*, *11*, 167.

Rafiki, A., Boulland, J. L., Halestrap, A. P., Ottersen, O. P., & Bergersen, L. (2003). Highly differential expression of the monocarboxylate transporters MCT2 and MCT4 in the developing rat brain. *Neuroscience*, *122*(3), 677–688.

Rajendran, V. M., Black, J., Ardito, T. A., Sangan, P., Alper, S. L., Schweinfest, C., et al. (2000). Regulation of DRA and AE1 in rat colon by dietary Na depletion. *American Journal of Physiology - Gastrointestinal and Liver Physiology*, *279*(5), G931–G942.

Rato, L., Alves, M. G., Socorro, S., Duarte, A. I., Cavaco, J. E., & Oliveira, P. F. (2012). Metabolic regulation is important for spermatogenesis. *Nature Reviews Urology*, *9*(6), 330–338.

Riley, M. V., & Winkler, B. S. (1990). Strong Pasteur effect in rabbit corneal endothelium preserves fluid transport under anaerobic conditions. *Journal of Physiology*, *426*, 81–93.

Rink, T. J., Tsien, R. Y., & Pozzan, T. (1982). Cytoplasmic pH and free Mg^{2+} in lymphocytes. *Journal of Cellular Biology*, *95*(1), 189–196.

Ritzhaupt, A., Wood, I. S., Ellis, A., Hosie, K. B., & Shirazi-Beechey, S. P. (1998). Identification and characterization of a monocarboxylate transporter (MCT1) in pig and human colon: its potential to transport L-lactate as well as butyrate. *Journal of Physiology*, *513*(Pt 3), 719–732.

Roth, D. A., & Brooks, G. A. (1990). Lactate and pyruvate transport is dominated by a pH gradient-sensitive carrier in rat skeletal muscle sarcolemmal vesicles. *Archives of Biochemistry and Biophysics*, *279*(2), 386–394.

Roth, D. E., Venta, P. J., Tashian, R. E., & Sly, W. S. (1992). Molecular basis of human carbonic anhydrase II deficiency. *Proceedings of the National Academy of Sciences of the United States of America*, *89*(5), 1804–1808.

Rouach, N., Koulakoff, A., Abudara, V., Willecke, K., & Giaume, C. (2008). Astroglial metabolic networks sustain hippocampal synaptic transmission. *Science*, *322*(5907), 1551–1555.

Scheppach, W. (1994). Effects of short chain fatty acids on gut morphology and function. *Gut*, *35*(Suppl. 1), S35–S38.

Schiaffino, S., & Reggiani, C. (2011). Fiber types in mammalian skeletal muscles. *Physiological Reviews*, *91*(4), 1447–1531.

Schmid, H., Scholz, M., Mall, A., Schmidt, U., Guder, W. G., & Dubach, U. C. (1977). Carbohydrate metabolism in rat kidney: heterogeneous distribution of glycolytic and gluconeogenic key enzymes. *Current Problems In Clinical Biochemistry*, *8*, 282–289.

Schueler, C., Becker, H. M., McKenna, R., & Deitmer, J. W. (2011). Transport activity of the sodium bicarbonate cotransporter NBCe1 is enhanced by different isoforms of carbonic anhydrase. *PLoS One*, *6*(11), e27167.

Schwartz, C. E., May, M. M., Carpenter, N. J., Rogers, R. C., Martin, J., Bialer, M. G., et al. (2005). Allan-Herndon-Dudley syndrome and the monocarboxylate transporter 8 (MCT8) gene. *American Journal of Human Genetics*, 77(1), 41–53.

Schweinfest, C. W., Spyropoulos, D. D., Henderson, K. W., Kim, J. H., Chapman, J. M., Barone, S., et al. (2006). slc26a3 (dra)-deficient mice display chloride-losing diarrhea, enhanced colonic proliferation, and distinct up-regulation of ion transporters in the colon. *Journal of Biological Chemistry*, *281*(49), 37962–37971.

Sellin, J. H., & De Soignie, R. (1998). Short-chain fatty acids have polarized effects on sodium transport and intracellular pH in rabbit proximal colon. *Gastroenterology, 114*(4), 737–747.

Semenza, G. L. (2010). HIF-1: upstream and downstream of cancer metabolism. *Current Opinion In Genetics & Development, 20*(1), 51–56.

Simpson, I. A., Carruthers, A., & Vannucci, S. J. (2007). Supply and demand in cerebral energy metabolism: the role of nutrient transporters. *Journal of Cerebral Blood Flow and Metabolism, 27*(11), 1766–1791.

Sly, W. S., Hewett-Emmett, D., Whyte, M. P., Yu, Y. S., & Tashian, R. E. (1983). Carbonic anhydrase II deficiency identified as the primary defect in the autosomal recessive syndrome of osteopetrosis with renal tubular acidosis and cerebral calcification. *Proceedings of the National Academy of Sciences of the United States of America, 80*(9), 2752–2756.

Sonveaux, P., Vegran, F., Schroeder, T., Wergin, M. C., Verrax, J., Rabbani, Z. N., et al. (2008). Targeting lactate-fueled respiration selectively kills hypoxic tumor cells in mice. *Journal of Clinical Investigation, 118*(12), 3930–3942.

Spamer, C., & Pette, D. (1977). Activity patterns of phosphofructokinase, glyceraldehydephosphate dehydrogenase, lactate dehydrogenase and malate dehydrogenase in microdissected fast and slow fibres from rabbit psoas and soleus muscle. *Histochemistry, 52*(3), 201–216.

Sterling, D., Alvarez, B.V., & Casey, J. R. (2002). The extracellular component of a transport metabolon. Extracellular loop 4 of the human AE1 Cl^-/HCO_3^- exchanger binds carbonic anhydrase IV. *Journal of Biological Chemistry, 277*(28), 25239–25246.

Sterling, D., Reithmeier, R. A., & Casey, J. R. (2001). A transport metabolon. Functional interaction of carbonic anhydrase II and chloride/bicarbonate exchangers. *Journal of Biological Chemistry, 276*(51), 47886–47894.

Stone, J., van Driel, D., Valter, K., Rees, S., & Provis, J. (2008). The locations of mitochondria in mammalian photoreceptors: relation to retinal vasculature. *Brain Research, 1189*, 58–69.

Strauss, O. (2005). The retinal pigment epithelium in visual function. *Physiological Reviews, 85*(3), 845–881.

Suhre, K., Shin, S.Y., Petersen, A. K., Mohney, R. P., Meredith, D., Wagele, B., et al. (2011). Human metabolic individuality in biomedical and pharmaceutical research. *Nature, 477*(7362), 54–60.

Supuran, C.T. (2008). Carbonic anhydrases: novel therapeutic applications for inhibitors and activators. *Nature Reviews Drug Discovery*, 7(2), 168–181.

Suzuki, A., Stern, S. A., Bozdagi, O., Huntley, G. W., Walker, R. H., Magistretti, P. J., et al. (2011). Astrocyte-neuron lactate transport is required for long-term memory formation. *Cell, 144*(5), 810–823.

Svichar, N., Waheed, A., Sly, W. S., Hennings, J. C., Hubner, C. A., & Chesler, M. (2009). Carbonic anhydrases CA4 and CA14 both enhance AE3-mediated $Cl^-HCO_3^-$ exchange in hippocampal neurons. *Journal of Neuroscience, 29*(10), 3252–3258.

Tamai, I., Sai, Y., Ono, A., Kido, Y., Yabuuchi, H., Takanaga, H., et al. (1999). Immunohistochemical and functional characterization of pH-dependent intestinal absorption of weak organic acids by the monocarboxylic acid transporter MCT1. *Journal of Pharmacy and Pharmacology, 51*(10), 1113–1121.

Tsacopoulos, M., & Magistretti, P. J. (1996). Metabolic coupling between glia and neurons. *Journal of Neuroscience, 16*(3), 877–885.

Ullah, M. S., Davies, A. J., & Halestrap, A. P. (2006). The plasma membrane lactate transporter MCT4, but not MCT1, is up-regulated by hypoxia through a HIF-1alpha-dependent mechanism. *Journal of Biological Chemistry, 281*(14), 9030–9037.

Vince, J. W., & Reithmeier, R. A. (2000). Identification of the carbonic anhydrase II binding site in the Cl(−)/HCO(3)(−) anion exchanger AE1. *Biochemistry, 39*(18), 5527–5533.

von Engelhardt, W., Gros, G., Burmester, M., Hansen, K., Becker, G., & Rechkemmer, G. (1994). Functional role of bicarbonate in propionate transport across guinea-pig isolated caecum and proximal colon. *Journal of Physiology*, *477*(Pt 2), 365–371.

Wang, L., Kondo, M., & Bill, A. (1997). Glucose metabolism in cat outer retina. Effects of light and hyperoxia. *Investigative Ophthalmology & Visual Science*, *38*(1), 48–55.

Wang, X., Levi, A. J., & Halestrap, A. P. (1994). Kinetics of the sarcolemmal lactate carrier in single heart cells using BCECF to measure pHi. *American Journal of Physiology*, *267*(5 Pt 2), H1759–H1769.

Wang, L., Tornquist, P., & Bill, A. (1997). Glucose metabolism in pig outer retina in light and darkness. *Acta Physiologica Scandinavica*, *160*(1), 75–81.

Wangsa-Wirawan, N. D., & Linsenmeier, R. A. (2003). Retinal oxygen: fundamental and clinical aspects. *Archives of Ophthalmology*, *121*(4), 547–557.

Whitaker-Menezes, D., Martinez-Outschoorn, U. E., Lin, Z., Ertel, A., Flomenberg, N., Witkiewicz, A. K., et al. (2011). Evidence for a stromal-epithelial "lactate shuttle" in human tumors: MCT4 is a marker of oxidative stress in cancer-associated fibroblasts. *Cell Cycle*, *10*(11), 1772–1783.

Wilson, M. C., Meredith, D., Bunnun, C., Sessions, R. B., & Halestrap, A. P. (2009). Studies on the DIDS-binding site of monocarboxylate transporter 1 suggest a homology model of the open conformation and a plausible translocation cycle. *Journal of Biological Chemistry*, *284*(30), 20011–20021.

Wilson, M. C., Meredith, D., Fox, J. E., Manoharan, C., Davies, A. J., & Halestrap, A. P. (2005). Basigin (CD147) is the target for organomercurial inhibition of monocarboxylate transporter isoforms 1 and 4: the ancillary protein for the insensitive MCT2 is EMBIGIN (gp70). *Journal of Biological Chemistry*, *280*(29), 27213–27221.

Winkler, B. S., Arnold, M. J., Brassell, M. A., & Puro, D. G. (2000). Energy metabolism in human retinal Muller cells. *Investigative Ophthalmology & Visual Science*, *41*(10), 3183–3190.

Winkler, B. S., Starnes, C. A., Twardy, B. S., Brault, D., & Taylor, R. C. (2008). Nuclear magnetic resonance and biochemical measurements of glucose utilization in the cone-dominant ground squirrel retina. *Investigative Ophthalmology & Visual Science*, *49*(10), 4613–4619.

Witkiewicz, A. K., Whitaker-Menezes, D., Dasgupta, A., Philp, N. J., Lin, Z., Gandara, R., et al. (2012). Using the "reverse Warburg effect" to identify high-risk breast cancer patients: stromal MCT4 predicts poor clinical outcome in triple-negative breast cancers. *Cell Cycle*, *11*(6), 1108–1117.

Zawacka-Pankau, J., Grinkevich, V. V., Hunten, S., Nikulenkov, F., Gluch, A., Li, H., et al. (2011). Inhibition of glycolytic enzymes mediated by pharmacologically activated p.53: targeting Warburg effect to fight cancer. *Journal of Biological Chemistry*, *286*(48), 41600–41615.

Zhang, Y., Li, J., Xie, Q., & Bonanno, J. A. (2006). Molecular expression and functional involvement of the bovine calcium-activated chloride channel 1 (bCLCA1) in apical HCO_3^- permeability of bovine corneal endothelium. *Experimental Eye Research*, *83*(5), 1215–1224.

Zhang, Y., Xie, Q., Sun, X. C., & Bonanno, J. A. (2002). Enhancement of HCO(3)(−) permeability across the apical membrane of bovine corneal endothelium by multiple signaling pathways. *Investigative Ophthalmology & Visual Science*, *43*(4), 1146–1153.

CHAPTER TEN

Phosphate Transport Kinetics and Structure-Function Relationships of SLC34 and SLC20 Proteins

Ian C. Forster[1], Nati Hernando, Jürg Biber, Heini Murer

Institute of Physiology and Zurich Center for Integrative Human Physiology, University of Zurich, Winterthurerstrasse, Zurich, Switzerland

[1]Corresponding author: E-mail: iforster@access.uzh.ch

Contents

Current Topics in Membranes, Volume 70
ISSN 1063-5823
http://dx.doi.org/10.1016/B978-0-12-394316-3.00010-7

Abstract

Transport of inorganic phosphate (P_i) is mediated by proteins belonging to two solute carrier families (SLC20 and SLC34). Members of both families transport P_i using the electrochemical gradient for Na^+. The role of the SLC34 members as essential players in mammalian P_i homeostasis is well established, whereas that of SLC20 proteins is less well defined. The SLC34 family comprises the following three isoforms that preferentially cotransport divalent P_i and are expressed in epithelial tissue: the renal NaPi-IIa and NaPi-IIc are responsible for reabsorbing P_i in the proximal tubule, whereas NaPi-IIb is more ubiquitously expressed, including the small intestine, where it mediates dietary P_i absorption. The SLC20 family comprises two members (PiT-1, PiT-2) that preferentially cotransport monovalent P_i and are expressed in epithelial as well as nonepithelial tissue. The transport kinetics of members of both families have been characterized in detail using heterologous expression in *Xenopus* oocytes. For the electrogenic NaPi-IIa/b, and PiT-1,-2, conventional electrophysiological techniques together with radiotracer methods have been applied, as well as time-resolved fluorometric measurements that allow new insights into local conformational changes of the protein during the cotransport cycle. For the electroneutral NaPi-IIc, conventional tracer uptake and fluorometry have been used to elucidate its transport properties. The 3-D structures of these proteins remain unresolved and structure–function studies have so far concentrated on defining the topology and identifying sites of functional importance.

1. INTRODUCTION

In mammals, phosphate plays essential structural, metabolic, synthesis and signaling roles. Dietary phosphate is absorbed in the small intestine and circulates in the blood for delivery to soft tissue and bone as required. In plasma, 50% of phosphate is ionized inorganic phosphate (P_i), and the remainder comprises diffusible phosphate complexes (40%) and nondiffusible protein-bound phosphate (10%). In the physiological range of pH, titratable P_i is distributed between monovalent ($H_2PO_4^-$) and divalent (HPO_4^{2-}) ions according to the pH. Under normal conditions, the daily intake of phosphate and loss in feces and urine are balanced and the total plasma P_i lies in the range of 0.8–1.5 mM (in humans). The principle locus of P_i homeostasis is the kidney. At the glomerulus, P_i is freely filtered from the blood and then reabsorbed from the primary urine according to the bodily requirements; any excess P_i is excreted in the urine. Approximately 80% of P_i reabsorption occurs along the proximal tubule of the nephron (e.g. (Berndt & Knox, 1992)) and the remaining 20% is reabsorbed along the distal tubule. Proximal tubular P_i reabsorption is under the control of circulating hormones such as parathyroid hormone (PTH), growth hormone, thyroid hormone, the plasma P_i level itself, as well as a number of circulating proteins, collectively referred to as phosphotonins (e.g. Berndt & Kumar, 2007). The transmembrane flux of anionic P_i (monovalent or divalent) into the cytosol against its electrochemical gradient requires active transport mechanisms. In mammals, this need is fulfilled by secondary-active carrier proteins that have a high specificity for either $H_2PO_4^-$ or HPO_4^{2-} and use the inwardly directly electrochemical gradient of Na^+ as a free energy source to drive transport.

The two protein families considered in this review are gene products of the SLC20 and SLC34 solute carrier families, designated according to the solute carrier classification established by the Human Genome Organization (Hediger et al., 2004), www.bioparadigms.org/slc. This nomenclature will be used also when referring to nonmammalian orthologs. The gene products of the SLC34 family, often referred to as type II Na^+/P_i cotransporters, (or NaPi-II), comprise three isoforms: NaPi-IIa, NaPi-IIb and NaPi-IIc (SLC34A1, SLC34A2, SLC34A3, respectively). They are the principal active transport proteins for P_i absorption in the gut and reabsorption in the kidney. Those of the SLC20 family, also known as type III Na^+/P_i cotransporters, are currently represented by two isoforms, PiT-1 and PiT-2 (SLC20A1, SLC20A2, respectively). Until recently, they were considered to fulfill a housekeeping role consistent with their ubiquitous expression. However, new evidence suggests they fulfill specific roles in certain organs as outlined below.

Members of a third transporter gene family, SLC17 (Reimer & Edwards, 2004) were originally referred to as type I Na^+-dependent P_i transporters because the first cloned member of this family (NaPi-1) was originally shown to be a Na^+-dependent P_i transporter, using the *Xenopus* oocyte expression system (Werner et al., 1991). This protein was later shown to transport organic anions (Broer, et al., 1998; Busch, et al., 1996). Another SLC17 protein, BNP1, was also first classified as a Na^+/P_i transporter, but was then found to mediate vesicular glutamate transport (Bellocchio, Reimer, Fremeau, & Edwards, 2000). Thus, given their nonspecificity for transporting P_i, SLC17 proteins are no longer considered exclusively Na^+/P_i cotransporters and they will not be considered further in this review.

Finally, it should be noted that the mechanism by which P_i effluxes across the basolateral membrane of epithelial cells still remains unknown, the general consensus being that a yet-to-be-identified anion exchanger is responsible (e.g. Barac-Nieto, Alfred, & Spitzer, 2002).

1.1. SLC34 Family

The SLC34 proteins are the most extensively characterized in terms of their function, structure and regulation. They play vital physiological roles in the kidney for maintaining P_i homeostasis (for review, see Forster, Hernando,

Biber, & Murer, 2006; Murer, Hernando, Forster, & Biber, 2000). Transport studies in the precloning era using membrane vesicles from renal and intestinal epithelial tissue documented secondary-active, Na^+-dependent P_i transport (Berner, Kinne, & Murer, 1976; Hoffmann, Thees, & Kinne et al., 1976). Further studies used both cloned transporters and native tissue to characterize the kinetics and regulation of this transport (for review, see Biber, Hernando, Forster, & Murer, 2009; Murer et al., 2000).

Figure 1A compares the phylogeny of representative members of the SLC34 family from different species, including a bacterial ancestor from *Vibrio*

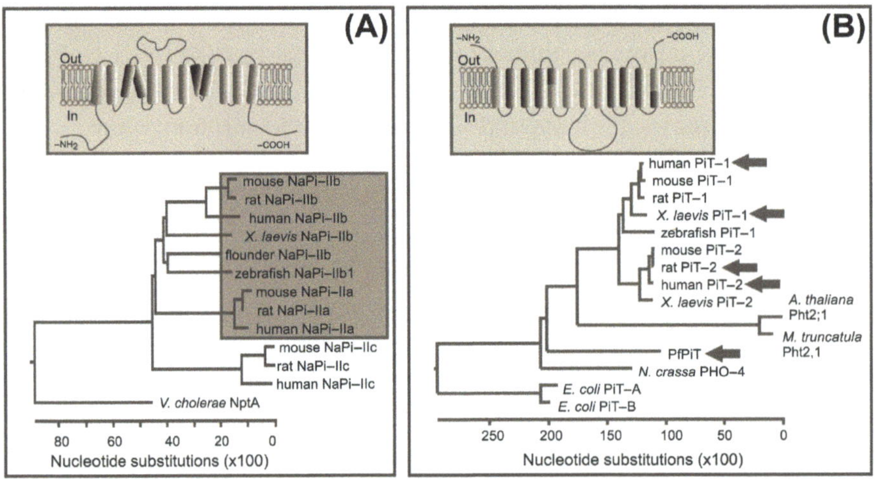

Figure 1 ***Phylogenetic trees of SLC34 and SLC20 families showing selected members.*** (A) SLC34 (type II Na^+/P_i transporters). The following sequences were used: mouse NaPi-IIb (AAC80007), rat NaPi-IIb (NP_445832), human NaPi-IIb (AAF31328), *X. laevis* NaPi-IIb (AAF21135), flounder NaPi-IIb (AAB16821), zebra fish NaPi-IIb (AF121796), mouse NaPi-IIa (AAC52361), rat NaPi-IIa (NP_037162), human NaPi-IIa (AAA36354), mouse NaPi-IIc (NP_543130), rat NaPi-IIc (NP_647554), human NaPi-IIc (NP_543153), *Vibrio cholerae* NptA (NP_230325). Boxed members are electrogenic. (B) SLC20 (type III Na^+/P_i transporters). The following sequences were used: human PiT-1 (NM_005415), mouse PiT-1 (AAB31458), rat PiT-1 (NM_031148), *X. laevis* PiT-1 (AAH59957), zebra fish PiT-1 (NM_213179), mouse PiT-2 (NM_011394), rat PiT-2 (NM_017223), human PiT-2 (NM_006749), *X. laevis* PiT-2 (BC084098), *Arabidopsis thaliana* Pht2; 1 (NM_113565), *Medicago truncatula* Pht2; 1 (AAN46087), *Plasmodium falciparum* PfPiT (AJ580003), *Neurospora crassa* Pho-4 (XM_954396), *Escherichia coli* PitA (NC_002655), *E. coli* PitB (NC_000913). Arrows indicate members with confirmed electrogenic behavior. Amino acid sequences were aligned using ClustalW. Genbank accession numbers are listed in parentheses. Insets in each panel depict consensus secondary topology of mammalian proteins. Inverted repeat regions indicated by dark shading. *(Data redrawn and modified from Virkki, Biber et al. (2007) with permission.)* See the color plate.

cholerae (NptA). All eukaryotic isoforms are predicted to have the same secondary topology (Fig. 1A, *inset*) comprising 12 α-helical units, cytosolic N and C termini and a common inverted repeat re-entrant motif. This latter feature is also predicted for NptA (Lebens, Lundquist, Soderlund, Todorovic, & Carlin, 2002), and is therefore thought to constitute the transport pathway and be involved in substrate coordination in all SLC34 proteins (Werner & Kinne, 2001). An important mechanistic subclassification of SLC34 family gene products concerns their electrogenicity. SLC34A1 and SLC34A22 (NaPi-IIa/b) are electrogenic, whereas SLC34A3 (NaPi-IIc) is electroneutral (Fig. 1A).

The first family member (SLC34A1or NaPi-IIa) was identified by expression cloning using *Xenopus* oocytes and a cDNA library derived from rat and human kidney (Magagnin et al., 1993). Immunohistochemistry confirmed its localization at the apical membrane of renal proximal tubular cells. NaPi-IIa has since been detected in other tissues such as brain, osteoclasts and osteoblast-like cells (Lundquist, Murer, & Biber, 2007) that suggests a potentially wider expression profile for this "renal" isoform. Nevertheless, the relative mRNA distribution for NaPi-IIa in humans is clearly weighted in favor of the kidney (Nishimura & Naito, 2008). Following the cloning of several NaPi-IIa isoforms, a Na^{+}-P_i cotransporter from the kidney and intestine of winter flounder was identified (Kohl et al., 1996; Werner, Murer, & Kinne, 1994). Its mammalian cousin (SLC34A2 or NaPi-IIb) was then cloned using a mouse embryonic EST clone (Hilfiker et al., 1998) and localized to the brush-border membrane of enterocytes and a number of other organs and tissues, including lung, colon, testes, salivary gland, thyroid gland mammary gland, uterus and liver (Frei et al., 2005; Hilfiker et al., 1998; Nishimura & Naito, 2008). In contrast to the flounder NaPi-IIb isoform, which is expressed in both kidney and intestine (Kohl et al., 1996), mammalian NaPi-IIb does not appear to be expressed at the protein level in the kidney. At the primary amino acid sequence level, NaPi-IIb isoforms are distinguished from the renal NaPi-IIa by a unique clustering of cysteine residues in their C-terminal region.

The third member of the SLC34 family to be identified (SLC34A3 or NaPi-IIc), similar to NaPi-IIa, is considered a renal-specific P_i transporter and is also apically expressed in the renal proximal tubule (Madjdpour, Bacic, Kaissling, Murer, & Biber, 2004; Picard et al., 2010; Segawa et al., 2002). It is distinguished from the other SLC34 proteins by its high expression in weaning animals, a cysteine cluster in the C-terminal region, similar to NaPi-IIb, and the distinctive functional hallmark of electroneutrality (Segawa et al., 2002).

In other respects, all three SLC34 members show a molecular identity of about 75% (Bacconi, Virki, Murer & Forster., 2005), if the N and C termini and the prominent extracellular loop that separates the protein into two halves are excluded (Section 3).

The expression and regulation of SLC34 proteins in their intestinal and renal context has been studied extensively because these organs represent the principle entry and exit points, respectively, for P_i in mammals. Renal regulation of NaPi-IIa has been the subject of extensive reviews (Forster et al., 2006; Miyamoto, Segawa, Ito, & Kuwahata, 2004; Murer et al., 2000; Tenenhouse, 2005; Tenenhouse & Murer, 2003; Werner & Kinne, 2001). The regulation of P_i reabsorption centers on controlling the number of transporter proteins in the proximal tubule brush-border membrane. So far, there is no evidence to suggest that the transport kinetics themselves are the target of regulatory factors.

The importance of SLC34 proteins in mammalian P_i handling is underscored by knockout animal studies and identification of naturally occurring mutations in patients with specific symptoms associated with impaired P_i homeostasis (for review, see Miyamoto et al., 2011). The critical role of NaPi-IIa for P_i homeostasis is further evidenced by the hyperphosphaturia phenotype documented in NaPi-IIa knockout mouse (Npt2a$^{-/-}$) (Beck et al., 1998). Dysregulation of NaPi-IIa abundance causes P_i-deficiency disorders, such as X-linked hypophosphatemia (XLH) and autosomal-dominant hypophosphatemic rickets (ADHR) (reviewed in Tenenhouse, 2005). No mutation in the NaPi-IIa gene has yet to be unequivocally linked to human hypophosphatemic disorders, and the present consensus is that the mutations alone are not responsible for the clinical disorders in the patients (Tenenhouse & Murer, 2003). On the other hand, naturally occurring mutations in the NaPi-IIc gene have been linked to the rare P_i-handling disease hereditary hypophosphatemic rickets with hypercalcuria (HHRH) (Bergwitz et al., 2006; Lorenz-Depiereux et al., 2006). Mutations in the NaPi-IIb gene have been proposed to cause pulmonary alveolar microlithiasis and possibly linked to testicular microlithiasis (Corut et al., 2006). In general, these mutations result in truncations and mistargeting to the plasma membrane rather than altered transport function of the protein itself.

1.2. SLC20 Family

SLC20 proteins show a widespread expression based primarily on the detection of mRNA in both epithelial and nonepithelial tissue (Bai, Collins, & Ghishan, 2000; Nishimura & Naito, 2008; Tatsumi et al., 1998; Uckert, Willimsky, Pedersen, Blankenstein, & Pedersen, 1998). Consequently, it was proposed that they fulfill a "housekeeping" role for P_i entry into most cells. They are represented by the two isoforms SLC20A1 (PiT-1) and SLC20A2 (PiT-2) that were originally identified as retroviral receptors that allow viral entry following a specific interaction with the receptor (Collins, Bai, & Ghishan, 2004). That the retroviral receptor Glvr-1, which renders cells susceptible to infection by the gibbon ape leukemia virus, was also a P_i transporter, was first proposed from its close homology to a phosphate permease (PHO-4) from the fungi *Neurospora crassa* (Johann, Gibbons, & O'Hara, 1992). A related protein, which renders cells susceptible to infection by amphotropic murine retrovirus (Ram-1) was subsequently identified (Miller & Miller, 1994; van Zeijl et al., 1994). Experimental evidence showing that retroviral receptors Glvr-1 and Ram-1 are electrogenic Na^+-dependent cotransporters then followed (Kavanaugh & Kabat, 1996; Kavanaugh et al., 1994; Olah, Lehel, Anderson, Eiden, & Wilson, 1994), and the receptors were thereafter renamed PiT-1 and PiT-2, respectively.

Proteins related to PiT-1 and PiT-2 are present in all phyla (Fig. 1B). However, one distinguishing feature concerns the driving cation. In prokaryotes and plants, P_i transport mediated by these proteins is coupled to the H^+ gradient (Daram et al., 1999; Harris, Webb, Howitt, & Cox, 2001; van Veen, 1997), whereas in animals and fungi, transport is coupled to the Na^+ gradient (Bai et al., 2000; Martinez & Persson, 1998; Tatsumi et al., 1998; Uckert et al., 1998; Versaw & Metzenberg, 1995). SLC20 proteins are presumed to have the same secondary structure (Fig. 1B, *inset*) with a common inverted repeat topology that differs significantly from that of SLC34 proteins. Moreover, it is assumed that SLC20 proteins are all electrogenic based on functional studies. Recent evidence also indicates that in addition to their housekeeping role, SLC20 proteins perform important physiological roles in bone P_i metabolism and in pathological conditions, such as hyperphosphatemia-induced calcification of vascular tissue (e.g. Ketteler & Giachelli, 2006), osteoarthritis (Cecil, Rose, Terkeltaub, & Liu-Bryan, 2005). Importantly, PiT-2 could play a significant role in diet-regulated P_i transport in the kidney, alongside NaPi-IIa and NaPi-IIc (Moe, 2009; Villa-Bellosta et al., 2009). The expression of both SLC20 isoforms at the mRNA level in human tissue confirms their ubiquity (Nishimura & Naito, 2008). Similar to SLC34 proteins, recent studies have further highlighted their developmental and physiological importance, pointing to novel and specific roles for this family. For example, loss of function mutations in the PiT-2 gene were reported that are implicated in disturbed brain P_i homeostasis (familial idiopathic basal ganglia calcification) (Wang et al., 2012). Furthermore, a PiT-1−/− mouse was reported to be embryonic fatal, suggested a critical role for PiT-1 in liver development (Beck et al., 2010).

This review focuses on the transport mechanism and structure–function relationships of SLC34 and SLC20 proteins. Given that it is the number of transporters in the membrane that determines the P_i flux, the characterization of the transport mechanisms and kinetic properties of these proteins are crucial to define the P_i flux contribution of a mixed population of transporters. Furthermore, the structure–function relationships can lead to greater insight into developing pharmaceuticals to target specific transporters. At the molecular level, 3-D crystal structures of SLC34 and SLC20 family members or their bacterial homologs are unavailable and therefore indirect methods (biophysical and biochemical) are the only means to correlate function with structure. Our current knowledge of transport mechanism and structure–function relationships has benefitted substantially from expression cloning of the different isoforms to allow investigation by heterologous expression, principally in *Xenopus* oocytes. Moreover, studies on the electrogenic isoforms can employ electrophysiology-based assays that allow resolution of the properties of the transport cycle in real time. We shall first review the proteins' kinetic properties, based primarily on investigation of the wild-type (WT) and engineered mutants expressed in *Xenopus* oocytes, and then consider their structure–function properties. For more detailed information on the roles that SLC34 and SLC20 proteins play in other aspects of P_i handling, the reader should consult extensive reviews on the physiological roles, regulation and pathophysiological implications (e.g. Biber et al., 2009; Forster, Hernando, Sorribas, & Werner, 2011; Miyamoto, Haito-Sugino et al., 2011; Tenenhouse, 2005).

2. TRANSPORT KINETICS

2.1. Steady-State Transport Properties

An overview of the steady-state properties of the two families is given in Table 1.

2.1.1. SLC34

2.1.1.1. Driving Forces

All SLC34 proteins use the inwardly directed Na^+ gradient to catalyze P_i transport (Fig. 2A). In addition, under voltage-clamp conditions, *Xenopus* oocytes that express the electrogenic isoforms show a strong dependence of P_i-induced current (I_{Pi}) on membrane potential. Typically, a change membrane of potential from 0 to −100 mV doubles I_{P_i} (Fig. 2A). On the basis of the finding of a 1:1 correlation between ^{32}P uptake and net charge translocated, I_{P_i} can then be used as a direct

Table 1 Kinetic properties of SLC34 and SCL20 proteins expressed in *Xenopus* oocytes

Family	SLC34			SLC20
Isoform	NaPi-IIa (SLC34A1)	NaPi-IIb (SLC34A2)	NaPi-IIc (SLC34A3)	PiT-1 (PiT-2) (SLC20A1/A2)
P_i species preferred	HPO_4^{2-}[a]	HPO_4^{2-}[a]	HPO_4^{2-}[b]	$H_2PO_4^{-}$[c]
Driving cations	Na^+ (Li^+[d])	Na^+ (Li^+[d])	Na^+[e]	Na^+, Li^+[c]
Electrogenicity (charge/cycle)	1[a]	1[a]	0[e]	1[c]
Stoichiometry ($Na:P_i$)	3:1[a,f]	3:1[a]	2:1[b]	2:1[c]
Presteady-state currents	Yes[g,f]	Yes[h,d,j]	No[b]	n.d.
$K_{0.5}^{P_i}$ (μM)	50[f], 54[g]	250,[k] 7[h], 29[i], 31[j]	70,[e] 80[l]	120[c], 24[m], 25[m]
$K_{0.5}^{Na}$ (mM)	40[f], 50[g]	25[h], 42[i], 46[j], 67[k]	48,[e] 43[l]	50[c]
pH dependence (pH↓)	Strong[g,f,n]	strong[h,j,k], weak[i]	Strong[e]	Weak[c]
Turnover (s^{-1}) (−60 mV)	4[g†]	9[o], 13[j]	n.d.	n.d.
Concentrating capacity (10:1 Na gradient, −60 mV)	10,000:1	10,000:1	100:1	1000:1
Arsenate (K_i^{As}) (mM)	1.08[p]	0.05[p]	1.01[p]	3.61[q] (3.95[q])
PFA (K_i^{PFA}) (mM)	1.05[p]	0.16[p]	0.9[p]	2.71[q] (4.63[q])
Leak (% of I_{P_i},1 mM P_i) (−100 mV)	~10[g,r]	~5[r]	n.d.	n.d.

n.d.: not determined; PFA: phosphonoformic acid.
[a]Rat NaPi-IIa, flounder NaPi-IIb (Forster, Loo et al. 1999).
[b]Mouse NaPi-IIc (Bacconi, Virkki et al. 2005).
[c]*Xenopus* P_iT-1 (Ravera, Virkki et al., 2007).
[d]Flounder NaPi-IIb (Andrini, Meinild et al., 2012).
[e]Human NaPi-IIc (Segawa, Kaneko et al., 2002).
[f]Human NaPi-IIa (Virkki et al., 2005a, 2005b).
[g]Rat NaPi-IIa (Forster, Hernando et al., 1998).
[h]Mouse NaPi-IIb (Forster, Virkki et al., 2006).
[i]Zebra fish NaPi-IIb2 (Graham, Nalbant et al., 2003).
[j]Flounder NaPi-IIb (Forster, Wagner et al., 1997).
[k]Zebra fish NaPi-IIb1 (Nalbant, Boehmer et al., 1999).
[l]Mouse NaPi-IIc (Ghezzi, Murer et al., 2009).
[m]Human PiT-1, rat P_iT-2 (Kavanaugh, Miller et al., 1994).
[n]Rat NaPi-IIa (Busch, Waldegger et al., 1994).
[o]L.V.Virkki, unpublished.
[p]Rat NaPi-IIa/b/c (Villa-Bellosta & Sorribas, 2008a, 2008b).
[q]Rat PiT-1,-2 (Villa-Bellosta, Bogaert et al., 2007).
[r]Rat NaPi-IIa, flounder NaPi-IIb (Andrini, Ghezzi et al., 2008).

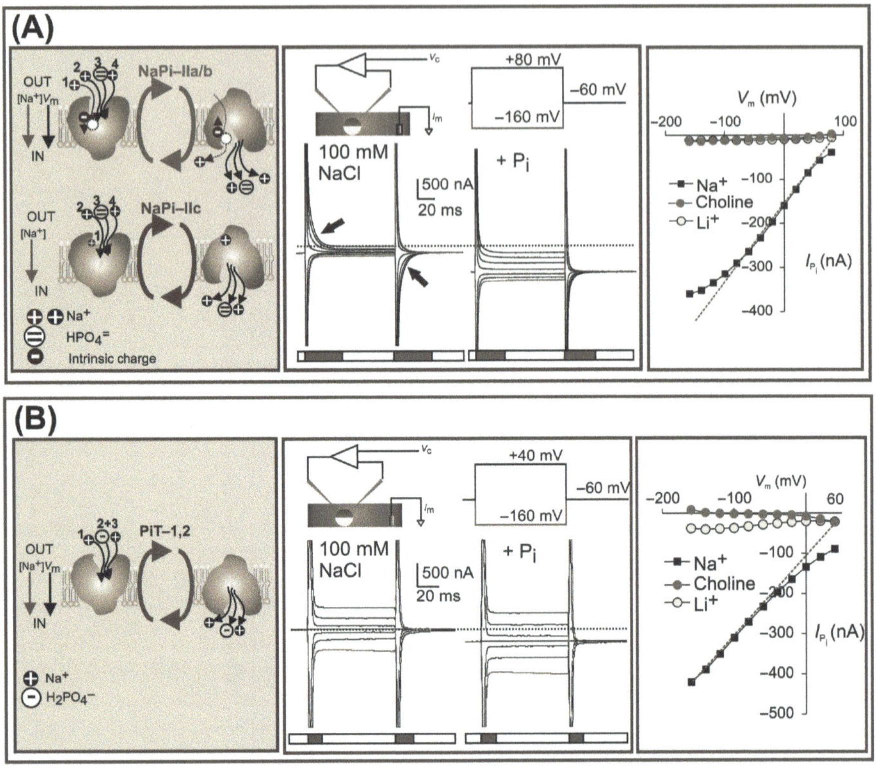

Figure 2 ***Overview of transport mechanisms of SLC34 and SLC20 proteins. (A) SLC34.*** *Left panel*: Cartoon of substrate interactions with electrogenic (NaPi-IIa/b) (*upper*) and electroneutral (NaPi-IIc) (*lower*) isoforms. Driving forces are indicated for the normal transport cycle. Substrates interact in an ordered sequence (indicated by numbers on cartoon) in which two Na^+ ions bind sequentially, followed by divalent P_i and a third Na^+ ion. For electrogenic NaPi-IIa/b, an effective negative charge, intrinsic to the protein confers voltage dependence to the transport cycle (Section 3.1.4). Movement of this charge leads to binding of the first Na^+ ion and subsequent translocation together with the other substrates, giving a 3:1 Na^+:P_i stoichiometry. After substrate release to the cytosol, the intrinsic charge again senses the transmembrane field, leading to a voltage-dependent reorientation of the empty carrier. For the electroneutral NaPi-IIc, a Na^+ ion first binds like NaPi-IIa/b, but is not translocated (Ghezzi, Murer et al., 2009), giving a 2:1 Na^+:P_i stoichiometry. *Center panel*: Original current recordings from a voltage-clamped oocyte expressing flounder NaPi-IIb in solution containing 100 mM Na^+ (*left*) and 100 mM Na^+ +1 mM P_i (*right*). The voltage was stepped from −160 to +80 mV in 40 mV increments. Arrows indicate presteady-state relaxations that are suppressed in the presence of P_i (see Section 2.2). Dashed line indicates zero current level. Bars indicate steady-state (open) and presteady-state (filled) regions of the records. *Right panel*: Steady-state P_i-dependent current (I_{Pi}) as a function of voltage for different external cations (at 100 mM) for an oocyte expressing flounder NaPi-IIb. Each data point is the difference

measure of the P_i transport rate (Forster, Loo, & Eskandari, 1999; Virkki, Forster, Bacconi, Biber, & Murer, 2005; Virkki, Forster, Biber, & Murer, 2005). For most isoforms, the voltage dependence of I_{P_i} also tends to saturate at the extremes of membrane potential that indicates the existence of rate limiting, voltage-independent partial reactions in the transport cycle (Section 2.2.1). In contrast, no I_{P_i} is detected for NaPi-IIc (Bacconi, Virkki, Biber, Murer, & Forster, 2005; Segawa et al., 2002). Moreover, ^{32}P uptake rates are independent of membrane potential (Bacconi et al., 2005, Ghezzi, Murer, & Forster, 2009) and cation interactions in the absence of P_i show no sensitivity to membrane potential (Ghezzi, et al., 2009).

2.1.1.2. Substrates

All SLC34 isoforms display Michaelian transport kinetics with respect to P_i (by radiotracer uptake or, where possible, by voltage-clamp assays), consistent with the translocation of one P_i molecule per transport cycle (e.g. Busch et al., 1994; Forster, Hernando, Biber, & Murer, 1998; Forster, Virkki, Bossi, Murer, & Biber, 2006; Forster et al., 1997; Ghezzi et al., 2009; Graham et al., 2003; Nalbant, Boehmer, Dehmelt, Wehner, & Werner, 1999; Segawa et al., 2002; Virkki, Forster, Bacconi, et al.,

between the oocyte holding current at the indicated membrane potential with and without 1 mM P_i (pH 7.4). No significant inward current is recorded when Na^+ is replaced with choline or Li^+. The dashed line indicates the deviation from linearity at extreme membrane potentials. (B) SLC20. *Left panel*: Cartoon of substrate interactions with PiT-1/2. Driving forces are indicated for the normal transport cycle. Substrate interactions comprise an initial binding of one Na^+ ion, followed by a random interaction of monovalent P_i and a second Na^+ ion (Ravera, Virkki et al., 2007). Reorientation of the loaded carrier leads to substrate release in the cytosol. Neither intrinsic charge movements nor voltage-dependent Na^+ ion interactions have been detected experimentally; therefore, the translocation step itself is hypothesized to confer voltage dependence. *Center panel*: Original current recordings from an oocyte expressing *Xenopus* PiT-1 superfused with 100 mM Na^+ (left) and 100 mM Na^+ +1 mM P_i (right). The voltage was stepped from −160 to +40 mV in 20 mV increments. Note that no presteady-state relaxations are visible. Dashed line indicates zero current level. Bars indicate steady-state (open) and presteady-state (filled) regions of current records. *Right panel*: Steady-state P_i-dependent current (I_{Pi}) plotted as a function of voltage for different driving cations (at 100 mM) for an oocyte expressing *Xenopus* PiT-1. No significant inward current is recorded when Na^+ is replaced with choline, whereas currents are reduced by ~85% for $V > 0$ mV when Na^+ is replaced by Li^+. *(Data redrawn and modified from* Virkki, Biber et al. (2007) *with permission.)* See the color plate.

2005, Virkki, Forster, Biber, et al., 2005). The apparent affinity constants for total P_i $\left(K_{0.5}^{P_i}\right)$ are typically <100 μM (pH 7.4 and $[Na^+]_o$ = 100 mM), with the exception of a zebra fish isoform (NaPi-IIb1) $\left(K_{0.5}^{P_i} = 250\ \mu m\right)$ (Graham et al., 2003). In contrast, the murine and rat NaPi-IIb exhibit exceptionally high apparent P_i affinities $\left(K_{0.5}^{P_i} \approx 10\ \mu M\right)$ (Forster et al., 2006; Villa-Bellosta & Sorribas, 2008a, 2008b).* Furthermore, all SLC34 proteins preferentially transport divalent P_i (HPO_4^{2-})(Bacconi et al., 2005; Forster et al., 1999). For the electrogenic NaPi-IIa and NaPi-IIb, preference for HPO_4^{2-} was established by simultaneously measuring the ratio of net charge translocated to P_i influx using ^{32}P as a tracer, on single voltage-clamped oocytes (Forster et al., 1999). This preference has also been confirmed by surface pH measurements on *Xenopus* oocytes using a pH-sensitive electrode (Ravera, Virkki, Murer, & Forster, 2007). For the electroneutral NaPi-IIc, the preference for HPO_4^{2-} is consistent with its 2:1 Na^+:P_i stoichiometry (Bacconi et al., 2005), and based on recent surface pH measurements using an ion-sensitive field effect transistor (ISFET) sensor (Schaffhauser et al., 2012).

With Na^+ as the variable substrate, the cotransport rate (by uptake or voltage clamp) shows sigmoidicity, with reported Hill coefficients ranging from 1.8 (Forster et al., 1999) to 3 (Forster et al., 1998). A Hill coefficient >1 indicates that >1 Na^+ ion is required to drive P_i transport and noninteger values reflect the degree of cooperativity of multiple cation interactions (e.g. Weiss, 1997). The apparent affinities for Na $\left(K_{0.5}^{Na}\right)$ are also reasonably consistent among isoforms, lying typically in the range 40–60 mM. The apparent affinities for Na and P_i indicate that under normal mammalian physiological conditions most SLC34 isoforms function at or near their maximum transport velocities.

Inorganic arsenate (As^V), with a similar structure to P_i, is also a potential substrate for SLC34 proteins. Arsenate is known to compete with P_i with an inhibition constant (K_i^{As}) typically ≈1 mM for NaPi-IIa/c but as low as ≈50 μM for NaPi-IIb (Villa-Bellosta & Sorribas, 2008a, 2008b), that is consistent with the higher P_i apparent affinity for this isoform. Given that the apparent affinity for As^V is close to K_i^{As}, it was therefore proposed that NaPi-IIb could play a significant role in the toxicokinetics of arsenic exposure (Beene, Halluer, Yoshinaga, Hamdi, & Liu, 2011; Villa-Bellosta & Sorribas, 2009; Villa-Bellosta & Sorribas, 2010).

* These species-specific differences in apparent P_i affinity may reflect different P_i-handling requirements. Note that the apparent affinity constants were determined by heterologous expression in *Xenopus* oocytes and determined at temperatures typically 20-24 °C that differ from the normal physiological conditions.

2.1.1.3. Stoichiometry and Concentrating Capacity

Simultaneous voltage-clamp measurement of net charge translocation and uptake (^{32}P or ^{22}Na) was used to resolve previous uncertainties concerning the stoichiometry of electrogenic Na-coupled P_i transport. For rat NaPi-IIa and flounder NaPi-IIb, a consistent value of 3Na:1P_i was obtained (Forster et al., 1999) with one net positive charge translocated per transport cycle. For NaPi-IIc, the predicted stoichiometry of 2:1 was established by dual tracer (^{32}P and ^{22}Na) assay (Bacconi et al., 2005). The stoichiometry difference between NaPi-IIa/b and NaPi-IIc means that the theoretical P_i concentrating capacity, assuming a 10:1 Na concentration gradient at a membrane potential of −60 mV, is approximately 100-fold higher for NaPi-IIa/b. With a 10-fold greater inward flux of Na^+ ions and net charge movement, compared to NaPi-IIc, this would significantly increase the energetic cost to a cell.

2.1.1.4. Cation Specificity and pH

Consistent with results from precloning studies on vesicles from native tissue, SLC34 proteins were considered exclusively Na^+-driven as little or no transport activity is observed when other cations fully replace external Na^+. However, it was recently shown that Li^+ ions are able to substitute for at least one of the three Na^+ ions in the transport cycle. Moreover, simultaneous electrophysiology and ^{22}Na uptake assays indicate that one Li^+ ion, together with two Na^+ions are transported per cycle (Andrini, Meinild, Ghezzi, Murer, & Forster, 2012). Unlike some other solute carriers, protons do not substitute for Na^+ as the driving cation (Forster, Biber, & Murer, 2000). However, the transport rate of all SLC34 proteins is strongly dependent on pH. Special care must be taken when assessing the effect of pH on SLC34 transport kinetics because P_i is titratable. Thus, if we assume a $pK_a = 6.8$ for P_i titration in low ionic strength physiological solutions, then at pH 7.4, $HPO_4^{2-}/H_2PO_4^- = 4$ and this ratio will be <1 for pH <6.8. In addition, protons alter the transport kinetics by interacting with voltage-sensitive partial reactions and competing for the third Na^+-binding interaction (Forster et al., 2000). The net result is a progressive reduction in maximum transport rate as external $[H^+]$ increases, together with a reduced apparent affinity for HPO_4^{2-}. At low pH (≈5), where P_i is almost only monovalent, P_i can still interact with the protein to block the uncoupled leak current (Ehnes, Forster, Bacconi, et al., 2004, Ehnes, Forster, Kohler, et al., 2004) (Section 2.1.1.6).

2.1.1.5. Binding Order

Early steady-state kinetic studies using native tissue established that Na-coupled P_i cotransport proceeds in an ordered manner with Na^+ binding first (e.g. Hoffmann et al., 1976). Studies one NaPi-IIa expressed in *Xenopus* oocytes refined this model by proposing that one Na^+ ion interacts with the protein before P_i binds (Forster, 2007; Forster et al., 1998), consistent with the finding that the Na^+-dependence of the uncoupled leak (see below) showed Michaelian kinetics (Forster et al., 1998). Moreover, with saturating P_i, the transport rate is strongly Na^+-dependent, which indicates that Na^+ is the last substrate to bind before translocation. The proposed "Na-Pi-Na-Na" model (Forster et al., 1998) has since been further revised to become the "Na-Na-Pi-Na" model based on (i) evidence from voltage-clamp fluorometry studies on NaPi-IIb with an engineered fluorophore labeling site (Virkki, Murer, & Forster, 2006a, 2006b) (Section 3.1.3), and (ii) detailed analysis of the wild-type (WT) presteady-state kinetics (e.g. Andrini et al., 2012) (Section 2.2.2). According to this model, two Na^+ ions bind sequentially followed by P_i binding. Finally, a third Na^+ binds before a proposed rate-limiting reorientation of the fully loaded carrier (Fig. 2A). Li^+ ions can compete with Na^+ for occupancy of the first cation-binding site (Andrini et al., 2012) (Fig. 2A). The order of release of substrates to the cytosol is still unknown.

For the electroneutral NaPi-IIc, evidence from fluorometric assays suggests that, similar to its electrogenic cousins, three Na^+ ions interact but only two are translocated and ultimately released to the cytosol to complete the transport cycle (Ghezzi et al., 2009) (Fig. 2A). The fate of the first bound Na^+ ion is critical in defining the electrogenic properties of the particular isoform. For NaPi-IIa/b, this Na^+ binds within the transmembrane field and is translocated; for NaPi-IIc, it may simply serve to increase the cooperativity of the second Na^+ interaction before P_i binding (Ghezzi et al., 2009).

2.1.1.6. Uncoupled Leak

Similar to many cation-coupled cotransporters, electrogenic SLC34 proteins exhibit an uncoupled leak current in the absence of the driven substrate (Andrini, Ghezzi, Murer, & Forster, 2008; Forster et al., 1998). No evidence for a Na^+-leak is found for the electroneutral NaPi-IIc (Bacconi et al., 2005), although naturally occurring mutations in NaPi-IIc have been reported to result in a significant Na^+-leak (Jaureguiberry, Carpenter, Forman, Juppner, & Bergwitz, 2008). The underlying mechanism at the molecular level has yet to be elucidated. Strong experimental evidence supports the notion that the leak observed in WT NaPi-IIa and NaPi-IIb is intrinsic to the protein (Andrini et al., 2008; Forster et al., 1998). The leak current is inhibited by phosphonoformic acid (PFA) (Section 2.1.3) but, unlike the cotransport activity, the leak is unaffected by external protons. Indeed, the inhibitory action of PFA on the leak current can be recapitulated using P_i at very low pH, where cotransport is fully suppressed (Ehnes, Forster, Bacconi, et al., 2004, Ehnes, Forster, Kohler, et al., 2004). This salient finding implies that monovalent P_i do interact with SLC34 proteins. On the basis of the leak current kinetics, it was proposed that the leak results from the movement of the first Na^+ ion through the protein in a uniport-like manner (Andrini et al., 2008; Forster et al., 1998) (Fig. 4A). This conclusion is also consistent with the high temperature dependence of the leak current (Bacconi, Ravera, Virkki, Murer, & Forster, 2007). In summary, the leak and cotransport modes are most likely intrinsically related and share the same partial reactions. Under normal physiological conditions with saturating P_i levels, the leak mode will make a negligible contribution to net Na^+ influx.

2.1.1.7. Turnover Rate

Transporter turnover rate defines the transport rate of a single protein. This can be determined from a measurement of the transport rate of a population of proteins under defined conditions (e.g. saturating substrate) if the number of active transporters is known. For electrogenic transporters, the steady-state substrate-induced current at a given voltage (V) and substrate concentration for N active transporters is given by:

$$I_S(V) = N\phi(V)n_t e \tag{10.1}$$

where $\phi(V)$ is the turnover rate (transport cycles, s^{-1}), n_t is the number of charges translocated per cycle and e is the electronic charge. One approach to estimate N uses the ratio of the total mobile charge (Q_{max}) to the apparent valence of each cotransporter functional unit (z) obtained from presteady-state analysis (Section 2.2.2). The turnover rate is then:

$$\phi(V) = I_S(V)z/(n_t Q_{max}) \tag{10.2}$$

Published estimates of $\phi(V)$ for SLC34 proteins lie in the range 4–13 s^{-1} (−60 mV, ~20 °C) (Forster et al., 2006). This range encompasses that reported for other electrogenic cotransporters and is typically <100 s^{-1} (e.g. SGLT1: 57 s^{-1} (Loo, Hazama, Supplisson, Turk, & Wright, 1993)); GAT1: 6–15 s^{-1} (Gonzales et al., 2007; Mager et al., 1993); EAAT2: 29 s^{-1} (Wadiche, Arriza, Amara, & Kavanaugh, 1995); NIS: 36 s^{-1} (Eskandari et al., 1997).

There are strong indications, based on freeze fracture studies of oocyte membranes, that z obtained from presteady-state data is an underestimate of the true charge/carrier (see Zampighi et al., 1995; Gonzales et al., 2007). Therefore, turnover rates using this approach may also be underestimated. Furthermore, SLC34 protein transport kinetics are strongly temperature-dependent (e.g. Bacconi et al., 2007). Thus, we would expect higher values under mammalian physiological conditions (37 °C), given that the estimates of ϕ were obtained from oocyte data (18–20 °C). If we assume that the temperature dependence of P_i transport for the flounder NaPi-IIb (Bacconi et al., 2007) is representative for all isoforms ($Q_{10} = 2.3$), then the turnover estimated experimentally for rat NaPi-IIa ($4.1 \pm 0.6\ s^{-1}$, at −60 mV) (Forster et al., 2006) would be $>10\ s^{-1}$ at 37 °C. The turnover rate for NaPi-IIc has not been experimentally determined.

2.1.2. SLC20

2.1.2.1. Driving Forces

SLC20 proteins also use the inwardly directed Na^+ gradient to catalyze P_i transport (Fig. 2B). From voltage-clamp studies, *Xenopus* oocytes expressing PiT-1 or PiT-2 display a curvilinear dependence of transport rate (P_i-induced current, I_{Pi}) on membrane potential. Typically, a change membrane potential from 0 to −100 mV doubles the transport rate (Fig. 2B). However, unlike most SLC34 isoforms, evidence for rate-limiting behavior at the hyperpolarizing extreme is less apparent (Kavanaugh & Kabat, 1996; Ravera et al., 2007).

2.1.2.2. Substrates

Similar to SLC34 proteins, SLC20 proteins show Michaelian kinetics with respect to P_i (by uptake or electrophysiology), consistent with the translocation of one P_i molecule per transport cycle (e.g. Bai et al., 2000; Bottger et al., 2006; Kavanaugh et al., 1994; Ravera et al., 2007). A cooperative P_i-dependence was reported for PiT-2 expressed in *Xenopus* oocytes under non-voltage-clamped conditions, a finding consistent with this isoform behaving as a functional dimer (Bottger et al., 2006). This finding has yet to be confirmed under voltage-clamp conditions. SLC20 proteins show an apparent affinity constant for total P_i ($K_{0.5}^{Pi}$) typically <100 μM (pH 7.4, $[Na^+] = 100$ mM) and they preferentially transport monovalent P_i ($H_2PO_4^-$) (Ravera et al., 2007; Saliba et al., 2006). The P_i species preference was established by combining electrophysiology and tracer flux assays as for the SLC34 isoforms (Ravera et al., 2007), and has also been confirmed by surface pH measurements on *Xenopus* oocytes (Ravera et al., 2007; Schaffhauser et al., 2012). The preference for monovalent P_i and relative insensitivity of transport kinetics to reduced pH means that SLC20 proteins could support P_i transport under acidotic conditions, where SLC34 proteins are functionally compromised. With Na as the variable substrate, the P_i activation kinetics for Pit-1 and Pit-2, determined electrophysiologically show no sigmoidicity. Given that the determined stoichiometry is 2:1, this indicates a weak cooperativity between Na interactions. However, for the malarial homolog (PfPIT), a strong cooperativity for Na was found (Saliba et al., 2006).

As with SLC34 proteins, inorganic arsenate (As^V) was also found to interact with PiT-1 (with $K_{0.5}{}^{As} = 0.83$ mM), but not for PiT-2 (Ravera et al., 2007), however, direct evidence for As^V as a cotransported species was lacking in that study. Recently, uptake assays using $^{73}As^V$ as a radiotracer (Villa-Bellosta & Sorribas, 2010) established that As^V is indeed transported by both SLC20 isoforms. The somewhat higher $K_{0.5}{}^{As} \approx 4$ mM may reflect the lack of controlled membrane potential in that study. It is unlikely that SLC20 proteins would provide a significant pathway for arsenic toxification in mammals given the relatively poor apparent affinity for As^V.

2.1.2.3. Stoichiometry and Concentrating Capacity

The Na^+:P_i stoichiometry for PiT-1 of 2:1 has been determined by simultaneous ^{22}Na and ^{32}P uptake under voltage-clamp conditions (Ravera et al., 2007) as well as by means of a static-head approach using malarial parasites expressing PfPiT (Saliba et al., 2006). These findings are consistent with the SLC20 preference for monovalent P_i.

2.1.2.4. Cation Specificity and pH

PiT-1 is significantly less sensitive to pH compared to the mammalian NaPi-IIa and flounder NaPi-IIb. The maximum transport rate is relatively constant over 3 pH units and $K_{0.5}^{P_i}$ increases significantly only at pH < 6. Unlike SLC34 proteins that are exclusively Na^+ driven, for PiT-1, Li^+ can fully replace Na^+ as the driving cation, albeit with a significantly reduced transport rate (Ravera et al., 2007). In the absence of Na^+, lowering pH from 7.5 to 6.0 induced significant P_i uptake in *Xenopus* oocytes that expressed PiT-2. This finding suggests that H^+ may also substitute for Na^+ for this isoform (Bottger et al., 2006; Villa-Bellosta, Bogaert, Levi, & Sorribas, 2007). The increased $K_{0.5}^{P_i}$ for PiT-1 with acidification suggests that protons may interact directly with the protein to reduce substrate affinity.

2.1.2.5. Binding Order

Interpretation of assays under voltage-clamp conditions with varying substrates suggests that Na^+ is the first ion to interact followed by an apparent random binding of Na^+ and P_i (Ravera et al., 2007).

2.1.3. Inhibitors for SLC34 and SLC20 Proteins

2.1.3.1. SLC34

Interest in inhibitors focuses on blockers for NaPi-IIb as a means to reduce intestinal P_i absorption in chronic kidney disease. Cotransport by all SLC34 isoforms is inhibited by PFA or foscarnet (Loghman-Adham, 1996), which competes for P_i binding. However, PFA has a higher K_i (~0.4–0.6 mM, e.g. Villa-Bellosta & Sorribas, 2009) than $K_{0.5}^{P_i}$ and displays nephrotoxic effects (Loghman-Adham, 1996). These characteristics make it unattractive as a means to limit dietary P_i absorption and prevent hyperphosphatemia in cases where kidney function is compromised (Wallin & Ryrfeldt, 1995). Other pharmacologically interesting compounds with significantly lower K_is compared to PFA have been the subject of commercial studies (Weinstock, 2004). These possibly act in a noncompetitive manner on SLC34 proteins, as reported for the compound JTP-59557 (Matsuo et al., 2005).

2.1.3.2. SLC20

PFA is a very poor inhibitor of SLC20-mediated transport activity (Ravera et al., 2007; Villa-Bellosta et al., 2007; Villa-Bellosta & Sorribas, 2009), and currently no other specific nontransportable inhibitors have been reported.

2.2. Presteady-State Kinetics and Voltage Dependence

2.2.1. Background

The voltage dependence of electrogenic cotransport is postulated to arise from transitions between conformational states that involve charges sensing the membrane potential. At extreme membrane potentials, these partial reactions in the cotransport cycle are no longer rate determining, thereby giving rise to the saturating effects seen in the steady-state I_{P_i} vs V relation (*right panel*, Fig. 2A). The charge movements reflect voltage-dependent conformational rearrangements that involve (i) charges intrinsic to the protein (i.e. associated with the empty carrier), and (ii) the movement of external cations to and from their respective binding sites within the protein. For cotransporters, the first experimental evidence for the existence of mobile charges was documented by applying rapid steps in membrane potential to *Xenopus* oocytes heterologously expressing SGLT1 (Loo et al., 1993) and GAT1 (Mager et al., 1993). With this approach, the time-dependent redistribution of mobile charges from a population of independently functioning transporters manifests itself macroscopically as transient decaying currents, termed presteady-state relaxations. For a single protein, these charge movements would most likely be discrete "shots" of current occurring with the change of conformational state—the probability distribution of which would follow Poisson statistics. They would be well below the detection limit of current instrumentation, for unlike ion channels, for which the current through a single channel is typically on the order of picoAmpere or more, the inherently slow turnover rate of cotransporters implies that for a single charge translocated at 100 s^{-1}, the current per protein will be ≈0.016 fA. However, a large population of proteins (typically $>10^9$ in the oocyte membrane) gives resolvable decaying currents comprising one or more exponential decays, depending on the number of partial reactions involved.

2.2.2. *SLC34*

2.2.2.1. Basic Properties

Presteady-state relaxations attributed to rat/human NaPi-IIa (Forster et al., 1998; Virkki, Forster, Bacconi, et al., 2005; Virkki, Forster, Biber, et al., 2005) and flounder/mouse NaPi-IIb (Forster et al., 1997; Forster et al., 2006; Andrini et al., 2012) give a macroscopic view of their molecular-scale movements in real time and have been instrumental in elucidating the underlying mechanism of the voltage-dependency of SLC34 proteins. The relaxations depend on the external [Na^+], are suppressed by P_i and the inhibitor PFA and are still detectable in the absence of external Na^+ (choline replacement) (Fig. 3A). Importantly, the amount of charge displaced for equal but opposite voltage steps is balanced, which confirms that they are capacitative in nature and arise from the displacement of a fixed number of mobile charges within the transmembrane electric field. Presteady-state relaxations are not detectable in *Xenopus* oocytes expressing functional NaPi-IIc, consistent with the notion that no voltage-dependent partial reactions contribute to its transport cycle.

2.2.2.2. Analysis and Interpretation

The oocyte endogenous capacitive charging component is first eliminated from the total current (e.g. by curve fitting, or by subtracting the data set obtained with PFA to suppress relaxations), and then the relaxations are

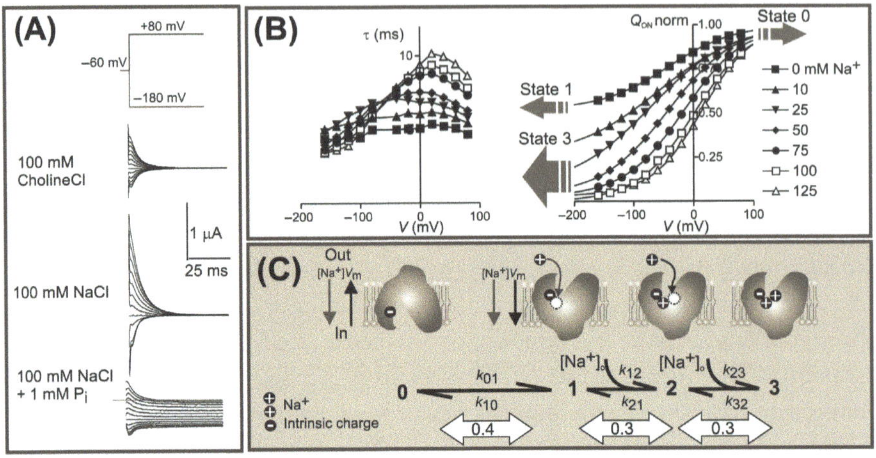

Figure 3 ***Presteady-state kinetics of electrogenic SLC34 proteins.*** (A) current recordings from a representative voltage-clamped oocyte expressing flounder NaPi-IIb in response to voltage steps from −60 mV to potentials in the range −180 mV to +80 mV in 20 mV increments for superfusion conditions indicated. Endogenous steady-state currents and capacitive transients were eliminated by baseline subtraction and curve fitting. (B) Relaxation time constant (τ) (*left*) and normalized charge (Q_{norm}) (*right*) plotted as a function of membrane potential (*V*) for a representative voltage-clamped oocyte expressing flounder NaPi-IIb with superfusion conditions indicated. Arrows on Q_{norm}–*V* data indicate the state occupancy at the voltage limits according to the scheme in (C). Continuous traces are fits using Eqn (10.3). (C) Cartoon depicting a four-state model that can account for the data in (B). Charge movements in response to a step change in membrane voltage arise from the movement of a lumped negative charge intrinsic to the protein and two Na^+ ions moving to their binding sites within the transmembrane electric field. At very positive potentials, all proteins occupy state 0, and at very negative potentials state 3 is occupied. Fitting the Q_{norm}–*V* data with a single Boltzmann function assumes a two-state model. At intermediate potentials, the proteins would be distributed among the four states according to the rate constants for the three partial reactions. Effective valence of each partial reaction is indicated, based on data analysis and simulations (Andrini, Meinild et al., 2012). See the color plate.

fit with a decaying exponential function. The relaxation time constant (τ) plotted as a function of the test (or "target") potential shows strong voltage dependence, as expected for a nonlinear dielectric. The characteristic depolarizing shift of the maximum τ with increasing [Na^+] (Fig. 3B) indicates that [Na^+] determines that rate of charge movement. Numerical integration of the relaxations gives the charge displaced (*Q*) as a function of *V*. For the WT NaPi-IIa/b, this relationship shows saturation at the potential extremes, as expected for a discrete number of mobile charges that move within the transmembrane electric field (Fig. 3B). The *Q*–*V* data can be parameterized by fitting with a single Boltzmann function of the form:

$$Q(V) = Q_{max}/\left(1 + \exp\left(V_{0.5} - V\right)ze/kT\right) + Q_{hyp} \qquad (10.3)$$

where $Q(V)$ is the charge moved for a voltage step from a given holding potential to V; z is the slope factor (interpreted as the apparent valency, associated with a lumped hypothetical charged entity that moves across the whole transmembrane electric field); Q_{max} is the total displaceable charge; Q_{hyp} is the charge at the hyperpolarizing extreme that depends on the holding or starting potential; $V_{0.5}$ is the voltage at which 50% of the total charge that has been displaced; e is the electronic charge; k is the Boltzmann's constant and T is the absolute temperature. If the charge distribution is first offset by Q_{hyp} and normalized to Q_{max} in maximum [Na^+], then a depolarizing shift with increasing [Na^+] is found. This is similar to the shift observed for the τ–V data (Fig. 3B). The intrinsic mobile charge of the protein (measured in the absence of external Na^+) accounts for ≈40% of the total detected with Na^+ present. A simple kinetic scheme that satisfactorily accounts for these data has four states corresponding to four conformations of the protein (Andrini et al., 2012) (Fig. 3C). The empty carrier is assumed to occupy one of the two states (state 0 or state 1) depending on the membrane potential. For a strong hyperpolarizing change in membrane potential, the protein assumes the "outward" facing conformation (state 1); two Na^+ ions can then interact sequentially so that the system occupies state 3 with two Na^+ ions bound (Andrini et al., 2012).

The dependence on external [Na^+] of the fit parameters (Q_{max}, $V_{0.5}$, z) (Eqn 10.3) reveals important properties that have been used for testing the validity of models for cation interaction. In particular, when the $V_{0.5}$ vs [Na^+] data for the WT flounder NaPi-IIb are plotted semilogarithmically, a linear relationship for [Na^+] > 25 mM was found with a limiting slope close to ≈120 mV/decade, consistent with the sequential interaction of two Na^+ ions ((Fig. 4B) (Andrini et al., 2012)). In contrast, when Li^+ replaced Na^+ in the external medium, the slope with respect to [Li^+] was reduced to ≈60 mV/decade, consistent with one Li^+ ion interacting with the protein (Andrini et al., 2012). Deviations from linearity are seen at low [Na^+] where the Q–V data can no longer be described in terms of a single Boltzmann function.

2.2.2.3. A 10-State Kinetic Model for Electrogenic SLC34

A 10-state kinetic model that describes the electrogenic SLC34 cotransport cycle as a sequence of partial reactions between 10 hypothetical conformational states of the protein has been proposed (Fig. 4A). Numerical simulations using this model predict the steady-state cotransport current (Fig. 4B) and presteady-state parameters (Fig. 4C) that match closely with those obtained from experimental data, thereby validating its application. In addition to providing basic mechanistic insight into the differences in voltage-dependent

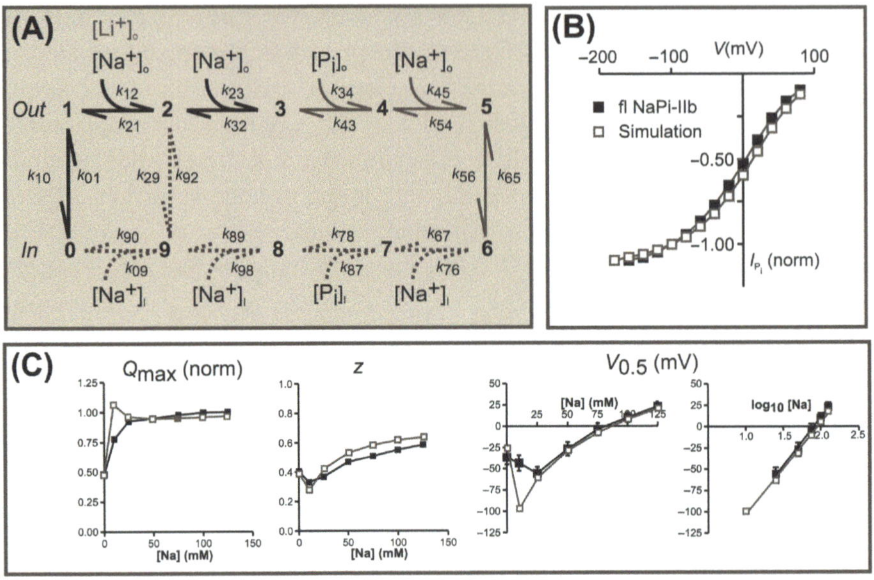

Figure 4 A 10-state model for simulating steady-state and presteady-state kinetics of electrogenic SLC34 proteins. (A) Scheme of 10-state model. The numbered "states" represent unique conformations of the protein that either favor substrate interactions from the external membrane face (1 ↔ 5) or on the inside (0 ↔ 6), consistent with an alternating access mechanism. The leak pathway is hypothesized to be 2 ↔ 9. Each transition is assumed to obey pseudo first-order kinetics, and for generality, nonrapid equilibrium conditions were assumed (see Parent, Supplisson et al., 1992; Loo, Eskandari et al., 2002). The scheme is symmetric, whereby substrate interactions from the external and internal membrane surfaces follow the same order, although this has not yet been established experimentally. To simplify the modeling, the transitions between state 6 and 0 are lumped into a single partial reaction (Andrini, Meinild et al., 2012). All voltage dependencies for the cotransport cycle can be accounted for by partial reactions that precede P_i binding (Fig. 3C). For the simulations, the second Na^+ interaction (2 ↔ 3) was assumed to be electroneutral. This is consistent with a proposed model for NaPi-IIc, in which the first Na^+ ion is not translocated and the lack of experimental evidence for voltage-dependent transitions (Ghezzi, Murer et al., 2009). More recent studies suggest that for the electrogenic NaPi-IIb, this partial reaction also involves charge movement (Andrini, Meinild et al., 2012). Li^+ ions compete for occupancy of the first Na^+-binding site, and in doing so, can replace one of the three Na^+ ions that participate in cotransport (Andrini, Meinild et al., 2012). Numerical simulations using this model were performed by solving the set of differential equations that describe the transitions between states. Rate constants for transitions involving charge movement were defined according to simple symmetrical barrier models based on Eyring transition state theory (see Ghezzi, Meinild et al., 2011; Andrini, Meinild et al., 2012 for more details). (B) Normalized steady-state I_{Pi} plotted as a function of membrane potential comparing representative data from an oocyte expressing flounder NaPi-IIb (filled squares) and simulated data (open squares). Data are normalized to I_{Pi} at −100 mV. (C)

kinetics for WT isoforms (e.g. Forster et al., 2006), and a mechanistic basis for cation interactions (Andrini et al., 2012; Forster et al., 2000), such a model can be used to identify critical partial reactions when the steady-state and presteady-state kinetics are altered by mutagenesis (see Section 3.1.4) (Bacconi et al., 2007; Ehnes, Forster, Bacconi, et al., 2004, Ehnes, Forster, Kohler, et al., 2004; Ghezzi et al., 2011). At present, the model is severely underdetermined because the kinetics of substrate interaction on the cytosolic side are unknown.

2.2.3. SLC20

Oocytes expressing PiT-1 do not exhibit readily detectable presteady-state relaxations superimposed on the oocyte capacitive charging transient (Fig. 2B). Efforts to resolve relaxations have involved subtraction of the step response in control superfusate from the response in the presence of saturating P_i (Ravera et al., 2007). Unlike NaPi-IIa/b, charge balance for the subtracted data was not found. This may reflect technical limitations in charge detection, or indicate the presence of another voltage-gated conductance that is sensitive to P_i. The lack of a specific inhibitor and the small magnitude of the associated charge movement have so far precluded more detailed analysis. It is also possible that PiT-1 transduces membrane potential in a different manner from NaPi-IIa/b, or that the charge movements are so fast that they cannot be resolved by the TEVC method. One possibility is that the voltage dependence arises from the translocation of the fully loaded carrier (net charge +1) in a mechanism similar to that proposed for neutral amino acid transporters of the SLC6 family (B^0AT1/SLC6A19) (O'Mara, Oakley, & Broer, 2006). The lack of evidence for presteady-state relaxations in the presence or absence of Na^+ would further support this possibility.

Comparison of Boltzmann fit parameters derived from simulations (open squares) of *Q–V* (in absence of P_i) with representative data obtained from oocytes expressing flounder NaPi-IIb (filled squares). The normalized Q_{max} is reasonably independent of [Na^+] except for low [Na^+] where the single Boltzmann assumption breaks down. The magnitude and variation of the apparent valence (slope factor of Boltzmann function (Eqn 10.3) with [Na^+] similarly reflects the fact that charge movement arises from three partial reactions. The limiting slope of the logarithmic plot of $V_{0.5}$ vs [Na^+] is ≈120 mV/decade, which is consistent with two Na^+ ions binding to the protein. See the color plate.

3. STRUCTURE–FUNCTION RELATIONSHIPS

3.1. SLC34

3.1.1. Molecular Features

SLC34 proteins belong to a unique class of membrane proteins and appear to share no homology with the other main cation-driven transporter families, even at the bacterial level. They are assumed to be functional monomers based on a study using the rat NaPi-IIa isoform (Kohler, Forster, Lambert, Biber, & Murer, 2000), but there is also evidence that they may oligomerzse as dimers or possibly tetramers in the membrane (Forster et al., 2006; Gisler et al., 2008; Haito-Sugino et al., 2012). The primary sequences of the mammalian isoforms vary from 599 amino acids (NaPi-IIc) to approximately 640 amino acids (NaPi-IIa and NaPi-IIb), with differences mainly in the intracellular C- and N-terminal regions and the signature large extracellular linker region (Fig. 5A). This linker contains 2 N-glycosylation sites and a disulfide bridge that links the two "halves" of the protein. Cleavage of the disulfide bond of expressed NaPi-IIa leads to loss of function (Lambert, Traebert, Biber, & Murer, 2000), but coexpression of the split protein (each containing one of the essential cysteines) results in functional re-assembly (Kohl, Wagner, Huelseweh, Busch, & Werner, 1998; Ehnes, Forster, Kohler, Biber, & Murer, 2002). The proposed topology of the eukaryotic isoforms comprises 12 α-helices (designated as transmembrane domains (TMDs)) (Fig. 5A). This model combines predictions from biochemical and biophysical studies including epitope labeling (Lambert, Traebert, Hernando, Biber & Murer, 1999), cysteine scanning mutagenesis (see Section 3.1.2) and in vitro glycosylation assays (Radanovic, Gisler, Biber, & Murer, 2006). The C-terminal region is important for targeting, hormonal regulation and protein–protein interactions. For example, the TRL motif in the C terminus plays a role as a PDZ-binding motif (Hernando, Karim-Jimenez, Biber, & Murer, 2001; Karim-Jimenez, Hernando, Biber, & Murer, 2001), and a KR motif located in an intracellular linker region (Fig. 5A) is critical for parathyroid hormone (PTH) sensitivity (Karim-Jimenez, Hernando, Biber, & Murer, 2000).

between TMD1 ↔ 2 is more accessible than the linker between TMD11 ↔ 12. Empty symbols indicate sites that were accessible, but resulted in unchanged transport behavior. Modified from (Ehnes et al., 2004b), with permission. *Inset*: Detail of region in TMD9 that shows the periodicity of an α-helix. From (Lambert, Forster et al., 2001) with permission. This region is hypothesized to form part of a re-entrant segment that physically associates with TMD4 (Ghezzi, Meinild et al., 2011). See the color plate.

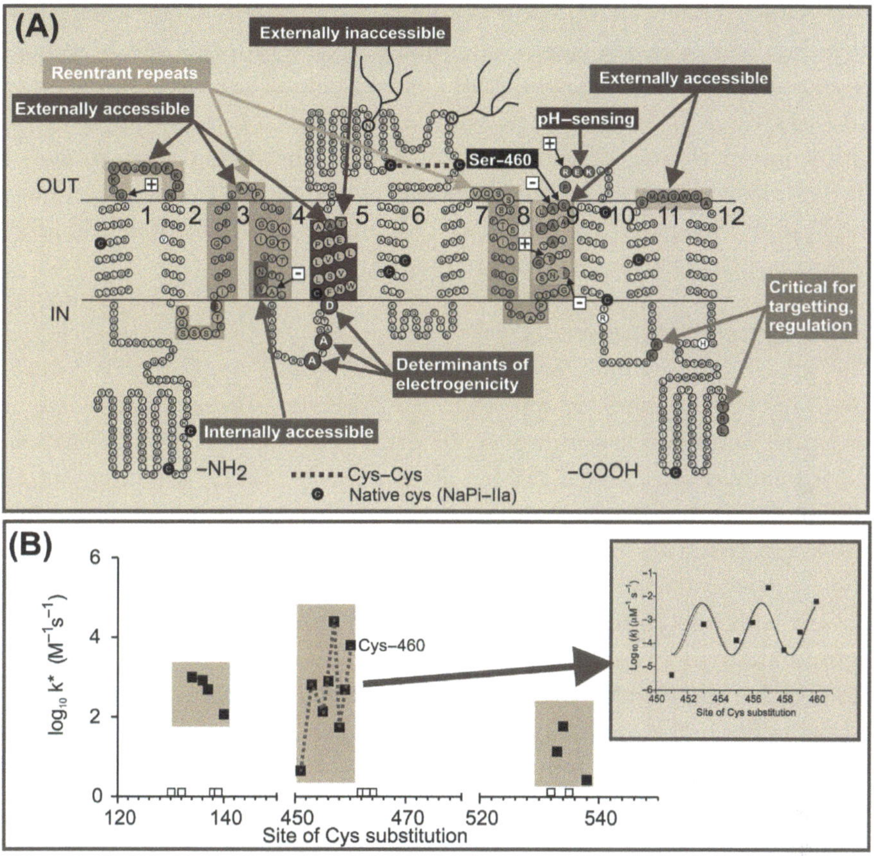

Figure 5 Topology and structure–function features of SLC34 proteins. (A) topology model based on rat NaPi-IIa sequence. Transmembrane domains (1–12) were assigned according to (Radanovic, Gisler et al., 2006). The model was drawn using TOPO2 software (http://www.sacs.ucsf.edu/TOPO/topo.html). Enlarged symbols indicate sites of significant structure–function importance. *Pink*: Identical residues in the re-entrant repeats; *green*: sites when mutated to a Cys that are externally accessible; *blue*: sites when mutated to a cysteine that are internally accessible; *purple*: sites externally inaccessible; *red*: sites critical for electrogenicity; *orange*: sites important for regulation and targeting; *black*: native Cys; *grey*: determinants of pH dependence. An essential disulfide bridge in the large extracellular loop is indicated (*dashed line*). Two N-glycosylation sites have been identified in the linker between TMD5 & 6. Boxed "–" and "+" refer to sites that result in mutant transporters with a hyperpolarizing or depolarizing shift, respectively, in their voltage dependence when a Cys is substituted. Modified and updated from (Virkki, Biber et al., 2007), with permission. (B) External accessibility determined by cysteine scanning. Summary of regions in the protein explored so far. Ordinate represents the modification rate (on logarithmic scale) of the Cys at the respective site and gives an indirect measure of its accessibility. These data suggest that the top of TMD9 is more accessible than the two outer linker regions (TMD1 ↔ 2 and TMD 11 ↔ 12), and the linker

A salient feature of all SLC34 proteins is the inverted repeat architecture (Fig. 5A), which is conserved among SLC34 isoforms and homologs in all phyla (Werner & Kinne, 2001). Similar motifs are also found in the 3-D structures of cotransporters with known architecture. Their existence can be interpreted as the structural basis for the alternating access model of carrier-mediated transport. Such repeats are considered to arise from gene duplication (Screpanti & Hunte, 2007; Abramson & Wright, 2009; Boudker & Verdon, 2010; Forrest, Kramer, & Ziegler, 2011).

3.1.2. Cysteine Mutagenesis

The substituted cysteine accessibility method (SCAM) (Karlin & Akabas, 1998) has been extensively applied to the NaPi-IIa isoform to elucidate its structure–function relationships (Lambert, Forster, Stange, Biber, & Murer, 1999; Lambert et al., 2001; Kohler, Forster, Stange, Biber, & Murer, 2002a, 2002b; Ehnes, Forster, Bacconi, et al., 2004; Ehnes, Forster, Kohler, et al., 2004; Virkki, Forster, Bacconi, et al., 2005; Virkki, Forster, Biber, et al., 2005). Cys are substituted at selected sites in the protein. If these can be specifically labeled with bulky methanethiosulfonate (MTS) reagents, and a change in transport activity occurs, then this is an indication of the functional importance of that site. By using MTS reagents with a known membrane permeability, sidedness of accessibility to the Cys can be determined. The validity of SCAM rests on the assumption that native Cys are inaccessible, or if labeled, do not alter function (Karlin & Akabas, 1998). Complete removal of all 12 native Cys in NaPi-IIa was not possible because of (i) the need to preserve one essential disulfide bridge, and (ii) a progressive decrease in functional activity. Nevertheless, eight native Cys were successfully removed and the behavior of one novel substituted Cys mutant was recapitulated (Kohler, Forster, Stange, Biber, & Murer, 2003), thereby validating using the WT backbone for SCAM. Moreover, using the cut-open oocyte voltage-clamp technique, MTS reagents were applied at the cytosolic interface and no change in WT activity was observed (unpublished experiment, K. Köhler, I. Gautschi, University of Lausanne). This confirmed that native Cys accessible only from the cytosol were also not functionally important. It is assumed that these findings hold for all other isoforms investigated.

So far, SCAM has been applied at sites within two predicted external linkers, one TMD and in the re-entrant repeat regions (Fig. 5A). Mutants were exposed to membrane permeable and impermeable MTS reagents and the transport properties assayed before and after exposure.

By determining the rate of Cys-modification, topological features were confirmed and identified, for example, an α-helical component of the second re-entrant segment (TMD9-10, Fig. 5B) (Lambert et al., 2001). In addition, functional changes—either caused by the Cys-mutagenesis itself or after MTS modification—were also observed, thereby giving further insight into functional roles played by different parts of the protein. For example, a Ser–Cys substitution at site 460 (at the top of TMD9, Fig. 5A) is well tolerated although thiol modification results in complete suppression of the cotransport. Nevertheless, the leak mode still operates, and moreover, P_i interacts with the protein (Lambert et al., 1999). The same loss of cotransport behavior seen after thiol modification of the equivalent mutation in the electroneutral NaPi-IIc (Ser-437) (Ghezzi et al., 2009), and NaPi-IIb (Ser-448) (Virkki et al., 2006a, 2006b) confirms the critical importance of this conserved site in all SLC34 members. In another study, reciprocal changes in the voltage dependence of I_{P_i} were found by Cys-mutagenesis and subsequent thiol modification in the linker regions between TMD1 &2 and TMD11 &12 (Ehnes, Forster, Bacconi, et al., 2004, Ehnes, Forster, Kohler, et al., 2004). This behavior suggested that specific residues in these linkers may provide flexibility to the protein as it undergoes voltage-dependent conformational changes during the transport cycle. Furthermore, the role of the predicted re-entrant segments as codeterminants of the leak and cotransport modes was elucidated by investigating mutants with Cys engineered in the respective repeat regions (Kohler et al., 2002a, 2002b). This study was recently extended using voltage-clamp fluorometry (VCF) (Section 3.1.3) and Cys cross-linking assays (Ghezzi, Meinild, Murer, & Forster, 2011) to establish that the inverted repeat segments very likely associate physically and can move relative to one another depending on substrate availability.

3.1.3. Voltage-Clamp Fluorometry

In contrast to presteady-state assays that report global rearrangements of charges within the protein, VCF allows investigation of local conformational changes by labeling engineered cysteines with thiol-reactive fluorophores. Labeled oocytes are voltage clamped and time-resolved fluorescence measurements can be made simultaneously (Fig. 6A,B). Changes in the emitted fluorescence intensity (ΔF), induced by substrate and steps in membrane potential, arise from changes in the microenvironment of the fluorophore

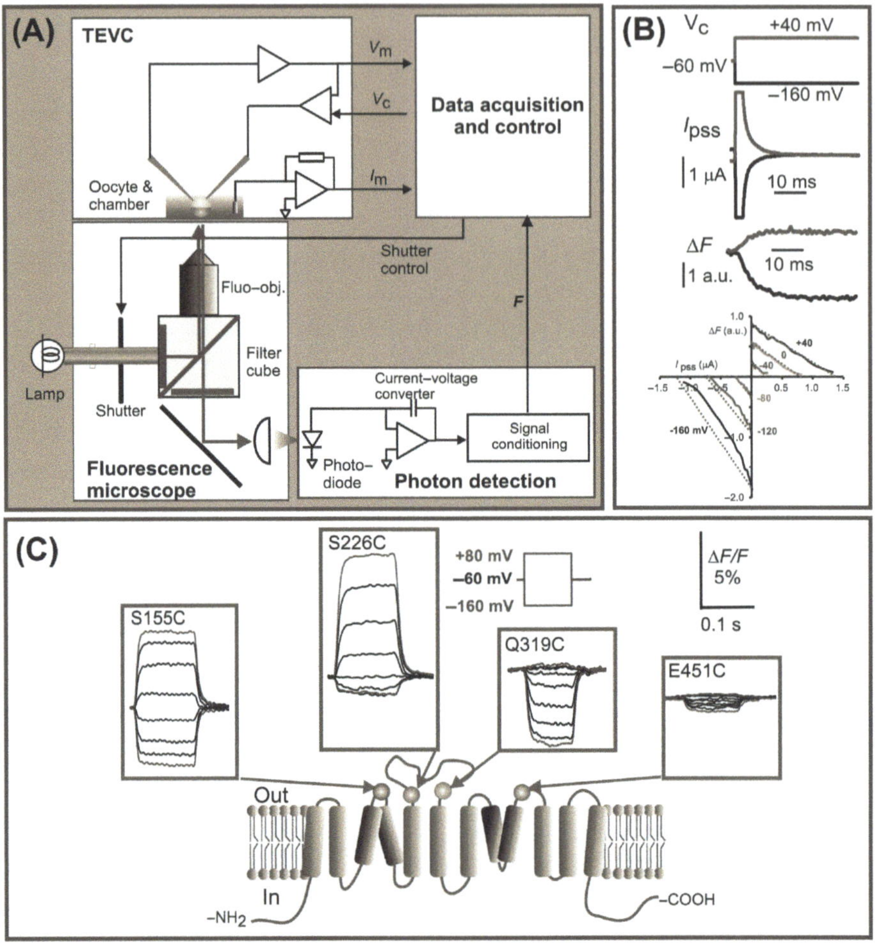

Figure 6 Application of voltage-clamp fluorometry to SLC34 structure–function studies. (A) Schematic of a typical voltage-clamp fluorometry (VCF) setup for *Xenopus* oocytes. The oocyte membrane is voltage clamped using a conventional two-electrode voltage clamp (TEVC), and placed in a chamber with a transparent bottom to allow fluorophores to be excited and their emission measured. For more technical information, see (Cha, Zerangue et al., 1998; Virkki et al., 2006a, 2006b). (B) *Upper panel*: Voltage steps evoke presteady-state relaxations (I_{pss}) and changes in emitted fluorescence intensity (ΔF) recorded simultaneously on the same oocyte expressing the flounder NaPi-IIb mutant (S448C) labeled with the fluorophore MTS-TAMRA. *Lower panel*: a parametric plot of I_{pss} against ΔF. The linear relationships found for $V > -40$ mV indicate that the time course of presteady-state relaxation and change in fluorescence intensity correlate. This suggests that the voltage-dependent charge displacement reported by I_{pss} and changes in the fluorophore microenvironment reported by ΔF reflect the same protein conformational changes in this membrane potential range. Modified and redrawn

and therefore report local conformational changes at or near the labeled site (Cha, Zerangue, Kavanaugh, & Bezanilla, 1998). The temporal resolution is limited only by the speed of charging the oocyte membrane. Substrate (Na^+, Li^+ and P_i) interactions have been explored using VCF applied to the S448C mutant in flounder NaPi-IIb (equivalent to S460C in NaPi-IIa, Fig. 5A) to establish that two Na^+ ions (but only one Li^+ ion) interact with the protein before P_i binding, and that in the presence of Na^+, P_i interaction induces a conformational change (Virkki et al., 2006a, 2006b). In this study, a direct correlation between the time course of presteady-state current and ΔF was found for voltage steps to potentials above −40 mV (Fig. 6B). This suggested that the charge movement and ΔF report the same physical process in that voltage range.

The power of the VCF technique was further demonstrated by attempting to "map" conformational changes in different regions of the protein. Four sites in the flounder NaPi-IIb isoform were identified that were labeled with the fluorophore *without* affecting the cotransport function (Virkki et al., 2006a, 2006b). These constructs permitted the characterization of substrate and voltage-dependent conformational changes associated with specific partial reactions in the complete cotransport cycle. ΔF was measured in (i) the absence of external Na (restricting the possible states to $0 \leftrightarrow 1$, Fig. 4A); (ii) 100 mM Na^+ (restricting the possible states to $0 \leftrightarrow 3$, Fig. 4A) and (iii) 100 mM Na^+ + 1 mM P_i (including all states in the cycle, Fig. 4A). The complementary ΔF obtained (Fig. 6C) suggested that, in response to the same membrane potential change, parts of the protein may experience complementary rearrangements relative to the aqueous medium (Fig. 6C). Similar to the presteady-state charge movements, changes in ΔF showed saturation at potential extremes and the ΔF–V data were fit with a Boltzmann function as for presteady-state charge analysis. The mutants yielded different ΔF–V relationships that depended on the substrate present and, based on the Boltzmann fit parameters, suggested that unique conformational changes occur when either one or two Na^+ ions are bound or when P_i was bound together with Na^+. Moreover, the

from (Virkki et al., 2006a, 2006b). (C) The voltage-dependent ΔF detected by labeling two sites in the N-terminal half of the NaPi-IIb protein (mutant S155C, S226C) are complementary to ΔF recorded from two sites in the C-terminal half (Q319C, E451C). This behavior suggests that complementary rearrangements of the protein occur relative to the aqueous medium *(Data adapted and redrawn from* Virkki et al., 2006a, 2006b). See the color plate.

dependence of ΔF on [P_i] at hyperpolarizing potentials was Michaelian and the $K_{0.5}^{P_i}$ values comparable to those found by electrophysiology.

Recently, the fluorescence technique has been applied to the electroneutral NaPi-IIc by engineering a Cys at the equivalent site (Ser-437) to Ser-460 in NaP$_i$-IIa (Fig. 5A) (Ghezzi et al., 2009). As expected, no voltage-dependent ΔF was detected. However, the emitted fluorescence intensity when the mutant was expressed was found to depend on the composition of the external solution, unlike the background fluorescence. With P_i as the variable substrate, the estimates of $K_{0.5}{}^{Pi}$ obtained from ^{32}P uptake and ΔF were in excellent agreement. This finding provided strong evidence that despite the lack of cotransport activity after labeling, P_i interactions were still intact and the partial reactions that determine the apparent affinity for P_i are independent of the substrate translocation. Importantly, this study also revealed that in the absence of P_i, most likely two Na^+ ions interact with NaPi-IIc as with the electrogenic isoforms (Eskandari, 2009; Ghezzi et al., 2009). In all these studies, a finding that no ΔF was detected when P_i was applied in the absence of external Na^+ confirms the strict dependence of P_i binding on the presence of Na^+.

3.1.4. Electrogenicity

The electroneutral NaPi-IIc (Segawa et al., 2002) provided an important protein tool to identify the molecular determinants of electrogenicity. A sequence comparison identified a conserved trio of amino acids (one of which is an aspartic acid) common to all electrogenic isoforms, located at the bottom of the fourth TMD (Fig. 5A). Substitution of these amino acids in the NaPi-IIc re-established electrogenicity, as evidenced by robust P_i-induced steady-state currents, presteady-state relaxations and the predicted 3:1 stoichiometry (Bacconi et al., 2005). However, despite the fact that electrogenic and electroneutral isoforms have similar substrate apparent affinities (Table 1), the molecular engineering of NaPi-IIc caused a significantly reduced apparent affinity for both Na^+ and P_i. This suggested that other differences in the amino acid sequences are also important for the kinetic profile. For the complementary experiment, in which the Asp was replaced with a Gly in the NaPi-IIa backbone, electroneutral cotransport was documented, thereby confirming the functional importance of the charged Asp in conferring electrogenicity (Virkki, Forster, Bacconi, et al., 2005; Virkki, Forster, Biber, et al., 2005).

Further insight into the determinants of electrogenicity has come from Cys scanning. Cys replacement at certain sites throughout the protein resulted in functional mutants that displayed either a hyperpolarizing or depolarizing shift in the voltage dependence relative to the WT (Fig. 5A). This was also seen as a concomitant shift in $V_{0.5}$ of the presteady-state Q–V. The mutagenesis did not significantly alter their other kinetic parameters (Lambert et al., 2001; Ghezzi et al., 2011). The similarity of their kinetic phenotypes may indicate that the respective sites are coordinated structurally. It is interesting that only one of these sites contains a charged residue (Fig. 5A). In one case (A175C in TMD4, Fig 5A), the hyperpolarizing shift could be simulated using the 10-state model (Fig. 4A) by setting a slower forward rate constant for the first Na^+ ion interaction (Ghezzi et al., 2011). Given its proximity to the critical Asp (Fig. 5A), this finding strongly supports the involvement of this part of the protein in voltage-dependent conformational changes associated with cation binding.

3.1.5. Proton Interactions

Several structure–function studies have focused on the intrinsic pH dependence of SLC34 proteins—first analyzed in detail for WT NaPi-IIa/b (Forster et al., 2000). A chimera approach was used to identify the region(s) responsible for conferring pH dependency to NaPi-IIa. This approach was based on the rationale that mouse NaPi-IIb displays a significantly reduced pH dependence compared to mouse NaPi-IIa (de la Horra et al., 2000), and led to the identification of a critical region between TMD 9 & 10 (Fig. 5A). This region contained a triad of neighboring, charged amino acids in NaPi-IIa (REK) that were substituted by an uncharged triad (GNT/A) in mammalian NaPi-IIb. Subsequent mutagenesis involved substitution of the REK triad into NaPi-IIb that restored the NaPi-IIa pH phenotype. Moreover, the pH dependence was abolished when the GNT triad was substituted in NaPi-IIa. These findings were also supported by the behavior of Cys-substituted NaPi-IIa mutants that displayed decreased pH sensitivity when any of the three residues was replaced by a Cys (Lambert et al., 2001). However, based on more recent studies and subsequent cloning of other SLC34 isoforms, it is unlikely that this motif represents the only proton interacting site in NaPi-IIa for several reasons. First, although the REK motif is conserved among all NaPi-IIa that display a strong pH dependence (Busch et al., 1994; Busch et al., 1995; Forster et al., 1998; Forster et al., 2000; Hartmann et al., 1995), it is present neither in the Zebra fish NaPi-IIb1 (Nalbant et al., 1999) nor NaPi-IIc, both of which show the phenotypically strong NaPi-IIa pH dependency (Segawa et al., 2002). Second, in the flounder NaPi-IIb that shows strong pH dependence (Forster et al., 2000) and the zebra fish NaPi-IIb2 that shows weaker pH dependence (Graham et al., 2003), the arginine is replaced by a neutral alanine. Third, an electrophysiological study on mouse NaPi-IIb (Forster et al., 2006) reported that $K_{0.5}^{P_i}$ was approximately fivefold smaller than for NaPi-IIa isoforms (Table 1). Therefore, for pH < 7, although the availability of divalent P_i is decreased, the low $K_{0.5}{}^{Pi}$ can explain the apparent insensitivity of the mouse NaPi-IIb isoform to acid pH. Given the variability of residues at the equivalent sites among the different isoforms, it is therefore likely that other residues also play a role in determining both pH sensitivity and apparent P_i affinity.

3.2. SLC20

3.2.1. Molecular Features

A 10-TMD topology was proposed for the human receptor of the gibbon ape virus (SLC20A1/PiT-1) based on hydrophobicity predictions (O'Hara et al., 1990). With subsequent sequencing of homologous receptors in other species, and the recognition that these proteins serve the dual function of viral receptor and P_i transporter, more detailed topological analyses led to the current secondary topology model of PiT-2 (SLC20A2) (Fig. 7). These studies used a combination of in vitro glycosylation assays, epitope tagging, cysteine scanning and sequence comparisons of all related family members (Salaun, Rodrigues, & Heard, 2001; Farrell, Tusnady et al., 2009; Bottger & Pedersen, 2011). The current model predicts a 12-TMD structure, but in contrast to SLC34 family proteins, the N and C termini are both extracellular. The topology is characterized by the large intracellular linker region between predicted TMD 7 & 8. An "inverted topology" containing amino acid blocks that are also conserved among all family members has been identified, and are referred to as PD1131 homology domains (Salaun et al., 2001) and comprise stretches of identical residues designated as PiT signature sequences (Bottger & Pedersen, 2005). Previously, much of the structure–function interest in these proteins has focused on identifying those parts responsible for viral receptor function (Pedersen, Johann, van Zeijl, Pedersen, & O'Hara, 1995; Pedersen, van Zeijl, Johann, & O'Hara, 1997; Farrell, Russ, Murthy, & Eiden, 2002; Feldman, Farrell, Murthy, Russ, & Eiden, 2004; Bottger & Pedersen, 2005; Farrell

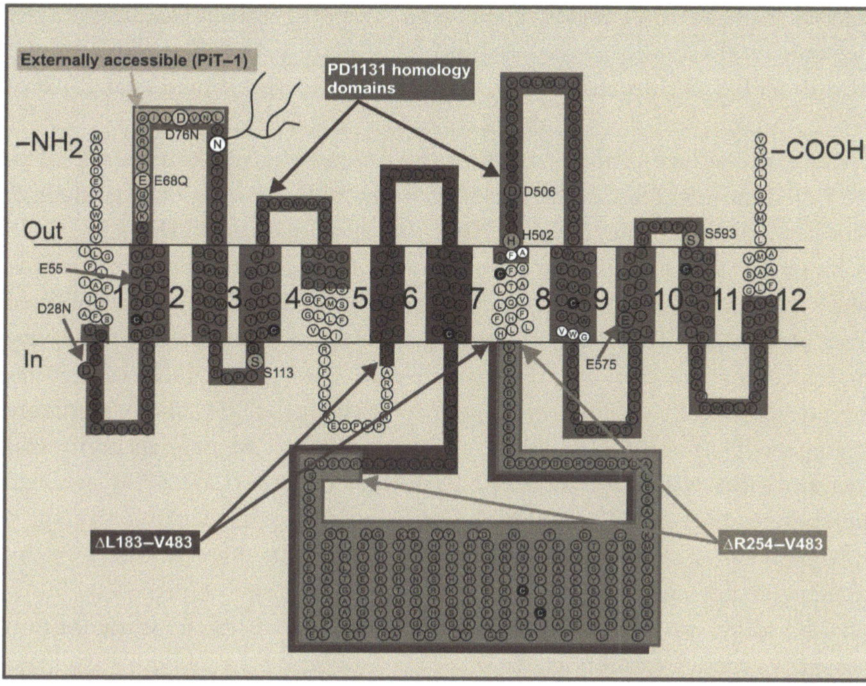

Figure 7 Topology and structure–function features of SLC20 proteins. This topology model for mammalian SLC20 proteins is based on the human PiT-2 sequence. The 12 predicted TMDs were assigned according to (Salaun, Rodrigues et al., 2001). The model was drawn using TOPO2 software (http://www.sacs.ucsf.edu/TOPO/topo.html). Residues important for SLC20 function that have been identified by mutagenesis (Bottger & Pedersen, 2004; Salaun, Marechal et al., 2004; Bottger & Pedersen, 2005) are shown enlarged. *Blue*: PD1131 homology domains (Salaun, Rodrigues et al., 2001). Two regions involving the large intracellular loop have been deleted. The ΔL183-V483 deletion causes loss of transport function with partially compromised viral receptor function (*green*) and the ΔR254-V483 leaves transport function intact but compromises viral receptor function (*orange*) (Bottger & Pedersen, 2011). Acidic residues important for transport function (D28, E55, D506, E575) (*red*) and those not critical for transport function (E68, D78, E91) (*pink*) are indicated. The protein is N-glycosylated at Asp-81. A region in the first extracellular loop of PiT-1 (also conserved in PiT-2) was subjected to SCAM, and sites accessible from the external medium have been identified (Ravera, S and Forster, I.C. manuscript in preparation). See the color plate.

et al., 2009; Bottger & Pedersen, 2011). For example, a PiT-2 mutant that lacked the large intracellular loop together with the preceding TMD 6 &7 still supported viral entry (Bottger & Pedersen, 2004). A recent study reported functional P_i transport when most of the intracellular loop was deleted (Bottger & Pedersen, 2011) (Fig. 7). PiT-1 and PiT-2 differ in the location of their putative virus-binding sites (e.g. the first extracellular loop of PiT-2 (Feldman et al., 2004), and the forth extracellular loop of PiT-1 (Johann, van Zeijl, Cekleniak, & O'Hara, 1993).

Several residues critical for transport function have been identified in the PiT-2 sequence by mutagenesis (Bottger & Pedersen, 2002; Bottger & Pedersen, 2005; Bottger et al., 2006). Two glutamate residues (E55, E575 in human PiT-2, Fig. 7) that are conserved in PiT-1 and PiT-2, as well as many related phosphate permeases from bacteria and fungi, were replaced with glutamines. Transport function was reduced significantly, while the viral receptor function remained unaffected (Bottger & Pedersen, 2002). In another study, two conserved aspartic acids in the PiT family signature sequences (D28 and D506, Fig. 7), were mutated to asparagines and transport function was again suppressed (Bottger & Pedersen, 2005), whereas mutations at E68 and D76 within the predicted first extracellular loop (Fig. 7) only partially affected function. Functional assays in these studies were by uptake and although membrane expression levels were confirmed by biotinylation assays, no detailed information on the effects of the mutations on transport kinetics was reported.

3.2.2. Cysteine Mutagenesis

An extensive Cys scanning of PiT-1 has been undertaken in the context of determining its secondary topology features and virus binding interaction (Farrell et al., 2009). The predicted topology from this study is very similar to the 12 TMD model for PiT-2 (Fig. 7). We have recently performed a Cys-scanning study on the first predicted extracellular loop of the *Xenopus* isoform of PiT-1 that contains a glutamine and aspartic acid previously identified to be critical for cotransport function (Bottger & Pedersen, 2005). Exposure of the WT PiT-1 to MTS reagents showed no change in electrogenic activity—a finding that confirms that none of the five native Cys was in a functionally sensitive position or readily accessible. Twelve sites were replaced individually with a Cys, and the mutants were expressed in oocytes as described above for SLC34. Whereas Cys substitution was well tolerated at all sites, ^{32}P uptake was significantly suppressed for six mutants after MTS exposure. Thus this linker region is possibly near the substrate-binding sites or associated with the transport pathway itself (manuscript in preparation, S. Ravera, I C. Forster).

4. FUTURE PERSPECTIVES

The characterization of the transport kinetics of WT SLC20 and SLC34 isoforms expressed in *Xenopus* oocytes has revealed fundamental differences in their transport kinetics. These differences are highlighted by the lack of detectable presteady-state charge movement for SLC20 proteins. For the electrogenic SLC34 isoforms (NaPi-IIa and NaPi-IIb), these relaxations display many of the characteristics found for electrogenic cotransporters from other SLC families (e.g. SGLT, NIS (SLC5) e.g. Loo et al., 1993; Eskandari et al., 1997; Loo, Eskandari, Hirayama, & Wright, 2002) and GAT (SLC6) (e.g. Mager et al., 1993; Sacher et al., 2002; Meinild & Forster, 2012). The similarities suggest that a common mechanism, which underlies the manner whereby the free energy of the membrane potential is transduced to drive cotransport, is shared among these families. For SLC20 proteins, the absence of detectable presteady-state charge movements suggests that the final translocation of the fully loaded protein is electrogenic. However, this notion requires further investigation. Moreover, further investigation of the SLC20 family kinetics would benefit from having a specific inhibitor.

At present, the kinetic model developed for electrogenic SLC34 proteins can successfully account for the forward mode of cotransport. However, the intact whole oocyte preparation severely limits the experimental manipulations necessary to characterize fully the reverse cycle and investigate substrate interactions at the cytosolic interface. Future work should therefore aim at characterizing the internal partial reactions using for example, excised giant membrane patches or the cut-open oocyte technique, as has been applied to SGLT1 (Eskandari, Wright, & Loo, 2005) and GAT1 (Hilgemann & Lu, 1998).

The structure–function studies carried out to date exemplify what can be achieved in the absence of 3-D structural information. The combination of a variety of approaches including bioinformatics, SCAM and VCF has provided important insights into the structure–function relationships of SLC34 proteins and should be extended to the SLC20 family. Structural predictions made by Cys scanning, taken together with TMD cross-linking assays, can identify features similar to those revealed in the 3-D structures, as exemplified by studies on lactose permease (LacY) (Kaback, Sahin-Toth, & Weinglass, 2001; Abramson et al., 2003) and the GAT1 bacterial homolog $LeuT_{Aa}$ (Yamashita, Singh, Kawate, Jin, & Gouaux, 2005; Kanner, 2006). This provides validation for the further use of these strategies for proteins with as-yet unresolved structures. At the same time, the kinetic analysis of engineered Cys-mutants has led to the identification of functionally important sites, underscoring the continued relevance of functional studies. In this context, the application of the fluorometric assays has already shown its potential as a tool to study the dynamics of transporters in their native environment and to complement conventional assays.

ABBREVIATIONS

ADHR autosomal-dominant hypophosphatemic rickets
HHRH hypophosphatemic rickets with hypercalciuria
MTS methane thiosulfonate
PFA phosphonoformic acid
PDZ post synaptic density protein, Drosophila disc large tumor suppressor, and Zonula occludens-1 protein
PTH parathyroid hormone
SCAM substituted cysteine accessibility method
TEVC two-electrode voltage clamp
TMD transmembrane domain
VCF voltage-clamp fluorometry

ACKNOWLEDGMENTS

Work from our own laboratories was supported in part by funding from the Swiss National Science Foundation. The authors thank the following past and present members for their invaluable contributions to many of the reported studies: Olga Andrini, Andrea Bacconi, Carmen de la Horra, Colin Ehnes, Chiara Ghezzi, Katja Köhler, Georg Lambert, Anne-Kristine Meinild, Monica Patti, Silvia Ravera and Leila Virkki.

REFERENCES

Abramson, J., Smirnova, I., Kasho, V., Verner, G., Iwata, S., & Kaback, H. R. (2003). The lactose permease of *Escherichia coli*: overall structure, the sugar-binding site and the alternating access model for transport. *FEBS Letters*, *555*(1), 96–101.

Abramson, J., & Wright, E. M. (2009). Structure and function of Na(+)-symporters with inverted repeats. *Current Opinion in Structural Biology*, *19*(4), 425–432.

Andrini, O., Ghezzi, C., Murer, H., & Forster, I. C. (2008). The leak mode of type II Na(+)-P(i) cotransporters. *Channels (Austin)*, *2*(5), 346–357.

Andrini, O., Meinild, A. K., Ghezzi, C., Murer, H., & Forster, I. C. (2012). Lithium interactions with Na+-coupled inorganic phosphate cotransporters: insights into the mechanism of sequential cation binding. *American Journal of Physiology – Cell Physiology*, *302*(3), C539–C554.

Bacconi, A., Ravera, S., Virkki, L. V., Murer, H., & Forster, I. C. (2007). Temperature dependence of steady-state and presteady-state kinetics of a type IIb Na+/P_i cotransporter. *Journal of Membrane Biology*, *215*(2–3), 81–92.

Bacconi, A., Virkki, L. V., Biber, J., Murer, H., & Forster, I. C. (2005). Renouncing electrogenicity is not free of charge: switching on electrogenicity in a Na^+-coupled phosphate cotransporter. *Proceedings of the National Academy of Sciences of the United States of America, 102*, 12606–12611.

Bai, L., Collins, J. F., & Ghishan, F. K. (2000). Cloning and characterization of a type III Na-dependent phosphate cotransporter from mouse intestine. *American Journal of Physiology, 279*(4), C1135–C1143.

Barac-Nieto, M., Alfred, M., & Spitzer, A. (2002). Basolateral phosphate transport in renal proximal-tubule-like OK cells. *Experimental Biology and Medicine (Maywood, N.J.), 227*(8), 626–631.

Beck, L., Karaplis, A. C., Amizuka, N., Hewson, A. S., Ozawa, H., & Tenenhouse, H. S. (1998). Targeted inactivation of Npt2 in mice leads to severe renal phosphate wasting, hypercalciuria, and skeletal abnormalities. *Proceedings of the National Academy of Sciences of the United States of America, 95*(9), 5372–5377.

Beck, L., Leroy, C., Beck-Cormier, S., Forand, A., Salaun, C., Paris, N., et al. (2010). The phosphate transporter PiT1 (Slc20a1) revealed as a new essential gene for mouse liver development. *PLoS ONE, 5*(2), e9148.

Beene, L. C., Halluer, J., Yoshinaga, M., Hamdi, M., & Liu, Z. (2011). Pentavalent arsenate transport by zebrafish phosphate transporter NaPi-IIb1. *Zebrafish, 8*(3), 125–131.

Bellocchio, E. E., Reimer, R. J., Fremeau, R. T., Jr., & Edwards, R. H. (2000). Uptake of glutamate into synaptic vesicles by an inorganic phosphate transporter. *Science, 289*(5481), 957–960.

Bergwitz, C., Roslin, N. M., Tieder, M., Loredo-Osti, J. C., Bastepe, M., Abu-Zahra, H., et al. (2006). SLC34A3 Mutations in patients with hereditary hypophosphatemic rickets with hypercalciuria predict a key role for the sodium-phosphate cotransporter NaPi-IIc in maintaining phosphate homeostasis. *American Journal of Human Genetics, 78*(2), 179–192.

Berndt, T. J., & Knox, F. G. (1992). Renal regulation of phosphate excretion. In D. W. Seldin, & G. Giebisch (Eds.), *The kidney: Physiology and pathophysiology* (pp. 2511–2526). New York: Raven Press.

Berndt, T., & Kumar, R. (2007). Phosphatonins and the regulation of phosphate homeostasis. *Annual Review of Physiology, 69*, 341–359.

Berner, W., Kinne, R., & Murer, H. (1976). Phosphate transport into brush-border membrane vesicles isolated from rat small intestine. *Biochemical Journal, 160*(3), 467–474.

Biber, J., Hernando, N., Forster, I., & Murer, H. (2009). Regulation of phosphate transport in proximal tubules. *Pflugers Archiv: European Journal of Physiology, 458*(1), 39–52.

Bottger, P., Hede, S. E., Grunnet, M., Hoyer, B., Klaerke, D. A., & Pedersen, L. (2006). Characterization of transport mechanisms and determinants critical for Na+-dependent P_i symport of the PiT-family paralogs, human PiT1 and PiT2. *American Journal of Physiology, 291*, C1377–C1387.

Bottger, P., & Pedersen, L. (2002). Two highly conserved glutamate residues critical for type III sodium-dependent phosphate transport revealed by uncoupling transport function from retroviral receptor function. *Journal of Biological Chemistry, 277*(45), 42741–42747.

Bottger, P., & Pedersen, L. (2004). The central half of PiT2 is not required for its function as a retroviral receptor. *Journal of Virology, 78*(17), 9564–9567.

Bottger, P., & Pedersen, L. (2005). Evolutionary and experimental analyses of inorganic phosphate transporter PiT family reveals two related signature sequences harboring highly conserved aspartic acids critical for sodium-dependent phosphate transport function of human PiT2. *FEBS Journal, 272*(12), 3060–3074.

Bottger, P., & Pedersen, L. (2011). Mapping of the minimal inorganic phosphate transporting unit of human PiT2 suggests a structure universal to PiT-related proteins from all kingdoms of life. *BMC Biochemistry, 12*, 21.

Boudker, O., & Verdon, G. (2010). Structural perspectives on secondary active transporters. *Trends in Pharmacological Sciences, 31*(9), 418–426.

Broer, S., Schuster, A., Wagner, C. A., Broer, A., Forster, I., Biber, J., et al. (1998). Chloride conductance and Pi transport are separate functions induced by the expression of NaPi-1 in *Xenopus* oocytes. *Journal of Membrane Biology, 164*(1), 71–77.

Busch, A. E., Schuster, A., Waldegger, S., Wagner, C. A., Zempel, G., Broer, S., et al. (1996). Expression of a renal type I sodium/phosphate transporter (NaPi-1) induces a conductance in *Xenopus* oocytes permeable for organic and inorganic anions. *Proceedings of the National Academy of Sciences of the United States of America, 93*(11), 5347–5351.

Busch, A. E., Wagner, C. A., Schuster, A., Waldegger, S., Biber, J., Murer, H., et al. (1995). Properties of electrogenic P_i transport by a human renal brush border Na^+/P_i transporter. *Journal of the American Society of Nephrology, 6*(6), 1547–1551.

Busch, A., Waldegger, S., Herzer, T., Biber, J., Markovich, D., Hayes, G., et al. (1994). Electrophysiological analysis of Na^+/P_i cotransport mediated by a transporter cloned from rat kidney and expressed in *Xenopus* oocytes. *Proceedings of the National Academy of Sciences of the United States of America, 91*(17), 8205–8208.

Cecil, D. L., Rose, D. M., Terkeltaub, R., & Liu-Bryan, R. (2005). Role of interleukin-8 in PiT-1 expression and CXCR1-mediated inorganic phosphate uptake in chondrocytes. *Arthritis and Rheumatism, 52*(1), 144–154.

Cha, A., Zerangue, N., Kavanaugh, M., & Bezanilla, F. (1998). Fluorescence techniques for studying cloned channels and transporters expressed in *Xenopus* oocytes. *Methods in Enzymology, 296*, 566–578.

Collins, J. F., Bai, L., & Ghishan, F. K. (2004). The SLC20 family of proteins: dual functions as sodium-phosphate cotransporters and viral receptors. *Pflugers Archiv: European Journal of Physiology, 447*(5), 647–652.

Corut, A., Senyigit, A., Ugur, S. A., Altin, S., Ozcelik, U., Calisir, H., et al. (2006). Mutations in SLC34A2 cause pulmonary alveolar microlithiasis and are possibly associated with testicular microlithiasis. *American Journal of Human Genetics, 79*(4), 650–656.

Daram, P., Brunner, S., Rausch, C., Steiner, C., Amrhein, N., & Bucher, M. (1999). Pht2;1 encodes a low-affinity phosphate transporter from *Arabidopsis*. *Plant Cell, 11*(11), 2153–2166.

Ehnes, C., Forster, I. C., Bacconi, A., Kohler, K., Biber, J., & Murer, H. (2004). Structure-function relations of the first and fourth extracellular linkers of the type IIa Na^+/P_i cotransporter: II. Substrate interaction and voltage dependency of two functionally important sites. *Journal of General Physiology, 124*, 489–503.

Ehnes, C., Forster, I. C., Kohler, K., Bacconi, A., Stange, G., Biber, J., et al. (2004). Structure-function relations of the first and fourth predicted extracellular linkers of the type IIa Na^+/Pi cotransporter: I. Cysteine scanning mutagenesis. *Journal of General Physiology, 124*(5), 475–488.

Ehnes, C., Forster, I. C., Kohler, K., Biber, J., & Murer, H. (2002). Functional studies on a split type II Na/Pi-cotransporter. *Journal of Membrane Biology, 188*(3), 227–236.

Eskandari, S. (2009). Remarkable commonalities of electrogenic and electroneutral Na+-phosphate cotransporters. *Journal of Physiology, 587*(Pt 17), 4131–4132.

Eskandari, S., Loo, D. D., Dai, G., Levy, O., Wright, E. M., & Carrasco, N. (1997). Thyroid Na^+/I^- symporter. Mechanism, stoichiometry, and specificity. *Journal of Biological Chemistry, 272*(43), 27230–27238.

Eskandari, S., Wright, E. M., & Loo, D. D. (2005). Kinetics of the reverse mode of the Na+/glucose cotransporter. *Journal of Membrane Biology, 204*(1), 23–32.

Farrell, K. B., Russ, J. L., Murthy, R. K., & Eiden, M.V. (2002). Reassessing the role of region A in Pit1-mediated viral entry. *Journal of Virology, 76*(15), 7683–7693.

Farrell, K. B., Tusnady, G. E., & Eiden, M. V. (2009). New structural arrangement of the extracellular regions of the phosphate transporter SLC20A1, the receptor for gibbon ape leukemia virus. *Journal of Biological Chemistry, 284*(43), 29979–29987.

Feldman, S. A., Farrell, K. B., Murthy, R. K., Russ, J. L., & Eiden, M. V. (2004). Identification of an extracellular domain within the human PiT2 receptor that is required for amphotropic murine leukemia virus binding. *Journal of Virology*, *78*(2), 595–602.

Forrest, L. R., Kramer, R., & Ziegler, C. (2011). The structural basis of secondary active transport mechanisms. *Biochimica et Biophysica Acta*, *1807*(2), 167–188.

Forster, I. (2007). *Electrogenic phosphate transport across renal epithelia: Mechanistic insights from experiments and simulations*. Workshop 5: The kidney: Cellular, tubular and vascular physiology, OSU: Mathematical Biosciences Institute.

Forster, I. C., Biber, J., & Murer, H. (2000). Proton-sensitive transitions of renal type II Na^+-coupled phosphate cotransporter kinetics. *Biophysical Journal*, *79*(1), 215–230.

Forster, I., Hernando, N., Biber, J., & Murer, H. (1998). The voltage dependence of a cloned mammalian renal type II Na^+/P_i cotransporter (NaPi-2). *Journal of General Physiology*, *112*(1), 1–18.

Forster, I. C., Hernando, N., Biber, J., & Murer, H. (2006). Proximal tubular handling of phosphate: A molecular perspective. *Kidney International*, *70*(9), 1548–1559.

Forster, I., Hernando, N., Sorribas, V., & Werner, A. (2011). Phosphate transporters in renal, gastrointestinal, and other tissues. *Advances in Chronic Kidney Disease*, *18*(2), 63–76.

Forster, I. C., Loo, D. D., & Eskandari, S. (1999). Stoichiometry and Na^+ binding cooperativity of rat and flounder renal type II Na^+-P_i cotransporters. *American Journal of Physiology*, *276*(4 Pt 2), F644–F649.

Forster, I. C., Virkki, L. V., Bossi, E., Murer, H., & Biber, J. (2006). Electrogenic kinetics of a mammalian intestinal Na^+/P_i-cotransporter. *Journal of Membrane Biology*, *212*, 177–190.

Forster, I. C., Wagner, C. A., Busch, A. E., Lang, F., Biber, J., Hernando, N., et al. (1997). Electrophysiological characterization of the flounder type II Na^+/P_i cotransporter (NaPi-5) expressed in *Xenopus laevis* oocytes. *Journal of Membrane Biology*, *160*(1), 9–25.

Frei, P., Gao, B., Hagenbuch, B., Mate, A., Biber, J., Murer, H., et al. (2005). Identification and localization of sodium-phosphate cotransporters in hepatocytes and cholangiocytes of rat liver. *American Journal of Physiology*, *288*(4), G771–G778.

Ghezzi, C., Meinild, A. K., Murer, H., & Forster, I. C. (2011). Voltage- and substrate-dependent interactions between sites in putative re-entrant domains of a Na(+)-coupled phosphate cotransporter. *Pflugers Archiv: European Journal of Physiology*, *461*(6), 645–663.

Ghezzi, C., Murer, H., & Forster, I. C. (2009). Substrate interactions of the electroneutral Na+-coupled inorganic phosphate cotransporter (NaPi-IIc). *Journal of Physiology*, *587*(Pt 17), 4293–4307.

Gisler, S. M., Kittanakom, S., Fuster, D., Wong, V., Bertic, M., Radanovic, T., et al. (2008). Monitoring protein-protein interactions between the mammalian integral membrane transporters and PDZ-interacting partners using a modified split-ubiquitin membrane yeast two-hybrid system. *Molecular and Cellular Proteomics*, 7(7), 1362–1377.

Gonzales, A. L., Lee, W., Spencer, S. R., Oropeza, R. A., Chapman, J. V., Ku, J. Y., et al. (2007). Turnover rate of the gamma-aminobutyric acid transporter GAT1. *Journal of Membrane Biology*, *220*(1–3), 33–51.

Graham, C., Nalbant, P., Scholermann, B., Hentschel, H., Kinne, R. K., & Werner, A. (2003). Characterization of a type IIb sodium-phosphate cotransporter from zebrafish (*Danio rerio*) kidney. *American Journal of Physiology – Renal Physiology*, *284*(4), F727–F736.

Haito-Sugino, S., Ito, M., Ohi, A., Shiozaki, Y., Kangawa, N., Nishiyama, T., et al. (2012). Processing and stability of type IIc sodium-dependent phosphate cotransporter mutations in patients with hereditary hypophosphatemic rickets with hypercalciuria. *American Journal of Physiology – Cell Physiology*, *302*(9), C1316–C1330.

Harris, R. M., Webb, D. C., Howitt, S. M., & Cox, G. B. (2001). Characterization of PitA and PitB from *Escherichia coli*. *Journal of Bacteriology*, *183*(17), 5008–5014.

Hartmann, C. M., Wagner, C. A., Busch, A. E., Markovich, D., Biber, J., Lang, F., et al. (1995). Transport characteristics of a murine renal Na/Pi-cotransporter. *Pflugers Archiv: European Journal of Physiology*, *430*(5), 830–836.

Hediger, M. A., Romero, M. F., Peng, J. B., Rolfs, A., Takanaga, H., & Bruford, E. A. (2004). The ABCs of solute carriers: physiological, pathological and therapeutic implications of human membrane transport proteins. Introduction. *Pflugers Archiv: European Journal of Physiology*, *447*(5), 465–468.

Hernando, N., Karim-Jimenez, Z., Biber, J., & Murer, H. (2001). Molecular determinants for apical expression and regulatory membrane retrieval of the type IIa Na/P_i cotransporter. *Kidney International*, *60*(2), 431–435.

Hilfiker, H., Hattenhauer, O., Traebert, M., Forster, I., Murer, H., & Biber, J. (1998). Characterization of a murine type II sodium-phosphate cotransporter expressed in mammalian small intestine. *Proceedings of the National Academy of Sciences of the United States of America*, *95*(24), 14564–14569.

Hilgemann, D. W., & Lu, C. C. (1998). Giant membrane patches: improvements and applications. *Methods in Enzymology*, *293*, 267–280.

Hoffmann, N., Thees, M., & Kinne, R. (1976). Phosphate transport by isolated renal brush border vesicles. *Pflugers Archiv: European Journal of Physiology*, *362*(2), 147–156.

de la Horra, C., Hernando, N., Lambert, G., Forster, I., Biber, J., & Murer, H. (2000). Molecular determinants of pH sensitivity of the type IIa Na/P_i cotransporter. *Journal of Biological Chemistry*, *275*(9), 6284–6287.

Jaureguiberry, G., Carpenter, T. O., Forman, S., Juppner, H., & Bergwitz, C. (2008). A novel missense mutation in SLC34A3 that causes hereditary hypophosphatemic rickets with hypercalciuria in humans identifies threonine 137 as an important determinant of sodium-phosphate cotransport in NaPi-IIc. *American Journal of Physiology – Renal Physiology*, *295*(2), F371–F379.

Johann, S. V., Gibbons, J. J., & O'Hara, B. (1992). GLVR1, a receptor for gibbon ape leukemia virus, is homologous to a phosphate permease of *Neurospora crassa* and is expressed at high levels in the brain and thymus. *Journal of Virology*, *66*(3), 1635–1640.

Johann, S. V., van Zeijl, M., Cekleniak, J., & O'Hara, B. (1993). Definition of a domain of GLVR1 which is necessary for infection by gibbon ape leukemia virus and which is highly polymorphic between species. *Journal of Virology*, *67*(11), 6733–6736.

Kaback, H. R., Sahin-Toth, M., & Weinglass, A. B. (2001). The kamikaze approach to membrane transport. *Nature Reviews – Molecular Cell Biology*, *2*(8), 610–620.

Kanner, B. I. (2006). Structure and function of sodium-coupled GABA and glutamate transporters. *Journal of Membrane Biology*, *213*(2), 89–100.

Karim-Jimenez, Z., Hernando, N., Biber, J., & Murer, H. (2000). A dibasic motif involved in parathyroid hormone-induced down-regulation of the type IIa NaPi cotransporter. *Proceedings of the National Academy of Sciences of the United States of America*, *97*(23), 12896–12901.

Karim-Jimenez, Z., Hernando, N., Biber, J., & Murer, H. (2001). Molecular determinants for apical expression of the renal type IIa Na+/Pi-cotransporter. *Pflugers Archiv: European Journal of Physiology*, *442*(5), 782–790.

Karlin, A., & Akabas, M. H. (1998). Substituted-cysteine accessibility method. *Methods in Enzymology*, *293*, 123–145.

Kavanaugh, M. P., & Kabat, D. (1996). Identification and characterization of a widely expressed phosphate transporter/retrovirus receptor family. *Kidney International*, *49*(4), 959–963.

Kavanaugh, M. P., Miller, D. G., Zhang, W., Law, W., Kozak, S. L., Kabat, D., et al. (1994). Cell-surface receptors for gibbon ape leukemia virus and amphotropic murine retrovirus are inducible sodium-dependent phosphate symporters. *Proceedings of the National Academy of Sciences of the United States of America*, *91*(15), 7071–7075.

Ketteler, M., & Giachelli, C. (2006). Novel insights into vascular calcification.. *Kidney International*, *105*(Suppl.), S5–S9.

Kohl, B., Herter, P., Hulseweh, B., Elger, M., Hentschel, H., Kinne, R. K., et al. (1996). Na-Pi cotransport in flounder: same transport system in kidney and intestine. *American Journal of Physiology, 270*(6 Pt 2), F937–F944.

Kohl, B., Wagner, C. A., Huelseweh, B., Busch, A. E., & Werner, A. (1998). The Na+-phosphate cotransport system (NaPi-II) with a cleaved protein backbone: implications on function and membrane insertion. *Journal of Physiology, 508*(Pt 2), 341–350.

Kohler, K., Forster, I. C., Lambert, G., Biber, J., & Murer, H. (2000). The functional unit of the renal type IIa Na+/Pi cotransporter is a monomer. *Journal of Biological Chemistry, 275*(34), 26113–26120.

Kohler, K., Forster, I. C., Stange, G., Biber, J., & Murer, H. (2002a). Identification of functionally important sites in the first intracellular loop of the NaPi-IIa cotransporter. *American Journal of Physiology, 282*(4), F687–F696.

Kohler, K., Forster, I. C., Stange, G., Biber, J., & Murer, H. (2002b). Transport function of the renal type IIa Na^+/P_i cotransporter is codetermined by residues in two opposing linker regions. *Journal of General Physiology, 120*, 693–703.

Kohler, K., Forster, I. C., Stange, G., Biber, J., & Murer, H. (2003). Essential cysteine residues of the type IIa Na^+/P_i cotransporter. *Pflugers Archiv: European Journal of Physiology, 446*(2), 203–210.

Lambert, G., Forster, I. C., Stange, G., Biber, J., & Murer, H. (1999). Properties of the mutant Ser-460-Cys implicate this site in a functionally important region of the type IIa Na^+/P_i cotransporter protein. *Journal of General Physiology, 114*(5), 637–652.

Lambert, G., Forster, I. C., Stange, G., Kohler, K., Biber, J., & Murer, H. (2001). Cysteine mutagenesis reveals novel structure-function features within the predicted third extracellular loop of the type IIa Na^+/P_i cotransporter. *Journal of General Physiology, 117*(6), 533–546.

Lambert, G., Traebert, M., Biber, J., & Murer, H. (2000). Cleavage of disulfide bonds leads to inactivation and degradation of the type IIa, but not type IIb sodium phosphate cotransporter expressed in *Xenopus laevis* oocytes. *Journal of Membrane Biology, 176*(2), 143–149.

Lambert, G., Traebert, M., Hernando, N., Biber, J., & Murer, H. (1999). Studies on the topology of the renal type II NaPi-cotransporter. Pflugers Archiv. *European Journal of Physiology, 437*(6), 972–978.

Lebens, M., Lundquist, P., Soderlund, L., Todorovic, M., & Carlin, N. I. (2002). The NptA gene of *Vibrio cholerae* encodes a functional sodium-dependent phosphate cotransporter homologous to the type II cotransporters of eukaryotes. *Journal of Bacteriology, 184*(16), 4466–4474.

Loghman-Adham, M. (1996). Use of phosphonocarboxylic acids as inhibitors of sodium-phosphate cotransport. *General Pharmacology, 27*(2), 305–312.

Loo, D. D., Eskandari, S., Hirayama, B. A., & Wright, E. M. (2002). A kinetic model for secondary active transport. In H. E. Layton, & A. M. Weinstein (Eds.), *Membrane transport and renal physiology* (pp. 65–84). : Springer.

Loo, D. D., Hazama, A., Supplisson, S., Turk, E., & Wright, E. M. (1993). Relaxation kinetics of the Na^+/glucose cotransporter. *Proceedings of the National Academy of Sciences of the United States of America, 90*(12), 5767–5771.

Lorenz-Depiereux, B., Benet-Pages, A., Eckstein, G., Tenenbaum-Rakover, Y., Wagenstaller, J., Tiosano, D., et al. (2006). Hereditary hypophosphatemic rickets with hypercalciuria is caused by mutations in the sodium-phosphate cotransporter gene SLC34A3. *American Journal of Human Genetics, 78*(2), 193–201.

Lundquist, P., Murer, H., & Biber, J. (2007). Type II Na+-Pi cotransporters in osteoblast mineral formation: regulation by inorganic phosphate. *Cellular Physiology and Biochemistry, 19*(1–4), 43–56.

Madjdpour, C., Bacic, D., Kaissling, B., Murer, H., & Biber, J. (2004). Segment-specific expression of sodium-phosphate cotransporters NaPi-IIa and -IIc and interacting proteins in mouse renal proximal tubules. *Pflugers Archiv: European Journal of Physiology, 448*(4), 402–410.

Magagnin, S., Werner, A., Markovich, D., Sorribas, V., Stange, G., Biber, J., et al. (1993). Expression cloning of human and rat renal cortex Na/P_i cotransport. *Proceedings of the National Academy of Sciences of the United States of America, 90*(13), 5979–5983.

Mager, S., Naeve, J., Quick, M., Labarca, C., Davidson, N., & Lester, H. A. (1993). Steady states, charge movements, and rates for a cloned GABA transporter expressed in *Xenopus* oocytes. *Neuron, 10*(2), 177–188.

Martinez, P., & Persson, B. L. (1998). Identification, cloning and characterization of a derepressible Na+-coupled phosphate transporter in Saccharomyces cerevisiae. *Molecular and General Genetics, 258*(6), 628–638.

Matsuo, A., Negoro, T., Seo, T., Kitao, Y., Shindo, M., Segawa, H., et al. (2005). Inhibitory effect of JTP-59557, a new triazole derivative, on intestinal phosphate transport in vitro and in vivo. *European Journal of Pharmacology, 517*(1–2), 111–119.

Meinild, A. K., & Forster, I. C. (2012). Using lithium to probe sequential cation interactions with GAT1. *American Journal of Physiology – Cell Physiology, 302*(11), C1661–C1675.

Miller, D. G., & Miller, A. D. (1994). A family of retroviruses that utilize related phosphate transporters for cell entry. *Journal of Virology, 68*(12), 8270–8276.

Miyamoto, K., Haito-Sugino, S., Kuwahara, S., Ohi, A., Nomura, K., Ito, M., et al. (2011). Sodium-dependent phosphate cotransporters: lessons from gene knockout and mutation studies. *Journal of Pharmaceutical Sciences, 100*(9), 3719–3730.

Miyamoto, K., Segawa, H., Ito, M., & Kuwahata, M. (2004). Physiological regulation of renal sodium-dependent phosphate cotransporters. *Japanese Journal of Physiology, 54*, 93–102.

Moe, O.W. (2009). PiT-2 coming out of the pits. *American Journal of Physiology – Renal Physiology, 296*(4), F689–F690.

Murer, H., Hernando, N., Forster, I., & Biber, J. (2000). Proximal tubular phosphate reabsorption: molecular mechanisms. *Physiological Reviews, 80*(4), 1373–1409.

Nalbant, P., Boehmer, C., Dehmelt, L., Wehner, F., & Werner, A. (1999). Functional characterization of a Na^+-phosphate cotransporter (NaPi-II) from zebrafish and identification of related transcripts. *Journal of Physiology, 520*(Pt 1), 79–89.

Nishimura, M., & Naito, S. (2008). Tissue-specific mRNA expression profiles of human solute carrier transporter superfamilies. *Drug Metabolism and Pharmacokinetics, 23*(1), 22–44.

O'Hara, B., Johann, S.V., Klinger, H. P., Blair, D. G., Rubinson, H., Dunn, K. J., et al. (1990). Characterization of a human gene conferring sensitivity to infection by gibbon ape leukemia virus. *Cell Growth and Differentiation, 1*(3), 119–127.

O'Mara, M., Oakley, A., & Broer, S. (2006). Mechanism and putative structure of B(0)-like neutral amino acid transporters. *Journal of Membrane Biology, 213*(2), 111–118.

Olah, Z., Lehel, C., Anderson, W. B., Eiden, M.V., & Wilson, C. A. (1994). The cellular receptor for gibbon ape leukemia virus is a novel high affinity sodium-dependent phosphate transporter. *Journal of Biological Chemistry, 269*(41), 25426–25431.

Parent, L., Supplisson, S., Loo, D. D., & Wright, E. M. (1992). Electrogenic properties of the cloned Na^+/glucose cotransporter: II. A transport model under nonrapid equilibrium conditions. [Erratum appears in J Membr Biol 1992 Nov;130(2):203]. *Journal of Membrane Biology, 125*(1), 63–79.

Pedersen, L., Johann, S. V., van Zeijl, M., Pedersen, F. S., & O'Hara, B. (1995). Chimeras of receptors for gibbon ape leukemia virus/feline leukemia virus B and amphotropic murine leukemia virus reveal different modes of receptor recognition by retrovirus. *Journal of Virology, 69*(4), 2401–2405.

Pedersen, L., van Zeijl, M., Johann, S.V., & O'Hara, B. (1997). Fungal phosphate transporter serves as a receptor backbone for gibbon ape leukemia virus. *Journal of Virology, 71*(10), 7619–7622.

Picard, N., Capuano, P., Stange, G., Mihailova, M., Kaissling, B., Murer, H., et al. (2010). Acute parathyroid hormone differentially regulates renal brush border membrane phosphate cotransporters. *Pflugers Archiv: European Journal of Physiology, 460*(3), 677–687.

Radanovic, T., Gisler, S. M., Biber, J., & Murer, H. (2006). Topology of the type IIa Na+/P(i) cotransporter. *Journal of Membrane Biology, 212*(1), 41–49.

Ravera, S., Virkki, L.V., Murer, H., & Forster, I. C. (2007). Deciphering PiT transport kinetics and substrate specificity using electrophysiology and flux measurements. *American Journal of Physiology – Cell Physiology, 293*(2), C606–C620.

Reimer, R. J., & Edwards, R. H. (2004). Organic anion transport is the primary function of the SLC17/type I phosphate transporter family. *Pfluegers Archiv: European Journal of Physiology, 447*(5), 629–635.

Sacher, A., Nelson, N., Ogi, J.T., Wright, E. M., Loo, D. D., & Eskandari, S. (2002). Presteady-state and steady-state kinetics and turnover rate of the mouse gamma-aminobutyric acid transporter (mGAT3). *Journal of Membrane Biology, 190*(1), 57–73.

Salaun, C., Marechal, V., & Heard, J. M. (2004). Transport-deficient Pit2 phosphate transporters still modify cell surface oligomers structure in response to inorganic phosphate. *Journal of Molecular Biology, 340*(1), 39–47.

Salaun, C., Rodrigues, P., & Heard, J. M. (2001). Transmembrane topology of PiT-2, a phosphate transporter-retrovirus receptor. *Journal of Virology, 75*(12), 5584–5592.

Saliba, K. J., Martin, R. E., Broer, A., Henry, R. I., McCarthy, C. S., Downie, M. J., et al. (2006). Sodium-dependent uptake of inorganic phosphate by the intracellular malaria parasite. *Nature, 443*(7111), 582–585.

Schaffhauser, D. F., Patti, M., Goda, T., Miyahara, Y., Forster, I. C., & Dittrich, P. S. (2012). An integrated field-effect microdevice for monitoring membrane transport in *Xenopus laevis* oocytes via lateral proton diffusion. *PLoS ONE*, 7(7), e39238.

Screpanti, E., & Hunte, C. (2007). Discontinuous membrane helices in transport proteins and their correlation with function. *Journal of Structural Biology, 159*(2), 261–267.

Segawa, H., Kaneko, I., Takahashi, A., Kuwahata, M., Ito, M., Ohkido, I., et al. (2002). Growth-related renal type II Na/Pi cotransporter. *Journal of Biological Chemistry, 277*(22), 19665–19672.

Tatsumi, S., Segawa, H., Morita, K., Haga, H., Kouda, T., Yamamoto, H., et al. (1998). Molecular cloning and hormonal regulation of PiT-1, a sodium-dependent phosphate cotransporter from rat parathyroid glands. *Endocrinology, 139*(4), 1692–1699.

Tenenhouse, H. S. (2005). Regulation of phosphorus homeostasis by the type IIa Na/phosphate cotransporter. *Annual Review of Nutrition, 25*, 197–214.

Tenenhouse, H. S., & Murer, H. (2003). Disorders of renal tubular phosphate transport. *Journal of the American Society of Nephrology, 14*(1), 240–248.

Uckert, W., Willimsky, G., Pedersen, F. S., Blankenstein, T., & Pedersen, L. (1998). RNA levels of human retrovirus receptors Pit1 and Pit2 do not correlate with infectibility by three retroviral vector pseudotypes. *Human Gene Therapy, 9*(17), 2619–2627.

van Veen, H. W. (1997). Phosphate transport in prokaryotes: molecules, mediators and mechanisms. *Antonie Van Leeuwenhoek, 72*(4), 299–315.

van Zeijl, M., Johann, S.V., Closs, E., Cunningham, J., Eddy, R., Shows, T. B., et al. (1994). A human amphotropic retrovirus receptor is a second member of the gibbon ape leukemia virus receptor family. *Proceedings of the National Academy of Sciences of the United States of America, 91*(3), 1168–1172.

Versaw, W. K., & Metzenberg, R. L. (1995). Repressible cation-phosphate symporters in Neurospora crassa. *Proceedings of the National Academy of Sciences of the United States of America, 92*(9), 3884–3887.

Villa-Bellosta, R., Bogaert, Y. E., Levi, M., & Sorribas, V. (2007). Characterization of phosphate transport in rat vascular smooth muscle cells: implications for vascular calcification. *Arteriosclerosis, Thrombosis, and Vascular Biology, 27*(5), 1030–1036.

Villa-Bellosta, R., Ravera, S., Sorribas, V., Stange, G., Levi, M., Murer, H., et al. (2009). The Na+-P_i cotransporter PiT-2 (SLC20A2) is expressed in the apical membrane of rat renal proximal tubules and regulated by dietary P_i. *American Journal of Physiology – Renal Physiology, 296*(4), F691–F699.

Villa-Bellosta, R., & Sorribas, V. (2008). Role of rat sodium/phosphate cotransporters in the cell membrane transport of arsenate. *Toxicology and Applied Pharmacology, 232*(1), 125–134.

Villa-Bellosta, R., & Sorribas, V. (2009). Different effects of arsenate and phosphonoformate on P(i) transport adaptation in opossum kidney cells. *American Journal of Physiology – Cell Physiology, 297*(3), C516–C525.

Villa-Bellosta, R., & Sorribas, V. (2010). Arsenate transport by sodium/phosphate cotransporter type IIb. *Toxicology and Applied Pharmacology, 247*(1), 36–40.

Virkki, L. V., Biber, J., Murer, H., & Forster, I. C. (2007). Phosphate transporters: a tale of two solute carrier families. *American Journal of Physiology – Renal Physiology, 293*(3), F643–F654.

Virkki, L.V., Forster, I. C., Bacconi, A., Biber, J., & Murer, H. (2005). Functionally important residues in the predicted 3rd transmembrane domain of the type IIa sodium-phosphate co-transporter (NaPi-IIa). *Journal of Membrane Biology, 206*(3), 227–238.

Virkki, L.V., Forster, I. C., Biber, J., & Murer, H. (2005). Substrate interactions in the human type IIa sodium-phosphate cotransporter (NaPi-IIa). *American Journal of Physiology, 288*, F969–F981.

Virkki, L.V., Murer, H., & Forster, I. C. (2006a). Mapping conformational changes of the type IIb Na^+/P_i cotransporter by voltage clamp fluorometry. *Journal of Biological Chemistry, 281*, 28837–28849.

Virkki, L.V., Murer, H., & Forster, I. C. (2006b). Voltage clamp fluorometric measurements on a type II Na^+-coupled P_i cotransporter: shedding light on substrate binding order. *Journal of General Physiology, 127*, 539–555.

Wadiche, J. I., Arriza, J. L., Amara, S. G., & Kavanaugh, M. P. (1995). Kinetics of a human glutamate transporter. *Neuron, 14*(5), 1019–1027.

Wallin, A., & Ryrfeldt, A. (1995). The toxicity of Foscarnet (phosphonoformic acid, trisodium salt; PFA) studied in cultured dog renal tubular cells. *Toxicol In Vitro, 9*(3), 237–244.

Wang, C., Li, Y., Shi, L., Ren, J., Patti, M., Wang, T., et al. (2012). Mutations in SLC20A2 link familial idiopathic basal ganglia calcification with phosphate homeostasis. *Nature Genetics, 44*(3), 254–256.

Weinstock, J. (2004). Inhibitors of sodium-dependent phosphate transport. *Expert Opinion Therapeutic Patents, 14*(1), 3.

Weiss, J. N. (1997). The Hill equation revisited: uses and misuses. *FASEB Journal, 11*(11), 835–841.

Werner, A., & Kinne, R. K. (2001). Evolution of the Na-P_i cotransport systems. *American Journal of Physiology, 280*(2), R301–R312.

Werner, A., Moore, M. L., Mantei, N., Biber, J., Semenza, G., & Murer, H. (1991). Cloning and expression of cDNA for a Na/Pi cotransport system of kidney cortex. *Proceedings of the National Academy of Sciences of the United States of America, 88*(21), 9608–9612.

Werner, A., Murer, H., & Kinne, R. K. (1994). Cloning and expression of a renal Na-P_i cotransport system from flounder. *American Journal of Physiology, 267*(2 Pt 2), F311–F317.

Yamashita, A., Singh, S. K., Kawate, T., Jin, Y., & Gouaux, E. (2005). Crystal structure of a bacterial homologue of Na+/Cl^- dependent neurotransmitter transporters. *Nature, 437*(7056), 215–223.

Zampighi, G. A., Kreman, M., Boorer, K. J., Loo, D. D., Bezanilla, F., Chandy, G., et al. (1995). A method for determining the unitary functional capacity of cloned channels and transporters expressed in *Xenopus laevis* oocytes. *Journal of Membrane Biology, 148*(1), 65–78.

CHAPTER ELEVEN

Functional and Physiological Role of Vitamin C Transporters

Marc Bürzle*,†, Matthias A. Hediger*,†,1
*Institute of Biochemistry and Molecular Medicine, University of Bern, Bern, Switzerland
†Swiss National Centre of Competence in Research, NCCR TransCure, University of Bern, Bern, Switzerland
[1]Corresponding author: E-mail: matthias.hediger@ibmm.unibe.ch

Contents

Abstract

Vitamin C (ascorbic acid) is required for the synthesis of collagen, carnitine, catecholamine and the neurotransmitter norepinephrine. Vitamin C also plays an important role in protection against oxidative stress. Transporters for vitamin C and its oxidized form dehydroascorbate (DHA) are crucial to keep vitamin concentrations optimal in the body. The human SLC23 family consists of the Na^+-dependent vitamin C transporters SVCT1 (SLC23A1) and SVCT2 (SLC23A2) and the orphan transporter SVCT3 (SLC23A3). Phylogenetically, the SLC23 family belongs to the nucleobase–ascorbate transporter family although no specificity for nucleobases has yet been demonstrated for the human members of this family. In fact, the SVCT1 and SVCT2 transporters are rather specific for ascorbic acid. SVCT1 is expressed in epithelial tissues such as intestine, where it contributes to the maintenance of whole-body ascorbic acid levels, whereas the expression of SVCT2 is relatively widespread either to protect metabolically active cells and specialized tissues from oxidative stress or to deliver ascorbic acid to tissues that are in high demand of the vitamin for enzymatic reactions. DHA, the oxidized form of ascorbic acid is taken up and distributed in the body by facilitated transport via members of the SLC2/GLUT family (GLUT1, GLUT3, and GLUT4). Although, the main focus of this review is on the SLC23 family of ascorbic acid transporters, transporters of DHA and nucleobases are also briefly discussed for completeness.

Current Topics in Membranes, Volume 70
ISSN 1063-5823
http://dx.doi.org/10.1016/B978-0-12-394316-3.00011-9

1. INTRODUCTION

Ascorbic acid plays a crucial role in health and disease. As an antioxidant and free-radical scavenger, it protects metabolically active cells from oxidative stress. It donates two electrons to prevent other compounds from being oxidized. The ascorbyl radical is formed and subsequently dehydroascorbic acid (DHA), the oxidized form of vitamin C. As a cofactor of enzymatic reactions, ascorbic acid acts as an electron donor in numerous enzymatic reactions such as collagen hydroxylation (adding hydroxyl groups to proline or lysine residues to strengthen the collagen triple helix), norepinephrine synthesis and peptide hormone amidation. In this function, ascorbic acid maintains the prosthetic metal ions of these enzymes (i.e. Fe^{2+} and Cu^{2+}) in their reduced forms (Englard & Seifter, 1986).

Most mammals, including mice, synthesize ascorbic acid from glucose, but humans and other primates are not able to synthesize ascorbic acid because of the lack of the *Gulo* gene, which is coding for the L-gulono-γ-lactone oxidase. This enzyme is required for the last step in the synthesis of ascorbic acid from glucose (Nishikimi, Fukuyama, Minoshima, Shimizu, & Yagi, 1994). Also, DHA cannot be completely reduced back to ascorbic acid in humans. Thus, humans need to absorb ascorbic acid from the diet via a specialized transporter to maintain appropriate levels in the body. Ascorbic acid is specifically transported by SVCTs encoded by the SLC23 family. The oxidized form of ascorbic acid (dehydroascorbic acid) can be taken up by facilitated diffusion via members of the GLUT family (GLUT1, GLUT3, and GLUT4) (Rumsey et al., 1997, 2000; Montel-Hagen et al., 2008).

The human SLC23 family consists of two Na^+-dependent vitamin C transporters, SVCT1 and SVCT2, and an orphan transporter SVCT3 (Tsukaguchi et al., 1999; Berger & Hediger, 2000; Berger et al., 2003; Castro et al., 2001; Clark, Rohrbaugh, Otterness, & Kraus, 2002; Daruwala, Song, Koh, Rumsey, & Levine, 1999; Dhariwal, Hartzell, & Levine, 1991). SVCT1 and SVCT2 are integral membrane proteins with 12 putative transmembrane domains (Faaland et al., 1998; Liang, Johnson, & Jarvis, 2001; Tsukaguchi et al., 1999; Wang et al., 1999, 2000). Both SVCT1 and SVCT2 transport ascorbic acid with high affinity. SVCT1 is responsible for absorption of vitamin C from the diet across the apical membrane of enterocytes. It is also expressed in lung, liver, kidney, and skin. SVCT2 is expressed more ubiquitously to provide cells with ascorbic acid for enzymatic function and to protect cells from oxidative stress (Clark et al., 2002; Lee et al., 2006; Rajan et al., 1999; Tsukaguchi et al., 1999; Wang et al., 2000).

It is known that the SLC23 family belongs to the nucleobase-ascorbate transporter (NAT) family, which consists of more than 500 proteins derived from bacteria, archea, diatoms, fungi, plants, and animals. The members of the NAT family are divided into three groups according to their substrate specificity. One group is specific for xanthine and/or uric acid, one for uracil, and the other for ascorbic acid. The first group is expressed in bacteria, fungi and plants with K_m values of 2–8 μM for xanthine and 7–130 μM for uric acid. The transport proteins specific for uracil are only found in bacteria and the ascorbic acid transporters only exist in vertebrates. SVCT1 and SVCT2 belong to this group of ascorbic acid transporters (de Koning & Diallinas, 2000).

SVCT3 (SLC23A3) is an orphan transporter that was originally cloned from mouse yolk sac (YSPL1) (Guimarães et al., 1995). Subsequent studies showed that SVCT3 is predominantly expressed in the kidney. However, the function of SVCT3 has not yet been reported thus far. It remains unknown whether SVCT3 is an ascorbic acid or nucleobase transporter. In rats, an additional isoform exists that is called SLC23A4. It was recently shown to transport nucleobases and it is predominantly expressed in the small intestine. In humans, however the human SLC23A4 gene is a pseudogene.

Human SVCT1 and SVCT2 are 66% identical at the amino acid sequence level. SVCT1 and SVCT3 are 34% and SVCT2 and SVCT3 are 31% identical (Fig. 1A,B).

2. NAT FAMILY

As already noted in Section 1, the ascorbic acid transporters of the SLC23 family belong to the NAT family, which is present in bacteria, fungi, plants and vertebrates. Members of the SLC23 family exhibit a remarkable sequence similarity to nucleobase transporters. Members of the NAT family are widely distributed and they play a diverse range of physiological roles. Purines and pyrimidines are required in nucleic acid and nucleoside synthesis (e.g. uracil) and serve as important nitrogen sources for plants and microorganisms. Uric acid and xanthine are converted to urea and ammonium and, therefore, used as nitrogen sources (Pantazopoulou & Diallinas, 2007).

(A)

```
                                                                         
hSVCT1   ------------MRAQEDLEGRTQHETTRDPSTPLPTEP--------------------- 27
hSVCT2   MMGIGKNTTSKSMEAGSSTEGKYEDEAKHPAFFTLPVVINGGATSSGEQDNEDTELMAIY 60

                                                          1
hSVCT1   -------------------------KFDMLYKIEDVPPWYLCILLGFQHYLTCFSGTIAV 62
hSVCT2   TTENGIAEKSSLAETLDSTGSLDPQRSDMIYTIEDVPPWYLCIFLGLQHYLTCFSGTIAV 120

                           2                             3
hSVCT1   PFLLAEALCVGHDQHMVSQLIGTIFTCVGITTLIQTTVGIRLPLFQASAFAFLVPAKAIL 122
hSVCT2   PFLLADAMCVGYDQWATSQLIGTIFFCVGITTLLQTTFGCRLPLFQASAFAFLAPARAIL 180

                                                             4
hSVCT1   ALERWKCPPEEEIYGN-WSLPLNTSHIWHPRIREVQGAIMVSSVVEVVIGLLGLPGALLN 181
hSVCT2   SLDKWKCNTTDVSVANGTAELLHTEHIWYPRIREIQGAIIMSSLIEVVIGLLGLPGALLK 240

                                        5
hSVCT1   YIGPLTVTPTVSLIGLSVFQAAGDRAGSHWGISACSILLIILFSQYLRNLTFLLPVYRWG 241
hSVCT2   YIGPLTITPTVALIGLSGFQAAGERAGKHWGIAMLTIFLVLLFSQYARNVKFPLPIYKSK 300

                             6
hSVCT1   KGLTLLRIQIFKMFPIMLAIMTVWLLCYVLTLTDVLPTDPKAYGFQARTDARGDIMAIAP 301
hSVCT2   KGWTAYKLQLFKMFPIILAILVSWLLCFIFTVTDVFPPDSTKYGFYARTDARQGVLLVAP 360

                           7
hSVCT1   WIRIPYPCQWGLPTVTAAAVLGMFSATLAGIIESIGDYYACARLAGAPPPPVHAINRGIF 361
hSVCT2   WFKVPYPFQWGLPTVSAAGVIGMLSAVVASIIESIGDYYACARLSCAPPPPIHAINRGIF 420

             8                                           9
hSVCT1   TEGICCIIAGLLGTGNGSTSSSPNIGVLGITKVGSRRVVQYGAAIMLVLGTIGKFTALFS 421
hSVCT2   VEGLSCVLDGIFGTGNGSTSSSPNIGVLGITKVGSRRVIQCGAALMLALGMIGKFSALFA 480

                     10                           11
hSVCT1   SLPDPILGGMFCTLFGMITAVGLSNLQFVDMNSSRNLFVLGFSMFFGLTLPNYLESNPGA 481
hSVCT2   SLPDPVLGALFCTLFGMITAVGLSNLQFIDLNSSRNLFVLGFSIFFGLVLPSYLRQNP-- 538

                       12
hSVCT1   INTGILEVDQILIVLLTTEMFVGGCLAFILDNTVPGSPEERGLIQWKAGAHANSDMSSSL 541
hSVCT2   LVTGITGIDQVLNVLLTTAMFVGGCVAFILDNTIPGTPEERGIRKWKKGVGKGNKSLDGM 598

hSVCT1   KSYDFPIGMGIVKRITFLKYIPICPVFKGFSSSSKDQIAIPEDTPENTETASVCTKV 598
hSVCT2   ESYNLPFGMNIIKKYRCFSYLPISPTFVGYTWKGLRKSDNSRSSDEDSQATG----- 650
```

(B)

		SVCT1
	SVCT2	66%
SVCT3	31%	34%

Figure 1 Amino acid sequence alignment and sequence identities. (A) Alignment of the amino acid sequences of the human SLC23 transporter proteins SVCT1 (Swiss-Prot: Q9UHI7) and SVCT2 (Swiss-Prot: Q9UGH3) with conserved histidine residues (yellow), PKC (red circle), and PKA (blue square) sites. Transmembrane domains are indicated in boxes. The alignment was generated using ClustalW. (B) Sequence identities of the human SLC23 transporter proteins were determined using the tool "BLAST 2 sequences". See the color plate.

Depending on the subtype, 12–14 putative transmembrane domains (TM) are predicted for members of the NAT family. For instance, UapA (*Aspergillus nidulans*) consists of 12 putative TM, whereas for UraA (*Escherichia coli*), 14 TM are predicted. In all NAT proteins, various conserved sequence segments have been found and the highly conserved signature sequence motif (**Q/E/P**)**N**X**G**XXXX**T**(**K/R/G**) was reported to be pivotal for function and specificity (Diallinas, Valdez, Sophianopoulou, Rosa, & Scazzocchio, 1998).

3. FUNCTIONAL PROPERTIES OF SVCT1 AND SVCT2 (SLC23A1 AND SLC23A2)

SVCT1 and SVCT2 cotransport Na^+ and ascorbic acid with a 2:1 stoichiometry down the electrochemical Na^+ gradient. Both transporters show a binding order of Na^+, ascorbic acid, and then Na^+ (Godoy et al., 2007; Mackenzie, Illing, & Hediger, 2008). Transport is electrogenic and the Hill coefficient for Na^+ was found to be near 2 under normal conditions (Tsukaguchi et al., 1999; Wang et al., 1999, 2000).

Studies by Godoy et al. revealed that SVCT2-mediated ascorbic acid transport is Ca^{2+}–Mg^{2+}-dependent, because there was no transport in the absence of Ca^{2+} and Mg^{2+}, even if Na^+ is present. The authors suggest that SVCT2 is converted into an inactive state when Ca^{2+} and Mg^{2+} are absent. Furthermore, there is a low affinity conformation of SVCT2 at low Na^+ concentrations (<20 mM), and there are two high affinity conformations when Na^+ concentrations are increased and cooperativity of Na^+ is modulated by ascorbic acid (Godoy et al., 2007).

SVCT1 and SVCT2 both are Na^+-dependent transporters that mediate the uptake of Na^+-coupled ascorbic acid (the reduced and biologically active form of vitamin C) with high affinity (Biondi et al., 2007; Corpe et al., 2005; Godoy et al., 2007; Luo, Wang, Kansara, Pal, & Mitra, 2008; Mackenzie et al., 2008; Wang et al., 2000; Wilson, 2005). Ascorbic acid transport by SVCT1 is found to be saturable with an apparent K_m of approximately 70 μM (Mackenzie et al., 2008), but values vary in the literature from 20 to 100 μM (Tsukaguchi et al., 1999; Wang et al., 1999, 2000). The process is electrogenic and pH sensitive, with an optimal pH at around 7.5. Transport of ascorbic acid is 50–60% reduced at pH 5.5 compared to pH 7.5 (Mackenzie et al., 2008; Tsukaguchi et al., 1999; Wang et al., 2000). SVCT2 has an apparent K_m for ascorbic acid of approximately 15 μM.

4. STRUCTURE, PHOSPHORYLATION AND *N*-GLYCOSYLATION OF VITAMIN C TRANSPORTERS

Topology studies have been done for SVCT1 and SVCT2, and hydropathy analyses of both proteins predict 12 transmembrane domains (TM) (Faaland et al., 1998; Liang et al., 2001; Tsukaguchi et al., 1999; Wang et al., 1999, 2000). Using antibodies to the N and C termini of human SVCT1, and to the C terminus of human SVCT2, it was shown that these regions are intracellularly located (Fig. 2) (Velho & Jarvis, 2009). Four conserved histidine residues were identified in SVCT1 and SVCT2. It was shown that the first histidine is within TM1 and the third histidine is within TM5; different positions are predicted for the second and fourth histidine residues (Fig. 1A) (Liang et al., 2001; Tsukaguchi et al., 1999; Varma, Campbell, & Kuo, 2008). To investigate the correlation between these conserved histidine residues and protein function and localization, histidines were mutated to alanine (Fig. 1A). Apical membrane localization was not influenced by mutating the histidines in SVCT1, but transport activity was reduced sixfold by the His51Ala mutation. Mutation of the corresponding residue in SVCT2 (His109Ala) led to the same observations. Membrane localization was not affected but there was a complete loss of ascorbate transport activity

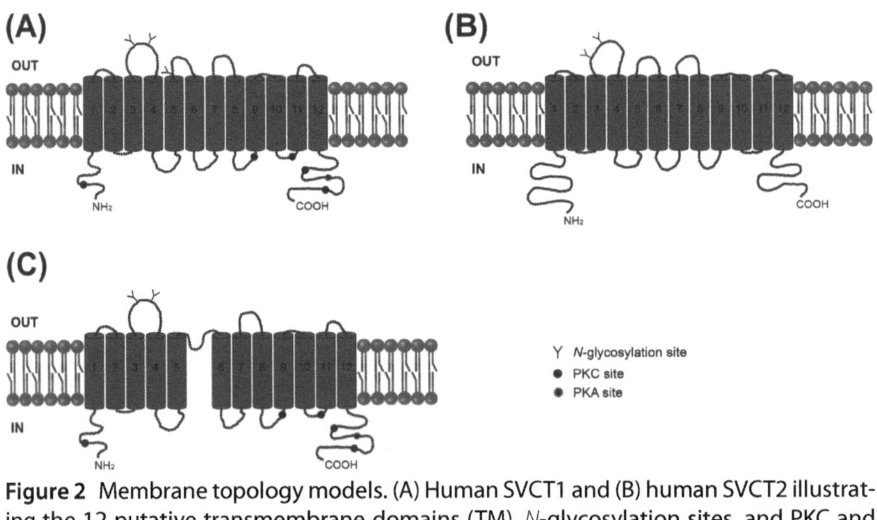

Figure 2 Membrane topology models. (A) Human SVCT1 and (B) human SVCT2 illustrating the 12 putative transmembrane domains (TM), *N*-glycosylation sites, and PKC and PKA sites for human SVCT1. (C) Velho and Jarvis, 2009 suggested a revised topological model of human SVCT1 with a re-entrant loop between TM 5 and TM 6. See the color plate.

in SVCT2. Thus, His51 in SVCT1 appears to play a crucial role in substrate binding (Ormazabal et al., 2010; Varma et al., 2008). In SVCT2, the five histidine residues His109, His203, His206, His269, and His413 were shown to be important for the function, pH sensitivity, transport kinetics, Na^+ cooperativity, conformational stability and localization of the protein in the plasma membrane (Ormazabal et al., 2010).

There are five protein kinase C (PKC)-dependent phosphorylation sites in the human SVCT1 sequence and six in the human SVCT2 sequence (Daruwala et al., 1999; Liang et al., 2001; Wang et al., 2000). Phorbol 12-myristate 13-acetate (PMA), a PKC-activating substrate, decreases the uptake of ascorbic acid in SVCT1 and SVCT2-expressing *Xenopus* oocytes, as well as in transfected COS-1 cells (40–60% reduction). It was shown that the reduced transport activity of SVCT1 in PMA-treated COS-1 cells was due to redistribution of SVCT1 from the cell surface to intracellular membranes, whereas there was no change in distribution of SVCT2 (Liang, Johnson, Ma, & Jarvis, 2002). These findings were confirmed in ascorbic acid uptake studies with HepG2 cells (Reidling, Subramanian, Dahhan, Sadat, & Said, 2008). In contrast, PMA increased the expression and function of SVCT2 in THP-1 monocytes during macrophage differentiation (Qiao & May, 2009).

Additionally, the human SVCT1 sequence contains one protein kinase A (PKA) consensus site, although there is no further evidence yet that it is functionally important (Daruwala et al., 1999; Liang et al., 2002; Wang et al., 2000).

Both human SVCT1 and SVCT2 have the following potential sites for *N*-glycosylation in the extracellular domains: Asn138, Asn144, and Asn230 in SVCT1, and Asn188 and Asn196 in SVCT2 (Subramanian, Marchant, Reidling, & Said, 2008; Velho & Jarvis, 2009; Wang et al., 2000). Asn230 in SVCT1 is not thought to be a functional glycosylation site (Velho & Jarvis, 2009). When Asn138 and Asn144 are mutated (N138Q and N144Q), ascorbic acid uptake is decreased in transfected COS-1 and HepG2 cells compared to that seen in wild-type SVCT1 transfected cells. The N230Q mutation showed no change in ascorbic acid uptake in COS-1 cells, but there was an impaired uptake in HepG2 cells (Subramanian et al., 2008; Velho & Jarvis, 2009). Mutation of Asn138 led to a significant retention of SVCT1 in intracellular membranes. Thus, Asn138 is important for cell-surface targeting (Subramanian et al., 2008). Mutations of Asn188 and Asn196 in SVCT2 (N188Q and N196Q) showed significant reduction in uptake function and expression of SVCT2 (Subramanian et al., 2008; Velho & Jarvis, 2009).

5. PHYSIOLOGICAL SIGNIFICANCE

5.1. Tissue Distribution

Human SVCT1 is expressed in epithelial tissues, such as intestine and kidney, and therefore, contributes to the maintenance of whole-body ascorbic acid levels (Fig. 3). In kidney, SVCT1 is expressed on the brush border of proximal straight tubules. In kidney and intestine, SVCT1 is responsible for the absorption of ascorbic acid on the apical membrane. It is still not known how ascorbic acid is transported out of the cell on the basolateral side and enters the blood (Lee et al., 2006; Tsukaguchi et al., 1999). Human SVCT2 is more widely distributed either to protect cells from oxidative stress or to deliver vitamin C into cells for specific, metal ion-dependent enzymatic reactions (Fig. 3) (Clark et al., 2002; Rajan et al., 1999; Tsukaguchi et al., 1999; Wang et al., 2000). SVCT1 and SVCT2 appear to function independently because SVCT2 deficiency affects neither the expression of SVCT1 nor the ascorbic acid levels in organs predominantly expressing SVCT1 (Kuo, MacLean, McCormick, & Wilson, 2004).

SVCT2 mRNA is expressed in different parts of the brain, such as hippocampus, cerebral cortex, cerebellum, and choroid plexus. As shown in Fig. 3, SVCT2 is expressed in the cells of choroid plexus where ascorbic acid is taken

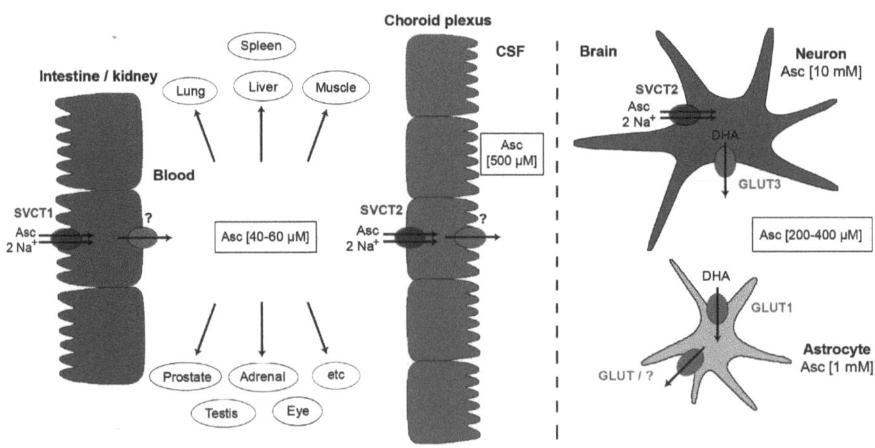

Figure 3 Whole body representation of the physiological roles of the SVCT1 and SVCT2 transporters. Expression and function of SVCT1 and SVCT2 is shown in different tissues and cell types of the human body. Transport pathways of vitamin C to brain and between specific cells of the brain are presented. See the color plate.

up from the blood via SVCT2-mediated transport. Thus far, it is unknown which pathway transports ascorbic acid out of the cells of the choroid plexus into the cerebrospinal fluid (CSF). SVCT2 is not expressed in the blood–brain barrier but, instead, the glucose transporter GLUT1 is present in the endothelial cells of the blood–brain barrier. Thus, a predominant transport pathway for this vitamin across the blood–brain barrier may be the facilitated transport of its oxidized form (DHA) via GLUT1 (SLC2A1) (Friedman & Zeidel, 1999; Margaret, 2000). In human blood, the concentration of DHA is about 2–10 μM (Dhariwal et al., 1991; Nakayama, Akiyama, Inagaki, Gotoh, & Oguchi, 2001). Given this relatively low blood concentration compared to that of ascorbic acid, it is still unclear whether DHA transport represents a major transport route for delivering the vitamin to the brain.

In situ hybridization experiments revealed that SVCT2 mRNA is, under normal conditions, expressed in neurons but not astrocytes (Astuya et al., 2005; Berger & Hediger, 2000). Also, intracellular levels of ascorbic acid were shown to be 10 times higher in neurons compared to glial cells (Fig. 3) (Margaret E 2000). These findings are in agreement with the concept that, after ascorbic acid is transported into neurons by SVCT2, the oxidized product, DHA, is transported via GLUT3 out of the neurons and then via GLUT1 into astrocytes where the vitamin is converted back into ascorbic acid (Fig. 3). Indeed, it was shown that astrocytes primarily transport DHA by the glucose transporter GLUT1 (Astuya et al., 2005). DHA can be reduced back to ascorbic acid by several mechanisms, such as different thiol transferases, NADPH-dependent reductases or glutathione reduction (Wells & Xu, 1994). However, as noted in the introduction, there is only partial recovery of DHA in humans, and explains why humans can get scurvy as a result of extreme vitamin C deficiency. Recovered ascorbic acid is then delivered back into neurons (Fig. 3). The cellular efflux pathways of vitamin C are still poorly understood.

Interestingly, focal ischemia induced by unilateral middle cerebral artery conclusion led to elevated SVCT2 mRNA expression in both neurons and astrocytes. This increase in SVCT2 expression presumably plays a neuroprotective role by protecting cells from oxidative stress (Berger et al., 2003). Qui et al. showed that SVCT2 is mainly expressed in the axons of mature hippocampal neurons in culture (Qiu, Li, Weeber, & May, 2007). The studies further revealed that impaired neurite outgrowth, less glutamate receptor clustering, reduced spontaneous neuronal activity, increased oxidative damage, and *N*-methyl-D-aspartate induced excitotoxicity are the consequences of SVCT2 deficiency.

5.2. Role of SVCTs in Different Tissues

SVCT1 knockout mice show a 50–70% reduction of plasma ascorbic acid levels compared to wild-type mice, and a threefold increased ascorbic acid urinary loss. Thus, even though mice do not depend on intestinal absorption of ascorbic acid from the diet via SVCT1 because they are capable of synthesizing it from glucose, SVCT1 plays a crucial role in renal ascorbic acid reabsorption (Corpe et al., 2010).

In 2002, Sotiriou et al. showed that SVCT2 knockout mice die within a few minutes after birth due to respiratory failure and hemorrhage in brain cortex and brain stem. Ascorbic acid levels in SVCT2 knockout mice were significantly decreased in SVCT2-expressing tissues, such as placenta, cortex, and lung, but not liver (Harrison et al., 2010; Sotiriou et al., 2002). The phenotype of the SVCT2 knockout highlights the vital importance of this vitamin C transporter. However, the causes of the rapid death and the brain hemorrhaging are still not completely understood.

A large amount of the whole-body ascorbic acid content is used in skeletal muscle to protect cells from oxidative stress and sustain enzyme activity by maintaining the prosthetic metal Fe^{2+} in the reduced form, which is important for carnitine biosynthesis. Therefore, ascorbic acid is transported into the muscle cells and stored in concentrations of 3–4 mg/100 g tissue weight (Peake, 2003; Sen, 2001). Skeletal muscles are also capable of taking up DHA because they express GLUT transporters (Korcok, Dixon, Lo, & Wilson, 2003). Expression of SVCT2 was found in mouse and rat muscle cell lines (C2C12 and L6C5), and preferentially in oxidative type I slow-twitched muscle fibers of chick embryonic skeletal muscle cells, as well as skeletal muscle cells from postnatal chick, rat, rabbit, guinea pig, and human (Low et al., 2009, 2011; Savini et al., 2005). During physical exercise, reactive oxygen species are produced in skeletal muscle cells and, interestingly, SVCT2 mRNA and protein expression is sensitive to the redox balance. When C2C12 myotubes are treated with H_2O_2, SVCT2 expression is increased. In contrast, treatment with the antioxidant lipoate led to a down-regulation of SVCT2 (Savini, Rossi et al., 2007).

In cultured human vascular smooth muscle cells, SVCT2 provides the cells with ascorbic acid, which is crucial for the synthesis and maturation of type I collagen by increasing intracellular generation of mature $\alpha1$(I) and $\alpha2$(I) subunits of type I procollagen (Qiao, Bell, Juliao, Li, & May, 2009; Rubin, Dey, & Reidling, 2005).

Uptake of ascorbic acid by SVCT2, which is expressed in human umbilical vein endothelial cells (HUVECs), is inhibited by inflammatory cytokines such as tumor necrosis factor-α and interleukin-1β (Seno et al., 2004). These findings are interesting because antioxidants such as ascorbic acid may be important in the prevention of cardiovascular diseases where oxidative stress is correlated with the pathogenesis of atherosclerosis (Seno et al., 2004).

Both SVCT1 and SVCT2 were found to be expressed in human skin. SVCT1 and SVCT2 are expressed by keratinocytes in the epidermis, and SVCT2 is also expressed by fibroblasts and microvascular endothelial cells in the dermis. Although SVCT2 is more highly expressed than SVCT1 in keratinocytes, SVCT1-mediated ascorbic acid uptake is the main transport pathway for ascorbic acid into these cells. mRNA expression of SVCT1, and subsequent ascorbic acid uptake in keratinocytes are significantly decreased by UVB irradiation in mice, whereas SVCT2 expression was not changed (Steiling et al., 2007). In contrast, Kang et al. demonstrated that in UVB-irradiated keratinocytes, SVCT1 is translocated from the cytosol to the membrane, thereby leading to enhanced ascorbic acid uptake and protection of the cells from UVB-induced oxidative stress (Kang et al., 2006). Steiling et al. assume that differences in the UV protocols may be the cause of these diverse results (Steiling et al., 2007).

SVCT2 expression was also found in the eye (Kannan, Stolz, Ji, Prasad, & Ganapathy, 2001; Tsukaguchi et al., 1999), where ascorbic acid is required to protect against radiation-induced damage (Rose & Bode, 1991). In the human lens epithelial cell line HLE-B3, SVCT2 was upregulated by exposure to the chemical oxidant tert-butylhydroperoxide that confirms the suggestion that SVCT2-mediated ascorbic acid uptake protects the eye from oxidative stress and UV radiation (Kannan et al., 2001).

Furthermore, SVCT-mediated ascorbic acid uptake plays a crucial role in catecholamine biosynthesis and adrenal steroidogenesis (Patak, Willenberg, & Bornstein, 2004; Wu et al., 2008), stimulates differentiation and mineralization of osteoblasts (Wu et al., 2004), promotes the development of male germ cells by ascorbic acid uptake of Sertoli cells (Angulo et al., 2008), protects macrophages from oxidative stress (May, Huang et al., 2005; May, Li et al., 2005), and contributes to platelet activation and aggregation, and thrombus stability (Savini, Catani et al., 2007).

6. REGULATION

Intracellular ascorbic acid is an important regulator of SVCT1 and SVCT2 expression. In cortex and cerebellum of late embryonic and early postnatal stages of normal mice, ascorbic acid levels were high although SVCT2 mRNA and protein levels were low. During adolescence, ascorbic acid levels decreased, whereas mRNA and protein levels of SVCT2 were increased. In contrast, in liver, both ascorbic acid and SVCT2 mRNA and protein levels were low in embryogenesis but increased after birth (Meredith, Harrison, & May, 2011).

In *Gulo−/−* mice unable to synthesize ascorbic acid, depletion of ascorbic acid led to enhanced SVCT1 and SVCT2 mRNA and protein expression in the liver, SVCT1 mRNA expression in the small intestine, and SVCT2 protein expression in cerebellum (Amano, Aigaki, Maruyama, & Ishigami, 2010; Meredith et al., 2011). Furthermore, ascorbic acid uptake was significantly increased in hepatocytes derived from *Gulo−/−* mice compared to hepatocytes from wild-type mice (Amano et al., 2010).

C and N termini of SVCT1 and SVCT2 are required for proper cell-surface targeting, membrane incorporation, and retention (Subramanian, Marchant, Boulware, & Said, 2004; Varma, Sobey, Campbell, & Kuo, 2009). These findings were confirmed in studies on human SVCT2 in human hepatocytes (HepG2 cells). Furthermore, the authors demonstrated that both microtubules and microfilaments regulate the cell-surface expression of SVCT2. Transport of SVCT2-containing vesicles throughout the cell was impaired by nocodazole, a microtubule depolarizing agent. It was suggested that microfilaments play a role in regulating the insertion and retrieval of SVCT2 because ascorbic acid uptake was increased by BDM and blebbistatin, inhibitors of the myosin-II ATPase. Inactivation of myosin probably impairs endocytosis of SVCT2 from the cell surface or insertion of SVCT2 in the plasma membrane (Subramanian, Marchant, & Said, 2010).

A naturally occurring short form of human SVCT2 that is missing 345 base pairs (nucleotides 482–827) was found to act as a dominant-negative inhibitor of ascorbic acid transport. The inhibitory effect of this nonfunctional protein is probably due to a protein–protein complex of full-length SVCT2 with the short form. The expression of this short isoform seems to be ubiquitous because it was detected in all tested tissues (Lutsenko, Carcamo, & Golde, 2004).

As already mentioned in Section 5, activation of the PKC intracellular pathway by phorbol 12-myristate 13-acetate (PMA) downregulates ascorbic acid uptake due to internalization of SVCT1. The membrane expression of SVCT2, however, is not affected by PKC-pathway activation (Liang et al., 2002; Reidling et al., 2008). Additionally, other pathways have been tested. Ascorbic acid uptake is significantly decreased by (i) activation of the PKA-mediated pathway, (ii) inhibition of the PTK-mediated pathway, and (iii) inhibition of the Ca^{2+}/CaM-mediated pathway (Reidling et al., 2008). However, PMA-induced differentiation of THP-1 monocytes shows increased SVCT2 mRNA and protein levels, and enhanced function through activation of PKCβI/II, MAP kinase, NADPH oxidase, and NF-κB (Qiao & May, 2009).

In osteoblast cells (MC3T3-E1), prostaglandin E2 (PGE2) increases the localization of SVCT2 in the plasma membrane and therefore, enhances ascorbic acid uptake. PEG2 associated effects in these cells require the activation of PKA (Wu, Zeng, Taniguchi, & Xie, 2007).

7. POLYMORPHISMS AND PATHOLOGICAL RELEVANCE

Several single nucleotide polymorphisms (SNPs) in SVCT1 and SVCT2, and their risk association with diseases were evaluated. A genetic variant (rs33972313) was found to be in exon 8 of the SLC23A1 locus. It is suggested that this missense causes a conformational change or protein failure that is responsible for lower concentrations of circulating ascorbic acid (−5.98 μmol/L) in the general population (Timpson et al., 2010). Furthermore, genetic variations in SLC23A1 and SLC23A2 were found to be associated with spontaneous preterm delivery. Especially, an SNP in intron 2 of SLC23A2 (rs6139591) revealed a 1.7-fold higher risk for spontaneous preterm birth in heterozygotes and a 2.7-fold higher risk for spontaneous preterm birth in homozygotes (Erichsen et al., 2006). Other studies on genetic variations showed that various SNPs in SLC23A2, one intronic SNP (rs12479919) and a haplotype that contained the common allele of the rs6139591, rs2681116, and rs14147458 SNPs are inversely associated with gastric cancer (Wright et al., 2009). Also, several SNPs and haplotypes in SLC23A1 and SLC23A2 influence the risk for lymphoma (Skibola et al., 2008).

An SNP in intron 8 of SLC23A2 (rs4987219) alters the risk of human papillomavirus type 16 (HPV16) associated head and neck squamous cell carcinomas (HNSCC). In the patients used for this study, the risk for HNSCC associated with HPV16 was highest for those with a wild-type allele of SLC23A2, whereas the risk significantly decreased for those with the homozygous variant (Chen et al., 2009).

8. OUTLOOK

Because vitamin C is required for crucial enzymatic pathways and given its function as an important antioxidant and free-radical scavenger, investigating the transport mechanisms, physiological roles, regulation of expression and pharmacology of vitamin C transporters is of great interest. Studying the therapeutic potential of SVCTs for the treatment of human diseases, such as cancer or cardiovascular diseases, is a relevant and promising field for the future. Likewise, it will be important to determine the functional role of SVCT3 and to identify the transported substrates. If SVCT3 turns out to be a nucleobase transporter, it may be a promising therapeutic target as well. Also, it may be possible to exploit nucleobase transporters as a drug delivery system. As a nucleobase transporter, it may affect the transport and metabolism of antiviral and antineoplastic drugs.

More data about the structure of the SLC23 family members (SVCT1-3) is required to generate better insight into the transport mechanisms and ligand binding sites of these proteins, and therefore to facilitate the development of specific inhibitors or activators for SLC23 family members. The structure of the uric acid-xanthine transporter UapA *(A. nidulans)* is one of the best characterized protein structures from the NAT family to which the SLC23 family is a subset. The UapA protein is rather distantly related to the SVCTs (23% amino acid sequence identity with SVCT1) while a structure of a member from the SLC23 family has not yet been determined.

ACKNOWLEDGMENTS

This publication was supported by the Swiss National Science Foundation (SNSF) through the National Centre of Competence in Research (NCCR) TransCure (www.transcure.org).

REFERENCES

Amano, A., Aigaki, T., Maruyama, N., & Ishigami, A. (2010). Ascorbic acid depletion enhances expression of the sodium-dependent vitamin C transporters, SVCT1 and SVCT2, and uptake of ascorbic acid in livers of SMP30/GNL knockout mice. *Archives of Biochemistry and Biophysics*, *496*(1), 38–44.

Angulo, C., Castro, M. A., Rivas, C. I., Segretain, D., Maldonado, R., Yañez, A. J., et al. (2008). Molecular identification and functional characterization of the vitamin C transporters expressed by Sertoli cells. *Journal of Cellular Physiology*, *217*(3), 708–716.

Astuya, A., Caprile, T., Castro, M., Salazar, K., García, M. d. l. A., Reinicke, K., et al. (2005). Vitamin C uptake and recycling among normal and tumor cells from the central nervous system. *Journal of Neuroscience Research*, *79*(1–2), 146–156.

Berger, U. V., & Hediger, M. A. (2000). The vitamin C transporter SVCT2 is expressed by astrocytes in culture but not in situ. *NeuroReport*, *11*(7), 1395–1399.

Berger, U. V., Lu, X.-C. M., Liu, W., Tang, Z., Slusher, B. S., & Hediger, M. A. (2003). Effect of middle cerebral artery occlusion on mRNA expression for the sodium-coupled vitamin C transporter SVCT2 in rat brain. *Journal of Neurochemistry*, *86*(4), 896–906.

Biondi, C., Pavan, B., Dalpiaz, A., Medici, S., Lunghi, L., & Vesce, F. (2007). Expression and characterization of vitamin C transporter in the human trophoblast cell line HTR-8/SVneo: effect of steroids, flavonoids and NSAIDs. *Molecular Human Reproduction*, *13*(1), 77–83.

Castro, M., Caprile, T., Astuya, A., Millán, C., Reinicke, K., Vera, J. C., et al. (2001). High-affinity sodium–vitamin C co-transporters (SVCT) expression in embryonic mouse neurons. *Journal of Neurochemistry*, *78*(4), 815–823.

Chen, A. A., Marsit, C. J., Christensen, B. C., Houseman, E. A., McClean, M. D., Smith, J. F., et al. (2009). Genetic variation in the vitamin C transporter, SLC23A2, modifies the risk of HPV16-associated head and neck cancer. *Carcinogenesis*, *30*(6), 977–981.

Clark, A. G., Rohrbaugh, A. L., Otterness, I., & Kraus, V. B. (2002). The effects of ascorbic acid on cartilage metabolism in guinea pig articular cartilage explants. *Matrix Biology*, *21*(2), 175–184.

Corpe, C. P., Lee, J. -H., Kwon, O., Eck, P., Narayanan, J., Kirk, K. L., et al. (2005). 6-Bromo-6-deoxy-l-ascorbic Acid. *Journal of Biological Chemistry*, *280*(7), 5211–5220.

Corpe, C. P., Tu, H., Eck, P., Wang, J., Faulhaber-Walter, R., Schnermann, J., et al. (2010). Vitamin C transporter Slc23a1 links renal reabsorption, vitamin C tissue accumulation, and perinatal survival in mice. *Journal of Clinical Investigation*, *120*(4), 1069–1083.

Daruwala, R., Song, J., Koh, W. S., Rumsey, S. C., & Levine, M. (1999). Cloning and functional characterization of the human sodium-dependent vitamin C transporters hSVCT1 and hSVCT2. *FEBS Letters*, *460*(3), 480–484.

Dhariwal, K., Hartzell, W., & Levine, M. (1991). Ascorbic acid and dehydroascorbic acid measurements in human plasma and serum. *The American Journal of Clinical Nutrition*, *54*(4), 712–716.

Diallinas, G., Valdez, J., Sophianopoulou, V., Rosa, A., & Scazzocchio, C. (1998). Chimeric purine transporters of *Aspergillus nidulans* define a domain critical for function and specificity conserved in bacterial, plant and metazoan homologues. *EMBO Journal*, *17*(14), 3827–3837.

Englard, S., & Seifter, S. (1986). The Biochemical functions of ascorbic acid. *Annual Review of Nutrition*, *6*(1), 365–406.

Erichsen, H. C., Engel, S. A. M., Eck, P. K., Welch, R., Yeager, M., Levine, M., et al. (2006). Genetic variation in the sodium-dependent vitamin C transporters, SLC23A1, and SLC23A2 and risk for preterm delivery. *American Journal of Epidemiology*, *163*(3), 245–254.

Faaland, C. A., Race, J. E., Ricken, G., Warner, F. J., Williams, W. J., & Holtzman, E. J. (1998). Molecular characterization of two novel transporters from human and mouse kidney and from LLC-PK1 cells reveals a novel conserved family that is homologous to bacterial and *Aspergillus* nucleobase transporters. *Biochimica et Biophysica Acta (BBA) – Gene Structure and Expression*, *1442*(2–3), 353–360.

Friedman, P. A., & Zeidel, M. L. (1999). Victory at C. *Natural Medicines*, *5*(6), 620–621.

Godoy, A., Ormazabal, V., Moraga-Cid, G., Zúñiga, F. A., Sotomayor, P., Barra, V., et al. (2007). Mechanistic insights and functional determinants of the transport cycle of the ascorbic acid transporter SVCT2. *Journal of Biological Chemistry*, *282*(1), 615–624.

Guimarães, M. J., Bazan, J. F., Zlotnik, A., Wiles, M. V., Grimaldi, J. C., Lee, F., & McClanahan, T. (1995). A new approach to the study of haematopoietic development in the yolk sac and embryoid bodies. *Development*, *121*(10), 3335–3346.

Harrison, F. E., Dawes, S. M., Meredith, M. E., Babaev, V. R., Li, L., & May, J. M. (2010). Low vitamin C and increased oxidative stress and cell death in mice that lack the sodium-dependent vitamin C transporter SVCT2. *Free Radical Biology and Medicine*, *49*(5), 821–829.

Kang, J. S., Kim, H. N., Jung, D. J., Kim, J. E., Mun, G. H., Kim, Y. S., et al. (2006). Regulation of UVB-Induced IL-8 and MCP-1 Production in skin keratinocytes by increasing vitamin C uptake via the redistribution of SVCT-1 from the cytosol to the membrane. *Journal of Investigative Dermatology*, *127*(3), 698–706.

Kannan, R., Stolz, A., Ji, Q., Prasad, P. D., & Ganapathy, V. (2001). Vitamin C transport in human lens epithelial cells: evidence for the presence of SVCT2. *Experimental Eye Research, 73*(2), 159–165.

de Koning, H., & Diallinas, G. (2000). Nucleobase transporters (Review). *Molecular Membrane Biology, 17*(2), 75–94.

Korcok, J., Dixon, S. J., Lo, T. C.Y., & Wilson, J. X. (2003). Differential effects of glucose on dehydroascorbic acid transport and intracellular ascorbate accumulation in astrocytes and skeletal myocytes. *Brain Research, 993*(1–2), 201–207.

Kuo, S.-M., MacLean, M. E., McCormick, K., & Wilson, J. X. (2004). Gender and sodium-ascorbate transporter isoforms determine ascorbate concentrations in mice. *The Journal of Nutrition, 134*(9), 2216–2221.

Lee, J., Oh, C., Mun, G., Kim, J., Chung, Y., Hwang, Y., et al. (2006). Immunohistochemical localization of sodium-dependent L-ascorbic acid transporter 1 protein in rat kidney. *Histochemistry and Cell Biology, 126*(4), 491–494.

Liang, W. J., Johnson, D., & Jarvis, S. M. (2001). Vitamin C transport systems of mammalian cells. *Molecular Membrane Biology, 18*(1), 87–95.

Liang, W. -J., Johnson, D., Ma, L.-S., & Jarvis, S. M. (2002). Regulation of the human vitamin C transporters expressed in COS-1 cells by protein kinase C. *American Journal of Physiology – Cell Physiology, 283*(6), C1696–C1704.

Low, M., Sandoval, D., Avilés, E., Pérez, F., Nualart, F., & Henríquez, J. (2009). The ascorbic acid transporter SVCT2 is expressed in slow-twitch skeletal muscle fibres. *Histochemistry and Cell Biology, 131*(5), 565–574.

Low, M., Sandoval, D., Morales, B., Nualart, F., & Henríquez, J. P. (2011). Up-regulation of the vitamin C transporter SVCT2 upon differentiation and depolarization of myotubes. *FEBS Letters, 585*(2), 390–396.

Luo, S., Wang, Z., Kansara, V., Pal, D., & Mitra, A. K. (2008). Activity of a sodium-dependent vitamin C transporter (SVCT) in MDCK-MDR1 cells and mechanism of ascorbate uptake. *International Journal of Pharmaceutics, 358*(1–2), 168–176.

Lutsenko, E. A., Carcamo, J. M., & Golde, D. W. (2004). A human sodium-dependent vitamin C transporter 2 isoform acts as a dominant-negative inhibitor of ascorbic acid transport. *Molecular and Cellular Biology, 24*(8), 3150–3156.

Mackenzie, B., Illing, A. C., & Hediger, M. A. (2008). Transport model of the human Na+-coupled L-ascorbic acid (vitamin C) transporter SVCT1. *American Journal of Physiology – Cell Physiology, 294*(2), C451–C459.

Margaret, E. R. (2000). Ascorbate regulation and its neuroprotective role in the brain. *Trends in Neurosciences, 23*(5), 209–216.

May, J. M., Huang, J., & Qu, Z.-c. (2005). Macrophage uptake and recycling of ascorbic acid: response to activation by lipopolysaccharide. *Free Radical Biology and Medicine, 39*(11), 1449–1459.

May, J. M., Li, L., Qu, Z.-c., & Huang, J. (2005). Ascorbate uptake and antioxidant function in peritoneal macrophages. *Archives of Biochemistry and Biophysics, 440*(2), 165–172.

Meredith, M. E., Harrison, F. E., & May, J. M. (2011). Differential regulation of the ascorbic acid transporter SVCT2 during development and in response to ascorbic acid depletion. *Biochemical and Biophysical Research Communications, 414*(4), 737–742.

Montel-Hagen, A., Kinet, S., Manel, N., Mongellaz, C., Prohaska, R., Battini, J. L., Delaunay, J., Sitbon, M., & Taylor, N. (2008). Erythrocyte Glut1 triggers dehydroascorbic acid uptake in mammals unable to synthesize vitamin C. *Cell, 132*(6), 1039–1048.

Nakayama, H., Akiyama, S., Inagaki, M., Gotoh, Y., & Oguchi, K. (2001). Dehydroascorbic acid and oxidative stress in haemodialysis patients. *Nephrology Dialysis Transplantation, 16*(3), 574–579.

Nishikimi, M., Fukuyama, R., Minoshima, S., Shimizu, N., & Yagi, K. (1994). Cloning and chromosomal mapping of the human nonfunctional gene for L-gulono-gamma-lactone oxidase, the enzyme for L-ascorbic acid biosynthesis missing in man. *Journal of Biological Chemistry*, *269*(18), 13685–13688.

Ormazabal, V., Zuñiga, F. A., Escobar, E., Aylwin, C., Salas-Burgos, A., Godoy, A., et al. (2010). Histidine residues in the Na+-coupled ascorbic acid Transporter-2 (SVCT2) are central regulators of SVCT2 function, modulating pH sensitivity, transporter kinetics, Na+ cooperativity, conformational stability, and Subcellular localization. *Journal of Biological Chemistry*, *285*(47), 36471–36485.

Pantazopoulou, A., & Diallinas, G. (2007). Fungal nucleobase transporters. *FEMS Microbiology Reviews*, *31*(6), 657–675.

Patak, P., Willenberg, H. S., & Bornstein, S. R. (2004). Vitamin C is an important cofactor for both adrenal cortex and adrenal medulla. *Endocrine Research*, *30*(4), 871–875.

Peake, J. M. (2003). Vitamin C: effects of exercise and requirements with training. *International Journal of Sport Nutrition and Exercise Metabolism*, *13*(2), 125–151.

Qiao, H., Bell, J., Juliao, S., Li, L., & May, J. M. (2009). Ascorbic acid uptake and regulation of type I collagen synthesis in cultured vascular smooth muscle cells. *Journal of Vascular Research*, *46*(1), 15–24.

Qiao, H., & May, J. M. (2009). Macrophage differentiation increases expression of the ascorbate transporter (SVCT2). *Free Radical Biology and Medicine*, *46*(8), 1221–1232.

Qiu, S., Li, L., Weeber, E. J., & May, J. M. (2007). Ascorbate transport by primary cultured neurons and its role in neuronal function and protection against excitotoxicity. *Journal of Neuroscience Research*, *85*(5), 1046–1056.

Rajan, D. P., Huang, W., Dutta, B., Devoe, L. D., Leibach, F. H., Ganapathy, V., et al. (1999). Human Placental sodium-dependent vitamin C transporter (SVCT2): molecular cloning and transport function. *Biochemical and Biophysical Research Communications*, *262*(3), 762–768.

Reidling, J. C., Subramanian, V. S., Dahhan, T., Sadat, M., & Said, H. M. (2008). Mechanisms and regulation of vitamin C uptake: studies of the hSVCT systems in human liver epithelial cells. *American Journal of Physiology – Gastrointestinal and Liver Physiology*, *295*(6), G1217–G1227.

Rose, R. C., & Bode, A. M. (1991). Ocular ascorbate transport and metabolism. *Comparative Biochemistry and Physiology A. Comparative Physiology*, *100*(2), 273–285.

Rubin, S. A., Dey, S., & Reidling, J. C. (2005). Functional analysis of two regulatory regions of the human Na+-dependent vitamin C transporter 2, SLC23A2, in human vascular smooth muscle cells. *Biochimica et Biophysica Acta (BBA) – Gene Structure and Expression*, *1732*(1–3), 76–81.

Rumsey, S. C., Daruwala, R., Al-Hasani, H., Zarnowski, M. J., Simpson, I. A., & Levine, M. (2000). Dehydroascorbic acid transport by GLUT4 in *Xenopus* oocytes and isolated rat adipocytes. *Journal of Biological Chemistry*, *275*(36), 28246–28253.

Rumsey, S. C., Kwon, O., Xu, G. W., Burant, C. F., Simpson, I., & Levine, M. (1997). Glucose transporter isoforms GLUT1 and GLUT3 transport dehydroascorbic acid. *Journal of Biological Chemistry*, *272*(30), 18982–18989.

Savini, I., Catani, M.V., Arnone, R., Rossi, A., Frega, G., Del Principe, D., et al. (2007). Translational control of the ascorbic acid transporter SVCT2 in human platelets. *Free Radical Biology and Medicine*, *42*(5), 608–616.

Savini, I., Rossi, A., Catani, M. V., Ceci, R., & Avigliano, L. (2007). Redox regulation of vitamin C transporter SVCT2 in C2C12 myotubes. *Biochemical and Biophysical Research Communications*, *361*(2), 385–390.

Savini, I., Valeria Catani, M., Duranti, G., Ceci, R., Sabatini, S., & Avigliano, L. (2005). Vitamin C homeostasis in skeletal muscle cells. *Free Radical Biology and Medicine*, *38*(7), 898–907.

Sen, C. K. (2001). Antioxidants in exercise Nutrition. *Sports Medicine*, *31*(13), 891–908.
Seno, T., Inoue, N., Matsui, K., Ejiri, J., Hirata, K., Kawashima, S., et al. (2004). Functional expression of sodium-dependent vitamin C transporter 2 in human endothelial cells. *Journal of Vascular Research*, *41*(4), 345–351.
Skibola, C. F., Bracci, P. M., Halperin, E., Nieters, A., Hubbard, A., Paynter, R. A., et al. (2008). Polymorphisms in the estrogen receptor 1 and vitamin C and matrix metalloproteinase gene families are associated with susceptibility to lymphoma. *PLoS ONE*, *3*(7), e2816.
Sotiriou, S., Gispert, S., Cheng, J., Wang, Y., Chen, A., Hoogstraten-Miller, S., et al. (2002). Ascorbic-acid transporter Slc23a1 is essential for vitamin C transport into the brain and for perinatal survival. *Natural Medicines*, *8*(5), 514–517.
Steiling, H., Longet, K., Moodycliffe, A., Mansourian, R., Bertschy, E., Smola, H., et al. (2007). Sodium-dependent vitamin C transporter isoforms in skin: distribution, kinetics, and effect of UVB-induced oxidative stress. *Free Radical Biology and Medicine*, *43*(5), 752–762.
Subramanian, V. S., Marchant, J. S., Boulware, M. J., & Said, H. M. (2004). A C-terminal region dictates the apical plasma membrane targeting of the human sodium-dependent vitamin C transporter-1 in polarized epithelia. *Journal of Biological Chemistry*, *279*(26), 27719–27728.
Subramanian, V. S., Marchant, J. S., & Said, H. M. (2010). Molecular determinants dictating cell surface expression of the human sodium-dependent vitamin C transporter-2 in human liver cells. *American Journal of Physiology – Gastrointestinal and Liver Physiology*, *298*(2), G267–G274.
Subramanian, V. S., Marchant, J. S., Reidling, J. C., & Said, H. M. (2008). N-Glycosylation is required for Na+-dependent vitamin C transporter functionality. *Biochemical and Biophysical Research Communications*, *374*(1), 123–127.
Timpson, N. J., Forouhi, N. G., Brion, M. -J., Harbord, R. M., Cook, D. G., Johnson, P., et al. (2010). Genetic variation at the SLC23A1 locus is associated with circulating concentrations of l-ascorbic acid (vitamin C): evidence from 5 independent studies with >15,000 participants. *The American Journal of Clinical Nutrition*, *92*(2), 375–382.
Tsukaguchi, H., Tokui, T., Mackenzie, B., Berger, U. V., Chen, X.-Z., Wang, Y., et al. (1999). A family of mammalian Na+-dependent L-ascorbic acid transporters. *Nature*, *399*(6731), 70–75.
Varma, S., Campbell, C. E., & Kuo, S. -M. (2008). Functional role of conserved transmembrane segment 1 residues in human sodium-dependent vitamin C transporters. *Biochemistry*, *47*(9), 2952–2960.
Varma, S., Sobey, K., Campbell, C. E., & Kuo, S.-M. (2009). Hierarchal contribution of N- and C-terminal sequences to the differential localization of homologous sodium-dependent vitamin C transporters, SVCT1 and SVCT2, in epithelial cells. *Biochemistry*, *48*(13), 2969–2980.
Velho, A. M., & Jarvis, S. M. (2009). Topological studies of hSVCT1, the human sodium-dependent vitamin C transporter and the influence of N-glycosylation on its intracellular targeting. *Experimental Cell Research*, *315*(13), 2312–2321.
Wang, H., Dutta, B., Huang, W., Devoe, L. D., Leibach, F. H., Ganapathy, V., et al. (1999). Human Na+-dependent vitamin C transporter 1 (hSVCT1): primary structure, functional characteristics and evidence for a non-functional splice variant. *Biochimica et Biophysica Acta (BBA) – Biomembranes*, *1461*(1), 1–9.
Wang, Y., Mackenzie, B., Tsukaguchi, H., Weremowicz, S., Morton, C. C., & Hediger, M. A. (2000). Human vitamin C (L-Ascorbic acid) transporter SVCT1. *Biochemical and Biophysical Research Communications*, *267*(2), 488–494.
Wells, W. W., & Xu, D. P. (1994). Dehydroascorbate reduction. *Journal of Bioenergetics and Biomembranes*, *26*(4), 369–377.

Wilson, J. X. (2005). Regulation of vitamin C transport. *Annual Review of Nutrition, 25*(1), 105–125.

Wright, M. E., Andreotti, G., Lissowska, J., Yeager, M., Zatonski, W., Chanock, S. J., et al. (2009). Genetic variation in sodium-dependent ascorbic acid transporters and risk of gastric cancer in Poland. *European Journal of Cancer, 45*(10), 1824–1830.

Wu, X., Iguchi, T., Itoh, N., Okamoto, K., Takagi, T., Tanaka, K., et al. (2008). Ascorbic acid transported by sodium-dependent vitamin C transporter 2 stimulates steroidogenesis in human choriocarcinoma cells. *Endocrinology, 149*(1), 73–83.

Wu, X., Itoh, N., Taniguchi, T., Hirano, J., Nakanishi, T., & Tanaka, K. (2004). Stimulation of differentiation in sodium-dependent vitamin C transporter 2 overexpressing MC3T3-E1 osteoblasts. *Biochemical and Biophysical Research Communications, 317*(4), 1159–1164.

Wu, X., Zeng, L. H., Taniguchi, T., & Xie, Q. M. (2007). Activation of PKA and phosphorylation of sodium-dependent vitamin C transporter 2 by prostaglandin E2 promote osteoblast-like differentiation in MC3T3-E1 cells. *Cell Death and Differentiation, 14*(10), 1792–1801.

INDEX

Note: Page numbers followed by "f" and "t" indicate figures and tables respectively.

C

I

N

T

U

V

W

X

Z

Color Plates

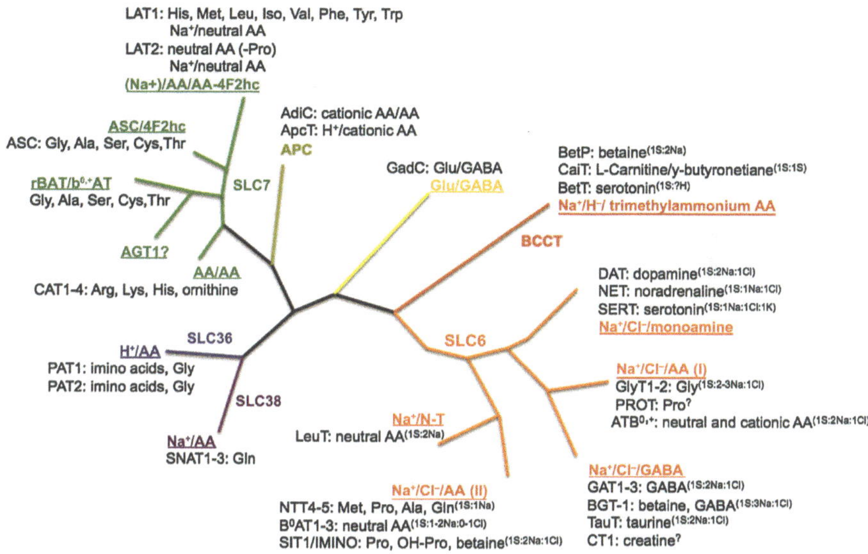

PLATE 1 (Figure 1 on page 4 of this volume)

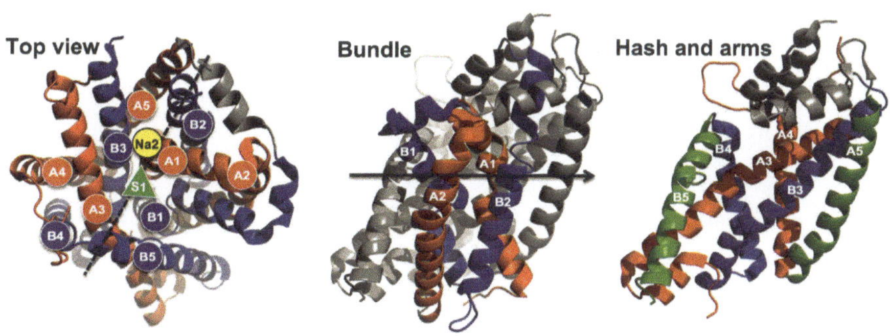

PLATE 2 (Figure 2 on page 4 of this volume)

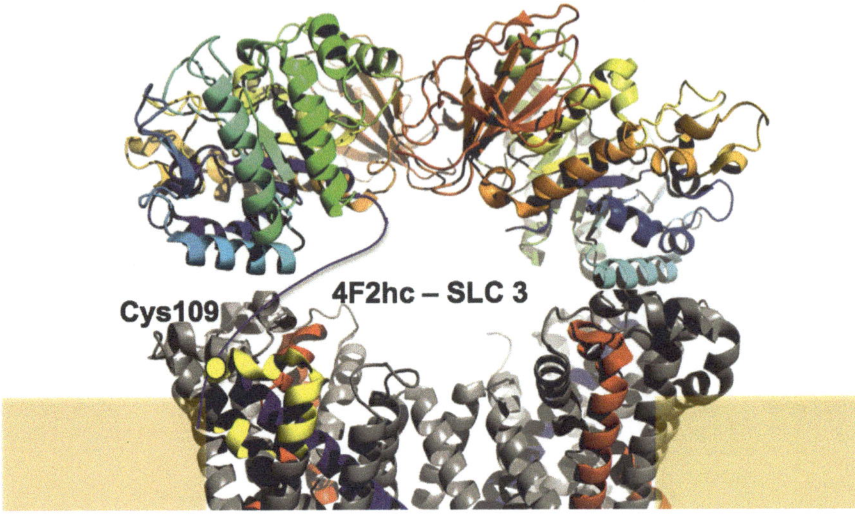

PLATE 3 (Figure 3 on page 9 of this volume)

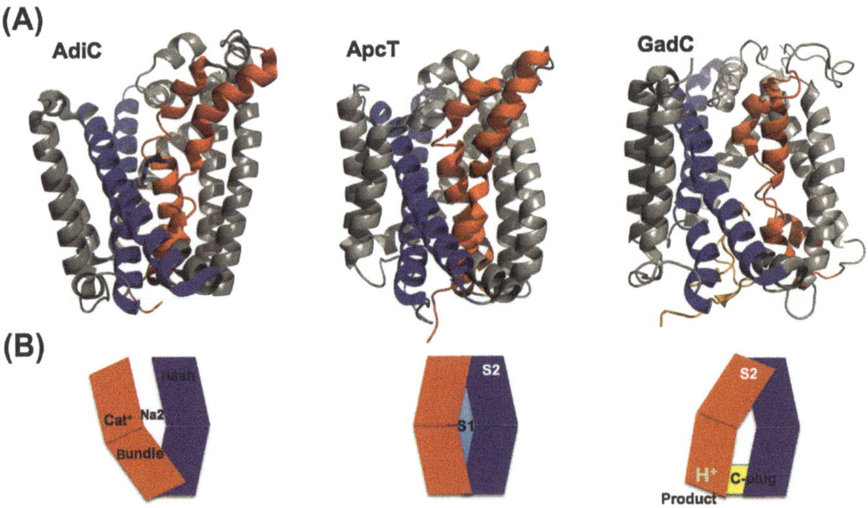

PLATE 4 (Figure 4 on page 15 of this volume)

```
                                                           ----------TM14-----------
hSGLT1    544   ISLLTKPIPDVHLYRLCWSLRNSKEERIDLDAEEEN-------IQEGPKETIEIETQ-----VPEK-KKGIFRRAYDLFC  610
hSGLT2    544   VSLCTAPIPRKHLHRLVFSLRHSKEEREDLDADEQQG--SSLPVQNGCPESA-MEMN-----EPQAPAPSLFRQCLLWFC  615
hSGLT3    544   ISLLTKPIPDVHLYRLCWVLRNSTEERIDIDAEEKS-------QEETDDG---VEED-----YPEK-SRGCLKKAYDLFC  607
hSGLT4    548   VSLCTTPIPEEQLTRLTWWTRNCPLSELEKEAHESTPEISERPAGECPAGGGAAENSSLGQEQPEAPSRSWGKLLWSWFC  627
hSGLT5    531   GSLLTPPPQSVQIENLTWWTL----------------------AQDVPLGTKAGDGQ-----------------------  565

                                              -----------TM14------------
hSGLT1    611   GLEQHGA---PKMTEEEEKAMKMKMTDTSEKPLWRTVLNVNGIILVTVAVFCHAYFA  664
hSGLT2    616   GMSRGGVGSPPPLTQEEAAAAARRLEDISEDPSWARVVNLNALLMMAVAVFLWGFYA  672
hSGLT3    608   GLQK--G---PKLTKEEEEALSKKLTDTSERPSWRTIVNINAILLLAVVVFIHGYYA  659
hSGLT4    628   GLSGTPEQA---LSPAEKAALEQKLTSIEEEPLWRHVCNINAVLLLAINIFLWGYFA  681
hSGLT5    566   ----TP----------------------QKHAFWARVCGFNAILLMCVNIFFYAYFA  596
```

PLATE 5 (Figure 5 on page 43 of this volume)

```
                                                                                           ---------TM2
Human SGLT1       1     MD-SSTWSPKTTAVTRPVETHELIRNAADISIIVIYFVVVMAVGLWAMFSTNRGTVGGFFLAGRSMVWWPIGASLFASNI  79
Xenopus SGLT-1    1     METSSQSSPQTTPGMEAFPKKS--LDTIDIVVLVLYFVFVLAVGLLSMCRTKRGTVKGYFLAGKDMAWWPVGASLFASNV  78
Human SMIT1       1     MR-A-------------------VLDTADIAIVALYFILVMCIGFFAMWKSNRSTVSGYFLAGRSMTWVTIGASLFVSNI  60

                        ------------     ----------TM3-----------                   ---------------TM4--
Human SGLT1       80    GSGHFVGLAGTGAASGIAIGGFEWNALVLVVVLGWLFVPIYIKAGVVTMPEYLRKRFGGQRIQVYLSLLSLLLYIFTKIS  159
Xenopus SGLT-1    79    GSGHFIGLAGSGAASGIAVTAYEWNGLFCVLALAWLFLPIYISAGVTTMPEYLQRRFGGKRIQIFLAILYLFIYIFTKIS  158
Human SMIT1       61    GSEHFIGLAGSGAASGFAVGAWEFNALLLLQLLGWVFIPIYIRSGVYTMPEYLSKRFGGHRIQVYFAALSLILYIFTKLS  140

                        -----------       ------TM5-----          -------TM6----------
Human SGLT1       160   ADIFSGAIFINLALGLNLYLAIFLLLAITALYTITGGLAAVIYTDTLQTVIMLVGSLILTGFAFHEVGGYDAFMEKYMKA  239
Xenopus SGLT-1    159   VDMYAGALFIQQALQWDLYVAVIGLLVITAIYTVAGGLAAVIYTDTLQTVIMIIGALILMAYSFIEIGGFEALQEKYFHA  238
Human SMIT1       141   VDLYSGALFIQESLGWNLYVSVILLIGMTALLTVTGGLVAVIYTDTLQALLMIIGALTLMIISIMEIGGFEEVKRRYMLA  220

                                                                  -------------TM7-----------   --------
Human SGLT1       240   IPTIVSD---GNTTFQEKC-YTPRADSFHIFRDPLTGDLPWPGFIFGMSILTLWYWCTDQVIVQRCLSAKNMSHVKGGCI  315
Xenopus SGLT-1    239   IPNTHSG--------NSTC-GIPREDAFHIFRDPVTSDLPWPGVLVGMTIPSLWYWCTDQVIVQRSLSAKNLSHAKAGSL  309
Human SMIT1       221   SPDVTSILLTYNLSNTNSCNVSPKKEALKMLRNPTDEDVPWPGFILGQTPASVWYWCADQVIVQRVLAAKNIAHAKGSTL  300
                        ------TM8---------------                                     --------------TM9--
Human SGLT1       316   LCGYLKLMPMFIMVMPGMISRILYTEKIACVVPSECEKYCGTKVGCTNIAYPTLVVELMPNGLRGLMLSVMLASLMSSLT  395
Xenopus SGLT-1    310   LAASLKVLPLFMMVLPGMISRVLFTDQVACADPELCKEICGNPSGCSDIAYPKMVIELLPTGLRGLMMSVMIAALMSSLT  389
Human SMIT1       301   MAGFLKLLPMFIIVVPGMISRILFTDDIACINPEHCMLVCGSRAGCSNIAYPRLVMKLVPVGLRGLMMAVMIAALMSDLD  380

                        -----------------       ------------TM10----------       -----------TM11--------
Human SGLT1       396   SIFNSASTLFTMDIYAKVRKRASEKELMIAGRLFILVLIGISIAWVPIVQSAQSGQLFDYIQSITSYLGPPIAAVFLLAI  475
Xenopus SGLT-1    390   SIFNSASTIFTMDLWRHIRPRSTEWELMIVGRVFVLVLVVVSILWIPLVQASQGGQLFIYIQSISSYLQPPVAVVFIAGC  469
Human SMIT1       381   SIFNSASTIFTLDVYKLIRKSASSRELMIVGRIFVAFMVVISIAWVPIIVEMQGGQMYLYIQEVADYLTPPVAALFLLAI  460

                        --    --------TM12------------                      --------TM13-----------
Human SGLT1       476   FWKRVNEPGAFWGLILGLLIGISRMITEFAYGTGSCMEPSNCPTIICGVHYLYFAIILFAISFITIVVISLLTKPIPDVH  555
Xenopus SGLT-1    470   FWKRTNEKGAFWGMTIGLVVGLIRMVLDFIYVAPQCDQPDTRPGVVKYVHYLYLSMILGLLTLLVVVAVSLWTEP-PSKQ  548
Human SMIT1       461   FWKRCNEQGAFYGGMAGFVLGAVRLILAFAYRAPECDQPDNRPGFIKDIHYMYVATGLFWVTGLITVIVSLLTPP-PTKE  539

Human SGLT1       556   LYRLC--WSLRN--------------SKEERIDLDAEEEN------IQEGPKETI-------------------------  588
Xenopus SGLT-1    549   MISRLTWFTRFDAEVEDPVETNHRP-----------AENGISVVEDISEEPHTTSTDAIYNDTT--------DDNPSSS-  608
Human SMIT1       540   QIRTTTFWSKKNLVVKENCSPKEEPYQMQEKSILRCSENNETINHIIPNGKSEDSIKGLQPEDVNLLVTCREEGNPVASL  619

                                                                                                 -------
Human SGLT1       589   ---EIETQVP-------------EKKK-----GIFRRAYDLFCGLEQHGAPKMTEE---EEKAMKMKMTDTSEKPLWRTV  644
Xenopus SGLT-1    609   --------------------------------SLLKKTILWLCGMDSRKGDKHDQAPPAPLEPAEVLLYE---RPLLKQV  653
Human SMIT1       620   GHSEAETPVDAYSNGQAALMGEKERKKETDDGGRYWKFIDWFCGFKSKSLSKRSLRDLMEEEAVCLQMLEETRQV--KVI  697
                        ----TM14------------
Human SGLT1       645   LNVNGIILVTVAVFCHAYFA-  664
Xenopus SGLT-1    654   LNTAVILCMSAGVFLWAYFG-  673
Human SMIT1       698   LNIGLFAVCSLGIFMFVYFSL  718
```

PLATE 6 (Figure 6 on page 44 of this volume)

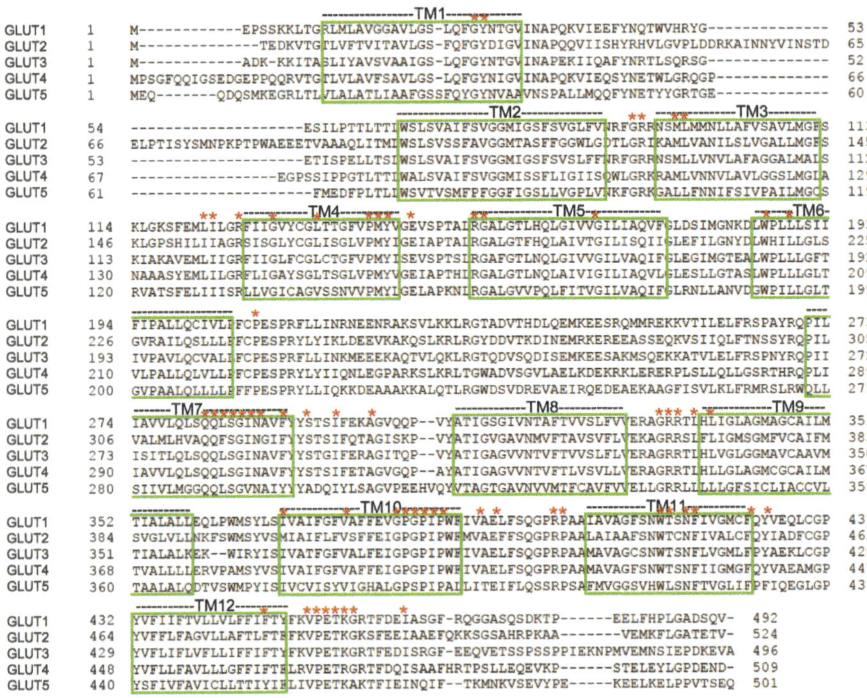

PLATE 7 (Figure 7 on page 46 of this volume)

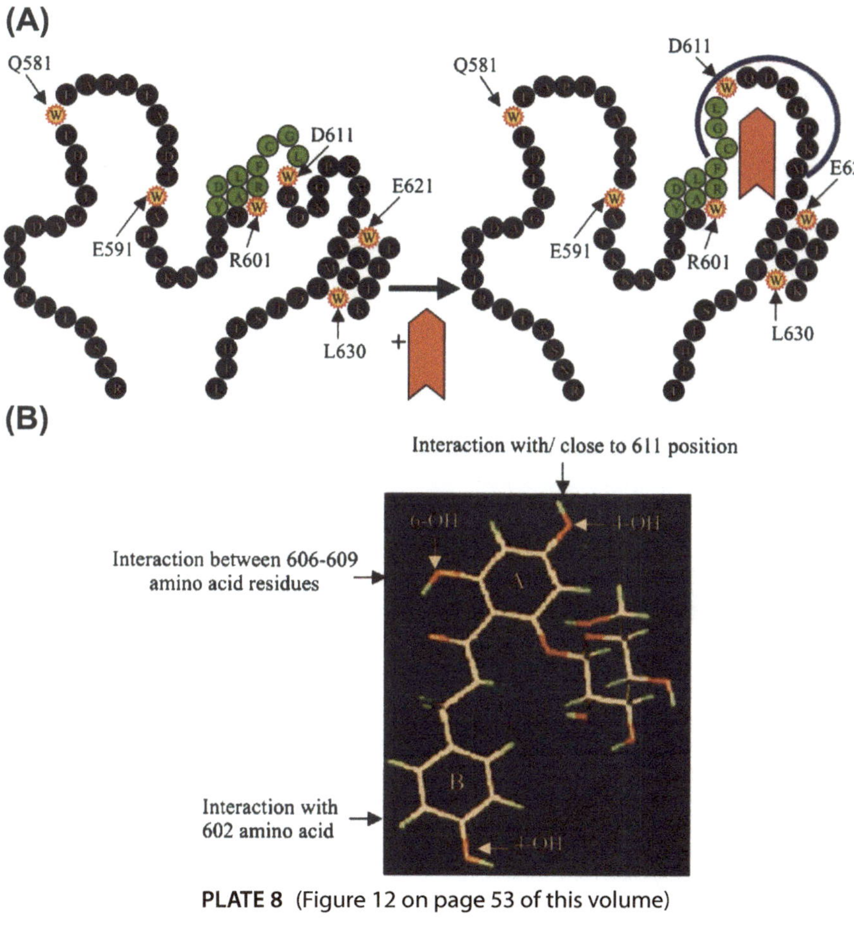

PLATE 8 (Figure 12 on page 53 of this volume)

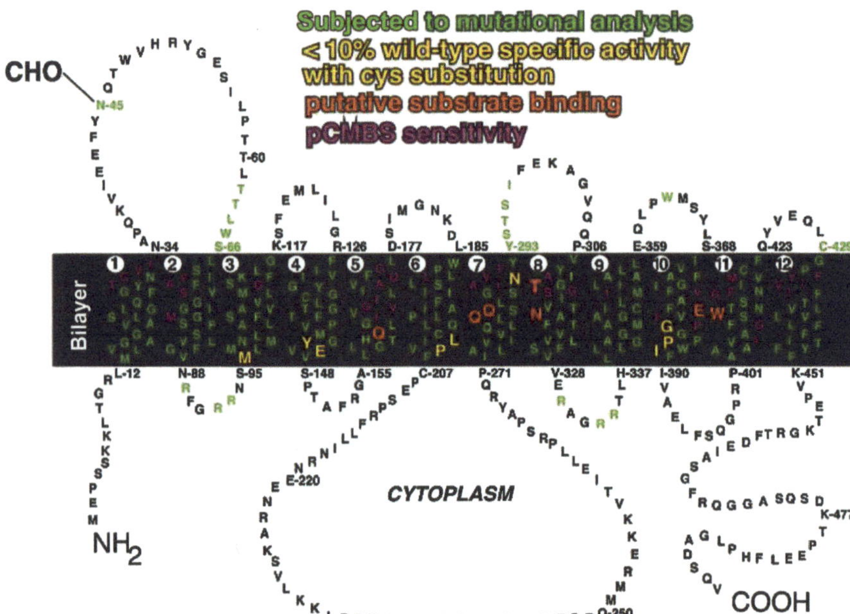

PLATE 9 (Figure 13 on page 56 of this volume)

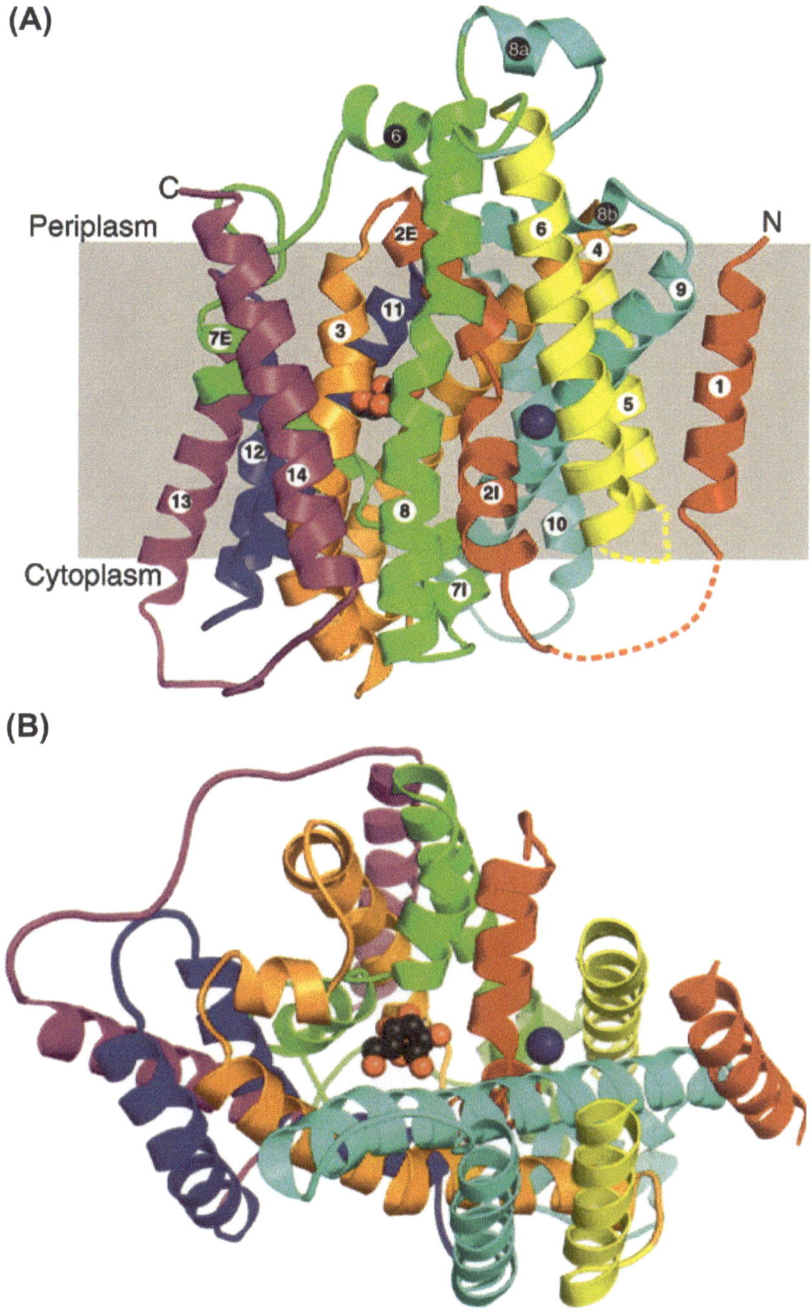

PLATE 10 (Figure 14 on page 57 of this volume)

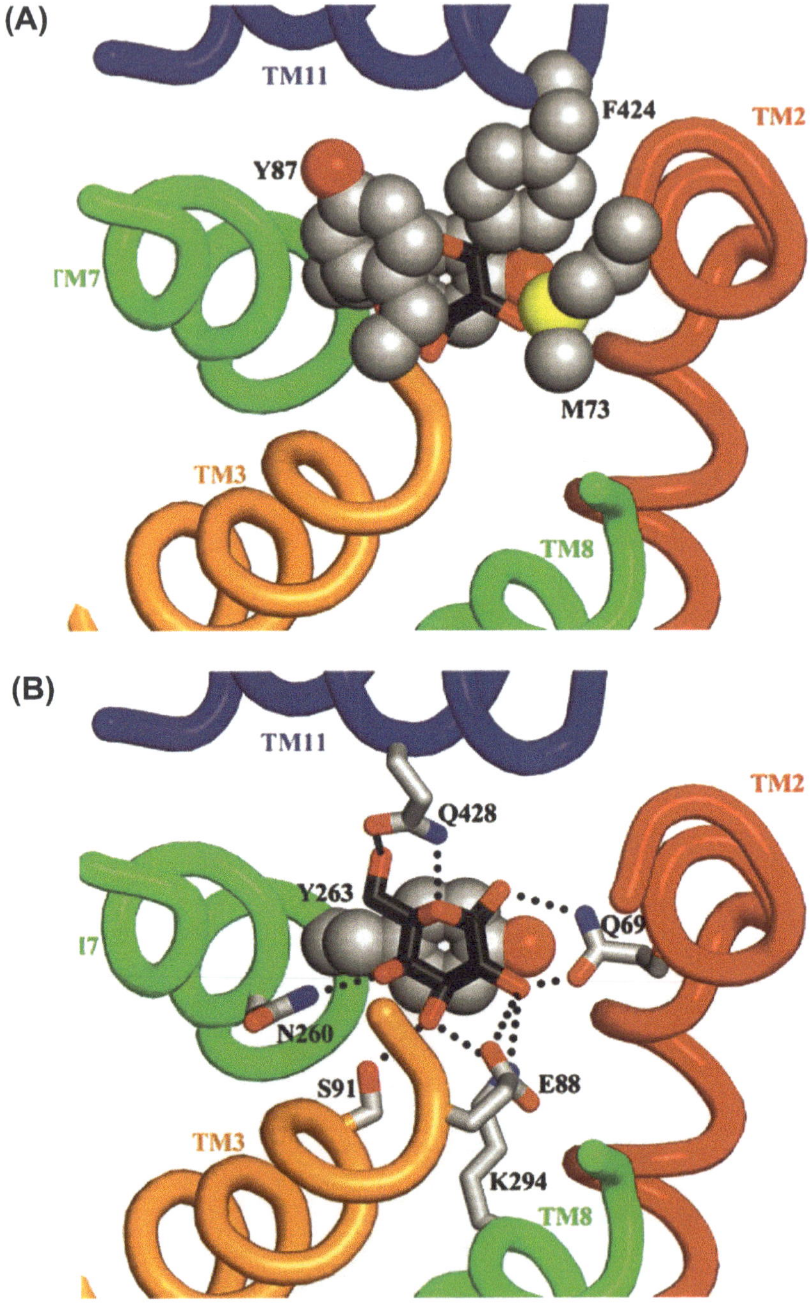

PLATE 11 (Figure 15 on page 59 of this volume)

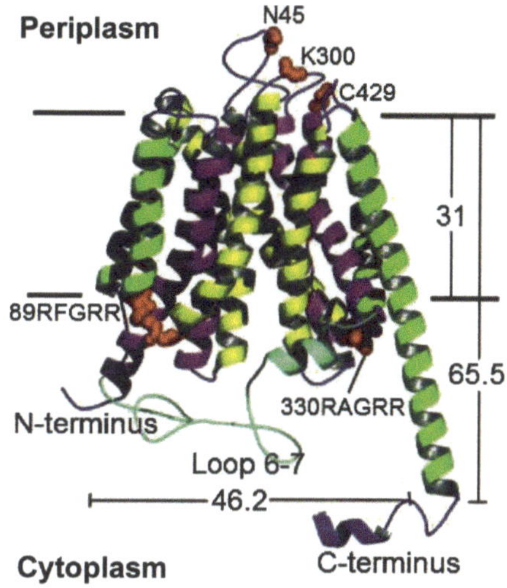

PLATE 12 (Figure 17 on page 62 of this volume)

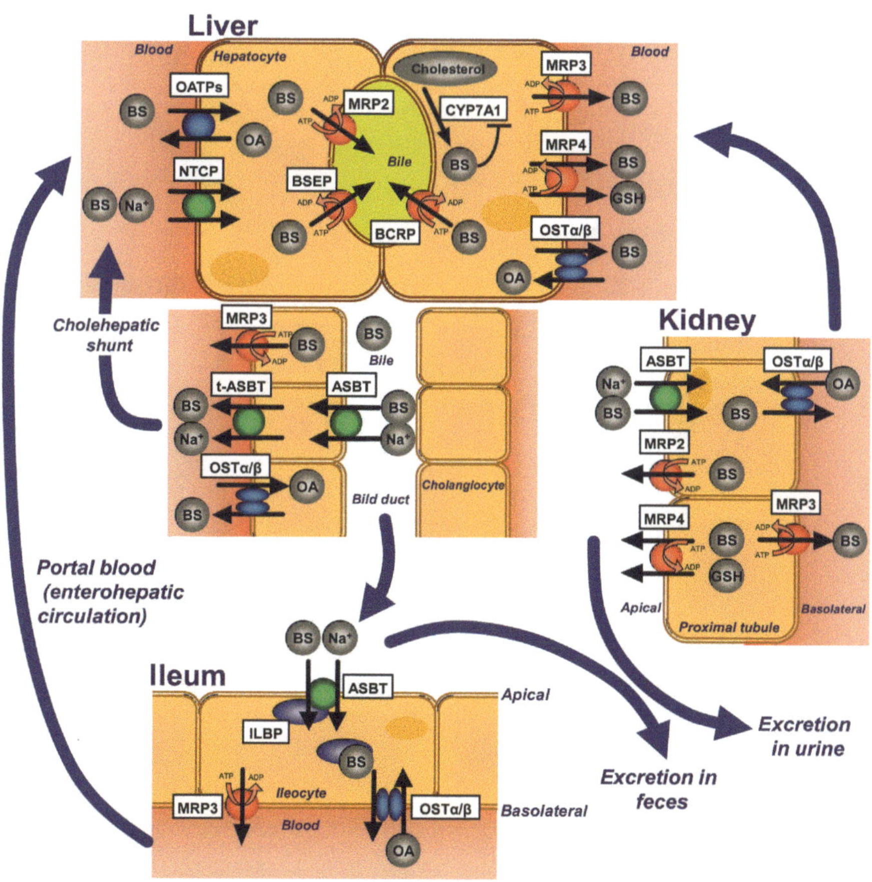

PLATE 13 (Figure 3 on page 114 of this volume)

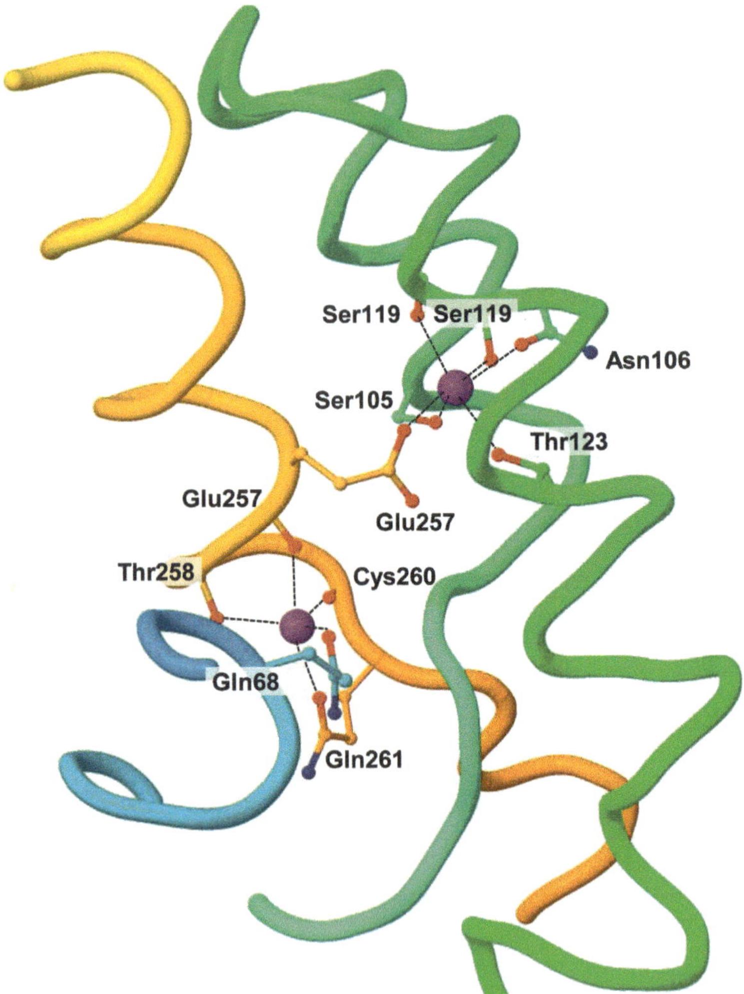

PLATE 14 (Figure 4 on page 120 of this volume)

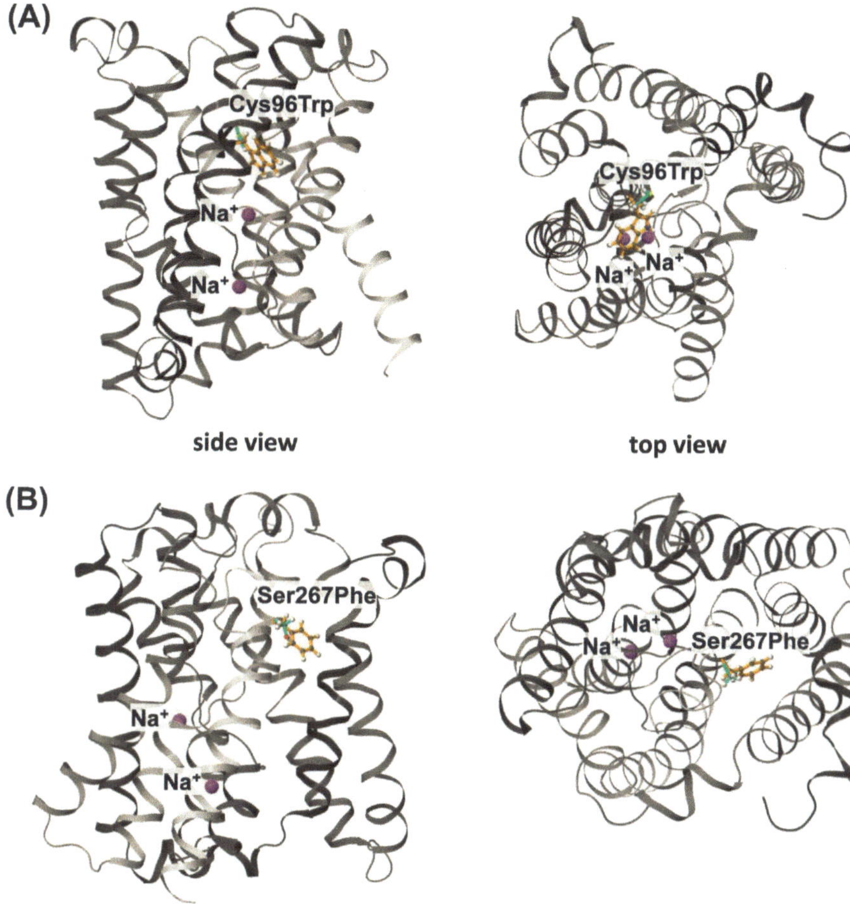

PLATE 15 (Figure 5 on page 133 of this volume)

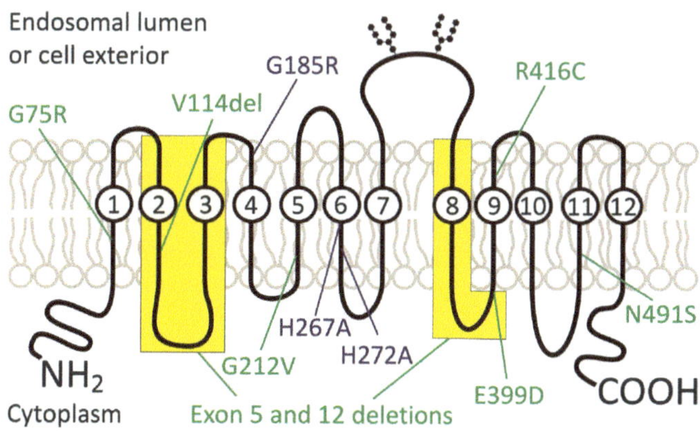

PLATE 16 (Figure 1 on page 172 of this volume)

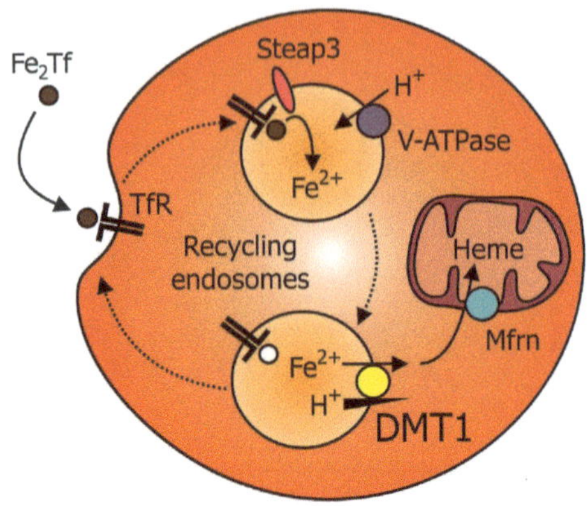

PLATE 17 (Figure 4 on page 192 of this volume)

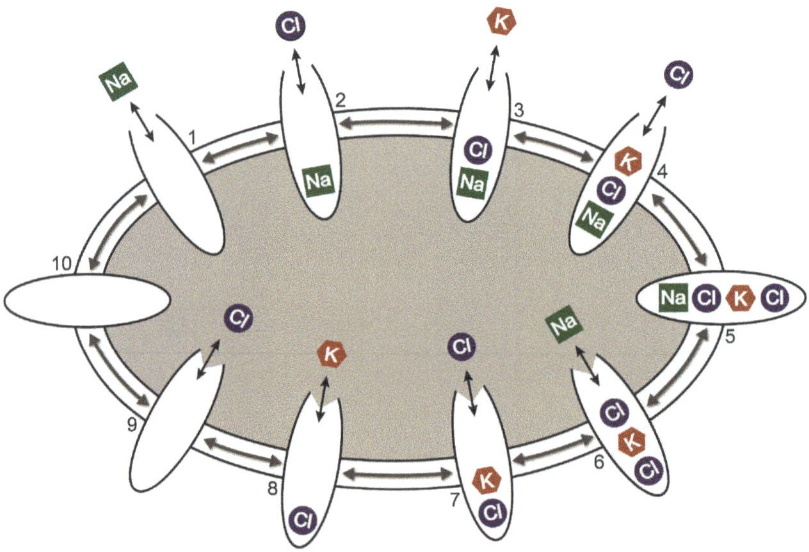

PLATE 18 (Figure 1 on page 218 of this volume)

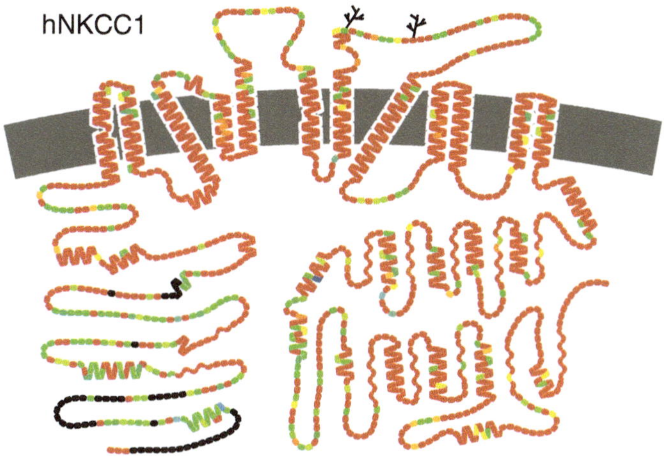

PLATE 19 (Figure 2 on page 220 of this volume)

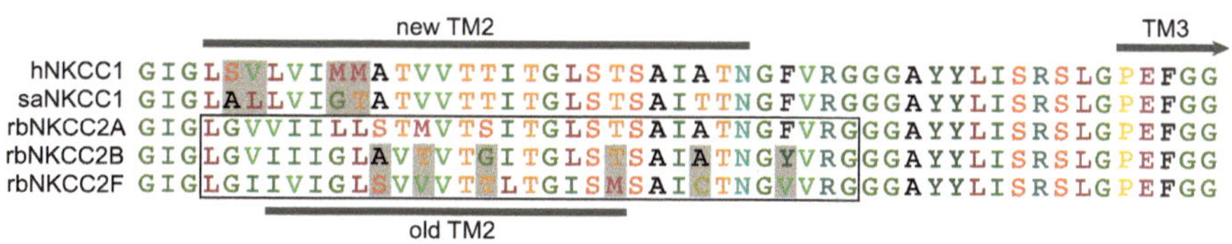

PLATE 20 (Figure 3 on page 222 of this volume)

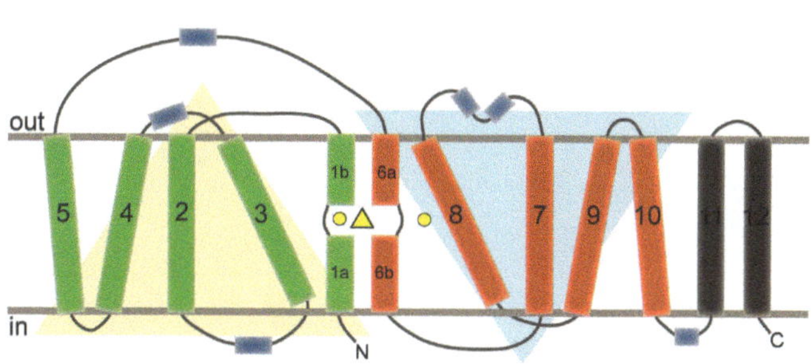

PLATE 21 (Figure 4 on page 225 of this volume)

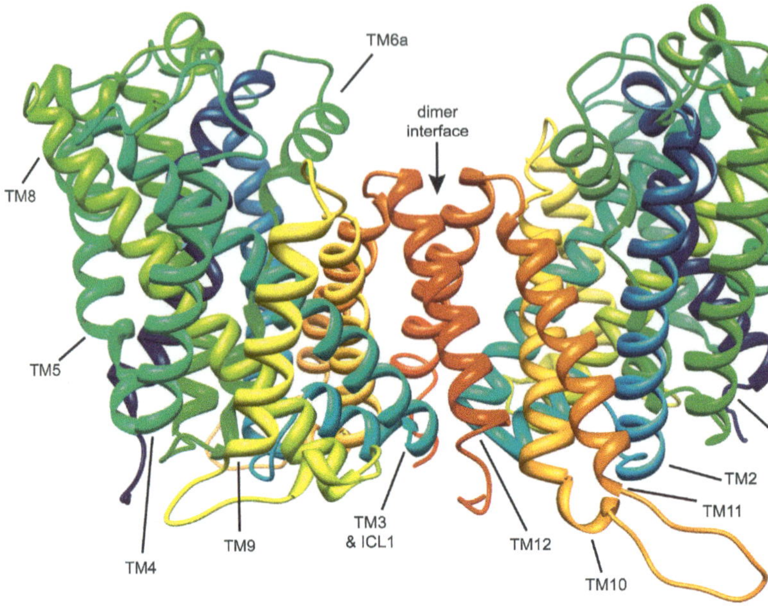

PLATE 22 (Figure 5 on page 227 of this volume)

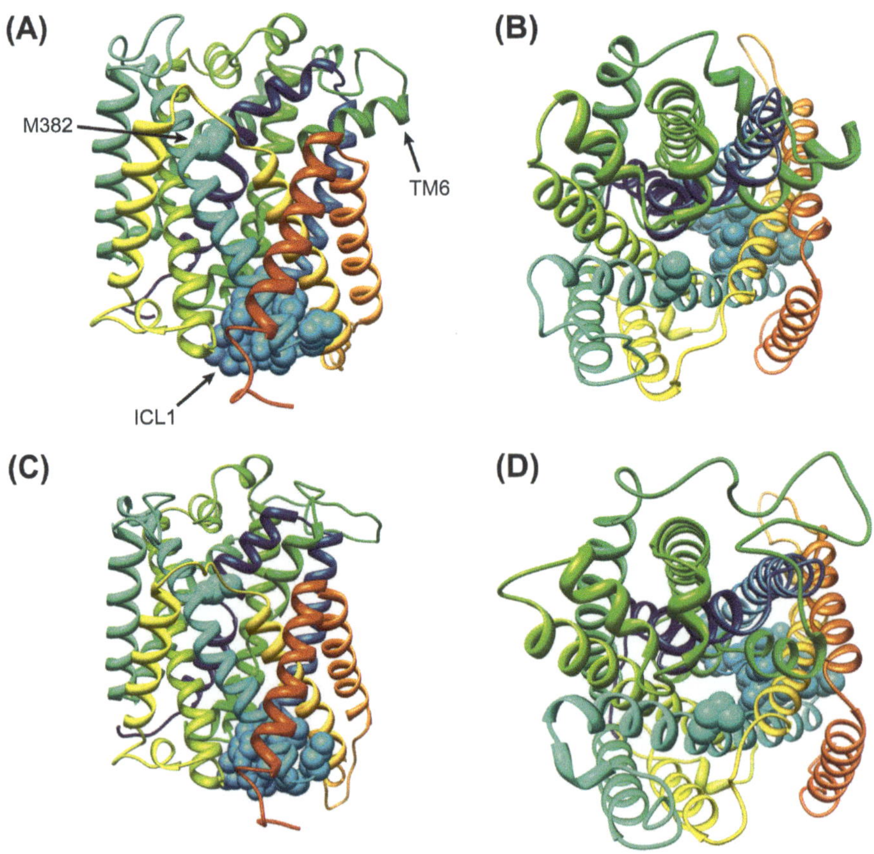

PLATE 23 (Figure 6 on page 228 of this volume)

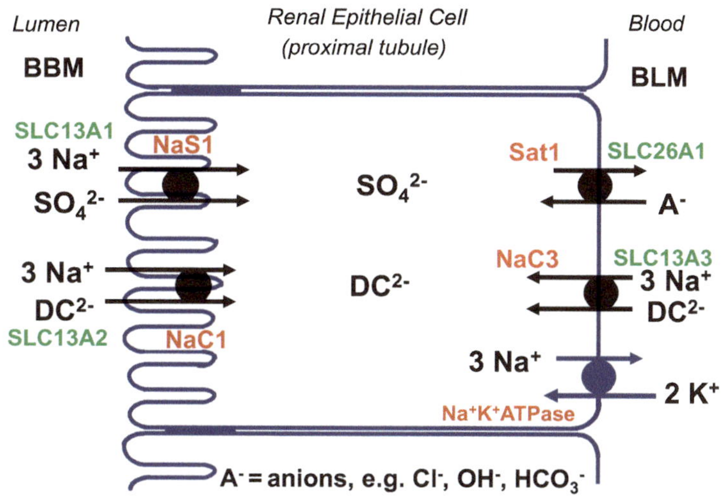

PLATE 24 (Figure 2 on page 249 of this volume)

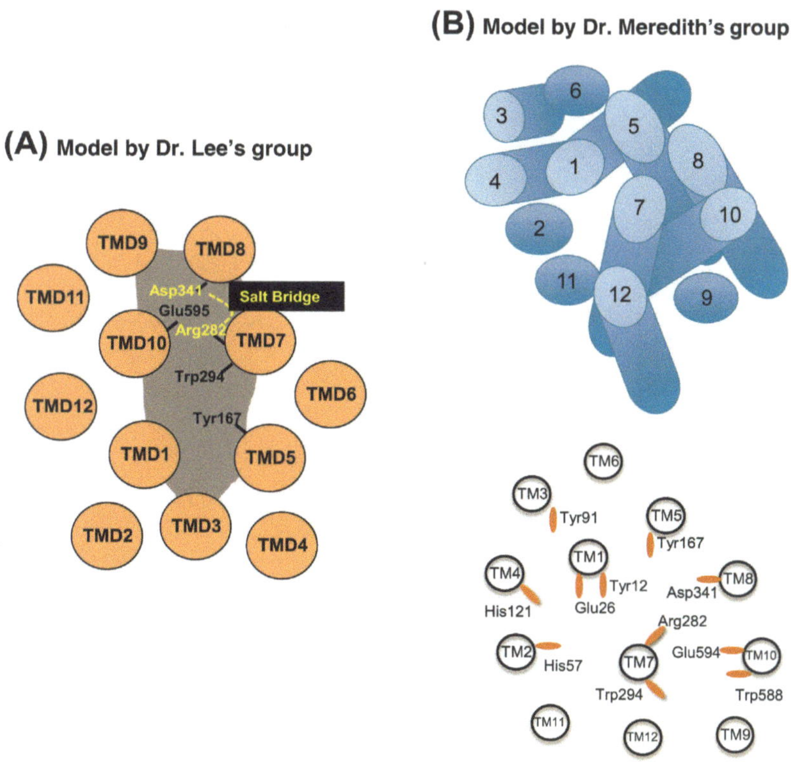

PLATE 25 (Figure 4 on page 265 of this volume)

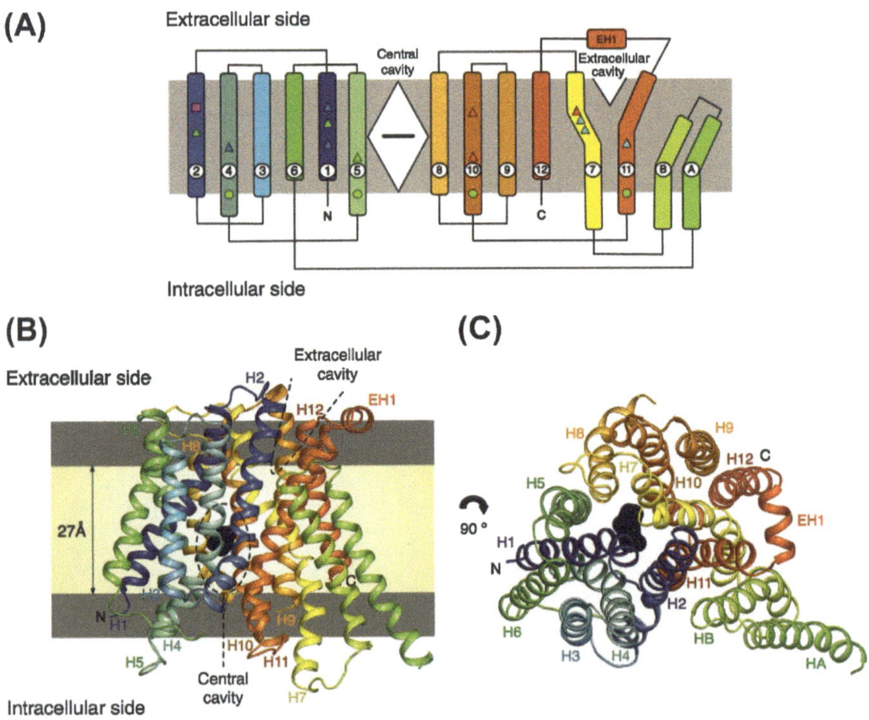

PLATE 26 (Figure 5 on page 267 of this volume)

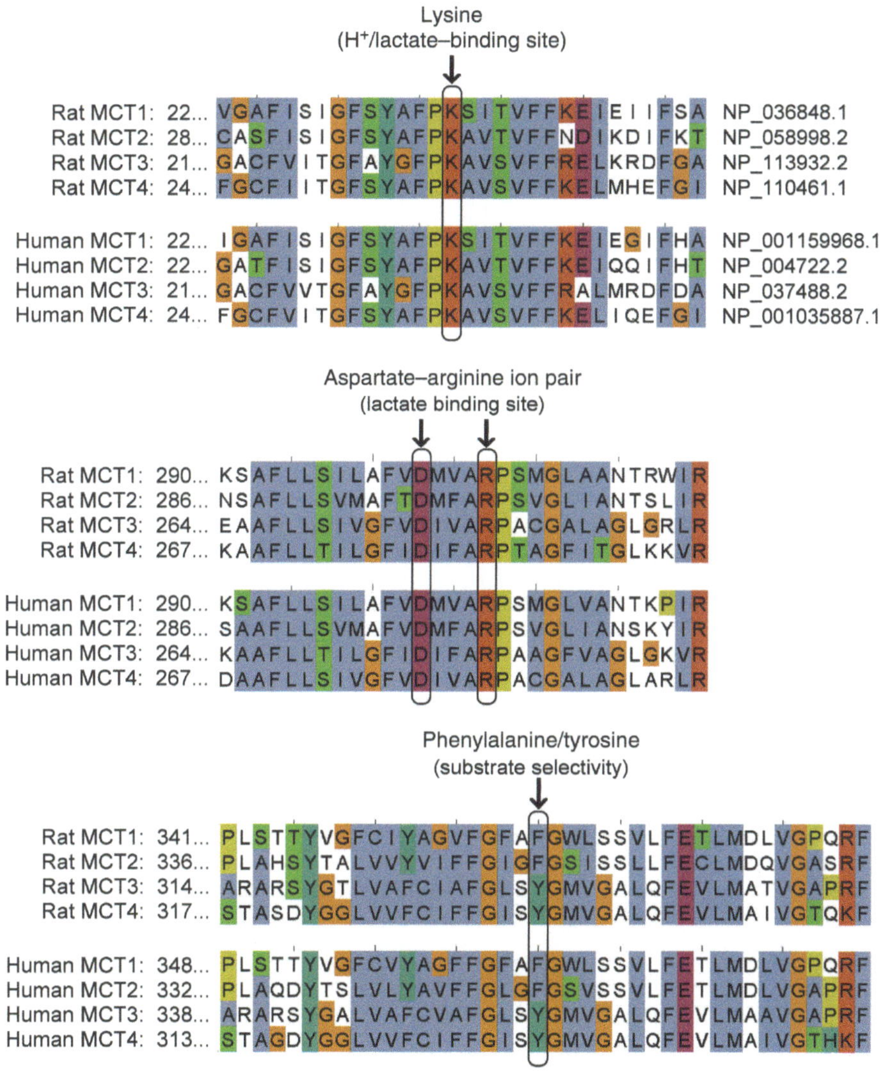

PLATE 27 (Figure 1 on page 278 of this volume)

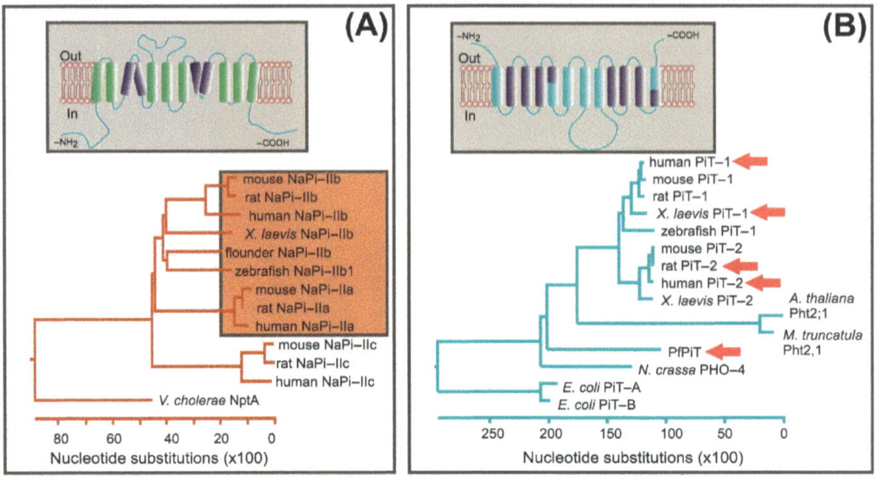

PLATE 28 (Figure 1 on page 316 of this volume)

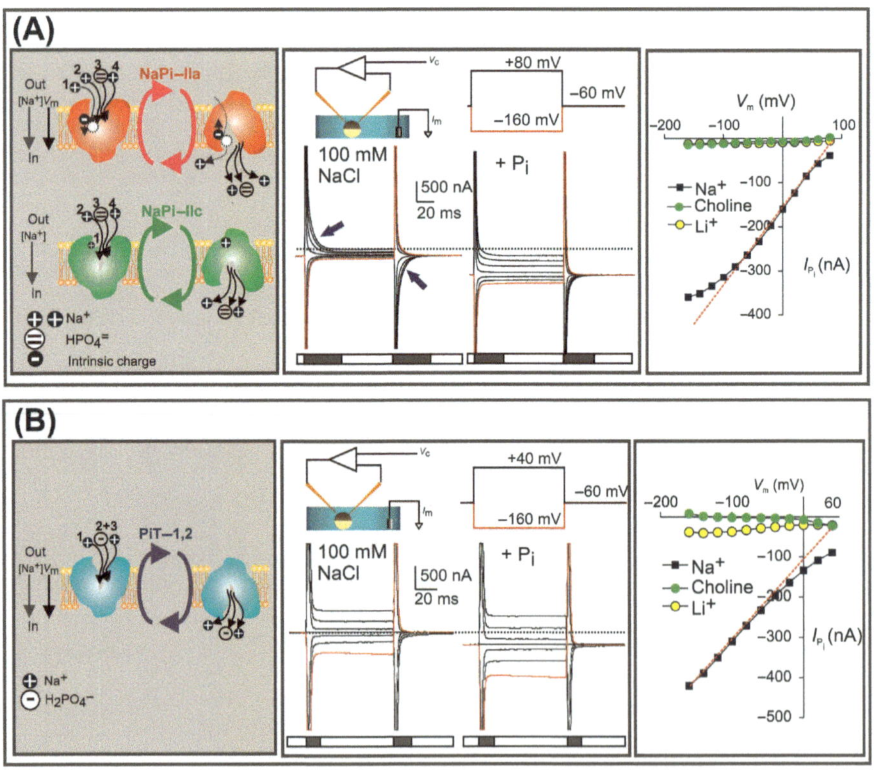

PLATE 29 (Figure 2 on page 322 of this volume)

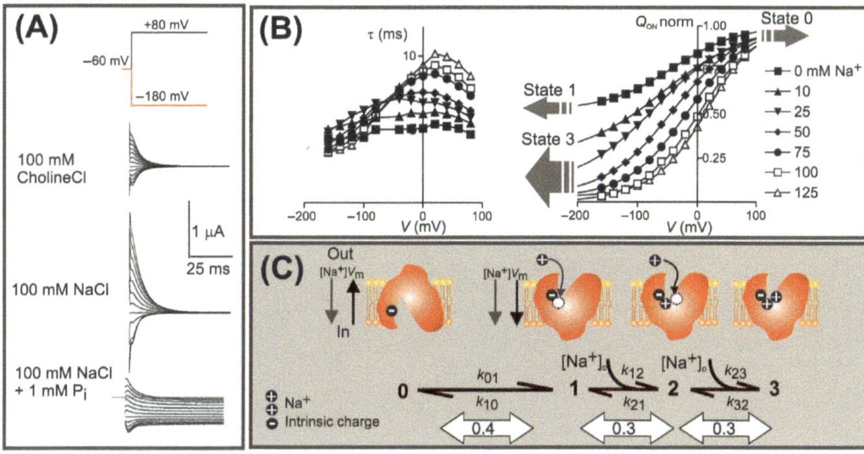

PLATE 30 (Figure 3 on page 332 of this volume)

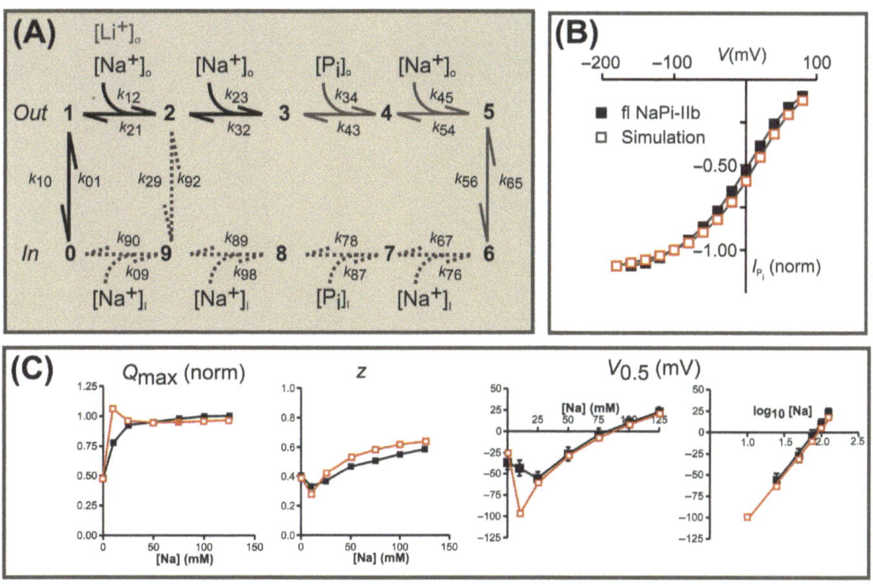

PLATE 31 (Figure 4 on page 334 of this volume)

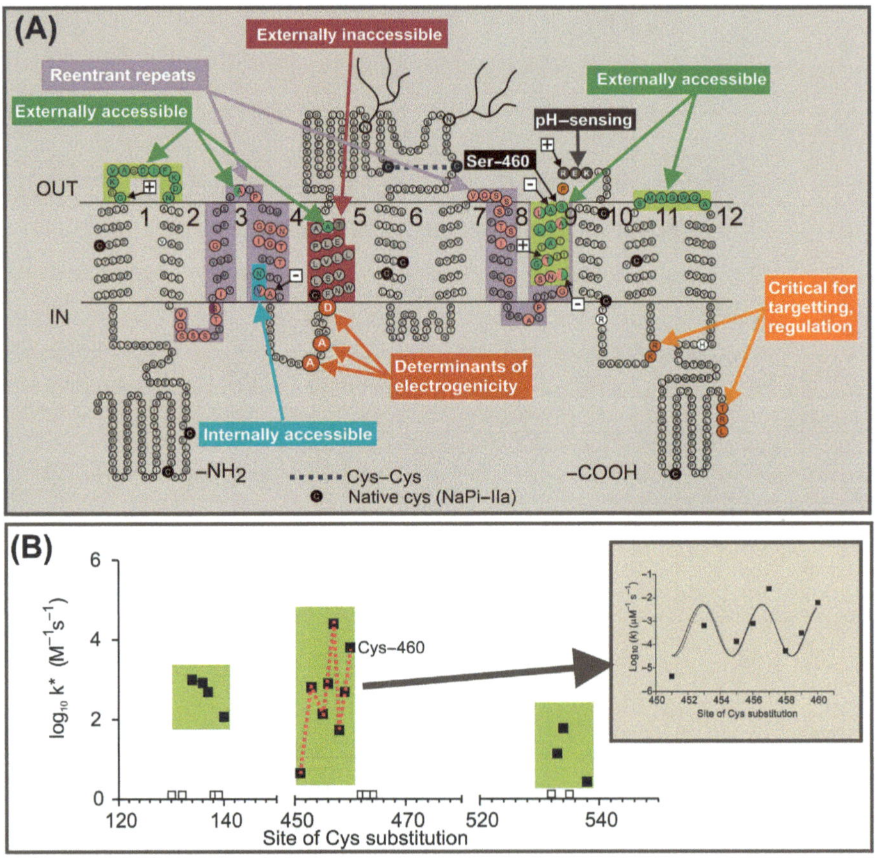

PLATE 32 (Figure 5 on page 337 of this volume)

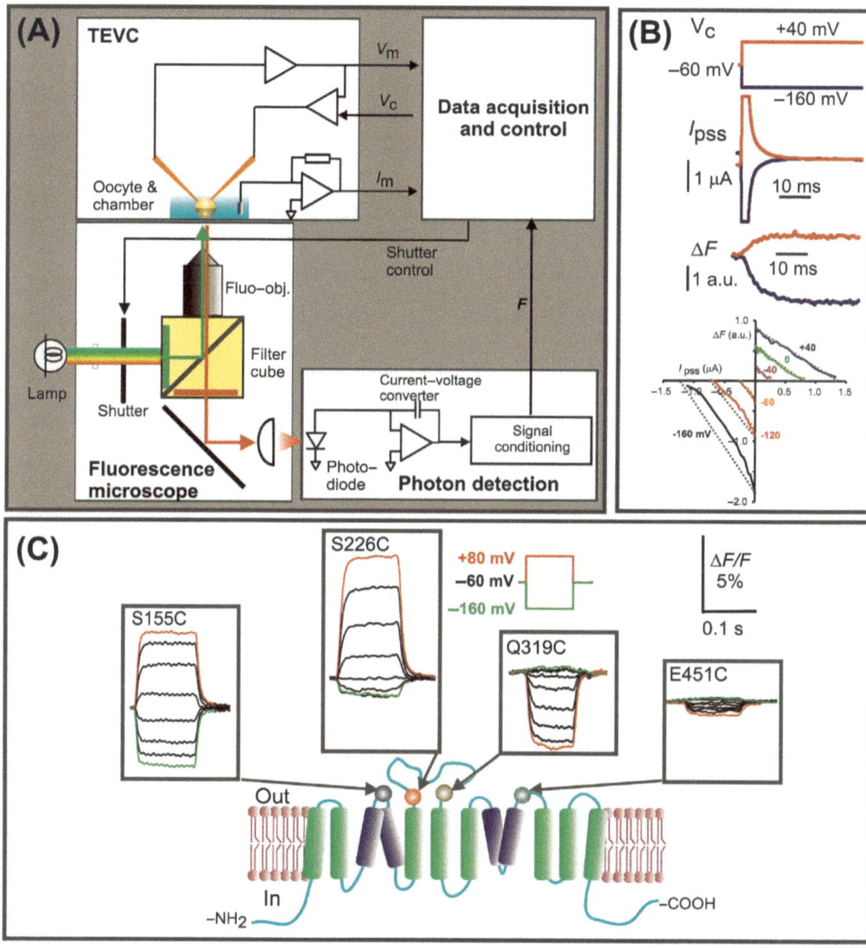

PLATE 33 (Figure 6 on page 340 of this volume)

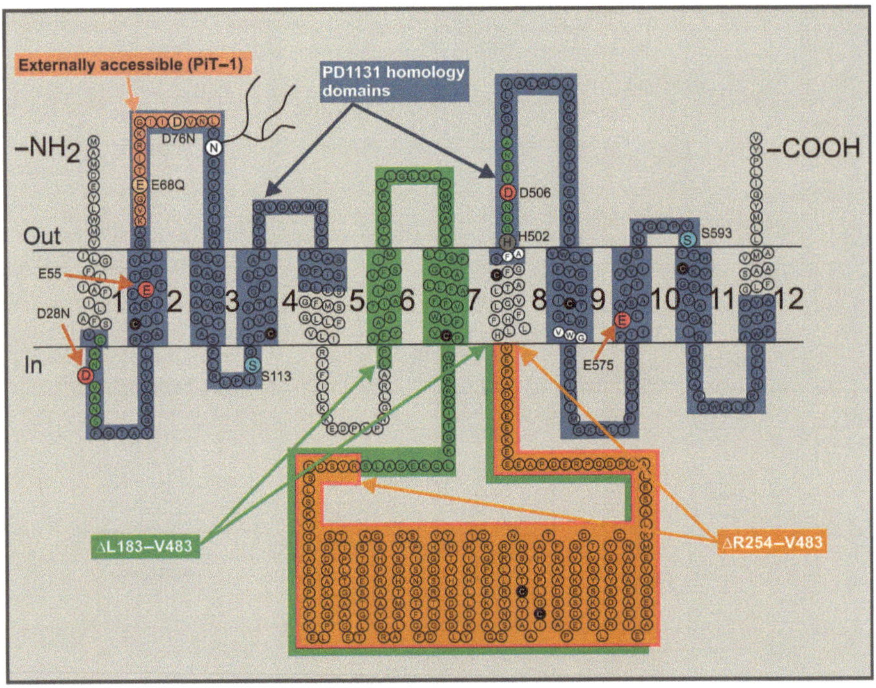

PLATE 34 (Figure 7 on page 345 of this volume)

(A)

```
hSVCT1   ------------MRAQEDLEGRTQHETTRDPSTPLPTEP--------------------- 27
hSVCT2   MMGIGKNTTSKSMEAGSSTEGKYEDEAKHPAFFTLPVVINGGATSSGEQDNEDTELMAIY 60

hSVCT1   -------------------------KFDMLYKIEDVPPWYLCILLGFQHYLTCFSGTIAV 62
hSVCT2   TTENGIAEKSSLAETLDSTGSLDPQRSDMIYTIEDVPPWYLCIFLGLQHYLTCFSGTIAV 120

hSVCT1   PFLLAEALCVGHDQHMVSQLIGTIFTCVGITTLIQTTVGIRLPLFQASAFAFLVPAKAIL 122
hSVCT2   PFLLADAMCVGYDQWATSQLIGTIFFCVGITTLLQTTFGCRLPLFQASAFAFLAPARAIL 180

hSVCT1   ALERWKCPPEEEIYGN-WSLPLNTSHIWHPRIREVQGAIMVSSVVEVVIGLLGLPGALLN 181
hSVCT2   SLDKWKCNTTDVSVANGTAELLHTEHIWYPRIREIQGAIIMSSLIEVVIGLLGLPGALLK 240

hSVCT1   YIGPLTVTPTVSLIGLSVFQAAGDRAGSHWGISACSILLIILFSQYLRNLTFLLPVYRWG 241
hSVCT2   YIGPLTITPTVALIGLSGFQAAGERAGKHWGIAMLTIFLVLLFSQYARNVKFPLPIYKSK 300

hSVCT1   KGLTLLRIQIFKMFPIMLAIMTVWLLCYVLTLTDVLPTDPKAYGFQARTDARGDIMAIAP 301
hSVCT2   KGWTAYKLQLFKMFPIILAILVSWLLCFIFTVTDVFPPDSTKYGFYARTDARQGVLLVAP 360

hSVCT1   WIRIPYPCQWGLPTVTAAAVLGMFSATLAGIIESIGDYYACARLAGAPPPPVHAINRGIF 361
hSVCT2   WFKVPYPFQWGLPTVSAAGVIGMLSAVVASIIESIGDYYACARLSCAPPPPIHAINRGIF 420

hSVCT1   TEGICCIIAGLLGTGNGSTSSSPNIGVLGITKVGSRRVVQYGAAIMLVLGTIGKFTALFS 421
hSVCT2   VEGLSCVLDGIFGTGNGSTSSSPNIGVLGITKVGSRRVIQCGAALMLALGMIGKFSALFA 480

hSVCT1   SLPDPILGGMFCTLFGMITAVGLSNLQFVDMNSSRNLFVLGFSMFFGLTLPNYLESNPGA 481
hSVCT2   SLPDPVLGALFCTLFGMITAVGLSNLQFIDLNSSRNLFVLGFSIFFGLVLPSYLRQNP-- 538

hSVCT1   INTGILEVDQILIVLLTTEMFVGGCLAFILDNTVPGSPEERGLIQWKAGAHANSDMSSSL 541
hSVCT2   LVTGITGIDQVLNVLLTTAMFVGGCVAFILDNTIPGTPEERGIRKWKKGVGKGNKSLDGM 598

hSVCT1   KSYDFPIGMGIVKRITFLKYIPICPVFKGFSSSSKDQIAIPEDTPENTETASVCTKV 598
hSVCT2   ESYNLPFGMNIIKKYRCFSYLPISPTFVGYTWKGLRKSDNSRSSDEDSQATG----- 650
```

(B)

		SVCT1
	SVCT2	66%
SVCT3	31%	34%

PLATE 35 (Figure 1 on page 360 of this volume)

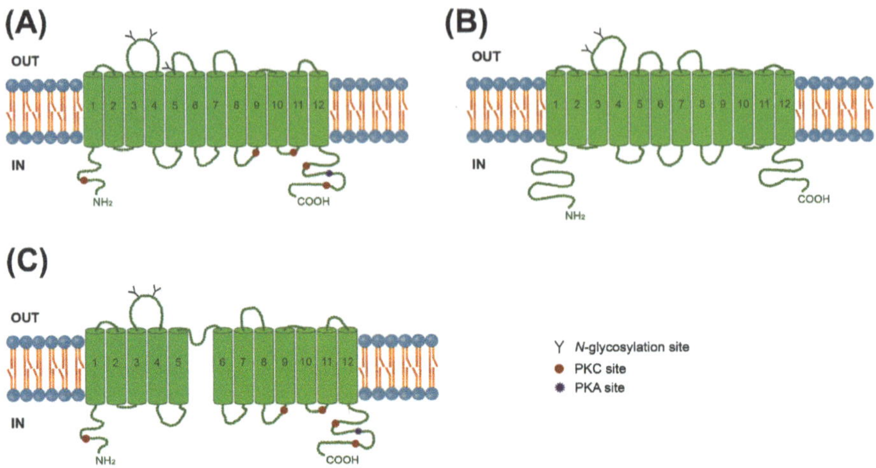

PLATE 36 (Figure 2 on page 362 of this volume)

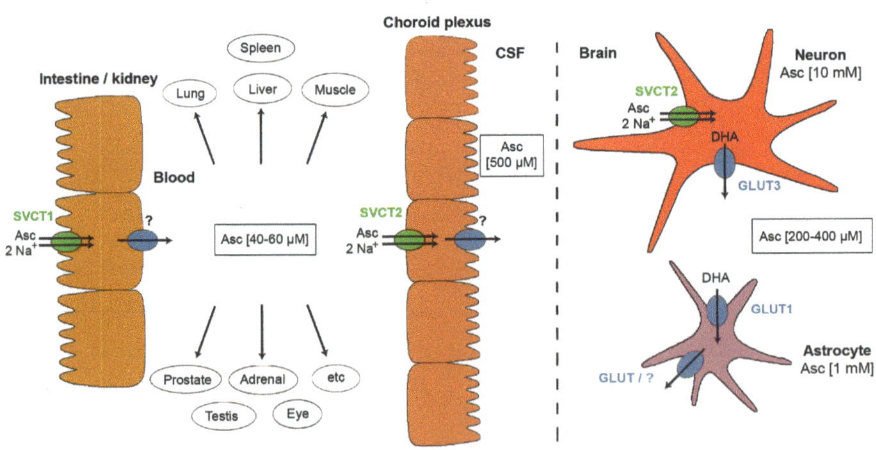

PLATE 37 (Figure 3 on page 364 of this volume)

FSC
www.fsc.org
MIX
Papier | Fördert gute Waldnutzung
FSC® C083411

Kolibri 360 GmbH
Ferdinand-Jühlke-Straße 7
99095 Erfurt, Deutschland
produktsicherheit@kolibri360.de